Abner J. Salgado, Steven M. Wise, Calvin Wong
**Multigrid Methods**

## Also of Interest

*Applied Nonlinear Functional Analysis*
*An Introduction*
Nikolaos S. Papageorgiou, Patrick Winkert, 2024
ISBN 978-3-11-128421-7, e-ISBN (PDF) 978-3-11-128695-2

*Applied Linear Analysis for Chemical Engineers*
*A Multi-scale Approach with Mathematica®*
Vemuri Balakotaiah, Ram R. Ratnakar, 2025
ISBN 978-3-11-159738-6, e-ISBN (PDF) 978-3-11-159805-5

*Numerical Methods with Python*
*for the Sciences*
William Miles, 2023
ISBN 978-3-11-077645-4, e-ISBN (PDF) 978-3-11-077664-5

*Numerical Analysis on Time Scales*
Svetlin G. Georgiev, Inci M. Erhan, 2022
ISBN 978-3-11-078725-2, e-ISBN (PDF) 978-3-11-078732-0

*Numerical Methods for Black-Box Software in Computational Continuum Mechanics*
*Parallel High-Performance Computing*
Sergey I. Martynenko, 2023
ISBN 978-3-11-131729-8, e-ISBN (PDF) 978-3-11-131956-8

Abner J. Salgado, Steven M. Wise, Calvin Wong

# Multigrid Methods

Axiomatic Convergence Theory for Linear and Weakly Nonlinear Problems

DE GRUYTER

**Mathematics Subject Classification 2020**
Primary: 65N55, 65-01, 65-04; Secondary: 65N22, 65F10

**Authors**
Prof. Abner J. Salgado
Department of Mathematics
University of Tennessee
Knoxville TN 37996-1300
USA
asalgad1@utk.edu

Calvin Wong
Department of Mathematics
University of Tennessee
Knoxville TN 37996-1300
USA
mwong4@vols.utk.edu

Prof. Steven M. Wise
Department of Mathematics
University of Tennessee
Knoxville TN 37996-1300
USA
swise1@utk.edu

ISBN 978-3-11-135419-4
e-ISBN (PDF) 978-3-11-135488-0
e-ISBN (EPUB) 978-3-11-135523-8

**Library of Congress Control Number: 2025943007**

**Bibliographic information published by the Deutsche Nationalbibliothek**
The Deutsche Nationalbibliothek lists this publication in the Deutsche Nationalbibliografie; detailed bibliographic data are available on the Internet at http://dnb.dnb.de.


Cover image: david_addimage / iStock / Getty Images Plus
Typesetting: VTeX UAB, Lithuania

www.degruyterbrill.com
Questions about General Product Safety Regulation:
productsafety@degruyterbrill.com

We dedicate this work to the memory of our esteemed colleague, mentor, teacher, and friend, Tim Schulze.

# Preface

The purpose of this book is to give a gentle introduction to the theory of multigrid methods for the fast numerical solution of linear and weakly nonlinear systems of equations resulting from the discretization of elliptic PDE. We use the finite element and finite difference methods to discretize the PDE problems. Solving the resulting linear and nonlinear systems of $n$ equations in $n$ variables represents the central computational challenge in the numerical approximation process. To analyze the convergence properties of the multigrid methods that we construct, we will mostly use the finite element framework, as the Galerkin structure it provides is the most natural choice for analysis. However, we do take a deep dive in the world of cell-centered finite differences. To analyze multigrid in that setting, the tools become a little more complicated. To our knowledge, the latter theory is not widely available in a textbook-like format.

The reader will get a rigorous, though not exhaustive, treatment of this topic as part of their tour through basic multigrid concepts. No previous exposure to numerical discretization methods is assumed. All that is required from the reader is a little curiosity and basic knowledge of matrix theory and the theory of finite-dimensional vector spaces at the level of, say, our first book [31]. We will provide most of the needed background for the more advanced topics later in the book, namely, Hilbert spaces and convex optimization.

Coding the multigrid method is notoriously difficult. The current book will additionally train the interested reader in the construction of multigrid codes using the simple MATLAB© programming environment. We use a mostly matrix-based approach in the book, both as a way of presenting the theory in a natural and simple setting and as a means for translating the theory into practical codes. In particular, the operators described in the text are typically transcribed into code with exactly the same names and functions. This makes reading the codes quite simple, even if the reader has had little exposure to software development. In this sense, the book is the software manual for the supplied codes. The whole process is designed so that the reader gets involved in translation of theory into implementation, and vice versa. All of the codes listed herein (and several more) can be downloaded from our GitHub repository

https://github.com/stevenmwise/MultigridCourse

We deviate a little from the matrix-based approach in the presentation of the framework for nonlinear problems in the later part of the book. This represents an area of current research. In fact, the book takes the reader all the way from the basics and simple implementation issues to the front lines of multigrid research. Although the book is not comprehensive in its treatment of multigrid methods and does not pretend to be, the reader will cover a lot of territory in the span of about 470 pages.

The book is meant either for a one- or two-semester course on the subject. For a one-semester (15-week) course, we recommend covering Chapters 1 and 3–7. The entire

https://doi.org/10.1515/9783111354880-203

book can be covered in two 15-week semesters or three 10-week quarters. The book is ideal for either self-study or an instructor-led classroom experience. In fact, the germ of this book can be found in a set of LaTeX typeset notes that a former student produced while attending SMW's lectures on the topic back in 2014. The notes were then used again several times to teach a course on the subject and have evolved into what you now hold in your hands.

There are lots of styles for teaching multigrid methods and a number of great books on the subject, including, but not limited to, [3, 6, 8, 12, 23, 36, 37, 38]. Our book is not better than these in terms of content. However, what we do very well, in our opinion, is that we give a modern take on the subject, with modern notation and important insights into the implementation of the methods in MATLAB©. Moreover, the present book is mathematically rigorous and can be used as a solid textbook, since it has many classroom tested exercises that can help students go deeper. We believe that this will be an excellent resource for any student or researcher wanting to go from the basics all the way to front lines of research while getting to practice with well-tested exercises and codes.

We use an axiomatic approach for our presentation, as in [8]. For the basic linear geometric multigrid analysis, we break assumptions (i. e., axioms) into three distinct types: Galerkin, approximation, and smoothing. We carry this a bit further when we discuss modern methods based on subspace decompositions in the context of convex optimization.

Some comments about organization and numbering conventions should be made. Each chapter is organized into sections, which are numbered sequentially within each chapter. Thus, “Section X.Y” refers to Section number Y in Chapter X. Named statements, specifically definitions, theorems, examples, remarks, et cetera, are numbered sequentially in each section, and share the same numbering. The same applies to equations, which have their own numbering. Thus “Statement X.Y.Z” refers to the named statement number Z in Section Y of Chapter X. “Equation (X.Y.Z)” is understood in a similar way. This may depart from established practice but, in our experience, it makes locating ideas and formulas easier. Figures are numbered sequentially within each chapter. Every named statement is meant to say something important. We have tried to encapsulate this by giving each such statement a short title.

The symbol □, as it is now customary, indicates the end of a proof. The symbols ■, ▲, and ◂ indicate the end of a definition, example, and remark, respectively. The end of an axiom is marked by the symbol •. The relation $\coloneqq$ indicates equality by definition. Thus $A \coloneqq B$ states that $A$ is defined to be equal to the (already defined) object $B$. Sometimes, during derivations and in the course of proofs, a relation is overset by a reference, say,

$$A \overset{(X.Y.Z)}{\leq} B.$$

This is used to state that said inequality holds because of the validity of formula or estimate (X.Y.Z). Finally, and this is a well-established tradition, for generic constants, we use the symbol $C$, and the value of such constant may change at each occurrence. If we want to make a constant specific, then we may decorate it with sub- or superindices, like $C_p$.

Finally, we must admit that no textbook is absolutely error free, and this text of ours is no exception. However, we hope that you will find it useful. If you do find it useful, then let us know — we will be glad to hear from you — and feel free to share your thoughts. If, while reading it carefully, you find typos, errors, and/or blunders, then kindly email them to us. We will be glad to fix the offending text and acknowledge your help in future iterations and editions. An updated errata file can be found in our GitHub repository.

Knoxville, TN, June 2025 AJS, SMW, CW

# Acknowledgment

We acknowledge funding from the National Science Foundation, which supported us in our research during the writing of the book. This project represents an attempt on our part to make the subject of multigrid and multilevel methods more accessible from the theoretical point of view and more interesting to a broader mathematics audience. At the same time, much of what we have learned at the frontiers of research in numerical partial differential equations and scientific computing has made it into the textbook.

We warmly thank our former PhD student Wenqiang Feng, who during his studies here at the University of Tennessee, typed the first draft of this manuscript from SMW's lecture notes, way back in the Fall of 2014. Many of the original figures were done by him in the LaTeX package `TikZ` and exist in the present text in only slightly modified forms. While this book has greatly expanded in content since those 2014 lecture notes, because he kindly shared his LaTeX files, we were able to jumpstart this book project a couple of years ago. So, Wenqiang, thanks for your generosity.

Thanks to Long Chen (University of California, Irvine) and Xiaozhe Hu (Tufts University) for granting us permission to use the LaTeX source for a previous paper [13] to begin our writing of the last two chapters. Even though we changed the LaTeX significantly, there may still be some words, phrases, and/or equations from [13] bleeding through. We thank Long Chen, separately and additionally, for making his `iFEM` [1] package readily available and for answering all our questions about it.

*AJS:* As a means of a thank you and acknowledgements, I would like to begin by telling a story. It was the Fall of 2003, and I was in the first semester of my senior year, thinking I wanted to pursue research in Functional Analysis and/or the Theory of Function Spaces. It was time to find an advisor, and at the suggestion of a former instructor, I went to talk to V.G. Korneev, who would become my BS and MS mentor. He was the first of many that showed me that in the Numerical Analysis of PDEs, there is a very neat synergy between PDEs, functional analysis, approximation theory, and many other subjects. More to the point, the very first paper he had me read was [29], which deals with the DD (a kind of subspace decomposition) preconditioning of the $p$-finite element method. This began my relationship with Multigrid and its variants. So, first of all, I would like to thank V.G. Korneev for being my first exposure to the topic. I also need to thank my PhD advisor J.-L. Guermond, who convinced Joe Pasciak to spend a semester trying to teach me the subject. Most importantly, I need to thank Steve Wise. He is the kindest human being I have ever met, and it was only through our conversations that I finally, kind-of, understood what Multigrid is about. Thank you Steve for bringing me along in this journey. We are one step closer to our grand goal. Finally, I cannot omit our MATLAB© guru and coauthor Calvin. You are destined for great things, and, hopefully, this will be the first step in that direction.

*SMW:* I warmly thank my postdoctoral office mate Junseok Kim, now a professor of mathematics at Korea University, who generously gave me his `C` source codes of several geometric multigrid solvers more than 20 years ago. I learned a lot about the struc-

https://doi.org/10.1515/9783111354880-204

ture of multigrid from those codes, and certainly some of the ideas from those lines of C have surely trickled down into the MATLAB© codes presented here. I also thank John S. Lowengrub, my postdoctoral advisor, for encouraging and nurturing my initial infatuation with multigrid all those years ago. I always wondered why the multigrid algorithms that we coded up, when we changed some operator here or there, worked sometimes better and often times worse. At first, I learned by trial and error only. As a professor, I set out to figure these things out theoretically by, naturally, acting like an expert and teaching courses on the subject. So a big thanks to my students, who endured several iterations of the lecture notes, which were really just the scribbles of someone trying to explain ideas to himself. Those lecture notes evolved into this book, and I hope that the explanations are finally clear (or a little clearer, at least). Thanks to my two coauthors and colaborers, Abner and Calvin. We have set out to make Tennessee a first-rate institution for numerical analysis training and research. The book is, as you say, Abner, a part of the grand goal. I think it turned out great.

*CW:* First, I would like to thank Dr. Wise, my coauthor and PhD advisor, for teaching me everything about multigrid — the most elegant method that incorporates modern numerical techniques to solve meaningful and large-scale problems. He is the kindest and coolest mathematician I've encountered, setting high standards for me and unconditionally helping me achieve them. I would also like to thank my other coauthor Dr. Salgado for meticulously writing and reviewing our work and for guiding me through many difficult aspects of this project. I extend my gratitude to Dr. Cory Hauck, my PhD co-advisor, for teaching me the importance of effective communication in scientific research. Special thanks to Pam Armentrout, the Graduate Coordinator of UTK Math, for her tremendous support, companionship, nurturing, and kindness throughout my PhD journey. My heartfelt appreciation goes to my classmates Jared Ohler, George Babus, and Evan Habbershaw for introducing me to American culture and sharing numerous enjoyable restaurant adventures. Lastly, I express my deepest love to my cats, Rocky and Stormy, for their unconditional love. The readers can even spot Stormy in Figure 7.8.

# Contents

**Preface — VII**

**Acknowledgment — XI**

**1 Classical iterative methods — 1**
1.1 Notation and basic facts — **1**
1.2 General linear iterative schemes (GLISs) — **6**
1.3 The spectral convergence theory — **7**
1.4 Splitting methods — **10**
1.4.1 Jacobi iterative method — **12**
1.4.2 The Gauss–Seidel method — **14**
1.4.3 Richardson's method — **19**
1.5 Additive, multiplicative, and symmetrized GLIS — **22**
1.5.1 Additive and multiplicative GLIS — **22**
1.5.2 Symmetrized GLIS — **23**
1.5.3 Symmetric Gauss–Seidel method — **25**
1.6 Convergence in the energy norm — **25**
1.6.1 The energy norm — **25**
1.6.2 Energy convergence of Gauss–Seidel methods — **29**
1.7 A computational experiment — **31**
1.8 Problems — **36**
Listings — **38**

**2 The preconditioned conjugate gradient method — 44**
2.1 Minimization of quadratic functions — **44**
2.2 The preconditioned steepest descent method — **46**
2.3 The zero-start preconditioned gradient descent method — **50**
2.4 Analysis of the zero-start PCG — **54**
2.5 The standard PCG — **56**
2.6 A computational experiment — **58**
2.7 Problems — **60**
Listings — **61**

**3 The two-grid algorithm — 70**
3.1 Components of the two-grid algorithm — **70**
3.2 The coarse-grid Ritz projection — **74**
3.3 The two-grid error transfer matrix — **77**
3.4 Problems — **80**

**4 Fourier analysis of the two-grid algorithm — 81**
4.1 Discretization of the model problem in 1D — **81**
4.1.1 The finite difference method — **81**
4.1.2 The finite element method — **82**
4.2 Properties of the stiffness matrix — **85**
4.3 The damped Jacobi smoother — **90**
4.4 The Gauss–Seidel smoother — **94**
4.5 Two-level grids and high-frequency smoothing — **95**
4.6 Prolongation and restriction operators — **99**
4.7 Other restriction matrices — **106**
4.8 Fourier error analysis — **107**
4.9 Some computational experiments — **117**
4.10 Problems — **126**
Listings — **127**

**5 An energy-norm analysis of the two-grid method — 135**
5.1 The approximation property — **135**
5.2 Richardson's smoother — **139**
5.3 The smoothing property — **140**
5.4 Convergence in the energy norm — **142**
5.5 Problems — **144**

**6 Multigrid: algorithms and axiomatic convergence theory — 145**
6.1 Basic building blocks of multigrid — **145**
6.2 Galerkin conditions — **152**
6.3 Strong and weak approximation properties — **157**
6.4 The first smoothing property — **158**
6.5 Convergence of the two-grid method revisited — **159**
6.6 Convergence of the W-cycle algorithm — **163**
6.7 Convergence of the simple symmetric V-cycle — **166**
6.8 The Braess–Hackbusch result — **172**
6.9 When the standard Galerkin condition fails — **175**
6.10 Problems — **186**

**7 Multigrid and the conforming finite element method — 187**
7.1 Description of the problem — **187**
7.2 Nested families of finite element spaces — **188**
7.2.1 Finite element approximation — **192**
7.2.2 Prolongation and restriction — **193**
7.3 Properties of the stiffness matrices — **195**
7.4 The strong approximation property — **201**
7.5 The full multigrid algorithm — **208**

7.6 Some computational experiments — 212
7.6.1 Numerical experiments in one dimension — 213
7.6.2 The iFEM package in MATLAB© — 213
7.7 Problems — 217
Listings — 218

**8 Cell-centered finite difference methods and multigrid — 230**
8.1 Discretization in one dimension — 230
8.2 The damped Jacobi and Richardson smoothers — 238
8.2.1 The damped Jacobi smoother — 238
8.2.2 The generic Richardson and Λ-Richardson smoothers — 239
8.2.3 The damped quasi-Jacobi method — 240
8.3 Smoothing properties — 241
8.4 Prolongation, restriction, and the imbalanced Galerkin condition — 246
8.5 Some additional multilevel tools — 249
8.6 The $\alpha$-weak approximation property — 252
8.7 The variable-diffusivity case — 258
8.8 Cell-centered multigrid in two dimensions — 261
8.8.1 The constant-diffusivity case — 265
8.8.2 The variable-diffusivity case — 270
8.8.3 Smoothers in two dimensions — 275
8.9 Some computational experiments — 278
8.9.1 A one-dimensional problem — 279
8.9.2 A two-dimensional elliptic problem with variable diffusivity — 281
8.10 Problems — 284
Listings — 285

**9 Multigrid as a multiplicative process — 311**
9.1 Multilevel matrices — 311
9.2 Properties of the multilevel matrices — 315
9.3 Multigrid error transfer matrices in multiplicative forms — 319
9.4 Problems — 324

**10 Additive preconditioners based on subspace decompositions — 325**
10.1 Subspace decompositions — 325
10.2 Hierarchical bases — 333
10.3 Hierarchical basis preconditioner — 336
10.4 The BPX preconditioner — 349
10.5 Some details on implementation — 355
10.6 A computational experiment — 357
10.7 Problems — 360
Listings — 361

**11 Convex optimization — 366**
11.1 Analysis on vector spaces — **366**
11.2 Convex sets and functions — **377**
11.3 Minimization of convex functionals — **381**
11.4 The gradient descent method — **387**
11.5 Problems — **391**

**12 Axiomatic methods of subspace corrections — 392**
12.1 Description of the problem and basic axioms — **392**
12.2 Successive subspace optimization methods — **394**
12.3 Full approximation storage method — **402**
12.4 The fast subspace descent (FASD) method — **405**
12.5 FASD with approximate line search — **417**
12.6 Problems — **423**

**13 FASD applied to nonlinear finite element problems — 424**
13.1 A model nonlinear problem — **424**
13.2 Finite element discretization and space decomposition — **427**
13.3 A one-dimensional implementation — **431**
13.4 Some computational experiments — **438**
13.5 Problems — **440**
Listings — **441**

**Bibliography — 449**

**Index — 451**

**List of Figures — 459**

**List of Tables — 461**

**List of Listings — 463**

**List of Symbols — 465**

# 1 Classical iterative methods

In this chapter, we discuss classical linear iterative methods for solving the linear system of equations

$$\mathsf{A}\boldsymbol{u} = \boldsymbol{f}.$$

We will usually be concerned with linear systems that arise from the approximation of boundary value problems (BVPs) for elliptic partial differential equations (PDEs). The prototypical example is the Poisson problem

$$-\Delta u = f \text{ in } \Omega, \quad u = 0 \text{ on } \partial\Omega.$$

The approximation will be carried out via the finite difference method (FDM) or finite element method (FEM). However, in this first chapter, we will not specifically take such a narrow view. Much of our discussion will remain at a general level. We only make the links to PDEs and discretization methods as applications of the general framework later in the book. We will always assume that $\mathsf{A} \in \mathbb{R}^{n\times n}$ is invertible, at least. For most of the text, we will assume that it is symmetric positive definite (SPD).

## 1.1 Notation and basic facts

The approach to multigrid that we follow in this text is a matrix-based one, at least for the first half of the book. For this reason, we begin by setting the notation for matrices and vectors. We will assume familiarity and maturity with normed vector spaces at the level of [31]. For example, we assume that the reader knows about the properties of eigenvalues and knows how to define an induced matrix norm.

We use the symbol $\mathbb{R}^n$ to denote the set of all vectors (or, more generally, $n$-tuples) of length $n$. This, of course, becomes a vector space with the usual, that is, componentwise, definitions of vector addition and scalar multiplication. We define the set

$$\mathbb{R}^n_\star := \{\boldsymbol{u} \in \mathbb{R}^n \mid \boldsymbol{u} \neq \boldsymbol{0}\}.$$

When we write $\boldsymbol{u} = [u_i] \in \mathbb{R}^n$, we will always understand this object as a column vector or, equivalently, as a matrix having $n$ rows and 1 column. To access the elements of a vector, we write

$$[\boldsymbol{u}]_j = u_j, \quad j \in \{1, \dots, n\}.$$

To simplify the notation, we will usually write $\boldsymbol{u}$ instead of $\boldsymbol{u} = [u_i]$, with the understanding that the elements of a vector are represented by the same letter. Vectors will be represented using bold typeface, whereas scalars will be written with the regular one.

https://doi.org/10.1515/9783111354880-001

To access the elements of a generic matrix $\mathsf{A} = [a_{i,j}] \in \mathbb{R}^{m\times n}$, we write

$$[\mathsf{A}]_{i,j} = a_{i,j}, \quad i \in \{1,\dots,m\},\ j \in \{1,\dots,n\}.$$

As with vectors, we will usually drop the $[a_{i,j}]$ part. Matrices will always be represented using capital letters in the sans serif typeface. Matrix–vector multiplication is defined as usual: for $\mathsf{A} \in \mathbb{R}^{m\times n}$ and $\boldsymbol{u} \in \mathbb{R}^n$, the product $\mathsf{A}\boldsymbol{u}$ is a vector in $\mathbb{R}^m$ with components

$$[\mathsf{A}\boldsymbol{u}]_i := \sum_{j=1}^{n} a_{i,j}u_j, \quad i \in \{1,\dots,m\}.$$

Matrix–matrix multiplication is also defined as usual: for $\mathsf{A} = [a_{i,j}] \in \mathbb{R}^{m\times p}$ and $\mathsf{B} = [b_{i,j}] \in \mathbb{R}^{p\times n}$, the product $\mathsf{AB}$ is a matrix in $\mathbb{R}^{m\times n}$ whose components are

$$[\mathsf{AB}]_{i,j} := \sum_{k=1}^{p} a_{i,k}b_{k,j}, \quad i \in \{1,\dots,m\},\ j \in \{1,\dots,n\}.$$

Suppose that $\mathsf{A} = [a_{i,j}] \in \mathbb{R}^{m\times n}$. The transpose of $\mathsf{A}$, written $\mathsf{A}^\top$, is a matrix in $\mathbb{R}^{n\times m}$ with components

$$[\mathsf{A}^\top]_{i,j} := a_{j,i}, \quad i \in \{1,\dots,n\},\ j \in \{1,\dots,m\}.$$

For $\boldsymbol{u}, \boldsymbol{v} \in \mathbb{R}^n$, the object

$$\boldsymbol{v}^\top\boldsymbol{u} = \sum_{i=1}^{n} v_i u_i$$

is a real number. This definition can be viewed in the context of matrix vector multiplication. On the other hand, $\boldsymbol{v}\boldsymbol{u}^\top$ is a matrix $\mathbb{R}^{n\times n}$ with components

$$[\boldsymbol{v}\boldsymbol{u}^\top]_{i,j} = v_i u_j, \quad i,j \in \{1,\dots,n\}.$$

This can also be deduced from matrix–matrix multiplication.

**Definition 1.1.1** (Euclidean inner product). Define the mapping $(\cdot,\cdot) : \mathbb{R}^n \times \mathbb{R}^n \to \mathbb{R}$ via

$$(\boldsymbol{u}, \boldsymbol{v}) := \boldsymbol{v}^\top\boldsymbol{u},$$

for all $\boldsymbol{u}, \boldsymbol{v} \in \mathbb{R}^n$. This object is called the **canonical inner product** or **Euclidean inner product.** ■

**Definition 1.1.2** (Euclidean norm). The mapping $\|\cdot\| : \mathbb{R}^n \to \mathbb{R}$ defined by

$$\|\boldsymbol{u}\| = \sqrt{(\boldsymbol{u}, \boldsymbol{u})}, \quad \forall\, \boldsymbol{u} \in \mathbb{R}^n,$$

is the **Euclidean or $\ell^2$-norm.** ■

The vector space $\mathbb{R}^n$ can be normed in many different ways. The so-called $\ell^p$-norms are often useful.

**Definition 1.1.3** ($\ell^p$-norm). Let $p \in [1, \infty)$. The $\ell^p$**-norm** of a vector $\boldsymbol{v} \in \mathbb{R}^n$ is defined as

$$\|\boldsymbol{v}\|_{\ell^p} := \left( \sum_{i=1}^{n} |v_i|^p \right)^{1/p}.$$

Similarly, the $\ell^\infty$**-norm** is

$$\|\boldsymbol{v}\|_{\ell^\infty} := \max_{i=1}^{n} |v_i|. \quad \blacksquare$$

For $p \in [1, \infty]$ and a matrix $\mathsf{A} \in \mathbb{R}^{n \times n}$, we denote by $\|\mathsf{A}\|_p$ the norm induced by the $\ell^p$-norm on $\mathbb{R}^n$. In what follows, unless expressly stated otherwise, whenever we write $\|\cdot\|$, we mean the Euclidean norm or the norm induced by the Euclidean one. Of course, we will quickly observe that

$$\|\cdot\| = \|\cdot\|_{\ell^2}.$$

Positive definite matrices will feature prominently in our discussions, so we discuss them in more detail.

**Definition 1.1.4** (Symmetric matrix). We say that $\mathsf{A} \in \mathbb{R}^{n \times n}$ is **symmetric** iff, for all $\boldsymbol{u}, \boldsymbol{v} \in \mathbb{R}^n$,

$$(\mathsf{A}\boldsymbol{u}, \boldsymbol{v}) = (\boldsymbol{u}, \mathsf{A}\boldsymbol{v}).$$

We say that A is **symmetric positive definite** (SPD) iff it is symmetric and, for all $\boldsymbol{u} \in \mathbb{R}^n_\star$,

$$\boldsymbol{u}^\top \mathsf{A}\boldsymbol{u} = (\mathsf{A}\boldsymbol{u}, \boldsymbol{u}) > 0. \quad \blacksquare$$

Let $\mathsf{A} \in \mathbb{R}^{n \times n}$. Define, for all $\boldsymbol{u}, \boldsymbol{v} \in \mathbb{R}^n$,

$$(\boldsymbol{u}, \boldsymbol{v})_{\mathsf{A}} := (\mathsf{A}\boldsymbol{u}, \boldsymbol{v}) = \boldsymbol{v}^\top \mathsf{A}\boldsymbol{u}. \tag{1.1.1}$$

The reader can easily show the following two results.

**Proposition 1.1.5** (Symmetry criterion). *A matrix* $\mathsf{A} \in \mathbb{R}^{n \times n}$ *is symmetric iff* $\mathsf{A} = \mathsf{A}^\top$.

*Proof.* Exercise; see Problem 1.1. □

**Proposition 1.1.6** (A-inner product). *Let* $\mathsf{A} \in \mathbb{R}^{n \times n}$ *be SPD. Then* (1.1.1) *defines an inner product on* $\mathbb{R}^n$.

*Proof.* Exercise; see Problem 1.2. □

We comment that, when A is SPD, we call the object in (1.1.1) the **A-inner product.** In this case, the **A-norm** is then defined as

$$\|\boldsymbol{w}\|_{\mathsf{A}} := \sqrt{(\boldsymbol{w}, \boldsymbol{w})_{\mathsf{A}}}, \quad \forall\, \boldsymbol{w} \in \mathbb{R}^n.$$

There are many possible inner products on $\mathbb{R}^n$. It may be that a matrix is not symmetric in the usual sense but is symmetric with respect to some other inner product.

**Definition 1.1.7** (General symmetry). Suppose that $(\cdot,\cdot)_\star : \mathbb{R}^n \times \mathbb{R}^n \to \mathbb{R}$ is an inner product, not necessarily Euclidean. We say that $\mathsf{A} \in \mathbb{R}^{n\times n}$ is **symmetric with respect to** $(\cdot,\cdot)_\star$ iff, for all $\boldsymbol{u}, \boldsymbol{v} \in \mathbb{R}^n$,

$$(\mathsf{A}\boldsymbol{u}, \boldsymbol{v})_\star = (\boldsymbol{u}, \mathsf{A}\boldsymbol{v})_\star.$$

We say that A is **symmetric positive definite with respect to** $(\cdot,\cdot)_\star$ iff it is symmetric with respect $(\cdot,\cdot)_\star$ and, for all $\boldsymbol{u} \in \mathbb{R}^n_\star$,

$$(\mathsf{A}\boldsymbol{u}, \boldsymbol{u})_\star > 0.$$

■

We will take advantage of the following keystone result several times in the text.

**Theorem 1.1.8** (Spectral decomposition theorem). *If* $\mathsf{A} \in \mathbb{R}^{n\times n}$ *is symmetric in the Euclidean sense, then all its eigenvalues are real. Moreover, there is an orthonormal (with respect to the Euclidean inner product) basis of* $\mathbb{R}^n$ *consisting of eigenvectors of* A.

*Proof.* See [31, Chap. 01]. □

This can be generalized, as you might imagine.

**Theorem 1.1.9** (General spectral decomposition theorem). *Suppose that* $(\cdot,\cdot)_\star : \mathbb{R}^n \times \mathbb{R}^n \to \mathbb{R}$ *is an inner product. If* $\mathsf{A} \in \mathbb{R}^{n\times n}$ *is symmetric with respect to* $(\cdot,\cdot)_\star$, *then all its eigenvalues are real. Moreover, there is an orthonormal (with respect to* $(\cdot,\cdot)_\star$*) basis of* $\mathbb{R}^n$ *consisting of eigenvectors of* A.

*Proof.* See [31, Chap. 07]. □

**Proposition 1.1.10** (SPD criterion). *A matrix* $\mathsf{A} \in \mathbb{R}^{n\times n}$ *is SPD iff* $\mathsf{A} = \mathsf{A}^\top$ *and all of its eigenvalues are positive.*

*Proof.* Exercise; see Problem 1.3. □

This last result generalizes as well.

**Proposition 1.1.11** (General SPD criterion). *Suppose that* $(\cdot,\cdot)_\star : \mathbb{R}^n \times \mathbb{R}^n \to \mathbb{R}$ *is an inner product. A matrix* $\mathsf{A} \in \mathbb{R}^{n\times n}$ *is SPD with respect to* $(\cdot,\cdot)_\star$ *iff* A *is symmetric with respect to* $(\cdot,\cdot)_\star$ *and all of its eigenvalues are positive.*

*Proof.* Exercise; see Problem 1.4. □

**Proposition 1.1.12** (Product of SPD matrices). *Suppose that* $\mathsf{A}, \mathsf{B} \in \mathbb{R}^{n\times n}$ *are symmetric positive definite matrices in the Euclidean sense. Then* $(\cdot,\cdot)_{\mathsf{A}}$ *and* $(\cdot,\cdot)_{\mathsf{B}^{-1}}$ *are inner products on* $\mathbb{R}^n$. *Moreover, the product matrix* $\mathsf{BA}$ *is symmetric and positive definite with respect to these inner products. Consequently, all of the eigenvalues of* $\mathsf{BA}$ *are positive, and there is an orthonormal (with respect to either of the new inner products) basis of eigenvectors.*

*Proof.* As Proposition 1.1.6 shows, $(\cdot,\cdot)_{\mathsf{A}}$ and $(\cdot,\cdot)_{\mathsf{B}^{-1}}$ are indeed inner products. We now show that the product matrix $\mathsf{BA}$ is SPD with respect to $(\cdot,\cdot)_{\mathsf{B}^{-1}}$, the proof of the fact that $\mathsf{BA}$ is SPD with respect to $(\cdot,\cdot)_{\mathsf{A}}$ being similar. Let $\boldsymbol{u}, \boldsymbol{v} \in \mathbb{R}^n$ be arbitrary. Then

$$\begin{aligned}(\mathsf{BA}\boldsymbol{u}, \boldsymbol{v})_{\mathsf{B}^{-1}} &= \boldsymbol{v}^\top \mathsf{B}^{-1}\mathsf{BA}\boldsymbol{u} \\ &= \boldsymbol{v}^\top \mathsf{A}\boldsymbol{u} \\ &= \boldsymbol{v}^\top \mathsf{ABB}^{-1}\boldsymbol{u} \\ &= \boldsymbol{v}^\top \mathsf{A}^\top \mathsf{B}^\top \mathsf{B}^{-1}\boldsymbol{u} \\ &= (\mathsf{BA}\boldsymbol{v})^\top \mathsf{B}^{-1}\boldsymbol{u} \\ &= (\boldsymbol{u}, \mathsf{BA}\boldsymbol{v})_{\mathsf{B}^{-1}}.\end{aligned}$$

This shows that $\mathsf{BA}$ is symmetric with respect to $(\cdot,\cdot)_{\mathsf{B}^{-1}}$. Now suppose that $\boldsymbol{u} \in \mathbb{R}^n_\star$. Then

$$(\mathsf{BA}\boldsymbol{u}, \boldsymbol{u})_{\mathsf{B}^{-1}} = \boldsymbol{u}^\top \mathsf{B}^{-1}\mathsf{BA}\boldsymbol{u} = \boldsymbol{u}^\top \mathsf{A}\boldsymbol{u} > 0.$$

This shows that $\mathsf{BA}$ is positive definite with respect to $(\cdot,\cdot)_{\mathsf{B}^{-1}}$.

The other points follow from the general spectral decomposition theorem and some simple calculations; see [31, Chap. 07]. □

The previous result, together with Proposition 1.1.6, suggests that SPD matrices and inner products are essentially the same objects, that is, they can be identified.

**Proposition 1.1.13** (General inner product). *Suppose that* $(\cdot,\cdot)_\star : \mathbb{R}^n \times \mathbb{R}^n \to \mathbb{R}$ *is an inner product. Then there exists a unique SPD matrix* $\mathsf{A} \in \mathbb{R}^{n\times n}$ *such that for all* $\boldsymbol{u}, \boldsymbol{v} \in \mathbb{R}^n$,

$$(\mathsf{A}\boldsymbol{u}, \boldsymbol{v}) = (\boldsymbol{u}, \boldsymbol{v})_\star.$$

*Proof.* Exercise; see Problem 1.5. □

Finally, we discuss the notion of condition number.

**Definition 1.1.14** (Condition number). Let $\|\cdot\|_\star$ be a norm on $\mathbb{R}^{n\times n}$. For any nonsingular matrix $\mathsf{A} \in \mathbb{R}^{n\times n}$, its $\|\cdot\|_\star$**-condition number** is defined as

$$\kappa_\star(\mathsf{A}) := \|\mathsf{A}\|_\star \|\mathsf{A}^{-1}\|_\star.$$

If the norm in question is induced by the Euclidean norm, then the corresponding condition number is called the **spectral condition number** and denoted simply by $\kappa(\mathsf{A})$. ■

For SPD matrices, the condition number can be computed as the ratio of eigenvalues.

**Proposition 1.1.15** (Condition number for SPD matrices). *Let* $\mathsf{A} \in \mathbb{R}^{n\times n}$ *be SPD with respect to the norm induced by some inner product* $(\cdot,\cdot)_\star$. *Then its* $\|\cdot\|_\star$*-condition number can be computed as*

$$\kappa_\star(\mathsf{A}) = \frac{\max\{\lambda | \lambda \in \sigma(\mathsf{A})\}}{\min\{\lambda | \lambda \in \sigma(\mathsf{A})\}}.$$

*Proof.* Exercise; see Problem 1.6. □

## 1.2 General linear iterative schemes (GLISs)

**Definition 1.2.1** (General linear iterative scheme (GLIS)). Suppose that $\mathsf{A} \in \mathbb{R}^{n\times n}$ is invertible, $\boldsymbol{f} \in \mathbb{R}^n$, and $\boldsymbol{u} = \mathsf{A}^{-1}\boldsymbol{f}$. Assume that $\mathsf{B} \in \mathbb{R}^{n\times n}$ is invertible and $\boldsymbol{u}^0 \in \mathbb{R}^n$. A **general linear iterative scheme (GLIS)** is an algorithm for determining the terms of the sequence $\{\boldsymbol{u}^k\}_{k=1}^{\infty} \subset \mathbb{R}^n$ via the recursive iteration

$$\boldsymbol{u}^{k+1} := \boldsymbol{u}^k + \mathsf{B}(\boldsymbol{f} - \mathsf{A}\boldsymbol{u}^k). \tag{1.2.1}$$

The matrix B is called the **iterator** of the GLIS. The **error sequence** is defined via

$$\boldsymbol{e}^k := \boldsymbol{u} - \boldsymbol{u}^k.$$

The **residual sequence** is defined via

$$\boldsymbol{r}^k := \boldsymbol{f} - \mathsf{A}\boldsymbol{u}^k = \mathsf{A}\boldsymbol{e}^k.$$

We say that the GLIS (1.2.1) **converges unconditionally** iff $\lim_{k\to\infty} \boldsymbol{e}^k = \boldsymbol{0}$, for all initial vectors $\boldsymbol{u}^0 \in \mathbb{R}^n$. ■

**Remark 1.2.2** (Properties of the iterator). The application of B should be simple and computationally inexpensive. In some sense, B should approximate $\mathsf{A}^{-1}$. In the extreme case that $\mathsf{B} = \mathsf{A}^{-1}$, the GLIS converges after one iteration. ◂

**Remark 1.2.3** (Preconditioner). In the GLIS (1.2.1), the matrix B is often called a *preconditioner* of A, especially, when A and B are both symmetric positive definite (SPD). Suppose $\boldsymbol{u}^0 = \boldsymbol{0}$. Then

$$\boldsymbol{u}^1 = \mathsf{B}\boldsymbol{f}.$$

Thus the action of the preconditioner B on a vector $\boldsymbol{f}$ is equivalent to one iteration of the GLIS with $\boldsymbol{u}^0 = \boldsymbol{0} \in \mathbb{R}^n$. A GLIS (where B is SPD) may be used as (or viewed as) a preconditioner for a Krylov method, like the conjugate gradient method; see Chapter 2. A "good" preconditioner has the property that

$$\kappa_\star(\mathsf{BA}) = \frac{\lambda_n(\mathsf{BA})}{\lambda_1(\mathsf{BA})} = \mathcal{O}(1),$$

where

$$0 < \lambda_1(\mathsf{BA}) \le \lambda_2(\mathsf{BA}) \le \cdots \le \lambda_n(\mathsf{BA})$$

are the ordered positive eigenvalues of the matrix BA. Note that this matrix is SPD with respect the inner products $(\cdot,\cdot)_{\mathsf{A}}$ and $(\cdot,\cdot)_{\mathsf{B}^{-1}}$, so we may use Proposition 1.1.15. BA is not typically even symmetric with respect to the usual Euclidean inner product. See [31, Chap. 07] for more detail. ◂

It is easy to see that $\boldsymbol{u} = \mathsf{A}^{-1}\boldsymbol{f}$ is a fixed point of the GLIS:

$$\boldsymbol{u} = \boldsymbol{u} + \mathsf{B}(\boldsymbol{f} - \mathsf{A}\boldsymbol{u}).$$

From the last equation let us subtract (1.2.1) to obtain

$$\boldsymbol{e}^{k+1} = \boldsymbol{e}^k - \mathsf{BA}\boldsymbol{e}^k = (\mathrm{I} - \mathsf{BA})\boldsymbol{e}^k.$$

The matrix $\mathrm{I} - \mathsf{BA}$ has an important role in the analysis of the GLIS. We give it a name.

**Definition 1.2.4** (Error transfer matrix). Suppose that $\mathsf{A} \in \mathbb{R}^{n\times n}$ is invertible, $\boldsymbol{f} \in \mathbb{R}^n$, $\boldsymbol{u} = \mathsf{A}^{-1}\boldsymbol{f}$, and $\boldsymbol{u}^k$ is the GLIS iteration at the $k$th step. The **GLIS error transfer matrix** is defined as

$$\mathsf{T} := \mathrm{I} - \mathsf{BA}. \tag{1.2.2}$$

It satisfies

$$\boldsymbol{e}^k = \mathsf{T}\boldsymbol{e}^{k-1},$$

for each $k \in \mathbb{N}$, where $\boldsymbol{e}^k = \boldsymbol{u} - \boldsymbol{u}^k$ is the error vector. ■

## 1.3 The spectral convergence theory

**Definition 1.3.1** (Convergent matrix). Let $\mathsf{C} \in \mathbb{R}^{n\times n}$. C is called **convergent** iff, for every $\epsilon > 0$, there exists $K \in \mathbb{N}$ such that, if $k \ge K$, then, for all $1 \le i,j \le n$,

$$\left|[\mathsf{C}^k]_{i,j}\right| \le \epsilon,$$

where

$$\mathsf{C}^k = \prod_{m=1}^{k} \mathsf{C}.$$ ■

**Theorem 1.3.2** (Convergence criteria). *The following statements are equivalent*

1. $\mathsf{C} \in \mathbb{R}^{n\times n}$ *is convergent.*
2. *There is an induced matrix norm* $\|\cdot\|_\star : \mathbb{R}^{n\times n} \to \mathbb{R}$ *for which*
$$\lim_{k\to\infty} \|\mathsf{C}^k\|_\star = 0.$$
3. *For any induced matrix norm* $\|\cdot\|_\star$,
$$\lim_{k\to\infty} \|\mathsf{C}^k\|_\star = 0.$$
4. $\rho(\mathsf{C}) < 1$, *where* $\rho(\mathsf{C})$ *is the spectral radius of* $\mathsf{C}$,
$$\rho(\mathsf{C}) := \max_{1\le i\le n}\{|\lambda_i| \mid \lambda_i \in \sigma(\mathsf{C})\},$$
*and* $\sigma(\mathsf{C})$ *is the spectrum of* $\mathsf{C}$.
5. *For every* $\boldsymbol{x} \in \mathbb{R}^n$,
$$\lim_{k\to\infty} \mathsf{C}^k\boldsymbol{x} = \boldsymbol{0}.$$

*Proof.* See [31, Chaps. 4 and 6]. □

We have the following convergence result.

**Theorem 1.3.3** (Convergence of GLIS). *Suppose that* $\mathsf{A}, \mathsf{B} \in \mathbb{R}^{n\times n}$ *are invertible,* $\boldsymbol{f} \in \mathbb{R}^n$, *and* $\boldsymbol{u} = \mathsf{A}^{-1}\boldsymbol{f}$. *The GLIS* (1.2.1) *converges unconditionally iff*

$$\rho(\mathsf{T}) < 1,$$

*where* $\mathsf{T} = \mathsf{I} - \mathsf{BA}$ *is the error transfer matrix.*

*Proof.* ($\Longrightarrow$): Suppose (1.2.1) converges for any $\boldsymbol{u}^0 \in \mathbb{R}^n$. (It must converge to $\boldsymbol{u} \in \mathbb{R}^n$, incidentally.) Then

$$\lim_{k\to\infty} \boldsymbol{e}^k = \boldsymbol{0}.$$

Let $(\lambda, \boldsymbol{w}) \in \mathbb{C} \times \mathbb{C}^n$ be an eigenpair of $\mathsf{T}$ with $\|\boldsymbol{w}\|_{\ell^\infty} = 1$, and set $\boldsymbol{e}^0 = \boldsymbol{w}$. (Note that we must resort to the complex vector space, generically, since a matrix with all real entries need not have real eigenvalues.) Then

$$\boldsymbol{e}^k = \mathsf{T}^k\boldsymbol{e}^0 = \lambda^k\boldsymbol{e}^0.$$

Thus

$$\|\boldsymbol{e}^k\|_{\ell^\infty} = |\lambda|^k \to 0.$$

This implies that $|\lambda| < 1$. Since $\lambda \in \sigma(\mathsf{T})$ was arbitrary,

$$\rho(\mathsf{T}) < 1.$$

($\Longleftarrow$): If $\rho(\mathsf{T}) < 1$, then by Theorem 1.3.2

$$\lim_{k\to\infty} \mathsf{T}^k \boldsymbol{x} = \mathbf{0},$$

for all $\boldsymbol{x} \in \mathbb{R}^n$. Hence

$$\lim_{k\to\infty} \boldsymbol{e}^k = \lim_{k\to\infty} \mathsf{T}^k \boldsymbol{e}^0 = \mathbf{0},$$

and the proof is complete. □

**Theorem 1.3.4** (Sufficient condition for convergence). *If there is an induced matrix norm $\|\cdot\|_\star : \mathbb{R}^{n\times n} \to \mathbb{R}$ for which $\|\mathsf{T}\|_\star < 1$, then the GLIS* (1.2.1) *converges for any starting point $\boldsymbol{u}^0 \in \mathbb{R}^n$. Furthermore, we have the error estimates*

$$\|\boldsymbol{u} - \boldsymbol{u}^k\|_\star \leq \|\mathsf{T}\|_\star^k \|\boldsymbol{u} - \boldsymbol{u}^0\|_\star$$

*and*

$$\|\boldsymbol{u} - \boldsymbol{u}^k\|_\star \leq \frac{\|\mathsf{T}\|_\star^k}{1 - \|\mathsf{T}\|_\star} \|\boldsymbol{u}^1 - \boldsymbol{u}^0\|_\star.$$

*Proof.* We use the well-known fact that for any induced matrix norm $\|\cdot\|_\star$,

$$\rho(\mathsf{T}) \leq \|\mathsf{T}\|_\star.$$

Thus, if $\|\mathsf{T}\|_\star < 1$, then $\rho(\mathsf{T}) < 1$. Using Theorem 1.3.3, we get the desired convergence result.

Next, let us establish the error estimates. Using the consistency and submultiplicativity of the induced norm, we have

$$\begin{aligned} \|\boldsymbol{e}^k\|_\star &= \|\mathsf{T}^k \boldsymbol{e}^0\|_\star \\ &\leq \|\mathsf{T}^k\|_\star \|\boldsymbol{e}^0\|_\star \\ &\leq \|\mathsf{T}\|_\star^k \|\boldsymbol{e}^0\|_\star, \end{aligned}$$

that is,

$$\|\boldsymbol{u} - \boldsymbol{u}^k\|_\star \leq \|\mathsf{T}\|_\star^k \|\boldsymbol{u} - \boldsymbol{u}^0\|_\star,$$

which is the first error estimate.

Now write

$$\begin{aligned}\|e^0\|_\star &= \|u - u^0\|_\star \\ &= \|u - u^1 + u^1 - u^0\|_\star \\ &\le \|u - u^1\|_\star + \|u^1 - u^0\|_\star.\end{aligned}$$

Since

$$u - u^1 = e^1 = \mathsf{T}e^0 = \mathsf{T}(u - u^0),$$

it follows that

$$\|u - u^1\|_\star \le \|\mathsf{T}\|_\star \|u - u^0\|_\star = \|\mathsf{T}\|_\star \|e^0\|_\star.$$

Therefore

$$\begin{aligned}\|e^0\|_\star &= \|u - u^1 + u^1 - u^0\|_\star \\ &\le \|u - u^1\|_\star + \|u^1 - u^0\|_\star \\ &\le \|\mathsf{T}\|_\star \|e^0\|_\star + \|u^1 - u^0\|_\star.\end{aligned}$$

Hence

$$\|e^0\|_\star \le \frac{1}{1 - \|\mathsf{T}\|_\star} \|u^1 - u^0\|_\star.$$

Substituting into the first estimate yields

$$\begin{aligned}\|e^k\|_\star &\le \|\mathsf{T}\|_\star^k \|u - u^0\|_\star \\ &= \|\mathsf{T}\|_\star^k \|e^0\|_\star \\ &\le \frac{\|\mathsf{T}\|_\star^k}{1 - \|\mathsf{T}\|_\star} \|u^1 - u^0\|_\star,\end{aligned}$$

which is the desired second error estimate. □

## 1.4 Splitting methods

In this section, we define a couple of the classical iterative methods that we meet in an elementary course on numerical analysis, namely, the Jacobi, Gauss–Seidel, and Richardson methods. We analyze them using the spectral theory introduced in the last section.

**Definition 1.4.1** (Canonical splitting). Let $\mathsf{A} \in \mathbb{R}^{n\times n}$. Consider

$$\mathsf{A} = \mathsf{D} - \mathsf{L} - \mathsf{U}, \tag{1.4.1}$$

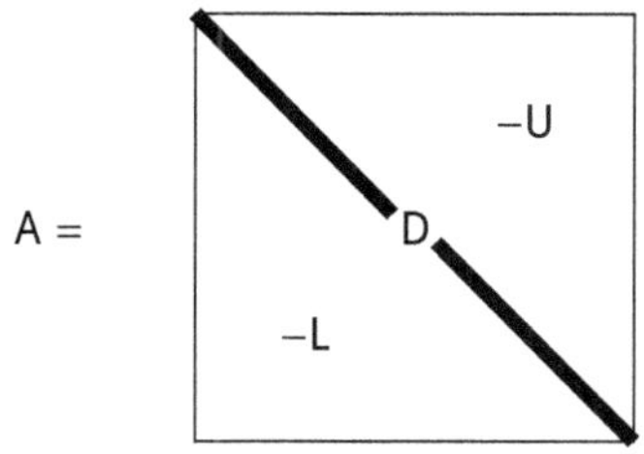


**Figure 1.1:** The canonical splitting of A.

where $D \in \mathbb{R}^{n\times n}$ is the diagonal of A, $-L \in \mathbb{R}^{n\times n}$ is the strictly lower triangular part of A, and $-U \in \mathbb{R}^{n\times n}$ is the strictly upper triangular part of A, respectively, as illustrated in Figure 1.1. Decomposition (1.4.1) is called the **canonical splitting of** A or **diagonal splitting of** A. ■

**Example 1.4.2** (Canonical splitting). Suppose that

$$A = \begin{bmatrix} 10 & -2 & -10 & 5 & -10 \\ 10 & 9 & 7 & -2 & -5 \\ 0 & 6 & 9 & 3 & -10 \\ 6 & 10 & 4 & -7 & -8 \\ -8 & 3 & 5 & 4 & 7 \end{bmatrix}.$$

Then

$$D = \begin{bmatrix} 10 & 0 & 0 & 0 & 0 \\ 0 & 9 & 0 & 0 & 0 \\ 0 & 0 & 9 & 0 & 0 \\ 0 & 0 & 0 & -7 & 0 \\ 0 & 0 & 0 & 0 & 7 \end{bmatrix},$$

$$U = \begin{bmatrix} 0 & 2 & 10 & -5 & 10 \\ 0 & 0 & -7 & 2 & 5 \\ 0 & 0 & 0 & -3 & 10 \\ 0 & 0 & 0 & 0 & 8 \\ 0 & 0 & 0 & 0 & 0 \end{bmatrix}, \quad L = \begin{bmatrix} 0 & 0 & 0 & 0 & 0 \\ -10 & 0 & 0 & 0 & 0 \\ 0 & -6 & 0 & 0 & 0 \\ -6 & -10 & -4 & 0 & 0 \\ 8 & -3 & -5 & -4 & 0 \end{bmatrix}.$$ ▲

**Proposition 1.4.3** (Splitting of SPD matrices). *If* A *is SPD, then* D *has only positive diagonal elements, and* $U = L^{\top}$.

*Proof.* Exercise; see Problem 1.7. □

### 1.4.1 Jacobi iterative method

Suppose that $A = D - L - U$ is the canonical splitting of A. Assume that D is invertible. Then the linear system $A\boldsymbol{u} = \boldsymbol{f}$ can be rewritten as

$$D\boldsymbol{u} = (U + L)\boldsymbol{u} + \boldsymbol{f}.$$

Jacobi's method results from the iteration

$$D\boldsymbol{u}^{k+1} := (U + L)\boldsymbol{u}^k + \boldsymbol{f}.$$

Put another way, we have the following:

**Definition 1.4.4** (The Jacobi method). Let $A = D - L - U$ be the canonical splitting of $A \in \mathbb{R}^{n\times n}$, where the diagonal matrix D is invertible. **The Jacobi method** is a GLIS with the iterator matrix

$$B = B_J := D^{-1}$$

and error transfer matrix

$$T = T_J := I - D^{-1}A.$$

The iteration sequence is defined by

$$\boldsymbol{u}^{k+1} := \boldsymbol{u}^k + D^{-1}(\boldsymbol{f} - A\boldsymbol{u}^k). \tag{1.4.2}$$

■

**Remark 1.4.5** (Error transfer). The error transfer matrix for the Jacobi method can be expressed as

$$\begin{aligned} T_J &= D^{-1}D - D^{-1}A \\ &= D^{-1}(D - D + U + L) \\ &= D^{-1}(U + L). \end{aligned}$$

◀

**Remark 1.4.6** (Component form). In component form, the Jacobi method can be written as

$$u_i^{k+1} = \frac{f_i - \sum_{j=1}^{i-1} a_{i,j}u_j^k - \sum_{j=i+1}^{n} a_{i,j}u_j^k}{a_{i,i}}.$$

Notice that the order in which we obtain the updated components is completely immaterial.

◀

**Definition 1.4.7** (Diagonal dominance). We say that a matrix $\mathsf{A} \in \mathbb{R}^{n\times n}$ is **diagonally dominant** or **rowwise diagonally dominant** if, for each $i \in \{1, \dots, n\}$,

$$\sum_{\substack{j=1\\j\neq i}}^{n} |a_{i,j}| < |a_{i,i}|. \qquad ■$$

**Proposition 1.4.8** (Diagonal dominant matrix). *If $\mathsf{A} \in \mathbb{R}^{n\times n}$ is diagonally dominant, then none of its diagonal elements is zero, and it is invertible.*

*Proof.* Exercise; see Problem 1.8. □

**Theorem 1.4.9** (Convergence of the Jacobi method). *If $\mathsf{A} \in \mathbb{R}^{n\times n}$ is diagonally dominant, then the Jacobi method is well defined and unconditionally convergent.*

*Proof.* We will show that $\|\mathsf{T}_\mathrm{J}\|_\infty < 1$. Since $\mathsf{A}$ is diagonally dominant, its diagonal elements are all nonzero. Thus the method is well defined because $\mathsf{D}$ is invertible. Now observe that

$$\mathsf{T}_\mathrm{J} = \begin{bmatrix} 1 & & 0 \\ & \ddots & \\ 0 & & 1 \end{bmatrix} - \begin{bmatrix} \frac{1}{a_{1,1}} & & 0 \\ & \ddots & \\ 0 & & \frac{1}{a_{n,n}} \end{bmatrix} \begin{bmatrix} a_{1,1} & \cdots & a_{1,n} \\ \vdots & \ddots & \vdots \\ a_{n,1} & \cdots & a_{n,n} \end{bmatrix}.$$

Thus, $\mathsf{T}_\mathrm{J} = [t_{i,j}]$ satisfies

$$t_{i,j} = \begin{cases} 0, & i = j, \\ -\dfrac{a_{i,j}}{a_{i,i}}, & i \neq j. \end{cases}$$

So

$$\|\mathsf{T}_\mathrm{J}\|_\infty = \max_{1\le i\le n} \sum_{j=1}^{n} |t_{i,j}| = \max_{1\le i\le n} \sum_{\substack{j=1\\j\neq i}}^{n} \left|\frac{a_{i,j}}{a_{i,i}}\right|.$$

Since $\mathsf{A}$ is diagonally dominant, there is $\delta > 0$ such that

$$|a_{i,i}| \ge \sum_{\substack{j=1\\j\neq i}}^{n} |a_{i,j}| + \delta$$

for each $i = 1, \dots, n$. Therefore, for each $i = 1, \dots, n$,

$$\sum_{\substack{j=1\\j\neq i}}^{n} \left|\frac{a_{i,j}}{a_{i,i}}\right| \le 1 - \frac{\delta}{|a_{i,i}|} \le 1 - \frac{\delta}{\alpha},$$

where

$$\alpha = \max_{1\le i\le n} |a_{i,i}|.$$

Hence

$$\|\mathsf{T}_J\|_\infty \le 1 - \frac{\delta}{\alpha} < 1. \qquad \square$$

**Definition 1.4.10** (Damped Jacobi method). Let $\mathsf{A} = \mathsf{D} - \mathsf{L} - \mathsf{U}$ be the canonical splitting of $\mathsf{A} \in \mathbb{R}^{n\times n}$, and let $\omega \in (0,1]$. Assume that D is invertible. The **damped Jacobi method**, or **weighted Jacobi method**, is defined by the iteration

$$\begin{aligned} \boldsymbol{z}^k &:= \boldsymbol{u}^k + \mathsf{D}^{-1}(\boldsymbol{f} - \mathsf{A}\boldsymbol{u}^k), \\ \boldsymbol{u}^{k+1} &:= \omega \boldsymbol{z}^k + (1-\omega)\boldsymbol{u}^k, \end{aligned}$$

where the starting value $\boldsymbol{u}^0 \in \mathbb{R}^n$ is given. ■

Thus the damped Jacobi method can be viewed as doing one step of the Jacobi method, followed by a damping, or weighting, step. We can eliminate $\boldsymbol{z}^k$ in the second equation to obtain the following:

**Proposition 1.4.11** (Damped Jacobi method). *With the same assumptions as in the last definition, the damped Jacobi method is a GLIS with iterator*

$$\mathsf{B} = \mathsf{B}_{J,\omega} := \omega \mathsf{D}^{-1}$$

*and error transfer matrix*

$$\mathsf{T} = \mathsf{T}_{J,\omega} := \mathrm{I} - \omega \mathsf{D}^{-1}\mathsf{A}.$$

*The iteration sequence can be expressed as*

$$\boldsymbol{u}^{k+1} = \boldsymbol{u}^k + \omega \mathsf{D}^{-1}(\boldsymbol{f} - \mathsf{A}\boldsymbol{u}^k). \tag{1.4.3}$$

*Proof.* Exercise; see Problem 1.9. □

We will come back to the damped Jacobi method in a later chapter.

### 1.4.2 The Gauss–Seidel method

Let $\mathsf{A} = \mathsf{D} - \mathsf{L} - \mathsf{U}$ be the canonical splitting of $\mathsf{A} \in \mathbb{R}^{n\times n}$. Then

$$(\mathsf{D} - \mathsf{L})\boldsymbol{u} = \mathsf{U}\boldsymbol{u} + \boldsymbol{f}.$$

The Gauss–Seidel method is simply stated as

$$(\mathsf{D}-\mathsf{L})\boldsymbol{u}^{k+1} := \mathsf{U}\boldsymbol{u}^{k} + \boldsymbol{f}.$$

Assuming that $\mathsf{D}-\mathsf{L}$ is invertible, some manipulations give

$$\begin{aligned}\boldsymbol{u}^{k+1} &= (\mathsf{D}-\mathsf{L})^{-1}\mathsf{U}\boldsymbol{u}^{k} + (\mathsf{D}-\mathsf{L})^{-1}\boldsymbol{f}\\ &= \boldsymbol{u}^{k} - (\mathsf{D}-\mathsf{L})^{-1}(\mathsf{D}-\mathsf{L})\boldsymbol{u}^{k} + (\mathsf{D}-\mathsf{L})^{-1}\mathsf{U}\boldsymbol{u}^{k} + (\mathsf{D}-\mathsf{L})^{-1}\boldsymbol{f}\\ &= \boldsymbol{u}^{k} + (\mathsf{D}-\mathsf{L})^{-1}(\boldsymbol{f} - \mathsf{A}\boldsymbol{u}^{k}).\end{aligned}$$

**Definition 1.4.12** (Forward Gauss–Seidel method). Suppose that $\mathsf{A} = \mathsf{D}-\mathsf{L}-\mathsf{U}$ is the canonical splitting of $\mathsf{A} \in \mathbb{R}^{n\times n}$ and $\mathsf{D}-\mathsf{L}$ is invertible. The **Gauss–Seidel method** – also called the **forward Gauss–Seidel method** – is a GLIS with iterator matrix

$$\mathsf{B} = \mathsf{B}_{\mathrm{GS}} := (\mathsf{D}-\mathsf{L})^{-1}$$

and error transfer matrix

$$\mathsf{T} = \mathsf{T}_{\mathrm{GS}} := \mathrm{I} - (\mathsf{D}-\mathsf{L})^{-1}\mathsf{A}.$$

The iteration sequence can be expressed as

$$\boldsymbol{u}^{k+1} := \boldsymbol{u}^{k} + (\mathsf{D}-\mathsf{L})^{-1}(\boldsymbol{f} - \mathsf{A}\boldsymbol{u}^{k}).$$ ■

**Remark 1.4.13** (Error transfer). The error transfer matrix for the Gauss–Seidel method can be expressed equivalently as

$$\begin{aligned}\mathsf{T}_{\mathrm{GS}} &= \mathrm{I} - (\mathsf{D}-\mathsf{L})^{-1}\mathsf{A}\\ &= (\mathsf{D}-\mathsf{L})^{-1}(\mathsf{D}-\mathsf{L}) - (\mathsf{D}-\mathsf{L})^{-1}(\mathsf{D}-\mathsf{L}-\mathsf{U})\\ &= (\mathsf{D}-\mathsf{L})^{-1}\mathsf{U}.\end{aligned}$$ ◀

In component form, we can express the forward Gauss–Seidel method as

$$\sum_{j=1}^{i} a_{i,j}u_j^{k+1} + \sum_{j=i+1}^{n} a_{i,j}u_j^{k} = f_i, \quad i = 1,\ldots,n.$$

To compute the update $\boldsymbol{u}^{k+1}$, we proceed from $i = 1$, in order, to $i = n$. So the first equation we solve is

$$u_1^{k+1} = \frac{f_1 - \sum_{j=2}^{n} a_{1,j}u_j^{k}}{a_{1,1}}.$$

Next, we have

$$u_2^{k+1} = \frac{f_2 - a_{2,1}u_1^{k+1} - \sum_{j=3}^{n} a_{2,j}u_j^k}{a_{2,2}}$$

and

$$u_3^{k+1} = \frac{f_3 - \sum_{j=1}^{2} a_{3,j}u_j^{k+1} - \sum_{j=4}^{n} a_{3,j}u_j^k}{a_{3,3}}.$$

We continue, in the same way, with $i = 4$, in order, to $i = n - 1$, and we end with

$$u_n^{k+1} = \frac{f_n - \sum_{j=1}^{n-1} a_{n,j}u_j^{k+1}}{a_{n,n}}.$$

The idea of the method is simple. Once we obtain an updated component $u_i^{k+1}$, we immediately use it to obtain $u_{i+1}^{k+1}$. One full pass from $i = 1$ to $i = n$ is called a *forward sweep.* The generic $i$th step is

$$u_i^{k+1} = \frac{f_i - \sum_{j=1}^{i-1} a_{i,j}u_j^{k+1} - \sum_{j=i+1}^{n} a_{i,j}u_j^k}{a_{i,i}}. \tag{1.4.4}$$

**Theorem 1.4.14** (Convergence of forward Gauss–Seidel). *Suppose that* $\mathsf{A} = \mathsf{D} - \mathsf{L} - \mathsf{U}$ *is the canonical splitting of* $\mathsf{A} \in \mathbb{R}^{n \times n}$. *If* $\mathsf{A}$ *is diagonally dominant, then the forward Gauss–Seidel method is well defined, that is,* $\mathsf{D} - \mathsf{L}$ *is invertible, and the method is unconditionally convergent.*

*Proof.* If $\mathsf{A}$ is diagonally dominant, then $\mathsf{D}$ has only nonzero diagonal entries. Hence $\mathsf{D} - \mathsf{L}$ is invertible.

For convergence, we again want to show that if $\mathsf{A}$ is diagonally dominant, then $\|\mathsf{T}_{\mathrm{GS}}\|_\infty < 1$. This will establish the unconditional convergence of the method. Since $\mathsf{A}$ is diagonally dominant, it follows that, for some $\delta > 0$,

$$|a_{i,i}| \geq \sum_{\substack{j=1 \\ j \neq i}}^{n} |a_{i,j}| + \delta = \sum_{j>i} |a_{i,j}| + \sum_{j<i} |a_{i,j}| + \delta.$$

Thus

$$|a_{i,i}| - \sum_{j<i} |a_{i,j}| \geq \sum_{j>i} |a_{i,j}| + \delta > \sum_{j>i} |a_{i,j}|,$$

which implies

$$\gamma := \max_{1 \leq i \leq n} \left\{ \frac{\sum_{j>i} |a_{i,j}|}{|a_{i,i}| - \sum_{j<i} |a_{i,j}|} \right\} < 1.$$

Now we will show that $\|\mathsf{T}_{\mathrm{GS}}\|_\infty \leq \gamma$.

Let $\boldsymbol{x} \in \mathbb{R}^n$ and $\boldsymbol{y} = \mathsf{T}_{\mathrm{GS}}\boldsymbol{x}$, that is,

$$\boldsymbol{y} = \mathsf{T}_{\mathrm{GS}}\boldsymbol{x} = (\mathsf{D} - \mathsf{L})^{-1}\mathsf{U}\boldsymbol{x}.$$

Let $k \in \{1, \dots, n\}$ be the index such that $\|\boldsymbol{y}\|_{\ell^\infty} = |y_k|$. Then we have

$$|[(\mathsf{D} - \mathsf{L})\boldsymbol{y}]_k| = |[\mathsf{U}\boldsymbol{x}]_k| = \left|\sum_{j>k} a_{k,j}x_j\right| \le \sum_{j>k} |a_{k,j}||x_j| \le \sum_{j>k} |a_{k,j}|\|\boldsymbol{x}\|_{\ell^\infty}.$$

Now notice that

$$\begin{aligned} |[(\mathsf{D} - \mathsf{L})\boldsymbol{y}]_k| &= \left|\sum_{j<k} a_{k,j}y_j + a_{k,k}y_k\right| \\ &\ge |a_{k,k}y_k| - \left|\sum_{j<k} a_{k,j}y_j\right| \\ &= |a_{k,k}|\|\boldsymbol{y}\|_{\ell^\infty} - \left|\sum_{j<k} a_{k,j}y_j\right| \\ &\ge |a_{k,k}|\|\boldsymbol{y}\|_{\ell^\infty} - \sum_{j<k} |a_{k,j}|\|\boldsymbol{y}\|_{\ell^\infty}. \end{aligned}$$

Therefore we have

$$|a_{k,k}|\|\boldsymbol{y}\|_{\ell^\infty} - \sum_{j<k} |a_{k,j}|\|\boldsymbol{y}\|_{\ell^\infty} \le \sum_{j>k} |a_{k,j}|\|\boldsymbol{x}\|_{\ell^\infty},$$

which implies

$$\|\boldsymbol{y}\|_{\ell^\infty} \le \frac{\sum_{j>k} |a_{k,j}|}{|a_{k,k}| - \sum_{j<k} |a_{k,j}|}\|\boldsymbol{x}\|_{\ell^\infty}.$$

So

$$\|\mathsf{T}_{\mathrm{GS}}\boldsymbol{x}\|_\infty \le \gamma\|\boldsymbol{x}\|_{\ell^\infty},$$

which implies

$$\|\mathsf{T}_{\mathrm{GS}}\|_\infty \le \gamma < 1.$$

□

**Definition 1.4.15** (Backward Gauss–Seidel method). Suppose that $\mathsf{A} \in \mathbb{R}^{n\times n}$ is invertible and $\mathsf{A} = \mathsf{D} - \mathsf{L} - \mathsf{U}$ is the canonical splitting of $\mathsf{A}$. Assume that $\mathsf{D} - \mathsf{U}$ is invertible. The **backward Gauss–Seidel method** is a GLIS with iterator matrix

$$\mathsf{B} = \mathsf{B}_{\mathrm{BGS}} \coloneqq (\mathsf{D} - \mathsf{U})^{-1}$$

and error transfer matrix

$$\mathsf{T} = \mathsf{T}_{\mathrm{BGS}} := \mathrm{I} - (\mathsf{D} - \mathsf{U})^{-1}\mathsf{A}.$$

The iteration scheme is precisely

$$\boldsymbol{u}^{k+1} := \boldsymbol{u}^{k} + (\mathsf{D} - \mathsf{U})^{-1}(\boldsymbol{f} - \mathsf{A}\boldsymbol{u}^{k}).$$ ■

**Remark 1.4.16** (Error transfer). The error transfer matrix for the backward Gauss–Seidel method can be reexpressed as

$$\begin{aligned}\mathsf{T}_{\mathrm{BGS}} &= I - (\mathsf{D} - \mathsf{U})^{-1}\mathsf{A} \\ &= \mathrm{I} - (\mathsf{D} - \mathsf{U})^{-1}(\mathsf{D} - \mathsf{L} - \mathsf{U}) \\ &= (\mathsf{D} - \mathsf{U})^{-1}\mathsf{L}.\end{aligned}$$ ◀

It is not hard to modify the proof of Theorem 1.4.14 to obtain the convergence of the backward Gauss–Seidel method.

**Theorem 1.4.17** (Convergence). *Suppose that* $\mathsf{A} = \mathsf{D} - \mathsf{L} - \mathsf{U}$ *is the canonical splitting of* $\mathsf{A}$. *If* $\mathsf{A}$ *is diagonally dominant, then the backward Gauss–Seidel method is well defined, that is,* $\mathsf{D} - \mathsf{U}$ *is invertible, and it is unconditionally convergent.*

*Proof.* Exercise; see Problem 1.10. □

In component form, we can express the backward Gauss–Seidel method as

$$\sum_{j=1}^{i-1} a_{i,j}u_j^k + \sum_{j=i}^{n} a_{i,j}u_j^{k+1} = f_i, \quad i = 1, \dots, n.$$

To compute the update $\boldsymbol{u}^{k+1}$, we proceed from $i = n$, in reverse order, to $i = 1$. We first find the $n$th component:

$$u_n^{k+1} = \frac{f_n - \sum_{j=1}^{n-1} a_{n,j}u_j^k}{a_{n,n}}.$$

Next, we have

$$u_{n-1}^{k+1} = \frac{f_{n-1} - \sum_{j=1}^{n-2} a_{n-1,j}u_j^k - a_{n-1,n}u_n^{k+1}}{a_{n-1,n-1}},$$

and then

$$u_{n-2}^{k+1} = \frac{f_{n-2} - \sum_{j=1}^{n-3} a_{n-2,j}u_j^k - \sum_{j=n-1}^{n} a_{n-2,j}u_j^{k+1}}{a_{n-2,n-2}}.$$

We continue, in the same way, with $i = n - 3$, in reverse order, to $i = 2$, and we end with

$$u_1^{k+1} = \frac{f_1 - \sum_{j=1}^{n-1} a_{1,j} u_j^{k+1}}{a_{1,1}}.$$

One full pass, from $i = n$, proceeding in reverse order, to $i = 1$ is called a *backward sweep*. The generic $i$th step is

$$u_i^{k+1} = \frac{f_i - \sum_{j=1}^{i-1} a_{i,j} u_j^{k} - \sum_{j=i+1}^{n} a_{i,j} u_j^{k+1}}{a_{i,i}}.$$

### 1.4.3 Richardson's method

Suppose $\omega \in \mathbb{R}_\star$ and consider the splitting

$$\mathsf{A} = \omega^{-1}\mathrm{I} + \mathsf{A} - \omega^{-1}\mathrm{I}.$$

Suppose

$$\mathsf{A}\boldsymbol{u} = \boldsymbol{f}.$$

Then

$$\omega^{-1}\boldsymbol{u} = (\omega^{-1}\mathrm{I} - \mathsf{A})\boldsymbol{u} + \boldsymbol{f}.$$

Richardson's method is, essentially,

$$\omega^{-1}\boldsymbol{u}^{k+1} := (\omega^{-1}\mathrm{I} - \mathsf{A})\boldsymbol{u}^{k} + \boldsymbol{f}.$$

Notice that this method is not based on the canonical diagonal splitting, but it is still quite useful and interesting.

**Definition 1.4.18** (Richardson's method). Suppose that $\mathsf{A} \in \mathbb{R}^{n\times n}$ is invertible and $\omega \in \mathbb{R}_\star$. **Richardson's method** is a GLIS with iterator matrix

$$\mathsf{B} = \mathsf{B}_{\mathrm{R}} := \omega\mathrm{I}$$

and error transfer matrix

$$\mathsf{T} = \mathsf{T}_{\mathrm{R}} := \mathrm{I} - \omega\mathsf{A}.$$

The iteration scheme is

$$\boldsymbol{u}^{k+1} := \boldsymbol{u}^{k} + \omega(\boldsymbol{f} - \mathsf{A}\boldsymbol{u}^{k}). \tag{1.4.5}$$

■

We make a slight departure from the now familiar approach and analyze this method using the Euclidean norm under the assumption that A is SPD.

**Theorem 1.4.19** (Convergence of Richardson's method). *Let* $\mathsf{A} \in \mathbb{R}^{n\times n}$ *be SPD, and let* $\sigma(\mathsf{A}) = \{\lambda_1, \ldots, \lambda_n\}$ *with* $0 < \lambda_1 \le \lambda_2 \le \cdots \le \lambda_n$. *Richardson's method converges unconditionally iff* $\omega \in (0, 2/\lambda_n)$. *In this case, we have the estimate*

$$\|\boldsymbol{e}^k\| \le \rho^k \|\boldsymbol{e}^0\|, \quad \rho = \rho(\omega) := \max\{|1 - \omega\lambda_n|, |1 - \omega\lambda_1|\}.$$

*From this it follows that, setting*

$$\omega = \omega_{\text{opt}} := \frac{2}{\lambda_1 + \lambda_n},$$

*we obtain the smallest possible value of* $\rho$, $\rho_{\text{opt}}$, *and*

$$\rho_{\text{opt}} := \rho(\omega_{\text{opt}}) = \frac{\lambda_n - \lambda_1}{\lambda_n + \lambda_1} = \frac{\kappa(\mathsf{A}) - 1}{\kappa(\mathsf{A}) + 1},$$

*where we recall that*

$$\kappa(\mathsf{A}) = \frac{\lambda_n}{\lambda_1}.$$

*Proof.* ($\Longleftarrow$): Since A is SPD, we know that the eigenvalues of A are positive real numbers and

$$\lambda_n = \|\mathsf{A}\|.$$

Notice also that $\mathsf{T}_{\mathrm{R}} = \mathsf{I}_n - \omega\mathsf{A} = \mathsf{T}_{\mathrm{R}}^{\mathsf{T}}$, which implies that the eigenvalues of $\mathsf{T}_{\mathrm{R}}$ are real. Observe that $(\lambda_i, \boldsymbol{w}_i)$ is an eigenpair of A iff $(\nu_i = 1 - \omega\lambda_i, \boldsymbol{w}_i)$ is an eigenpair of $\mathsf{T}_{\mathrm{R}}$. Assume that $0 < \omega < 2/\lambda_n$. Then

$$0 < \lambda_i\omega < 2\frac{\lambda_i}{\lambda_n}, \quad i = 1, \ldots, n,$$

which implies

$$1 > 1 - \lambda_i\omega > 1 - 2\frac{\lambda_i}{\lambda_n} \ge -1, \quad i = 1, \ldots, n.$$

It follows that

$$1 > \nu_1 \ge \cdots \ge \nu_n > -1, \quad \nu_i = 1 - \omega\lambda_i.$$

This guarantees that $\|\mathsf{T}_{\mathrm{R}}\| = \rho(\mathsf{T}_{\mathrm{R}}) < 1$, which implies convergence.

( $\Longrightarrow$ ): Conversely, if $\omega \notin (0, 2/\lambda_n)$, then $\rho(\mathrm{T_R}) \geq 1$, and the method cannot converge unconditionally.
(Error estimate): By consistency,

$$\|e^k\| = \|\mathrm{T}_\mathrm{R}^k e^0\| \leq \rho^k \|e^0\|.$$

Of course, it is easy to see that

$$\rho = \rho(\mathrm{T_R}) = \max\{|\nu_1|, |\nu_n|\} = \max\{|1 - \omega\lambda_n|, |1 - \omega\lambda_1|\}.$$

(Optimality): Finally, showing optimality amounts to minimizing $\rho$. See Figure 1.2. From this we see that the minimum of $\rho$ is attained when

$$|1 - \omega\lambda_1| = |1 - \omega\lambda_n|$$

or

$$1 - \omega\lambda_1 = \omega\lambda_n - 1,$$

which implies

$$\omega_{\mathrm{opt}} = \frac{2}{\lambda_1 + \lambda_n}.$$

Therefore

$$\rho_{\mathrm{opt}} = 1 - \omega_{\mathrm{opt}}\lambda_1 = \frac{\lambda_1 + \lambda_n - 2\lambda_1}{\lambda_1 + \lambda_n} = \frac{\lambda_n - \lambda_1}{\lambda_1 + \lambda_n} = \frac{\kappa(\mathrm{A}) - 1}{\kappa(\mathrm{A}) + 1}. \qquad \square$$

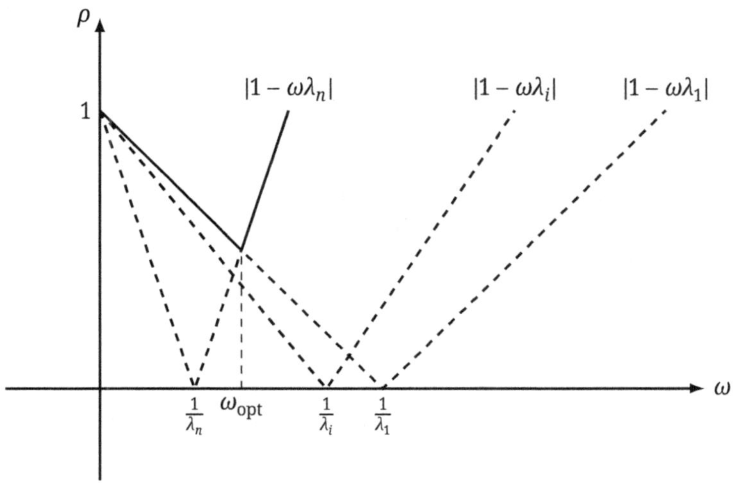


**Figure 1.2:** The curve $\rho(\mathrm{T_R})$ (in solid black) as a function of $\omega$.

## 1.5 Additive, multiplicative, and symmetrized GLIS

A GLIS is uniquely defined by its iterator matrix. If we have a collection of iterator matrices, then these can be combined in various ways to create a new GLIS.

### 1.5.1 Additive and multiplicative GLIS

**Definition 1.5.1** (Additive and multiplicative GLIS). Suppose $\mathsf{A} \in \mathbb{R}^{n\times n}$ is invertible and $\boldsymbol{f} \in \mathbb{R}^n$. Consider a family of iterator matrices $\mathsf{B}_i \in \mathbb{R}^{n\times n}$, $i = 1, \dots, r$. An iterative scheme is called **additive with respect to** $\{\mathsf{B}_i\}_{i=1}^r$ if it can be written in the form

$$\begin{aligned}
\boldsymbol{z}_1^k &:= \boldsymbol{u}^k + \mathsf{B}_1(\boldsymbol{f} - \mathsf{A}\boldsymbol{u}^k),\\
\boldsymbol{z}_2^k &:= \boldsymbol{z}_1^k + \mathsf{B}_2(\boldsymbol{f} - \mathsf{A}\boldsymbol{u}^k),\\
\boldsymbol{z}_3^k &:= \boldsymbol{z}_2^k + \mathsf{B}_3(\boldsymbol{f} - \mathsf{A}\boldsymbol{u}^k),\\
&\vdots\\
\boldsymbol{z}_{r-1}^k &:= \boldsymbol{z}_{r-2}^k + \mathsf{B}_{r-1}(\boldsymbol{f} - \mathsf{A}\boldsymbol{u}^k),\\
\boldsymbol{u}^{k+1} &:= \boldsymbol{z}_{r-1}^k + \mathsf{B}_r(\boldsymbol{f} - \mathsf{A}\boldsymbol{u}^k).
\end{aligned} \tag{1.5.1}$$

An iterative scheme is called **multiplicative with respect to** $\{\mathsf{B}_i\}_{i=1}^r$ if it can be written as

$$\begin{aligned}
\boldsymbol{z}_1^k &:= \boldsymbol{u}^k + \mathsf{B}_1(\boldsymbol{f} - \mathsf{A}\boldsymbol{u}^k),\\
\boldsymbol{z}_2^k &:= \boldsymbol{z}_1^k + \mathsf{B}_2(\boldsymbol{f} - \mathsf{A}\boldsymbol{z}_1^k),\\
\boldsymbol{z}_3^k &:= \boldsymbol{z}_2^k + \mathsf{B}_3(\boldsymbol{f} - \mathsf{A}\boldsymbol{z}_2^k),\\
&\vdots\\
\boldsymbol{z}_{r-1}^k &:= \boldsymbol{z}_{r-2}^k + \mathsf{B}_{r-1}(\boldsymbol{f} - \mathsf{A}\boldsymbol{z}_{r-2}^k),\\
\boldsymbol{u}^{k+1} &:= \boldsymbol{z}_{r-1}^k + \mathsf{B}_r(\boldsymbol{f} - \mathsf{A}\boldsymbol{z}_{r-1}^k).
\end{aligned} \tag{1.5.2}$$

■

**Proposition 1.5.2** (Iterator representation). *Suppose* $\mathsf{A} \in \mathbb{R}^{n\times n}$ *is invertible. Consider a family of iterator matrices* $\mathsf{B}_i \in \mathbb{R}^{n\times n}$, $i = 1, \dots, r$. *An additive iterative scheme with respect to* $\{\mathsf{B}_i\}_{i=1}^r$ *is a GLIS with iterator*

$$\mathsf{B} = \sum_{i=1}^r \mathsf{B}_i.$$

*A multiplicative iterative scheme with respect to* $\{\mathsf{B}_i\}_{i=1}^r$ *is a GLIS with the following recursively defined iterator:*

$$\mathsf{B} = \tilde{\mathsf{B}}_r,$$

*where* $\tilde{\mathsf{B}}_1 = \mathsf{B}_1$, *and for* $2 \le i \le r$,

$$\tilde{\mathsf{B}}_i = \tilde{\mathsf{B}}_{i-1} + \mathsf{B}_i - \mathsf{B}_i \mathsf{A} \tilde{\mathsf{B}}_{i-1}.$$

*Proof.* Exercise; see Problem 1.11. □

**Proposition 1.5.3** (Error transfer representation). *Suppose* $\mathsf{A} \in \mathbb{R}^{n\times n}$ *is invertible and* $\{\mathsf{B}_i\}_{i=1}^r \subset \mathbb{R}^{n\times n}$ *is a family of iterator matrices. The error transfer matrix for the additive GLIS with respect to* $\{\mathsf{B}_i\}_{i=1}^r$ *can be expressed as*

$$\mathsf{T} = \mathrm{I} - \left(\sum_{i=1}^{r} \mathsf{B}_i\right)\mathsf{A}.$$

*The error transfer matrix for the multiplicative GLIS with respect to* $\{\mathsf{B}_i\}_{i=1}^r$ *can be expressed as*

$$\mathsf{T} = (\mathrm{I} - \mathsf{B}_r\mathsf{A}) \cdots (\mathrm{I} - \mathsf{B}_1\mathsf{A}).$$

*Proof.* Exercise; see Problem 1.12. □

### 1.5.2 Symmetrized GLIS

It is often important in a GLIS to have a symmetric iterator matrix. In this section, we show how to symmetrize a generic iterative scheme using the multiplicative framework.

**Definition 1.5.4** (Symmetrized multiplicative GLIS). Suppose $\mathsf{A} \in \mathbb{R}^{n\times n}$ is invertible and $\boldsymbol{f} \in \mathbb{R}^n$. Consider a generic GLIS with iterator $\mathsf{B} \in \mathbb{R}^{n\times n}$. The **symmetrized multiplicative GLIS (SMGLIS) with respect to** $\mathsf{B}$ is defined as follows: given $\boldsymbol{u}^0$, find $\{\boldsymbol{u}^k\}_{k=1}^\infty$ via

$$\boldsymbol{z}^k := \boldsymbol{u}^k + \mathsf{B}(\boldsymbol{f} - \mathsf{A}\boldsymbol{u}^k), \tag{1.5.3}$$
$$\boldsymbol{u}^{k+1} := \boldsymbol{z}^k + \mathsf{B}^\top(\boldsymbol{f} - \mathsf{A}\boldsymbol{z}^k). \tag{1.5.4}$$

■

**Lemma 1.5.5** (SMGLIS representation). *Suppose* $\mathsf{A} \in \mathbb{R}^{n\times n}$ *is invertible and* $\boldsymbol{f} \in \mathbb{R}^n$. *Assume that* $\mathsf{B}$ *is invertible. The SMGLIS with respect to* $\mathsf{B}$ *can be written as*

$$\boldsymbol{u}^{k+1} = \boldsymbol{u}^k + \mathsf{B}_{\mathrm{SM}}(\boldsymbol{f} - \mathsf{A}\boldsymbol{u}^k), \tag{1.5.5}$$

*where*

$$\mathsf{B}_{\mathrm{SM}} := \mathsf{B} + \mathsf{B}^\top - \mathsf{B}^\top\mathsf{A}\mathsf{B}. \tag{1.5.6}$$

*In other words, the SMGLIS with respect to* $\mathsf{B}$ *is a GLIS with iterator* $\mathsf{B}_{\mathrm{SM}}$. *If* $\mathsf{B}$ *is invertible, then the iterator may be expressed as*

$$\mathsf{B}_{\mathrm{SM}} = \mathsf{B}^{\top}(\mathsf{B}^{-\top} + \mathsf{B}^{-1} - \mathsf{A})\mathsf{B}. \tag{1.5.7}$$

*If* $\mathsf{A}$ *is symmetric, then* $\mathsf{B}_{\mathrm{SM}}$ *is as well.*

*Proof.* Substituting (1.5.3) into (1.5.4) yields

$$\begin{aligned}
\boldsymbol{u}^{k+1} &= \boldsymbol{u}^k + \mathsf{B}(\boldsymbol{f} - \mathsf{A}\boldsymbol{u}^k) + \mathsf{B}^{\top}(\boldsymbol{f} - \mathsf{A}[\boldsymbol{u}^k + \mathsf{B}(\boldsymbol{f} - \mathsf{A}\boldsymbol{u}^k)]) \\
&= \boldsymbol{u}^k + \mathsf{B}(\boldsymbol{f} - \mathsf{A}\boldsymbol{u}^k) + \mathsf{B}^{\top}(\boldsymbol{f} - \mathsf{A}\boldsymbol{u}^k - \mathsf{A}\mathsf{B}(\boldsymbol{f} - \mathsf{A}\boldsymbol{u}^k)) \\
&= \boldsymbol{u}^k + \mathsf{B}(\boldsymbol{f} - \mathsf{A}\boldsymbol{u}^k) + \mathsf{B}^{\top}(\boldsymbol{f} - \mathsf{A}\boldsymbol{u}^k) - \mathsf{B}^{\top}\mathsf{A}\mathsf{B}(\boldsymbol{f} - \mathsf{A}\boldsymbol{u}^k) \\
&= \boldsymbol{u}^k + (\mathsf{B} + \mathsf{B}^{\top} - \mathsf{B}^{\top}\mathsf{A}\mathsf{B})(\boldsymbol{f} - \mathsf{A}\boldsymbol{u}^k),
\end{aligned}$$

which shows the SMGLIS with respect to $\mathsf{B}$ is a GLIS with iterator

$$\mathsf{B} = \mathsf{B}_{\mathrm{SM}} = \mathsf{B} + \mathsf{B}^{\top} - \mathsf{B}^{\top}\mathsf{A}\mathsf{B}.$$

If $\mathsf{B}$ is invertible, then

$$\begin{aligned}
\mathsf{B}_{\mathrm{SM}} &= \mathsf{B} + \mathsf{B}^{\top} - \mathsf{B}^{\top}\mathsf{A}\mathsf{B} \\
&= (\mathrm{I} + \mathsf{B}^{\top}\mathsf{B}^{-1} - \mathsf{B}^{\top}\mathsf{A})\mathsf{B} \\
&= (\mathsf{B}^{\top}\mathsf{B}^{-\top} + \mathsf{B}^{\top}\mathsf{B}^{-1} - \mathsf{B}^{\top}\mathsf{A})\mathsf{B} \\
&= \mathsf{B}^{\top}(\mathsf{B}^{-\top} + \mathsf{B}^{-1} - \mathsf{A})\mathsf{B}.
\end{aligned}$$

If $\mathsf{A}$ is symmetric, then $\mathsf{B}_{\mathrm{SM}}$ is symmetric, since

$$\begin{aligned}
\mathsf{B}_{\mathrm{SM}}^{\top} &= \mathsf{B}^{\top} + \mathsf{B} - \mathsf{B}^{\top}\mathsf{A}^{\top}\mathsf{B} \\
&= \mathsf{B}^{\top} + \mathsf{B} - \mathsf{B}^{\top}\mathsf{A}\mathsf{B} \\
&= \mathsf{B}_{\mathrm{SM}}.
\end{aligned}$$

□

**Remark 1.5.6** (Symmetrized additive method). We could also consider a symmetrized additive method:

$$\boldsymbol{z}^k := \boldsymbol{u}^k + \mathsf{B}(\boldsymbol{f} - \mathsf{A}\boldsymbol{u}^k), \tag{1.5.8}$$

$$\boldsymbol{u}^{k+1} := \boldsymbol{z}^k + \mathsf{B}^{\top}(\boldsymbol{f} - \mathsf{A}\boldsymbol{u}^k). \tag{1.5.9}$$

This is a GLIS with iterator matrix

$$\mathsf{B} = \mathsf{B}_{\mathrm{SA}} := \mathsf{B} + \mathsf{B}^{\top}.$$

However, it turns out that this is not so useful to us in the multigrid setting. ◂

### 1.5.3 Symmetric Gauss–Seidel method

The forward Gauss–Seidel method is the best example of a GLIS where the iterator is nonsymmetric, even when A is symmetric. However, we can use the methodology presented in the last section to symmetrize it.

**Definition 1.5.7** (Symmetric Gauss–Seidel method). Suppose that $\mathsf{A} \in \mathbb{R}^{n\times n}$ is invertible and $\mathsf{A} = \mathsf{D} - \mathsf{L} - \mathsf{U}$ is the canonical splitting of A. Assume that $\mathsf{D} - \mathsf{L}$ is invertible, and consider the Gauss–Seidel method characterized by the iterator

$$\mathsf{B}_{\mathrm{GS}} = (\mathsf{D} - \mathsf{L})^{-1}.$$

The **symmetric Gauss–Seidel method** is the SMGLIS with respect to $\mathsf{B}_{\mathrm{GS}}$ and is therefore a GLIS with iterator matrix

$$\mathsf{B}_{\mathrm{SGS}} := \mathsf{B}_{\mathrm{GS}} + \mathsf{B}_{\mathrm{GS}}^{\top} - \mathsf{B}_{\mathrm{GS}}^{\top}\mathsf{A}\mathsf{B}_{\mathrm{GS}}. \quad ■$$

**Proposition 1.5.8** (Symmetric Gauss–Seidel method). *Suppose that* $\mathsf{A} \in \mathbb{R}^{n\times n}$ *is symmetric and invertible and* $\mathsf{A} = \mathsf{D} - \mathsf{L} - \mathsf{U}$ *is the canonical splitting of* A. *Assume that* $\mathsf{D} - \mathsf{L}$ *is invertible. The iterator for the symmetric Gauss Seidel method can be written as*

$$\mathsf{B}_{\mathrm{SGS}} = (\mathsf{D} - \mathsf{U})^{-1}\mathsf{D}(\mathsf{D} - \mathsf{L})^{-1},$$

*and it is symmetric.*

*Proof.* Using equation (1.5.6) from Lemma 1.5.5, we find

$$\begin{aligned}
\mathsf{B}_{\mathrm{SGS}} &= \mathsf{B}_{\mathrm{GS}}^{\top}(\mathsf{B}_{\mathrm{GS}}^{-\top} + \mathsf{B}_{\mathrm{GS}}^{-1} - \mathsf{A})\mathsf{B}_{\mathrm{GS}} \\
&= (\mathsf{D}^{\top} - \mathsf{L}^{\top})^{-1}(\mathsf{D}^{\top} - \mathsf{L}^{\top} + \mathsf{D} - \mathsf{L} - \mathsf{A})(\mathsf{D} - \mathsf{L})^{-1} \\
&= (\mathsf{D} - \mathsf{U})^{-1}(\mathsf{D} - \mathsf{U} + \mathsf{D} - \mathsf{L} - \mathsf{A})(\mathsf{D} - \mathsf{L})^{-1} \\
&= (\mathsf{D} - \mathsf{U})^{-1}\mathsf{D}(\mathsf{D} - \mathsf{L})^{-1}.
\end{aligned} \quad □$$

## 1.6 Convergence in the energy norm

If the coefficient matrix A is SPD, then we can use it to construct a useful norm, one which will be the basis of measuring convergence in a new way.

### 1.6.1 The energy norm

**Definition 1.6.1** (Energy norm). Suppose that A is SPD. The **energy norm** associated with A is

$$\|\boldsymbol{u}\|_{\mathsf{A}} = \sqrt{(\boldsymbol{u}, \boldsymbol{u})_{\mathsf{A}}}. \quad ■$$

The energy norm is a natural metric for the convergence of the GLIS when A is SPD, as the next few results show.

**Theorem 1.6.2** (Convergence in energy norm). *Suppose* $\mathsf{A} \in \mathbb{R}^{n\times n}$ *is SPD and* $\mathsf{B} \in \mathbb{R}^{n\times n}$ *is symmetric and invertible. Define*

$$\mathsf{Q} := \mathsf{B}^{-1} - \frac{1}{2}\mathsf{A}.$$

*If* Q *is SPD, then the GLIS* (1.2.1) *is unconditionally convergent with respect to the* A*-norm, that is,*

$$\|\boldsymbol{e}^k\|_{\mathsf{A}} \to 0$$

*as* $k \to \infty$*. Moreover, the convergence is always monotonic in the sense that*

$$\|\boldsymbol{e}^{k+1}\|_{\mathsf{A}}^2 \le \|\boldsymbol{e}^k\|_{\mathsf{A}}^2, \quad \forall k \in \mathbb{N}_0 := \mathbb{N} \cup \{0\}.$$

*Proof.* Let $\boldsymbol{u}$ be the exact solution, and let $\boldsymbol{u}^k$ be the GLIS approximation at the $k$th step. Then

$$\boldsymbol{e}^{k+1} = \mathsf{T}\boldsymbol{e}^k = \boldsymbol{e}^k - \mathsf{B}\mathsf{A}\boldsymbol{e}^k.$$

Define $\boldsymbol{v}^{k+1} := \boldsymbol{e}^{k+1} - \boldsymbol{e}^k$. Then

$$\mathsf{B}^{-1}\boldsymbol{v}^{k+1} + \mathsf{A}\boldsymbol{e}^k = \mathbf{0}. \tag{1.6.1}$$

Taking the Euclidean inner product of the last equation with $\boldsymbol{v}^{k+1}$ gives

$$(\mathsf{B}^{-1}\boldsymbol{v}^{k+1}, \boldsymbol{v}^{k+1}) + (\mathsf{A}\boldsymbol{e}^k, \boldsymbol{v}^{k+1}) = 0.$$

Now observe that

$$\boldsymbol{e}^k = \frac{1}{2}(\boldsymbol{e}^{k+1} + \boldsymbol{e}^k) - \frac{1}{2}(\boldsymbol{e}^{k+1} - \boldsymbol{e}^k) = \frac{1}{2}(\boldsymbol{e}^{k+1} + \boldsymbol{e}^k) - \frac{1}{2}\boldsymbol{v}^{k+1}.$$

Then

$$\begin{aligned}
0 &= (\mathsf{B}^{-1}\boldsymbol{v}^{k+1}, \boldsymbol{v}^{k+1}) + (\mathsf{A}\boldsymbol{e}^k, \boldsymbol{v}^{k+1}) \\
&= (\mathsf{B}^{-1}\boldsymbol{v}^{k+1}, \boldsymbol{v}^{k+1}) + \frac{1}{2}(\mathsf{A}(\boldsymbol{e}^{k+1} + \boldsymbol{e}^k), \boldsymbol{v}^{k+1}) - \frac{1}{2}(\mathsf{A}\boldsymbol{v}^{k+1}, \boldsymbol{v}^{k+1}) \\
&= \left(\left(\mathsf{B}^{-1} - \frac{1}{2}\mathsf{A}\right)\boldsymbol{v}^{k+1}, \boldsymbol{v}^{k+1}\right) + \frac{1}{2}(A(\boldsymbol{e}^{k+1} + \boldsymbol{e}^k), \boldsymbol{v}^{k+1}) \\
&= \left(\left(\mathsf{B}^{-1} - \frac{1}{2}\mathsf{A}\right)\boldsymbol{v}^{k+1}, \boldsymbol{v}^{k+1}\right) + \frac{1}{2}(\mathsf{A}(\boldsymbol{e}^{k+1} + \boldsymbol{e}^k), \boldsymbol{e}^{k+1} - \boldsymbol{e}^k) \qquad (1.6.2) \\
&= \left(\left(\mathsf{B}^{-1} - \frac{1}{2}\mathsf{A}\right)\boldsymbol{v}^{k+1}, \boldsymbol{v}^{k+1}\right) + \frac{1}{2}(\|\boldsymbol{e}^{k+1}\|_{\mathsf{A}}^2 - \|\boldsymbol{e}^k\|_{\mathsf{A}}^2)
\end{aligned}$$

$$= (\mathsf{Q}\boldsymbol{v}^{k+1}, \boldsymbol{v}^{k+1}) + \frac{1}{2}(\|\boldsymbol{e}^{k+1}\|_{\mathsf{A}}^2 - \|\boldsymbol{e}^k\|_{\mathsf{A}}^2)$$
$$\geq \frac{1}{2}(\|\boldsymbol{e}^{k+1}\|_{\mathsf{A}}^2 - \|\boldsymbol{e}^k\|_{\mathsf{A}}^2).$$

Here we used the assumption that Q is SPD. Hence

$$\|\boldsymbol{e}^{k+1}\|_{\mathsf{A}}^2 \leq \|\boldsymbol{e}^k\|_{\mathsf{A}}^2,$$

and

$$\|\boldsymbol{e}^{k+1}\|_{\mathsf{A}} \leq \|\boldsymbol{e}^k\|_{\mathsf{A}}.$$

Thus $\{\|\boldsymbol{e}^k\|_{\mathsf{A}}\}_{k=0}^{\infty}$ is a nonincreasing sequence of nonnegative numbers. By the monotone convergence theorem there is $\gamma \geq 0$ such that, as $k \to \infty$,

$$\|\boldsymbol{e}^k\|_{\mathsf{A}} \to \gamma \geq 0.$$

Passing to the limit as $k \to \infty$ in (1.6.2), we get

$$\lim_{k\to\infty} (\mathsf{Q}\boldsymbol{v}^{k+1}, \boldsymbol{v}^{k+1}) = 0,$$

and this, using again that Q is SPD, in turn implies that $\boldsymbol{v}^{k+1} \to \mathbf{0}$. Using (1.6.1),

$$\boldsymbol{e}^k = -\mathsf{A}^{-1}\mathsf{B}^{-1}\boldsymbol{v}^{k+1},$$

we must have $\|\boldsymbol{e}^k\|_{\mathsf{A}} \to 0$. □

We can generalize the last result. First, we need the following:

**Lemma 1.6.3** (Q-norm). *Suppose that* $\mathsf{Q} \in \mathbb{R}^{n\times n}$ *is positive definite in the sense that*

$$(\mathsf{Q}\boldsymbol{y}, \boldsymbol{y}) > 0, \quad \forall \boldsymbol{y} \in \mathbb{R}^n_\star,$$

*but* Q *is not necessarily symmetric. Then*

$$\|\boldsymbol{w}\|_{\mathsf{Q}} = \sqrt{(\mathsf{Q}\boldsymbol{w}, \boldsymbol{w})}, \quad \forall \boldsymbol{w} \in \mathbb{R}^n,$$

*defines a norm.*

*Proof.* Suppose that Q is not symmetric, to avoid the simple case. Then

$$\mathsf{Q} = \mathsf{Q}_{\mathsf{S}} + \mathsf{Q}_{\mathsf{A}},$$

where

$$\mathsf{Q}_{\mathsf{S}} \coloneqq \frac{1}{2}(\mathsf{Q} + \mathsf{Q}^\top) \quad \text{and} \quad \mathsf{Q}_{\mathsf{A}} \coloneqq \frac{1}{2}(\mathsf{Q} - \mathsf{Q}^\top)$$

are the symmetric and antisymmetric parts, respectively. Observe that $\mathsf{Q}_\mathrm{S}^\top = \mathsf{Q}_\mathrm{S}$ and $\mathsf{Q}_\mathrm{A}^\top = -\mathsf{Q}_\mathrm{A}$. It follows that $(\mathsf{Q}_\mathrm{A}\boldsymbol{y},\boldsymbol{y})$ is zero, for all $\boldsymbol{y} \in \mathbb{R}^n$, because

$$(\mathsf{Q}_\mathrm{A}\boldsymbol{y},\boldsymbol{y}) = \boldsymbol{y}^\top \mathsf{Q}_\mathrm{A}\boldsymbol{y} = (\mathsf{Q}_\mathrm{A}^\top\boldsymbol{y})^\top \boldsymbol{y} = -(\mathsf{Q}_\mathrm{A}\boldsymbol{y})^\top\boldsymbol{y} = \boldsymbol{y}^\top\mathsf{Q}_\mathrm{A}\boldsymbol{y} = -(\mathsf{Q}_\mathrm{A}\boldsymbol{y},\boldsymbol{y}).$$

Therefore, for all $\boldsymbol{y} \in \mathbb{R}^n_\star$,

$$0 < (\mathsf{Q}\boldsymbol{y},\boldsymbol{y}) = (\mathsf{Q}_\mathrm{S}\boldsymbol{y},\boldsymbol{y}) + (\mathsf{Q}_\mathrm{A}\boldsymbol{y},\boldsymbol{y}) = (\mathsf{Q}_\mathrm{S}\boldsymbol{y},\boldsymbol{y}).$$

Thus $\mathsf{Q}_\mathrm{S}$ is SPD. Since

$$\|\boldsymbol{w}\|_{\mathsf{Q}_\mathrm{S}} = \sqrt{(\mathsf{Q}_\mathrm{S}\boldsymbol{w},\boldsymbol{w})}, \quad \forall \boldsymbol{w} \in \mathbb{R}^n,$$

defines a norm, the result follows. □

**Corollary 1.6.4** (Convergence in energy norm). *Suppose* A *is SPD and* B *is invertible. Define*

$$\mathsf{Q} := \mathsf{B}^{-1} - \frac{1}{2}\mathsf{A},$$

*and assume that* Q *is positive definite in the sense that*

$$(\mathsf{Q}\boldsymbol{y},\boldsymbol{y}) > 0, \quad \forall \boldsymbol{y} \in \mathbb{R}^n_\star,$$

*but is not necessarily symmetric. Then the GLIS* (1.2.1) *is convergent with respect to the* A*-norm, that is,*

$$\|\boldsymbol{e}^k\|_\mathsf{A} \to 0$$

*as $k \to \infty$. Moreover, the convergence is always monotonic:*

$$\|\boldsymbol{e}^{k+1}\|_\mathsf{A}^2 \le \|\boldsymbol{e}^k\|_\mathsf{A}^2, \quad \forall k \in \mathbb{N}_0.$$

*Proof.* Exercise; see Problem 1.13. □

**Theorem 1.6.5** (Householder–John). *Suppose that* $\mathsf{A} \in \mathbb{R}^{n\times n}$ *is nonsingular and symmetric and* $\mathsf{B} \in \mathbb{R}^{n\times n}$ *is invertible. Assume that*

$$\mathsf{Q} = \mathsf{B}^{-1} + \mathsf{B}^{-\top} - \mathsf{A}$$

*is SPD. Then the GLIS* (1.2.1)*, with error transfer matrix* $\mathsf{T} = \mathsf{I} - \mathsf{BA}$*, converges unconditionally iff* A *is SPD. Moreover, if* A *is SPD, then the convergence is always monotonic in the energy norm:*

$$\|\boldsymbol{e}^{k+1}\|_\mathsf{A}^2 \le \|\boldsymbol{e}^k\|_\mathsf{A}^2, \quad \forall k \in \mathbb{N}_0.$$

*Proof.* ( $\Longleftarrow$ ): Suppose that A is SPD. Recall that $\|\cdot\|_A$ defines a norm on $\mathbb{R}^n$. The error equation is precisely

$$\boldsymbol{e}^{k+1} = (\mathrm{I} - \mathsf{BA})\boldsymbol{e}^k.$$

Then

$$\begin{aligned}
\|\boldsymbol{e}^{k+1}\|_A^2 &= ((\mathrm{I} - \mathsf{BA})\boldsymbol{e}^k)^\top \mathsf{A}((\mathrm{I} - \mathsf{BA})\boldsymbol{e}^k) \\
&= (\boldsymbol{e}^{k\top}(\mathrm{I} - \mathsf{AB}^\top))\mathsf{A}((\mathrm{I} - \mathsf{BA})\boldsymbol{e}^k) \\
&= (\boldsymbol{e}^{k\top} - \boldsymbol{e}^{k\top}\mathsf{AB}^\top)(\mathsf{A}\boldsymbol{e}^k - \mathsf{ABA}\boldsymbol{e}^k) \\
&= \boldsymbol{e}^{k\top}\mathsf{A}\boldsymbol{e}^k - \boldsymbol{e}^{k\top}\mathsf{AB}^\top\mathsf{A}\boldsymbol{e}^k - \boldsymbol{e}^{k\top}\mathsf{ABA}\boldsymbol{e}^k + \boldsymbol{e}^{k\top}\mathsf{AB}^\top\mathsf{ABA}\boldsymbol{e}^k \\
&= \|\boldsymbol{e}^k\|_A^2 - \boldsymbol{e}^{k\top}\mathsf{A}(\mathsf{B}^\top + \mathsf{B} - \mathsf{B}^\top\mathsf{AB})\mathsf{A}\boldsymbol{e}^k \\
&= \|\boldsymbol{e}^k\|_A^2 - \boldsymbol{e}^{k\top}\mathsf{AB}^\top(\mathsf{B}^{-1} + \mathsf{B}^{-\top} - \mathsf{A})\mathsf{BA}\boldsymbol{e}^k \\
&= \|\boldsymbol{e}^k\|_A^2 - (\mathsf{BA}\boldsymbol{e}^k)^\top(\mathsf{B}^{-1} + \mathsf{B}^{-\top} - \mathsf{A})\mathsf{BA}\boldsymbol{e}^k.
\end{aligned}$$

Since $\mathsf{B}^{-1} + \mathsf{B}^{-\top} - \mathsf{A}$ is SPD and $\mathsf{BA}\boldsymbol{e}^k \neq \mathbf{0}$, in general, it follows that

$$\|\boldsymbol{e}^{k+1}\|_A^2 + \|\mathsf{BA}\boldsymbol{e}^k\|_Q^2 = \|\boldsymbol{e}^k\|_A^2,$$

and $\|\boldsymbol{e}^k\|_A$ is a decreasing sequence. Therefore, by the monotone convergence theorem, $\|\boldsymbol{e}^k\|_A$ converges, that is, there is $\alpha \in [0, \infty)$ such that

$$\lim_{k\to\infty} \|\boldsymbol{e}^k\|_A = \alpha = \lim_{k\to\infty} \|\boldsymbol{e}^{k+1}\|_A.$$

This implies

$$\lim_{k\to\infty} \|\mathsf{BA}\boldsymbol{e}^k\|_Q = 0,$$

which, in turn, implies that $\boldsymbol{e}^k \to \mathbf{0}$ as $k \to \infty$.
( $\Longrightarrow$ ): This direction is left as an exercise; see Problem 1.14. □

### 1.6.2 Energy convergence of Gauss–Seidel methods

The energy methods described earlier prove to be a more useful tool for establishing the convergence of the Gauss–Seidel methods than the spectral approach when the coefficient matrix is SPD.

**Theorem 1.6.6** (Energy convergence of Gauss–Seidel). *Suppose that* $\mathsf{A} \in \mathbb{R}^{n\times n}$ *is SPD. Then the forward, backward, and symmetric Gauss–Seidel methods are well-defined and converge unconditionally.*

*Proof.* Let us use Corollary 1.6.4 and Theorem 1.6.5.
(Forward): For the forward Gauss–Seidel method, consider

$$\begin{aligned} Q_1 &= B_{GS}^{-1} + B_{GS}^{-T} - A \\ &= D - L + D - U - D + L + U \\ &= D. \end{aligned}$$

Since A is SPD, its diagonal elements are positive. Thus D is SPD. Now we can apply the Householder–John Theorem 1.6.5 to see that the Gauss–Seidel method converges unconditionally.

We can also use Corollary 1.6.4. For that, set

$$\begin{aligned} Q_2 &= B_{GS}^{-1} - \frac{1}{2}A. \\ &= D - L - \frac{1}{2}D + \frac{1}{2}L + \frac{1}{2}U \\ &= \frac{1}{2}D - \frac{1}{2}L + \frac{1}{2}U. \end{aligned}$$

Now let us consider the symmetric part of $Q_2$:

$$\begin{aligned} Q_{2,S} &= \frac{1}{2}(Q_2 + Q_2^T) \\ &= \frac{1}{2}\left(\frac{1}{2}D - \frac{1}{2}L + \frac{1}{2}U + \frac{1}{2}D - \frac{1}{2}U + \frac{1}{2}L\right) \\ &= \frac{1}{2}D. \end{aligned}$$

Since $Q_{2,S}$ is SPD, Corollary 1.6.4 implies that the Gauss–Seidel method converges.
(Backward): The same techniques can be used to show that the backward Gauss–Seidel method converges.
(Symmetric): For the symmetric Gauss–Seidel method, we have

$$B_{SGS} = (D - U)^{-1}D(D - L)^{-1}.$$

Observe that the matrix

$$Q = B_{SGS}^{-1} + B_{SGS}^{-T} - A = A + 2U^TD^{-1}U$$

is SPD, since A is SPD. Now we can apply the Householder–John Theorem 1.6.5 to see that the symmetric Gauss–Seidel method converges. □

## 1.7 A computational experiment

In this brief section, we perform a comparison of the Jacobi, Gauss–Seidel, and SOR methods on a prototypical problem. SOR stands for successive overrelaxation, and it is a simple variant of the Gauss–Seidel method. We will describe it below.

We wish to approximate the solution of the following one-dimensional boundary value problem:

$$\begin{cases} u(0) = 0, \\ -u''(x) = f(x), \quad 0 < x < 1, \\ u(1) = 0. \end{cases} \tag{1.7.1}$$

To effect this approximation, we use the finite difference method with a uniformly spaced grid. Choose $n \in \mathbb{N}$ and set

$$h := \frac{1}{n+1}.$$

The set of grid points is the collection $\mathcal{G}_h = \{x_i\}_{i=0}^{n+1}$, where

$$x_i := ih, \quad i = 0, 1, 2, \ldots, n+1. \tag{1.7.2}$$

We call functions that map $\mathcal{G}_h$ to $\mathbb{R}$ **grid functions.** If $v : \mathcal{G}_h \to \mathbb{R}$ is a grid function, then we set $v_i = v(x_i)$ for $i = 0, \ldots, n+1$.

The finite difference scheme seeks to find the grid function $w : \mathcal{G}_h \to \mathbb{R}$ such that

$$\begin{cases} w_0 = 0, \\ \dfrac{-w_{i-1} + 2w_i - w_{i+1}}{h^2} = f_i := f(x_i), \quad i = 1, \ldots, n, \\ w_{n+1} = 0. \end{cases} \tag{1.7.3}$$

Those who are familiar with the finite difference method may recall that the idea of this scheme is that the grid function approximates the exact solution of our problem in the sense that $w_i \approx u(x_i)$. In fact, under reasonable assumptions, it is possible to show that

$$\max_{1 \le i \le n} |u(x_i) - w_i| \le Ch^2,$$

where $C > 0$ is a constant independent of $h$; see, for example, [31, Chap. 24] for details.

Let us set

$$\boldsymbol{w} := \begin{bmatrix} w_1 \\ w_2 \\ \vdots \\ w_{n-1} \\ w_n \end{bmatrix}, \quad \boldsymbol{f} := \begin{bmatrix} f_1 \\ f_2 \\ \vdots \\ f_{n-1} \\ f_n \end{bmatrix} \in \mathbb{R}^n,$$

and

$$\mathsf{A} = \frac{1}{h^2}\begin{bmatrix} 2 & -1 & 0 & \cdots & 0 \\ -1 & 2 & \ddots & & \vdots \\ 0 & \ddots & \ddots & -1 & 0 \\ \vdots & & -1 & 2 & -1 \\ 0 & \cdots & 0 & -1 & 2 \end{bmatrix} \in \mathbb{R}^{n\times n}. \tag{1.7.4}$$

Then in the matrix form the finite difference method is expressed as follows: find $\boldsymbol{w} \in \mathbb{R}^n$ such that

$$\mathsf{A}\boldsymbol{w} = \boldsymbol{f}. \tag{1.7.5}$$

The matrix A is called the finite difference stiffness matrix. As we will learn, this matrix is SPD, and therefore this problem is uniquely solvable; see Problem 1.15.

The SOR method is just the weighted Gauss–Seidel method with an overrelaxation parameter $\omega > 1$, chosen so as to achieve an optimal rate of convergence in the Euclidean norm. See, for example, [31, Chap. 6]. Following (1.4.4), the generic $i$th step of forward Gauss–Seidel is expressed as

$$w_i^{k+1} = \frac{f_i - \sum_{j=1}^{i-1} a_{i,j} w_j^{k+1} - \sum_{j=i+1}^{n} a_{i,j} w_j^k}{a_{i,i}},$$

where we step, in order, from $i = 1$ to $i = n$. The SOR method with a weighting parameter $\omega > 1$ can be expressed as

$$\begin{aligned} z_i^k &:= \frac{f_i - \sum_{j=1}^{i-1} a_{i,j} w_j^{k+1} - \sum_{j=i+1}^{n} a_{i,j} w_j^k}{a_{i,i}}, \\ w_i^{k+1} &:= \omega z_i^k + (1-\omega) w_i^k, \end{aligned}$$

where, again, we step, in order, from $i = 1$ to $i = n$.

In the present example, using the entries of the stiffness matrix defined in (1.7.4), the respective methods become

$$\begin{aligned} w_1^{k+1} &= \frac{h^2 f_1 + w_2^k}{2}, \\ w_i^{k+1} &= \frac{h^2 f_i + w_{i-1}^{k+1} + w_{i+1}^k}{2}, \quad 2 \le i \le n-1, \\ w_n^{k+1} &= \frac{f_n + w_{n-1}^{k+1}}{2}, \end{aligned}$$

and

$$
\begin{aligned}
z_1^k &= \frac{h^2 f_1 + w_2^k}{2}, & w_1^{k+1} &= \omega z_1^k + (1-\omega) w_1^k, \\
z_i^k &= \frac{h^2 f_i + w_{i-1}^{k+1} + w_{i+1}^k}{2}, & w_i^{k+1} &= \omega z_i^k + (1-\omega) w_i^k, \quad 2 \le i \le n-1, \\
z_n^k &= \frac{f_n + w_{n-1}^{k+1}}{2}, & w_n^{k+1} &= \omega z_n^k + (1-\omega) w_n^k.
\end{aligned}
$$

**Listing 1.1:** The main Gauss–Seidel iteration block. The full code is provided in Listing 1.5.

```
w(1) = (h2*f(1)+w(2))/2.0;
for i = 2:n-1
  w(i) = (h2*f(i)+w(i+1)+w(i-1))/2.0;
end
w(n) = (h2*f(n)+w(n-1))/2.0;
```

In either case, we immediately use an updated entry in the computation of the next entry. Thus, in the MATLAB© implementation, we can over-write entries to save storage. In short, we do not need to create separate storage spaces for the vectors $\boldsymbol{w}^{k+1}$ and $\boldsymbol{w}^k$.

**Listing 1.2:** Code for one pass of SOR iteration. The full code is provided in Listing 1.6.

```
tmp = (h2*f(1)+w(2))/2.0;
w(1) = omega*tmp+omOmega*w(1);
for i = 2:n-1
  tmp = (h2*f(i)+w(i+1)+w(i-1))/2.0;
  w(i) = omega*tmp+omOmega*w(i);
end
tmp = (h2*f(n)+w(n-1))/2.0;
w(n) = omega*tmp+omOmega*w(n);
```

**Example 1.7.1** (Comparison for $n = 31$). With the framework as above, let us suppose that $n = 31$ and $\boldsymbol{f} \in \mathbb{R}^{31}$ is constructed so that the exact solution $\boldsymbol{w} \in \mathbb{R}^{31}$ has the components

$$w_i = \exp(\sin(3\pi x_i)) - 1.$$

The code presented in Listing 1.3 solves the matrix problem $\mathsf{A}\boldsymbol{w} = \boldsymbol{f}$ using the Jacobi, Gauss–Seidel, and SOR linear iterative schemes, and it compares their rates of convergence. For instance, for the Jacobi method,

$$\|\boldsymbol{w} - \boldsymbol{w}_{\mathrm{J}}^{k+1}\| \approx \gamma_{\mathrm{J}} \|\boldsymbol{w} - \boldsymbol{w}_{\mathrm{J}}^k\|$$

for sufficiently large $k$, and similarly for the other cases; see Figure 1.3.

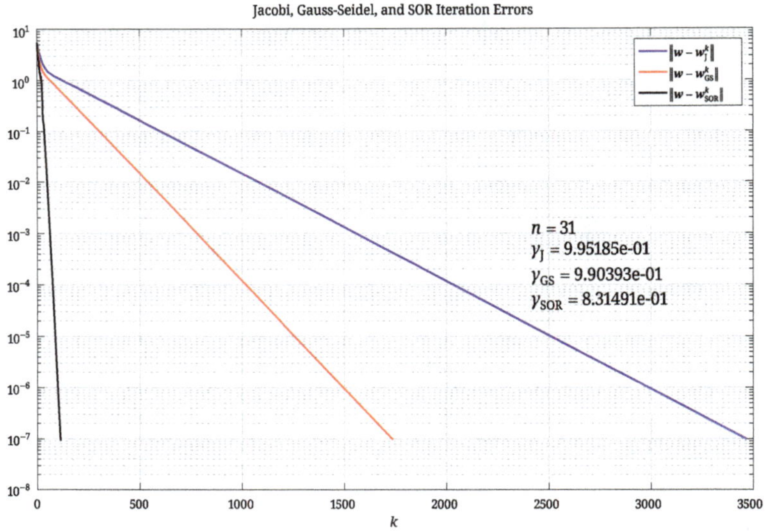


**Figure 1.3:** The error reduction rates for the Jacobi, Gauss–Seidel, and SOR methods with $n = 31$. The stopping tolerance is equal to $10^{-7}$. The computed error reduction factors $\gamma_J$, $\gamma_{GS}$, and $\gamma_{SOR}$ for the respective methods are shown.

The optimal value for $\omega > 1$ in the successive overrelaxation (SOR) method is chosen using the theory in [31, Example 6.4]. In fact, we confirm, to a good accuracy, the contraction rates given in that reference. The SOR method performs the best; it has the smallest contraction rate, as predicted by the theory. The Jacobi method is the worst of the three in that measure. To reach a stopping tolerance of $10^{-7}$, it requires almost 3500 iterations of the Jacobi method. ▲

**Example 1.7.2** (Comparison for $n = 63$). Suppose now that we take $n = 63$. We again assume that $\boldsymbol{f} \in \mathbb{R}^{63}$ is constructed so that the exact solution $\boldsymbol{w} \in \mathbb{R}^{63}$ has the components

$$w_i = \exp(\sin(3\pi x_i)) - 1.$$

The error contraction rates are given in Figure 1.4. Note that all the methods are degraded with the larger value of $n = 63$, compared to the previous case of $n = 31$. For example, reaching a stopping tolerance of $10^{-7}$ requires almost 14000 iterations of the Jacobi method. ▲

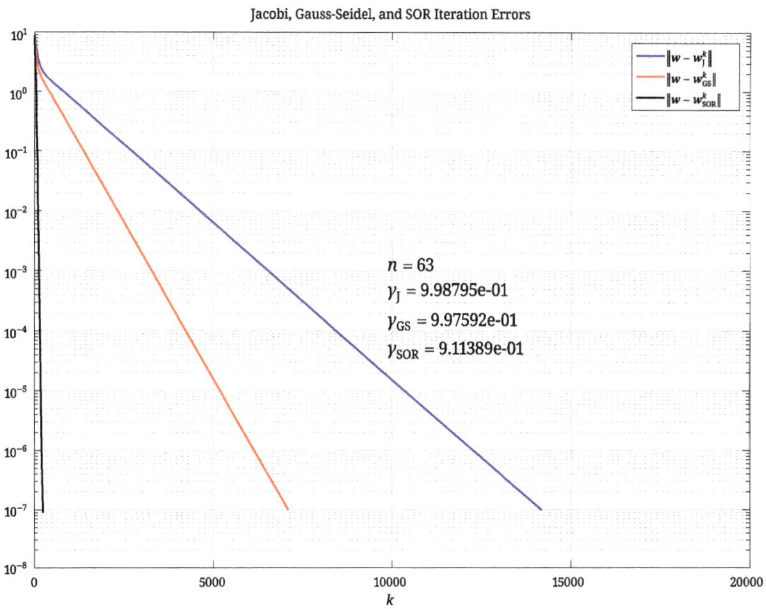


**Figure 1.4:** The error reduction rates for the Jacobi, Gauss–Seidel, and SOR methods with $n = 63$. The stopping tolerance is equal to $10^{-7}$. The computed error reduction factors $\gamma_J$, $\gamma_{GS}$, and $\gamma_{SOR}$ for the respective methods are shown. Observe how much more slowly the methods converge for the larger value of $n = 63$, compared to the case $n = 31$ shown in Figure 1.3.

## 1.8 Problems

1.1 Prove Proposition 1.1.5.
1.2 Prove Proposition 1.1.6.
1.3 Prove Proposition 1.1.10.
1.4 Prove Proposition 1.1.11.
1.5 Prove Proposition 1.1.13.
1.6 Prove Proposition 1.1.15.
1.7 Prove Proposition 1.4.3.
1.8 Prove Proposition 1.4.8.
1.9 Prove Proposition 1.4.11.
1.10 Prove Theorem 1.4.17.
1.11 Prove Proposition 1.5.2.
1.12 Prove Proposition 1.5.3.
1.13 Prove Corollary 1.6.4.
1.14 Complete the proof of the Householder–John Theorem 1.6.5.
1.15 Define $\mathsf{A} \in \mathbb{R}^{n\times n}$ via

$$\mathsf{A} = \begin{bmatrix} 2 & -1 & 0 & \cdots & 0 \\ -1 & 2 & \ddots & & \vdots \\ 0 & \ddots & \ddots & -1 & 0 \\ \vdots & & -1 & 2 & -1 \\ 0 & \cdots & 0 & -1 & 2 \end{bmatrix}.$$

(a) Find the eigenvalues and eigenvectors of A.
(b) Find the eigenvalues and eigenvectors of $\mathsf{T}_\mathrm{J}$ and $\mathsf{T}_\mathrm{GS}$, the error transfer matrices for the Jacobi and Gauss–Seidel methods, respectively.
(c) Find $\rho(\mathsf{T}_\mathrm{J})$ and $\rho(\mathsf{T}_\mathrm{GS})$.
(d) Construct the iterator and error transfer matrices $\mathsf{B}_{\mathrm{J},\omega}$ and $\mathsf{T}_{\mathrm{J},\omega}$, respectively, for the damped Jacobi method.
(e) Find the eigenvalues and eigenvectors of $\mathsf{T}_{\mathrm{J},\omega}$.
(f) Let $\mu_k(\omega)$ be the $k$th eigenvalue of $\mathsf{T}_{\mathrm{J},\omega}$ with the ordering

$$\mu_1 \geq \mu_2 \geq \cdots \geq \mu_n.$$

(You will observe in this problem that we may extend the definition of the eigenvalues in a natural way, so that $\mu_0 = 1$ and $\mu_{n+1} = 1 - 2\omega$.) Assume that $n + 1$ is even for simplicity. Prove that the quantity

$$S(\omega) := \max_{\frac{n+1}{2} \leq k \leq n+1} |\mu_k(\omega)|$$

is minimized by $\omega = \omega_0 = \frac{2}{3}$ and

$$|\mu_k(\omega_0)| \le \frac{1}{3},$$

for all $\frac{n+1}{2} \le k \le n+1$.

## Listings

```
./JacobiGSSOR/
    JGSSORDriver.m
    Jacobi.m
    GaussSeidel.m
    SOR.m
    contractionRates.m
```

**Listing 1.3:** `JGSORDriver.m`: MATLAB© code for comparing the Jacobi, Gauss–Seidel, and SOR iteration methods for solving the equation in Example 1.7.1.

```
%
% JGSSORDriver.m
%
% Calling script for solving the 1D Poisson equation
%
%   -u'' = f
%
% on an interval with homogeneous Dirichlet boundary
% conditions using three classical iterative methods:
% Jacobi, Gauss-Seidel, and Successive Over-Relaxation.
% The script compares convergence rates and iteration
% counts for all three methods.
%
%% Initialization and Parameter

% Clear command line window and workspace.
clear; clc;

n = 31;
h = 1/(n+1);
h2 = h*h;
kMax = 30000;
tol = 1e-07;

errVals = zeros(kMax,3);
kStop = zeros(3,1);

% Store variables in a parameter structure.
iterParam.n = n;
iterParam.h = h;
```

```
iterParam.h2 = h2;
iterParam.kMax = kMax;
iterParam.tol = tol;

% Construct the right-hand-side vector, f.
x = linspace(h,1.0-h,n)';
A = gallery('tridiag',n,-1.0/h2,2.0/h2,-1.0/h2);
wExact = exp(sin(3.0*pi*x))-1.0;
f = A*wExact;

% Initialize w.
SEED = 1234;
rng(SEED);
w = rand(n,1)-0.5;

% Jacobi iteration loop.
[w,errVals(:,1),kStop(1)] = Jacobi(f,w,wExact,iterParam);

% Plot Jacobi errors.
figure(1)
clf
semilogy(errVals(1:kStop(1),1),'b-','LineWidth',1.5)
hold on

% Re-initialize w.
SEED = 1234;
rng(SEED);
w = rand(n,1)-0.5;

% Gauss-Seidel iteration loop.
[w,errVals(:,2),kStop(2)] = GaussSeidel(f,w,wExact,...
                                        iterParam);

% Plot Gauss-Seidel errors.
figure(1)
semilogy(errVals(1:kStop(2),2),'r-','LineWidth',1.5)
hold on

% Re-initialize w.
SEED = 1234;
rng(SEED);
w = rand(n,1)-0.5;
```

```

% SOR iteration loop.
[w,errVals(:,3),kStop(3)] = SOR(f,w,wExact,iterParam);

% Plot SOR errors.
figure(1)
semilogy(errVals(1:kStop(3),3),'k-','LineWidth',1.5)
hold on

% Get contraction rates.
[rate] = contractionRates(errVals,kStop);

% Plotting the errors
xlabel('$k$','Interpreter','latex');
title(['Jacobi, Gauss-Seidel, and SOR ',...
       'Iteration Errors'],'Interpreter','latex');
legend(...
  '$\left\|{\bf w}-{\bf w}_{\rm J}^k\right\|$',...
  '$\left\|{\bf w}-{\bf w}_{\rm GS}^k \right\|$',...
  '$\left\|{\bf w}-{\bf w}_{\rm SOR}^k \right\|$',...
  'Interpreter','latex');
text(0.7*kStop(1),13500*tol,strcat(...
  '$n =\hspace{.1cm}$',...
  num2str(n)),'FontSize',14,...
  'Interpreter','latex')
text(0.7*kStop(1),4500*tol,strcat(...
  '$\gamma_{\rm J} =\hspace{.1cm}$',...
  num2str(rate(1),'%10.5e')),'FontSize',14,...
  'Interpreter','latex')
text(0.7*kStop(1),1500*tol,strcat(...
  '$\gamma_{\rm GS} =\hspace{.1cm}$',...
  num2str(rate(2),'%10.5e')),'FontSize',14,...
  'Interpreter','latex')
text(0.7*kStop(1),500*tol,strcat(...
  '$\gamma_{\rm SOR} =\hspace{.1cm}$',...
  num2str(rate(3),'%10.5e')),'FontSize',14,...
  'Interpreter','latex')
printstr = strcat('Err_JGSSOR_',num2str(n),'.pdf');
exportgraphics(gca, printstr)

hold off
```

**Listing 1.4:** `Jacobi.m`: The Jacobi iterative solver for the equation in Example 1.7.1.

```
function [w,errVals,kStop] = Jacobi(f,w,wExact,iterParam)

n = iterParam.n;
h2 = iterParam.h2;
kMax = iterParam.kMax;
tol = iterParam.tol;

errVals = zeros(kMax,1);
z = zeros(n,1);

k = 0;
err = tol+1.0;

% Jacobi loop.
while ((k < kMax) && (err > tol))

  z(1) = (h2*f(1)+w(2))/2.0;
  for i = 2:n-1
    z(i) = (h2*f(i)+w(i+1)+w(i-1))/2.0;
  end
  z(n) = (h2*f(n)+w(n-1))/2.0;

  w = z;
  err = norm(wExact-w,2);
  k = k+1;
  errVals(k) = err;

end

kStop = k;

end
```

**Listing 1.5:** `GaussSeidel.m`: The Gauss–Seidel iterative solver for the equation in Example 1.7.1.

```
function [w,errVals,kStop] = GaussSeidel(f,w,wExact,...
                                          iterParam)

n = iterParam.n;
h2 = iterParam.h2;
kMax = iterParam.kMax;
```

```
tol = iterParam.tol;

errVals = zeros(kMax,1);

k = 0;
err = tol+1.0;

% Gauss-Seidel loop.
while ((k < kMax) && (err > tol))

  w(1) = (h2*f(1)+w(2))/2.0;
  for i = 2:n-1
    w(i) = (h2*f(i)+w(i+1)+w(i-1))/2.0;
  end
  w(n) = (h2*f(n)+w(n-1))/2.0;

  err = norm(wExact-w,2);
  k = k+1;
  errVals(k) = err;

end

kStop = k;

end
```

**Listing 1.6:** SOR.m: The SOR iterative solver for the equation in Example 1.7.1.

```
function [w,errVals,kStop] = SOR(f,w,wExact,iterParam)

n = iterParam.n;
h = iterParam.h;
h2 = iterParam.h2;
kMax = iterParam.kMax;
tol = iterParam.tol;
sig = cos(pi*h);
omega = 2/(1+sqrt(1-sig*sig));
omOmega = 1.0-omega;

errVals = zeros(kMax,1);

k = 0;
err = tol+1.0;
```

```

% SOR loop.
while ((k < kMax) && (err > tol))

  tmp = (h2*f(1)+w(2))/2.0;
  w(1) = omega*tmp+omOmega*w(1);
  for i = 2:n-1
    tmp = (h2*f(i)+w(i+1)+w(i-1))/2.0;
    w(i) = omega*tmp+omOmega*w(i);
  end
  tmp = (h2*f(n)+w(n-1))/2.0;
  w(n) = omega*tmp+omOmega*w(n);

  err = norm(wExact-w,2);
  k = k+1;
  errVals(k) = err;

end

kStop = k;

end
```

**Listing 1.7:** `contractionRates.m`: MATLAB© code for estimating the contraction rates of various iterative solvers.

```
function [rate] = contractionRates(errVals,kStop)

kStop(:) = kStop(:)-mod(kStop(:),4);
rate = zeros(3,1);

% Estimate the rates of contraction:
if min(kStop) >= 4
  for i = 1:3
    kv = [kStop(i)/2,kStop(i)/2+kStop(i)/4,kStop(i)];
    le = log(errVals(kv,i));
    p1 = polyfit(kv,le,1);
    rate(i) = exp(p1(1));
  end
end

end
```

# 2 The preconditioned conjugate gradient method

No description of iterative schemes for the solution of linear systems of equations is complete without a description of Krylov subspace methods. Here we briefly present the preconditioned conjugate gradient method (PCG). We refer the reader to [31, Chap. 07] for a more detailed description of this method and to [30] for a complete presentation of Krylov subspace methods.

## 2.1 Minimization of quadratic functions

The methods we discuss in this chapter are suitable for the solution of linear systems with a matrix that is SPD. There is a deep connection between the solution of linear systems and the minimization of quadratic functions of several variables. We will discuss the minimization of functions with much more generality in Chapter 11.

In what follows, we assume that $(\cdot,\cdot)_\star$ is an inner product on $\mathbb{R}^n$, $\mathsf{M} \in \mathbb{R}^{n\times n}$ is SPD with respect to this inner product, and $\boldsymbol{g} \in \mathbb{R}^n$. Define the function

$$\Phi(\boldsymbol{w}) := \frac{1}{2}(\mathsf{M}\boldsymbol{w}, \boldsymbol{w})_\star - (\boldsymbol{w}, \boldsymbol{g})_\star. \tag{2.1.1}$$

For future reference, we record here the derivative of this function.

**Lemma 2.1.1** (Differentiability). *The function $\Phi$, defined in* (2.1.1) *is continuously differentiable. If $(\cdot,\cdot)_\star$ is the energy inner product of some SPD matrix* $\mathsf{B}$, *then*

$$\mathrm{D}\Phi(\boldsymbol{v}) := \mathsf{B}(\mathsf{M}\boldsymbol{v} - \boldsymbol{g}).$$

*Proof.* Let $\boldsymbol{v}, \boldsymbol{d} \in \mathbb{R}^n$ and recall that we must show that

$$\left|\Phi(\boldsymbol{v}+\boldsymbol{d}) - \Phi(\boldsymbol{v}) - \mathrm{D}\Phi(\boldsymbol{v})^\top \boldsymbol{d}\right| = \varepsilon(\|\boldsymbol{d}\|)$$

with

$$\frac{\varepsilon(\|\boldsymbol{d}\|)}{\|\boldsymbol{d}\|} \to 0$$

as $\boldsymbol{d} \to \boldsymbol{0}$.

The proof is a simple expansion, which we show for completeness. Indeed, notice that

$$\begin{aligned}\Phi(\boldsymbol{v}+\boldsymbol{d}) &= \frac{1}{2}(\mathsf{BM}(\boldsymbol{v}+\boldsymbol{d}), \boldsymbol{v}+\boldsymbol{d}) - (\mathsf{B}(\boldsymbol{v}+\boldsymbol{d}), \boldsymbol{g}) \\ &= \Phi(\boldsymbol{v}) + (\mathsf{B}(\mathsf{M}\boldsymbol{v} - \boldsymbol{g}), \boldsymbol{d}) + \frac{1}{2}(\mathsf{BM}\boldsymbol{d}, \boldsymbol{d}),\end{aligned}$$

https://doi.org/10.1515/9783111354880-002

where we used that M is symmetric with respect to the B-inner product. In summary,

$$|\Phi(\boldsymbol{v}+\boldsymbol{d}) - \Phi(\boldsymbol{v}) - D\Phi(\boldsymbol{v})^\top \boldsymbol{d}| = \frac{1}{2}(\mathsf{BM}\boldsymbol{d}, \boldsymbol{d}).$$

This not only shows the continuous differentiability, but also provides a formula for the derivative. □

We recall that by Proposition 1.1.13 no generality with respect to the inner product is lost in the previous result.

**Proposition 2.1.2** (Equivalence). *The function $\Phi$ given in* (2.1.1) *has a unique global minimum attained at* $\boldsymbol{v}^{\mathrm{E}} := \mathsf{M}^{-1}\boldsymbol{g}$.

*Proof.* Exercise; see Problem 2.1. □

**Proposition 2.1.3** (Minimization on subspaces). *Let $\Phi$ be defined in* (2.1.1), *and let $V \subset \mathbb{R}^n$ be a subspace. There is a unique $\boldsymbol{v}_V \in V$ that minimizes $\Phi$ over $V$. This minimizer is uniquely characterized by any of the following properties:*

- *The residual $\boldsymbol{r} = \boldsymbol{g} - \mathsf{M}\boldsymbol{v}_V$ is orthogonal, in the $(\cdot,\cdot)_\star$ inner product, to $V$, that is,*

$$(\boldsymbol{r}, \boldsymbol{w})_\star = 0, \quad \forall \boldsymbol{w} \in V.$$

- *Let $\boldsymbol{v}^{\mathrm{E}}$ be the global minimizer of $\Phi$. Define the error $\boldsymbol{e} = \boldsymbol{v}^{\mathrm{E}} - \boldsymbol{v}_V$. We have the following Galerkin orthogonality:*

$$(\mathsf{M}\boldsymbol{e}, \boldsymbol{w})_\star = 0, \quad \forall \boldsymbol{w} \in V.$$

- *The error is optimal, that is,*

$$(\mathsf{M}\boldsymbol{e}, \boldsymbol{e})_\star \le (\mathsf{M}(\boldsymbol{v}^{\mathrm{E}} - \boldsymbol{w}), \boldsymbol{v}^{\mathrm{E}} - \boldsymbol{w})_\star, \quad \forall \boldsymbol{w} \in V.$$

*Proof.* Exercise; see Problem 2.2. □

It turns out that minimization over subspaces is a recurring theme, and thus we give it a name.

**Definition 2.1.4** (Ritz approximation). Let $\Phi$ be defined in (2.1.1), let $\boldsymbol{v}^{\mathrm{E}}$ be its global minimizer, and let $V$ be a subspace of $\mathbb{R}^n$. The (necessarily) unique $\boldsymbol{v}_V \in V$ that minimizes $\Phi$ over $V$ is called the **Ritz approximation** of $\boldsymbol{v}^{\mathrm{E}}$ on $V$. ■

Proposition 2.1.3 shows that to approximate the solution of $\mathsf{A}\boldsymbol{u} = \boldsymbol{f}$ we can compute its Ritz approximation. Different choices of subspaces give rise to different methods.

## 2.2 The preconditioned steepest descent method

In the previous section, we realized that instead of solving a linear system of equations, we can find the minimum of a quadratic function, and that this minimum can be approximated over subspaces. Here we will describe one of the simplest of these strategies. Let $\boldsymbol{p} \in \mathbb{R}^n_\star$, which we call a *descent direction*, and consider $V = \operatorname{span}\{\boldsymbol{p}\}$. Over this subspace the function $\Phi$ defined in (2.1.1) simplifies to

$$\begin{aligned}\phi(\lambda) &:= \Phi(\lambda \boldsymbol{p}) \\ &= \frac{\lambda^2}{2}(\mathsf{M}\boldsymbol{p}, \boldsymbol{p})_\star - \lambda(\boldsymbol{p}, \boldsymbol{g})_\star.\end{aligned}$$

We can easily find the minimum of this quadratic function by computing

$$\phi'(\lambda) = \lambda(\mathsf{M}\boldsymbol{p}, \boldsymbol{p})_\star - (\boldsymbol{p}, \boldsymbol{g})_\star$$

and requiring $\phi'(\lambda) = 0$ to obtain

$$\lambda_\star := \frac{(\boldsymbol{p}, \boldsymbol{g})_\star}{(\mathsf{M}\boldsymbol{p}, \boldsymbol{p})_\star}, \tag{2.2.1}$$

which, since M is SPD with respect to $(\cdot, \cdot)_\star$, is meaningful as long as $\boldsymbol{p} \neq \mathbf{0}$.

This can be exploited to construct a numerical scheme.

**Definition 2.2.1** (Preconditioned steepest descent). Let matrices $\mathsf{A}, \mathsf{B} \in \mathbb{R}^{n\times n}$ be SPD, and let $\boldsymbol{f} \in \mathbb{R}^n$. Define $\boldsymbol{u}^{\mathrm{E}} = \mathsf{A}^{-1}\boldsymbol{f}$. The **preconditioned steepest descent (PSD)** method (with preconditioner B) is an iterative procedure to approximate $\boldsymbol{u}^{\mathrm{E}}$ that, starting from an arbitrary $\boldsymbol{u}^0 \in \mathbb{R}^n$, computes a sequence $\{\boldsymbol{u}^k\}_{k=0}^\infty \subset \mathbb{R}^n$ as follows. Assume that $\boldsymbol{u}^k$ has been defined. Then

- Define the **residual** and the **search direction**, respectively, as

$$\begin{aligned}\boldsymbol{r}^k &:= \boldsymbol{f} - \mathsf{A}\boldsymbol{u}^k, \\ \boldsymbol{p}^k &:= \mathsf{B}\boldsymbol{r}^k.\end{aligned}$$

- In (2.1.1), set $\mathsf{M} = \mathsf{BA}$, let $(\cdot, \cdot)_\star$ be the $\mathsf{B}^{-1}$-inner product, and let $\boldsymbol{g} = \boldsymbol{p}^k$.
- Define the **correction** $\boldsymbol{u}'$ as the minimizer of $\Phi$ over $V = \operatorname{span}\{\boldsymbol{p}^k\}$.
- The next iteration is

$$\boldsymbol{u}^{k+1} := \boldsymbol{u}^k + \boldsymbol{u}'.$$

■

The reason for the name of this method comes from a basic fact from calculus: the derivative points in the direction of maximal growth. Thus its negative indicates the direction of *steepest descent* or maximal decrease. According to Lemma 2.1.1, the derivative of the function

$$v \mapsto \frac{1}{2}(\mathsf{BA}v, v)_{\mathsf{B}^{-1}} - (v, \mathsf{B}f)_{\mathsf{B}^{-1}}$$

at the point $u^k$ is given by

$$\mathsf{B}^{-1}(\mathsf{BA}u^k - \mathsf{B}f) = -r^k.$$

In other words, the residual points in the direction of steepest descent.

**Remark 2.2.2** (Implementation). Due to (2.2.1), we can implement the PSD method as follows. Given $u^k$, compute

$$\begin{aligned} r^k &:= f - \mathsf{A}u^k, \\ p^k &:= \mathsf{B}r^k, \\ \lambda_k &:= \frac{(\mathsf{B}^{-1}p^k, p^k)}{(\mathsf{B}^{-1}\mathsf{BA}p^k, p^k)} \\ &= \frac{(r^k, p^k)}{(\mathsf{A}p^k, p^k)} \end{aligned}$$

and

$$u^{k+1} := u^k + \lambda_k p^k.$$ ◀

An important property of the steepest descent method, which can be used to provide a rate of convergence, is the following result.

**Lemma 2.2.3** (Error minimization). *At every step of the PSD method, the choice of $\lambda_k$ minimizes the* A*-norm of the error $e^{k+1} = u^{\mathrm{E}} - u^{k+1}$.*

*Proof.* The $(k+1)$th iterate can be written as

$$u^{k+1} = u^k + \lambda_k \mathsf{B}r^k.$$

Subtracting $u^{\mathrm{E}}$ from both sides of the equation and using that $\mathsf{A}e^k = r^k$, we get the equation that controls the error:

$$e^{k+1} = e^k - \lambda_k \mathsf{BA}e^k. \tag{2.2.2}$$

Compute the norms

$$\|e^{k+1}\|_{\mathsf{A}}^2 = \|e^k\|_{\mathsf{A}}^2 - 2\lambda_k(\mathsf{BA}e^k, e^k)_{\mathsf{A}} + \lambda_k^2\|\mathsf{BA}e^k\|_{\mathsf{A}}^2.$$

Consider the quadratic function

$$\psi(\lambda) = \|e^k\|_{\mathsf{A}}^2 - 2\lambda(\mathsf{BA}e^k, e^k)_{\mathsf{A}} + \lambda^2\|\mathsf{BA}e^k\|_{\mathsf{A}}^2.$$

We have

$$\psi'(\lambda) = -2(\mathsf{BA}\boldsymbol{e}^k, \boldsymbol{e}^k)_\mathsf{A} + 2\lambda \|\mathsf{BA}\boldsymbol{e}^k\|_\mathsf{A}^2,$$

so that $\psi$ is minimized at the point

$$\lambda_\star := \frac{(\mathsf{BA}\boldsymbol{e}^k, \boldsymbol{e}^k)_\mathsf{A}}{\|\mathsf{BA}\boldsymbol{e}^k\|_\mathsf{A}^2}.$$

Consider now the numerator of this expression. Using that $\mathsf{A}\boldsymbol{e}^k = \boldsymbol{r}^k$ and that the search direction is $\boldsymbol{p}^k = \mathsf{B}\boldsymbol{r}^k$, we get

$$\begin{aligned}(\mathsf{BA}\boldsymbol{e}^k, \boldsymbol{e}^k)_\mathsf{A} &= (\mathsf{BA}\boldsymbol{e}^k, \mathsf{A}\boldsymbol{e}^k) \\ &= (\boldsymbol{r}^k, \boldsymbol{p}^k).\end{aligned}$$

Similarly, the denominator reads

$$\begin{aligned}\|\mathsf{BA}\boldsymbol{e}^k\|_\mathsf{A}^2 &= \|\boldsymbol{p}^k\|_\mathsf{A}^2 \\ &= (\mathsf{A}\boldsymbol{p}^k, \boldsymbol{p}^k).\end{aligned}$$

Thus $\lambda_\star = \lambda_k$, and $\|\boldsymbol{e}^{k+1}\|_\mathsf{A}^2 = \psi(\lambda_\star)$. □

With this result in hand, we can estimate the convergence rate of PSD.

**Theorem 2.2.4** (Convergence of PSD). *Let $\mathsf{A}, \mathsf{B} \in \mathbb{R}^{n\times n}$ be SPD, let $\boldsymbol{f} \in \mathbb{R}^n$, and assume that the sequence $\{\boldsymbol{u}^k\}_{k=0}^\infty \subset \mathbb{R}^n$ was obtained by the PSD method with preconditioner $\mathsf{B}$. Then for every $k \geq 1$, we have*

$$\|\boldsymbol{e}^k\|_\mathsf{A} \leq \left(\frac{\kappa - 1}{\kappa + 1}\right)^k \|\boldsymbol{e}^0\|_\mathsf{A},$$

*where $\boldsymbol{u}^\mathrm{E} = \mathsf{A}^{-1}\boldsymbol{f}$, $\boldsymbol{e}^k = \boldsymbol{u}^\mathrm{E} - \boldsymbol{u}^k$ is the error, and*

$$\kappa = \kappa(\mathsf{BA}) = \frac{\max \sigma(\mathsf{BA})}{\min \sigma(\mathsf{BA})}$$

*is the condition number of the preconditioned system.*

*Proof.* All we need is already in the proof of the previous result. Indeed, (2.2.2) and the optimality condition can be rewritten as

$$\|\boldsymbol{e}^{k+1}\|_\mathsf{A} \leq \|\mathrm{I} - \lambda\mathsf{BA}\|_\mathsf{A} \|\boldsymbol{e}^k\|_\mathsf{A}, \quad \forall \lambda \geq 0.$$

In particular, we can choose, following the proof of Theorem 1.4.19,

$$\lambda_{\text{opt}} = \frac{2}{\min \sigma(\mathsf{BA}) + \max \sigma(\mathsf{BA})}.$$

This choice yields

$$\|\mathrm{I} - \lambda_{\text{opt}}\mathsf{BA}\|_{\mathsf{A}} \leq \frac{\kappa - 1}{\kappa + 1},$$

which implies the result. The details are left to the reader; see Exercise 2.3. □

An important property of PSD is given in the following result.

**Proposition 2.2.5** (Orthogonality). *Consider the PSD with* $\mathsf{B} = \mathrm{I}$. *Then every two consecutive residuals are orthogonal in the Euclidean sense.*

*Proof.* Exercise; see Problem 2.4. □

**Example 2.2.6** (Slow convergence). Theorem 2.2.4 has shown that PSD will converge at least as fast as the (preconditioned) Richardson method with optimal parameter, without the need of any tuning! We do not need to know bounds on the spectrum of BA, the method will find those.

However, in this example, we wish to illustrate that not all may be well and, even for simple systems, steepest descent may take many steps to converge. We consider $n = 2$ (the simplest possible case) and

$$\mathsf{A} = \mathsf{Q}^{\top} \begin{bmatrix} 1 & 0 \\ 0 & 10 \end{bmatrix} \mathsf{Q}, \quad \mathsf{Q} = \begin{bmatrix} \cos(\theta) & \sin(\theta) \\ -\sin(\theta) & \cos(\theta) \end{bmatrix},$$

with $\theta = 0.2000116726443145$. The matrix Q is orthogonal, and its effect is just a rotation by $\theta$ radians. We choose the right-hand side to be $\boldsymbol{f} = [1, 1]^{\top}$, and $\boldsymbol{u}^{\mathrm{E}} = \mathsf{A}^{-1}\boldsymbol{f}$.

Clearly, the matrix A is SPD. We implemented the PSD method with $\mathsf{B} = \mathrm{I}$; see Listing 2.2. The full script for this example is shown in Listing 2.1. We set $\boldsymbol{u}^0 = [2, 1]^{\top}$ and run the PSD method until the test

$$\|\boldsymbol{u}^{\mathrm{E}} - \boldsymbol{u}^k\| \leq 10^{-6}$$

is satisfied. Table 2.1 shows the convergence history, and Figure 2.1 shows the path that the iterates take. As we see, it takes quite a large number of iterations to achieve con-

**Table 2.1:** Convergence history of the PSD method in Example 2.2.6. The number of iterations is denoted by **Its**, whereas **Err** denotes $\|\boldsymbol{u}^{\mathrm{E}} - \boldsymbol{u}^k\|$.

| **Its** | 1 | 2 | 3 | 4 | 5 |
|---|---|---|---|---|---|
| **Err** | 0.8789 | 0.2436 | 0.1942 | 0.0538 | 0.04294 |
| ... | | | | | |
| **Its** | 16 | 17 | 18 | 19 | 20 |
| **Err** | 6.2850e-06 | 5.0112e-06 | 1.3893e-06 | 1.1077e-06 | 3.0712e-07 |

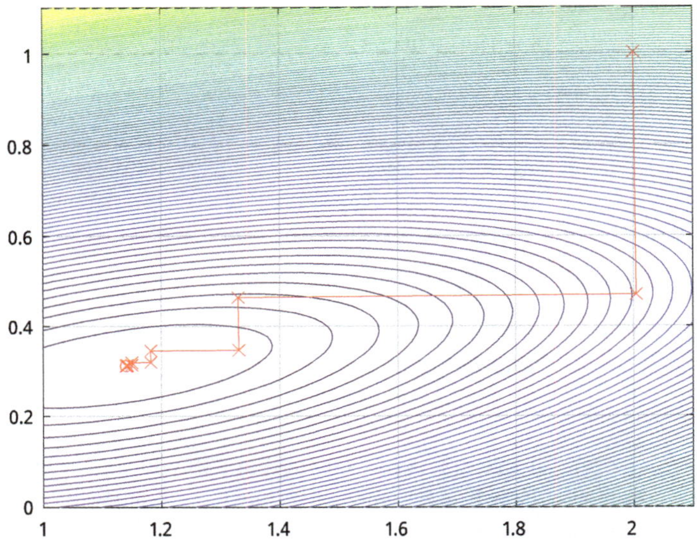


**Figure 2.1:** Convergence path for the PSD in Example 2.2.6. As we see, the difference between iterates is always aligned with one coordinate axis. This leads to slow convergence.

vergence, and it seems that the method can only move along the coordinate directions. This last "zig-zag" effect is explained by Proposition 2.2.5 and is partly the reason for the slow convergence in this example. ▲

## 2.3 The zero-start preconditioned gradient descent method

We begin by defining the method.

**Definition 2.3.1** (Zero-start PCG). Assume that A, B $\in \mathbb{R}^{n\times n}$ are SPD and $\boldsymbol{f} \in \mathbb{R}^n$. The **zero-start preconditioned conjugate gradient (PCG)** method (with preconditioner B) is an iterative procedure to approximate $\boldsymbol{u}^{\mathrm{E}} = \mathsf{A}^{-1}\boldsymbol{f}$ that, starting from $\boldsymbol{u}^0 = \mathbf{0}$, computes $\{\boldsymbol{u}^k\}_{k=0}^{\infty}$ as follows. Define

$$\boldsymbol{r}^0 := \boldsymbol{f}, \quad \boldsymbol{p}^0 := \mathsf{B}\boldsymbol{r}^0. \tag{2.3.1}$$

For $k \geq 0$:
– Update the iterate:

$$\alpha_k := \frac{(\mathsf{B}\boldsymbol{r}^k, \boldsymbol{r}^k)}{(\mathsf{A}\boldsymbol{p}^k, \boldsymbol{p}^k)}, \tag{2.3.2}$$

and

$$u^{k+1} := u^k + \alpha_k p^k. \tag{2.3.3}$$

– Update the residual:

$$r^{k+1} := r^k - \alpha_k \mathsf{A} p^k. \tag{2.3.4}$$

– Update the search direction:

$$\beta_k := \frac{(\mathsf{B} r^{k+1}, r^{k+1})}{(\mathsf{B} r^k, r^k)}, \tag{2.3.5}$$

and

$$p^{k+1} := \mathsf{B} r^{k+1} + \beta_k p^k. \tag{2.3.6}$$

■

Our next step is to understand the origin and properties of this method. As it turns out, every iterate is a Ritz approximation of the exact solution over a very specific collection of subspaces.

**Definition 2.3.2** (Krylov subspace). Let $\mathsf{M} \in \mathbb{R}^{n \times n}$ and $\mathbf{z} \in \mathbb{R}^n$. For $m \in \mathbb{N}$, the **Krylov subspace** of degree $m$ is

$$\mathcal{K}_m(\mathsf{M}, \mathbf{z}) := \operatorname{span}\{\mathsf{M}^k \mathbf{z}\}_{k=0}^{m-1}.$$

■

Let us present some immediate and simple properties of the Krylov subspaces.

**Proposition 2.3.3** (Properties of Krylov subspaces). *Let $\mathsf{M} \in \mathbb{R}^{n \times n}$ and $\mathbf{z} \in \mathbb{R}^n$. For $j, m \in \mathbb{N}$ such that $j \le m$, we have*

$$\mathcal{K}_j(\mathsf{M}, \mathbf{z}) \subset \mathcal{K}_m(\mathsf{M}, \mathbf{z}).$$

*In addition,*

$$\dim \mathcal{K}_j(\mathsf{M}, \mathbf{z}) \le j,$$

*and*

$$0 \le \dim \mathcal{K}_{j+1}(\mathsf{M}, \mathbf{z}) - \mathcal{K}_j(\mathsf{M}, \mathbf{z}) \le 1.$$

*Proof.* Exercise; see Problem 2.5. □

The following result shows the essence of PCG. It computes the Ritz approximations to $u^{\mathrm{E}}$ over Krylov subspaces of increasing degree.

**Theorem 2.3.4** (PCG is minimization). *The PCG method of Definition* 2.3.1 *is obtained by setting* $\mathsf{M} = \mathsf{BA}$, $\boldsymbol{g} = \mathsf{B}\boldsymbol{f}$, *and* $(\cdot,\cdot)_\star = (\cdot,\cdot)_{\mathsf{B}^{-1}}$ *in* (2.1.1). *Then* $\boldsymbol{u}^k$ *is the Ritz approximation of* $\boldsymbol{u}^{\mathrm{E}} := \mathsf{A}^{-1}\boldsymbol{f}$ *over* $\mathcal{K}_k(\mathsf{BA}, \mathsf{B}\boldsymbol{f})$.

*Proof.* In what follows, to shorten the notation, we set $\mathcal{K}_m := \mathcal{K}_m(\mathsf{BA}, \mathsf{B}\boldsymbol{f})$, $m \in \mathbb{N}$. We will proceed by induction and during the discussion, obtain several interesting properties of PCG. We begin by noticing that $\boldsymbol{p}^0 = \mathsf{B}\boldsymbol{f} \in \mathcal{K}_1$.

Suppose $\boldsymbol{u}^k$ is found. By construction $\boldsymbol{u}^k \in \mathcal{K}_k$, so that $\mathsf{BA}\boldsymbol{u}^k \in \mathcal{K}_{k+1}$. Define now $\bar{\boldsymbol{r}}^k := \boldsymbol{f} - \mathsf{A}\boldsymbol{u}^k$ and $\bar{\boldsymbol{z}}^k := \mathsf{B}\bar{\boldsymbol{r}}^k$. If we assume that $\bar{\boldsymbol{r}}^k \neq \boldsymbol{0}$, then

$$\begin{aligned}\bar{\boldsymbol{z}}^k &= \mathsf{B}(\boldsymbol{f} - \mathsf{A}\boldsymbol{u}^k)\\ &= \mathsf{B}\boldsymbol{f} - \mathsf{BA}\boldsymbol{u}^k\\ &= \boldsymbol{p}^0 - \mathsf{BA}\boldsymbol{u}^k\\ &\in \mathcal{K}_{k+1}.\end{aligned}$$

Moreover, the Galerkin orthogonality property of Proposition 2.1.2 shows that

$$(\mathsf{BA}(\boldsymbol{u}^{\mathrm{E}} - \boldsymbol{u}^k), \boldsymbol{w})_{\mathsf{B}^{-1}} = 0$$

for all $\boldsymbol{w} \in \mathcal{K}_k$. This can be rearranged to obtain, for all $\boldsymbol{w} \in \mathcal{K}_k$,

$$\begin{aligned}(\bar{\boldsymbol{z}}^k, \boldsymbol{w})_{\mathsf{B}^{-1}} &= (\mathsf{B}\boldsymbol{f} - \mathsf{BA}\boldsymbol{u}^k, \boldsymbol{w})_{\mathsf{B}^{-1}}\\ &= 0.\end{aligned}$$

In other words, $\bar{\boldsymbol{z}}^k \in \mathcal{K}_{k+1} \setminus \mathcal{K}_k$, and it is $(\cdot,\cdot)_{\mathsf{B}^{-1}}$-orthogonal to $\mathcal{K}_k$. Consequently, $\bar{\boldsymbol{r}}^k$ is orthogonal, in the Euclidean sense, to $\mathcal{K}_k$.

We have come to a very interesting and important observation. As long as $\bar{\boldsymbol{r}}^k \neq \boldsymbol{0}$, we have that $\dim \mathcal{K}_{k+1} = \dim \mathcal{K}_k + 1$. From this we can infer that there is $n_\star \leq n$ such that $\boldsymbol{u}^k = \boldsymbol{u}^{\mathrm{E}}$ for $k \geq n_\star$; see Exercise 2.6.

We continue the proof by denoting

$$\mathcal{K}_k^\perp := \{\boldsymbol{v} \in \mathcal{K}_{k+1} \mid (\boldsymbol{v}, \boldsymbol{w})_{\mathsf{A}} = 0,\ \forall\, \boldsymbol{w} \in \mathcal{K}_k\}$$

and observing that $\mathcal{K}_k^\perp \neq \{\boldsymbol{0}\}$ since $\bar{\boldsymbol{z}}^k \in \mathcal{K}_{k+1} \setminus \mathcal{K}_k$. In fact, $\dim \mathcal{K}_k^\perp = 1$, and

$$\mathcal{K}_{k+1} = \mathcal{K}_k \overset{\perp_{\mathsf{A}}}{\oplus} \mathcal{K}_k^\perp,$$

where we indicated that the sum is orthogonal in the A-inner product.

Next, let $\boldsymbol{u}^{k+1}$ be the minimizer of $\Phi$ over $\mathcal{K}_{k+1}$. By orthogonality we split this minimization into minimizing over $\mathcal{K}_k$, which yields $\boldsymbol{u}^k$, and minimizing over the one-dimensional subspace $\mathcal{K}_k^\perp$. Suppose now that $\boldsymbol{0} \neq \bar{\boldsymbol{p}}^k \in \mathcal{K}_k^\perp$ is chosen. Then $\mathcal{K}_k^\perp = \operatorname{span}\{\bar{\boldsymbol{p}}^k\}$, and

$$\boldsymbol{u}^{k+1} = \boldsymbol{u}^k + \lambda\bar{\boldsymbol{p}}^k.$$

The scalar $\lambda$ is chosen so as to minimize

$$\Phi(\boldsymbol{u}^{k+1} + \lambda\bar{\boldsymbol{p}}^k).$$

This is the same computation that led to (2.2.1); see Exercise 2.7. Thus the minimum is attained for

$$\bar{\alpha}_k = \frac{(\bar{\boldsymbol{r}}^k, \bar{\boldsymbol{p}}^k)}{(\mathsf{A}\bar{\boldsymbol{p}}^k, \bar{\boldsymbol{p}}^k)}.$$

The residual of the $(k+1)$th iteration is

$$\bar{\boldsymbol{r}}^{k+1} = \boldsymbol{f} - \mathsf{A}\boldsymbol{u}^{k+1} = \boldsymbol{f} - \mathsf{A}(\boldsymbol{u}^k + \alpha_k\bar{\boldsymbol{p}}^k) = \bar{\boldsymbol{r}}^k - \alpha_k\mathsf{A}\bar{\boldsymbol{p}}^k,$$

which coincides with (2.3.4), the formula to update the residual in Definition 2.3.1.

Let us now, for the time being, assume that $\bar{\boldsymbol{p}}^k$ was computed with (2.3.6). If that is the case, then

$$\begin{aligned}
\bar{\alpha}_k &= \frac{(\bar{\boldsymbol{r}}^k, \bar{\boldsymbol{p}}^k)}{(\mathsf{A}\bar{\boldsymbol{p}}^k, \bar{\boldsymbol{p}}^k)} \\
&= \frac{(\bar{\boldsymbol{r}}^k, \mathsf{B}\bar{\boldsymbol{r}}^k)}{(\mathsf{A}\bar{\boldsymbol{p}}^k, \bar{\boldsymbol{p}}^k)} + \beta_{k-1}\frac{(\bar{\boldsymbol{r}}^k, \bar{\boldsymbol{p}}^{k-1})}{(\mathsf{A}\bar{\boldsymbol{p}}^k, \bar{\boldsymbol{p}}^k)} \\
&= \frac{(\mathsf{B}\bar{\boldsymbol{r}}^k, \bar{\boldsymbol{r}}^k)}{(\mathsf{A}\bar{\boldsymbol{p}}^k, \bar{\boldsymbol{p}}^k)},
\end{aligned}$$

where in the last step, we used that $\bar{\boldsymbol{p}}^{k-1} \in \mathcal{K}_k$, and the residual $\bar{\boldsymbol{r}}^k$ is orthogonal, in the Euclidean sense, to $\mathcal{K}_k$. This coincides with (2.3.2), and all the computational formulas coincide. The sequence $\{\boldsymbol{u}^k\}_{k=0}^{\infty}$ is obtained via Ritz approximations over Krylov subspaces, and $\boldsymbol{r}^k = \bar{\boldsymbol{r}}^k$ and $\boldsymbol{p}^k = \bar{\boldsymbol{p}}^k$ for all $k \in \mathbb{N}_0$.

Let us now find $\bar{\boldsymbol{p}}^k$. By the way the vectors $\bar{\boldsymbol{p}}^0, \ldots, \bar{\boldsymbol{p}}^{k-1}$ are constructed, they are A-orthogonal, and thus they form a basis of $\mathcal{K}_k$. Recall that we have shown that $\bar{\boldsymbol{z}}^k = \mathsf{B}\bar{\boldsymbol{r}}^k$ has a component in $\mathcal{K}_k^{\perp}$. Thus we may define $\bar{\boldsymbol{p}}^k$ as the A-orthogonal projection of $\bar{\boldsymbol{z}}^k$ onto $\mathcal{K}_k^{\perp}$, that is,

$$\bar{\boldsymbol{p}}^k = \bar{\boldsymbol{z}}^k - \sum_{i=0}^{k-1} \frac{(\mathsf{A}\bar{\boldsymbol{z}}^k, \bar{\boldsymbol{p}}^i)}{(\mathsf{A}\bar{\boldsymbol{p}}^i, \bar{\boldsymbol{p}}^i)}\bar{\boldsymbol{p}}^i.$$

Notice now that

$$(\mathsf{A}\bar{\boldsymbol{z}}^k, \bar{\boldsymbol{p}}^i) = (\mathsf{B}\bar{\boldsymbol{r}}^k, \mathsf{A}\bar{\boldsymbol{p}}^i) = (\bar{\boldsymbol{r}}^k, \mathsf{B}\mathsf{A}\bar{\boldsymbol{p}}^i).$$

However, since $\bar{\boldsymbol{p}}^i \in \mathcal{K}_{i+1}$, we have that $\mathsf{B}\mathsf{A}\bar{\boldsymbol{p}}^i \in \mathcal{K}_{i+2}$, and then for $i + 2 \leq k$, we must have

$$(\mathsf{A}\bar{z}^k, \bar{p}^i) = 0.$$

This reduces the formula that defines $\bar{\boldsymbol{p}}^k$ to

$$\bar{\boldsymbol{p}}^k = \mathsf{B}\bar{\boldsymbol{r}}^k - \frac{(\mathsf{A}\bar{\boldsymbol{z}}^k, \bar{\boldsymbol{p}}^{k-1})}{(\mathsf{A}\bar{\boldsymbol{p}}^{k-1}, \bar{\boldsymbol{p}}^{k-1})}\bar{\boldsymbol{p}}^{k-1}.$$

Up to the definition of $\beta_k$, this is (2.3.6). Notice now that this coefficient can also be rewritten as

$$\begin{aligned}\frac{(\mathsf{A}\bar{\boldsymbol{z}}^k, \bar{\boldsymbol{p}}^{k-1})}{(\mathsf{A}\bar{\boldsymbol{p}}^{k-1}, \bar{\boldsymbol{p}}^{k-1})} &= \frac{1}{\bar{\alpha}_{k-1}} \frac{(\mathsf{B}\bar{\boldsymbol{r}}^k, \bar{\boldsymbol{r}}^k - \bar{\boldsymbol{r}}^{k-1})}{(\mathsf{A}\bar{\boldsymbol{p}}^{k-1}, \bar{\boldsymbol{p}}^{k-1})} \\ &= \frac{(\mathsf{A}\bar{\boldsymbol{p}}^{k-1}, \bar{\boldsymbol{p}}^{k-1})}{(\mathsf{B}\bar{\boldsymbol{r}}^{k-1}, \bar{\boldsymbol{r}}^{k-1})} \frac{(\mathsf{B}\bar{\boldsymbol{r}}^k, \bar{\boldsymbol{r}}^k - \bar{\boldsymbol{r}}^{k-1})}{(\mathsf{A}\bar{\boldsymbol{p}}^{k-1}, \bar{\boldsymbol{p}}^{k-1})} \\ &= \frac{(\mathsf{B}\bar{\boldsymbol{r}}^k, \bar{\boldsymbol{r}}^k - \bar{\boldsymbol{r}}^{k-1})}{(\mathsf{B}\bar{\boldsymbol{r}}^{k-1}, \bar{\boldsymbol{r}}^{k-1})},\end{aligned}$$

so that if we show that $(\mathsf{B}\bar{\boldsymbol{r}}^k, \bar{\boldsymbol{r}}^{k-1}) = 0$, then we indeed get

$$\frac{(\mathsf{A}\bar{\boldsymbol{z}}^k, \bar{\boldsymbol{p}}^{k-1})}{(\mathsf{A}\bar{\boldsymbol{p}}^{k-1}, \bar{\boldsymbol{p}}^{k-1})} = \beta_{k-1}.$$

All the necessary equivalences have been proved. To conclude, observe that $\mathsf{B}\bar{\boldsymbol{r}}^{k-1} = \bar{\boldsymbol{z}}^{k-1} \in \mathcal{K}_k$ and $\bar{\boldsymbol{r}}^k$ is orthogonal, in the Euclidean sense, to $\mathcal{K}_k$. □

## 2.4 Analysis of the zero-start PCG

Having understood the origin of the zero-start PCG method, let us study its convergence properties. First of all, during the proof of the previous result, we established that this method, at least in exact arithmetic, must converge in at most $n$ steps. In practice, however, the dimension of our space is very large, and we may not wish to perform that large number of iterations. Even if $n$ is moderate, rounding errors may affect our method, and thus we will not have convergence in a finite number of steps. For this reason, the zero-start PCG method is often considered as an iterative scheme. Its rate of convergence is the content of the next result.

**Theorem 2.4.1** (Convergence of zero-start PCG). *Let* $\mathsf{A}, \mathsf{B} \in \mathbb{R}^{n\times n}$ *be SPD, and let* $\boldsymbol{f} \in \mathbb{R}^n$. *Define* $\boldsymbol{u}^{\mathrm{E}} = \mathsf{A}^{-1}\boldsymbol{f}$. *Assume that the sequence* $\{\boldsymbol{u}^k\}_{k=0}^{\infty} \subset \mathbb{R}^n$ *was obtained by the zero-start PCG method with preconditioner* B. *Then, for every* $k \geq 1$, *we have*

$$\|\boldsymbol{e}^k\|_{\mathsf{A}} \leq 2\left(\frac{\sqrt{\kappa}-1}{\sqrt{\kappa}+1}\right)^k \|\boldsymbol{e}^0\|_{\mathsf{A}},$$

*where* $\boldsymbol{e}^k = \boldsymbol{u}^{\mathrm{E}} - \boldsymbol{u}^k$ *is the error, and*

$$\kappa = \kappa(\mathsf{BA}) = \frac{\max \sigma(\mathsf{BA})}{\min \sigma(\mathsf{BA})}$$

*is the condition number of the preconditioned system.*

*Proof.* We will continue using the notation established in the proof of Theorem 2.3.4. The optimality of Proposition 2.1.3 shows that

$$\|\boldsymbol{e}^k\|_{\mathsf{A}}^2 \le (\mathsf{A}(\boldsymbol{e}^0 - \boldsymbol{w}), (\boldsymbol{e}^0 - \boldsymbol{w})), \quad \forall \boldsymbol{w} \in \mathcal{K}_k.$$

Now each $\boldsymbol{w} \in \mathcal{K}_k$ can be written for some $\{c_i\}_{i=0}^{k-1} \subset \mathbb{R}$ as

$$\begin{aligned}\boldsymbol{w} &= \sum_{i=0}^{k-1} c_i(\mathsf{BA})^i \boldsymbol{p}^0 \\ &= \mathcal{P}(\mathsf{BA})\mathsf{B}\boldsymbol{r}^0 \\ &= \mathcal{P}(\mathsf{BA})\mathsf{BA}\boldsymbol{e}^0,\end{aligned}$$

where $\mathcal{P}(t) = \sum_{i=0}^{k-1} c_i t^i$ is a polynomial of degree at most $k-1$. We have thus created a bijection between $\mathcal{K}_k$ and $\mathbb{P}_{k-1}$, the space of polynomials of degree at most $k-1$. This means that

$$\begin{aligned}\boldsymbol{e}^0 - \boldsymbol{w} &= (\mathrm{I} - \mathcal{P}(\mathsf{BA})\mathsf{BA})\boldsymbol{e}^0 \\ &= \mathcal{Q}(\mathsf{BA})\boldsymbol{e}^0,\end{aligned}$$

where now $\mathcal{Q}$ is any polynomial of degree at most $k$ but such that $\mathcal{Q}(0) = 1$. In summary, we can rewrite the optimality condition as

$$\begin{aligned}\|\boldsymbol{e}^k\|_{\mathsf{A}}^2 &\le (\mathsf{A}\mathcal{Q}(\mathsf{BA})\boldsymbol{e}^0, \mathcal{Q}(\mathsf{BA})\boldsymbol{e}^0) \\ &\le \|\mathcal{Q}(\mathsf{BA})\|_{\mathsf{A}}^2 \|\boldsymbol{e}^0\|_{\mathsf{A}}^2.\end{aligned}$$

Here, for a matrix $\mathsf{M}$, we denoted by $\|\mathsf{M}\|_{\mathsf{A}}$ the induced matrix norm that comes from the vector norm with the same symbol, that is,

$$\|\mathsf{M}\|_{\mathsf{A}} := \sup_{\boldsymbol{v} \in \mathbb{R}^n_\star} \frac{\|\mathsf{M}\boldsymbol{v}\|_{\mathsf{A}}}{\|\boldsymbol{v}\|_{\mathsf{A}}}.$$

Notice now $\mathsf{BA}$ is self-adjoint and positive definite with respect to the A-inner product. Thus all the eigenvalues of $\mathsf{BA}$ are real and positive. Moreover,

$$\begin{aligned}\|\mathcal{Q}(\mathsf{BA})\|_{\mathsf{A}}^2 &= \max_{\lambda \in \sigma(\mathsf{BA})} |\mathcal{Q}(\lambda)| \\ &\le \sup_{t \in [\lambda_1, \lambda_n]} |\mathcal{Q}(t)|,\end{aligned}$$

where $\lambda_1 = \min \sigma(\mathsf{BA})$ and $\lambda_n = \max \sigma(\mathsf{BA})$.

Denote

$$\mathbb{P}_k^\star := \{p \in \mathbb{P}_k \mid p(0) = 1\}.$$

Then all the previous considerations lead us to

$$\|e^k\|_{\mathsf{A}} \leq \inf_{Q \in \mathbb{P}_k^\star} \sup_{t \in [\lambda_1, \lambda_n]} |Q(t)| \|e^0\|_{\mathsf{A}}.$$

The convergence properties of PCG are encoded in

$$\inf_{Q \in \mathbb{P}_k^\star} \sup_{t \in [\lambda_1, \lambda_n]} |Q(t)|.$$

The solution of this min-max problem is classical, and it can be found in many texts; see, for instance, [31, Chap. 07] and [30]. Here we just mention that this requires the use of the so-called Chebyshev polynomials

$$T_m(t) = \begin{cases} \cos(m \arccos(t)), & |t| \leq 1, \\ \cosh(m \cosh^{-1}(t)), & |t| \geq 1, \end{cases} \qquad m \in \mathbb{N}_0. \tag{2.4.1}$$

Indeed, it is possible to prove that

$$\inf_{Q \in \mathbb{P}_k^\star} \sup_{t \in [\lambda_1, \lambda_n]} |Q(t)| = \frac{1}{|T_k(\frac{1}{\rho})|}$$

with

$$\rho := \frac{\lambda_n - \lambda_1}{\lambda_n + \lambda_1}.$$

Properties of $T_k$ and the fact that $0 < \rho \leq 1$ can now be used to get

$$\frac{1}{|T_k(\frac{1}{\rho})|} \leq 2\left(\frac{\sqrt{\kappa} - 1}{\sqrt{\kappa} + 1}\right)^k. \tag{2.4.2}$$

The claimed bound immediately follows. □

## 2.5 The standard PCG

In the previous sections, we fully described the PCG scheme for a zero starting vector. There are situations where this is sufficient. However, sometimes, we have some knowledge about the solution to the system, and this can be used to furnish a suitable initial guess. Thus we need to consider the case of a nonzero starting vector.

**Definition 2.5.1** (Standard PCG). Assume that $\mathsf{A}, \mathsf{B} \in \mathbb{R}^{n\times n}$ are SPD and $\boldsymbol{f} \in \mathbb{R}^n$. The **standard preconditioned conjugate gradient (PCG)** method (with preconditioner B) is an iterative procedure to approximate $\boldsymbol{u}^{\mathrm{E}} = \mathsf{A}^{-1}\boldsymbol{f}$ that, starting from an arbitrary $\boldsymbol{u}^0 \in \mathbb{R}^n$, computes $\{\boldsymbol{u}^k\}_{k=0}^{\infty}$ as follows. Define

$$\boldsymbol{r}^0 := \boldsymbol{f} - \mathsf{A}\boldsymbol{u}^0, \quad \boldsymbol{p}^0 := \mathsf{B}\boldsymbol{r}^0. \tag{2.5.1}$$

For $k \geq 0$:
- Update the iterate:

$$\alpha_k := \frac{(\mathsf{B}\boldsymbol{r}^k, \boldsymbol{r}^k)}{(\mathsf{A}\boldsymbol{p}^k, \boldsymbol{p}^k)}, \tag{2.5.2}$$

  and

$$\boldsymbol{u}^{k+1} := \boldsymbol{u}^k + \alpha_k \boldsymbol{p}^k. \tag{2.5.3}$$

- Update the residual:

$$\boldsymbol{r}^{k+1} := \boldsymbol{r}^k - \alpha_k \mathsf{A}\boldsymbol{p}^k. \tag{2.5.4}$$

- Update the search direction:

$$\beta_k := \frac{(\mathsf{B}\boldsymbol{r}^{k+1}, \boldsymbol{r}^{k+1})}{(\mathsf{B}\boldsymbol{r}^k, \boldsymbol{r}^k)}, \tag{2.5.5}$$

  and

$$\boldsymbol{p}^{k+1} := \mathsf{B}\boldsymbol{r}^{k+1} + \beta_k \boldsymbol{p}^k. \tag{2.5.6}$$

■

As we see, the only difference between the zero-start PCG and the standard one is in the definition of the initial residual; see (2.3.1) and (2.5.1), respectively. It is no surprise then that the general PCG method is also connected to the minimization of a function $\Phi$ of the form (2.1.1).

To state the connection, we introduce affine sets.

**Definition 2.5.2.** If $\boldsymbol{v} \in \mathbb{R}^n$ is a fixed vector and $L \subset \mathbb{R}^n$ is a subspace, we define

$$\mathfrak{A}(\boldsymbol{v}; L) := \{\boldsymbol{v}\} + L := \{\boldsymbol{v} + \boldsymbol{z} \mid \boldsymbol{z} \in L\}.$$

$\mathfrak{A}(\boldsymbol{v}; L)$ is called the **affine set** parallel to subspace $L$ relative to the translation vector $\boldsymbol{v}$.

■

**Theorem 2.5.3** (Standard PCG). *The standard PCG method introduced in Definition 2.5.1 is obtained by setting* $\mathsf{M} = \mathsf{BA}$, $\boldsymbol{g} = \mathsf{B}\boldsymbol{f}$, *and* $(\cdot,\cdot)_\star = (\cdot,\cdot)_{\mathsf{B}^{-1}}$ *in (2.1.1). Then every iterate is of the form*

$$\boldsymbol{u}^k = \boldsymbol{u}^0 + \boldsymbol{z}^k,$$

*where* $\boldsymbol{z}^k$ *minimizes* $\Phi$ *over the affine set* $\{\boldsymbol{u}^0\} + \mathcal{K}_k(\mathsf{BA}, \boldsymbol{g})$. *In addition, for every* $k \geq 1$, *we have*

$$\|\boldsymbol{e}^k\|_{\mathsf{A}} \leq 2\left(\frac{\sqrt{\kappa} - 1}{\sqrt{\kappa} + 1}\right)^k \|\boldsymbol{e}^0\|_{\mathsf{A}},$$

*where* $\boldsymbol{e}^k = \boldsymbol{u}^{\mathrm{E}} - \boldsymbol{u}^k$ *is the error, and*

$$\kappa = \kappa(\mathsf{BA}) = \frac{\max \sigma(\mathsf{BA})}{\min \sigma(\mathsf{BA})}$$

*is the condition number of the preconditioned system.*

*Proof.* Exercise; see Problem 2.10. □

## 2.6 A computational experiment

To conclude this chapter, we compare the performance of the CG method with that of Richardson and SD. In Listing 2.4, we present an implementation of the PCG method. The driver for the full example is presented in Listing 2.5. An implementation of Richardson's method is presented in Listing 2.6, and that for PSD is in Listing 2.2. We consider the same setting as in Section 1.7 with $n = 63$. For the PCG and PSD codes, we are currently setting the preconditioner to be $\mathsf{B} = \mathrm{I}$ for simplicity. In other words, currently, both methods are run without preconditioning. However, the interested readers can easily write their own preconditioner and substitute the identity preconditioner code easily for further exploration.

Figure 2.2 shows the convergence history for all three schemes with $\mathsf{B} = \mathrm{I}$ for PCG and PSD and $\omega = \frac{h_L^2}{4}$ for Richardson. We immediately see how CG (even without preconditioning) outperforms the other two schemes by a large factor. We see, in addition, that CG seems to converge in $n$ steps as predicted by the theory.

Another worthy observation from Figure 2.2 is how close SD and Richardson seem to be. This is not surprising, as the damping factor $\omega$ for Richardson is chosen very close to its optimal value.

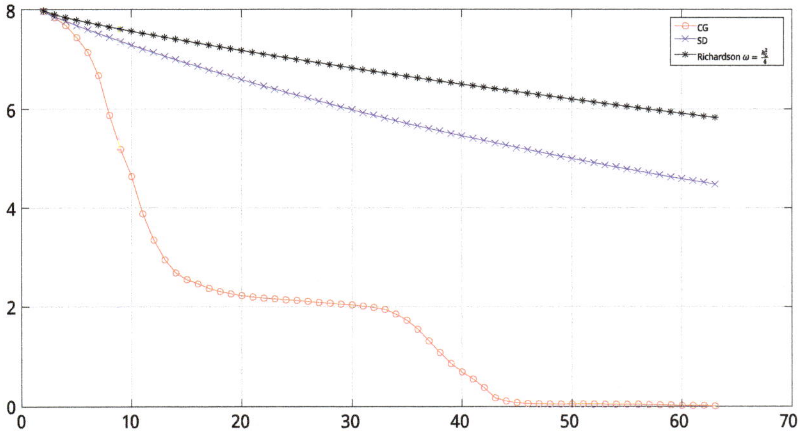


**Figure 2.2:** Comparison of the performance of CG, SD, and Richardson. The horizontal axis denotes the number of iterations, whereas the vertical is the error at each iteration. Here $n = 63$.

## 2.7 Problems

2.1 Prove Proposition 2.1.2.

2.2 Prove Proposition 2.1.3.

2.3 Complete the details of the proof of Theorem 2.2.4.

2.4 Prove Proposition 2.2.5.

2.5 Prove Proposition 2.3.3.

2.6 Prove that the PCG method of Definition 2.3.1 must converge in at most $n$ iterations.

2.7 Show that in the context of the proof of Theorem 2.3.4, the one-dimensional function

$$\begin{aligned}\phi(\lambda) &:= \Phi(\boldsymbol{u}^k + \lambda\bar{\boldsymbol{p}}^k)\\ &= \frac{1}{2}(\mathsf{BA}(\boldsymbol{u}^k + \lambda\bar{\boldsymbol{p}}^k), \boldsymbol{u}^k + \lambda\bar{\boldsymbol{p}}^k)_{\mathsf{B}^{-1}} - (\boldsymbol{u}^k + \lambda\bar{\boldsymbol{p}}^k, \mathsf{B}\boldsymbol{f})_{\mathsf{B}^{-1}}\end{aligned}$$

attains its global minimum at

$$\bar{\alpha}_k = \frac{(\bar{\boldsymbol{r}}^k, \bar{\boldsymbol{p}}^k)}{(\mathsf{A}\bar{\boldsymbol{p}}^k, \bar{\boldsymbol{p}}^k)}.$$

2.8 Prove using induction that for every $m \in \mathbb{N}_0$, (2.4.1) indeed defines a polynomial.

2.9 Use (2.4.1) and properties of hyperbolic trigonometric functions to derive (2.4.2).

2.10 Prove Theorem 2.5.3.

## Listings

./SteepestDescent/
    DriverDescent.m
    PSD.m
    PrecIdentity.m

./PCG/
    PCG.m
    DriverCompare.m
    Richardson.m

**Listing 2.1:** `DriverDescent.m`: Driver code for Example 2.2.6.

```
%
% DriverDescent.m
%
% Demonstration script for solving the linear system
%
%   Ax = f
%
% using the Preconditioned Steepest Descent (PSD) method.
%
%% Initialization and Parameter

% Clear command line window and workspace.
clear;
clc;


% Define rotation angle.
theta = 2.000116726443145e-01;

% Define rotation matrix.
rot = [cos(theta),sin(theta);-sin(theta),cos(theta)];

% Define the matrix A for the quadratic form A*x = f.
A = [1,0;0,10];
A = rot'*A*rot;

% Define the right-hand side vector f.
f = [1;1];
```

```
% Calculate the exact solution.
uExact = A\f;

% Define the quadratic function
% phi(x,y) = 0.5*x'*A*x - f'*x.
phi = @(x,y,A,f)...
      0.5*A(1,1)*x.^2+0.5*A(2,2)*y.^2+A(1,2)*x.*y...
      -f(1)*x-f(2)*y;

%% Plotting and Visualization Setup

% Create mesh for contour plot.
[X,Y] = meshgrid(1:0.005:2.1,0:0.005:1.1);

% Evaluate the quadratic function on the mesh.
contourVals = phi(X,Y,A,f);

% Create figure window and plot contours.
fg = figure(1);
contour(X,Y,contourVals,100);
grid on;
hold on;

% Plot the exact solution.
plot(uExact(1),uExact(2),'ro');
hold on;

%% Iterative Solution (Steepest Descent)

% Initial guess for the solution.
u = [2;1];

% Maximum number of iterations.
maxIter = 2000;

% Tolerance for convergence.
tol = 1e-06;

% Flag to indicate if solution converged.
converged = false;

```

```
for i = 1:maxIter

  % Perform one step of Preconditioned Steepest Descent.
  uNew = PSD(A,u,f,uExact,1,1e-3,@PrecIdentity);

  % Plot the step taken in this iteration.
  plot([u(1),uNew(1)],[u(2),uNew(2)],'rx-');
  hold on;

  % Calculate and print the norm of the error.
  nError = norm(uNew-uExact);
  fprintf('Iteration %d: Error Norm = %.4e\n',i,nError);

  % Check for convergence.
  if nError<tol
    fprintf('Solution found after %d iterations!\n',i);
    converged = true;
    u = uNew;
    break;
  end

  % Update u for the next iteration.
  u = uNew;
end

% Final error calculation and reporting.
if converged == false
  fprintf(['Solution did not converge after %d ',...
           'iterations.\n'], maxIter);
end
finalError = norm(u-uExact);
fprintf('\nFinal error norm = %g\n',finalError);

%% Save Plot and Clean Up

% Save the figure as a PDF.
print(fg,'SteepestDescent.pdf','-dpdf');

hold off;
```

**Listing 2.2:** PSD.m: PSD solver code for Example 2.2.6.

```
function [x,iStop,err] = PSD(A,x0,f,xExact,...
                             maxIterIn,tolIn,Prec)

x = x0;

% Pre-allocate error array for maxIterIn iterations.
err = zeros(maxIterIn,1);

for k = 1:maxIterIn
  r = f-A*x;
  p = Prec(r);
  Ap = A*p;
  lambda = (r'*p)/(p'*Ap);

  % Update solution.
  x = x+lambda*p;

  % Error after the k-th iteration.
  currentNorm = norm(xExact-x);
  err(k) = currentNorm;

  if currentNorm<tolIn
    iStop = k;
    err = err(1:iStop);
    return;

  end
end

% Reached if maxIterIn iterations are completed without
% meeting tolIn.
iStop = maxIterIn;

end
```

**Listing 2.3:** PrecIdentity.m: Identity preconditioner for the PSD solver.

```
function p = PrecIdentity(z)

% Identity preconditioner: returns the input vector
```

```
% unchanged.
p = z;

end
```

**Listing 2.4:** PCG.m: The PCG method.

```
function [x,iStop,err] = PCG(A,xInitial,f,xExact,...
                              maxIter,tolConvergence,Prec)

% Preallocate error history
% (err(1) for initial guess error).
errorHistory = zeros(maxIter+1,1);

x = xInitial;
r = f-A*x;
z = Prec(r);
p = z;

% Error of the initial guess.
errorHistory(1) = norm(xExact-x);

% PCG iterations. Loop runs maxIter times.
for k = 2:maxIter+1
  Ap = A*p;
  denom = p'*Ap;
  num = r'*z;
  alpha = num/denom;

  x = x+alpha*p;
  r = r-alpha*Ap;
  zNew = Prec(r);

  denomBeta = num;
  numBeta = zNew'*r;
  beta = numBeta/denomBeta;

  p = zNew+beta*p;
  z = zNew;

  currentErrorNorm = norm(xExact-x);
```

```
  errorHistory(k) = currentErrorNorm;

  if currentErrorNorm<tolConvergence
    iStop = k;
    err = errorHistory(1:iStop);
    return;

  end
end

% If maxIter iterations are performed without converging.
iStop = maxIter;
err = errorHistory;

end
```

**Listing 2.5:** `DriverCompare.m`: Code for the computational example of Section 2.6.

```
% Clear command line window and workspace.
clear;
clc;

%% Initialization and Parameters

% Discretization size.
n = 63;

% Grid spacing.
h = 1/(n+1);
hSquared = h*h;

% Maximum iterations for solvers.
maxIter = n;

% Tolerance for convergence.
tol = 1e-07;

%% Problem Setup

% Spatial grid (interior points).
x = linspace(h,1.0-h,n)';
```

```

% Discrete Laplacian matrix A.
A = gallery('tridiag',n,-1.0/hSquared,...
            2.0/hSquared,-1.0/hSquared);

% Exact solution.
wExact = exp(sin(3.0*pi*x))-1.0;

% Right-hand side vector f = A*wExact.
f = A*wExact;

% Initialize starting guess w (random).
SEED = 1234;
rng(SEED);
w = rand(n,1)-0.5;

%% Running Solvers

% Conjugate Gradient (CG).
[xCG,iStopCG,errCG] = PCG(A,w,f,wExact,maxIter,tol,...
                          @PrecIdentity);

% Steepest Descent (SD).
[xSD,iStopSD,errSD] = PSD(A,w,f,wExact,maxIter,tol,...
                          @PrecIdentity);

% Richardson iteration.
omega = 0.25*hSquared;
[xR,iStopR,errR] = Richardson(A,w,f,wExact,...
                              maxIter,tol,omega);

%% Plotting Convergence History

% Plotting range.
plotRange = maxIter;

% Create figure and plot.
figure(1);
plot(2:plotRange,errCG(2:plotRange),'ro-',...
     2:plotRange,errSD(2:plotRange),'bx-',...
     2:plotRange,errR(2:plotRange),'k*-');

```

```
legend('CG','SD','Richardson $\omega = h^2/4$',...
  'Interpreter','latex');
grid on;

%% Save Plot

% Define output filename.
printStr = 'ComparePCG.pdf';
print(figure(1),printStr,'-dpdf');
```

**Listing 2.6:** `Richardson.m`: Richardson method.

```
function [x,iStop,err] = Richardson(A,xInitial,f,...
                                   xExact,maxIter,...
                                   tolConvergence,omega)

% Preallocate error history
% (err(1) for initial guess value).
errorHistory = zeros(maxIter+1,1);
x = xInitial;

% Initial residual norm.
errorHistory(1) = norm(f-A*x);

% Richardson iterations. Loop runs maxIter times.
for k = 2:maxIter+1
  r = f-A*x;
  x = x+omega*r;

  % True error norm.
  currentErrorNorm = norm(xExact-x);
  errorHistory(k) = currentErrorNorm;

  if currentErrorNorm<tolConvergence
    iStop = k;
    err = errorHistory(1:iStop);
    return;

  end
end

```

```
% If maxIter iterations are performed without converging.
iStop = maxIter;
err = errorHistory;

end
```

# 3 The two-grid algorithm

In this chapter, we give the basics of the two-grid algorithm. Multigrid is basically a recursive form of the two-grid algorithm. Thus understanding the two-grid algorithm is almost essential for understanding multigrid methods. Our approach is based on matrices rather than operators. This helps the reader focus on implementation issues and dimensional considerations. This chapter presents the material within an otherwise generic framework. There is no connection to elliptic PDEs yet. This will be done in later chapters.

We build the two-grid algorithm upon the GLIS framework that we discussed in Chapter 1. Though it will not be obvious at first glance, we will show that, remarkably, the two-grid algorithm is a more complicated example of a particular GLIS.

## 3.1 Components of the two-grid algorithm

**Definition 3.1.1** (Galerkin condition). Suppose $n_0, n_1 \in \mathbb{N}$, $n_1 \geq n_0$, and $\mathsf{A}_1 \in \mathbb{R}^{n_1 \times n_1}$ and $\mathsf{A}_0 \in \mathbb{R}^{n_0 \times n_0}$ are both SPD. $\mathsf{A}_1$ is called the **fine grid stiffness matrix** or **level-1 stiffness matrix**, and $\mathsf{A}_0$ is called the **coarse grid stiffness matrix** or **level-0 stiffness matrix**. Any matrix $\mathsf{R}_0 \in \mathbb{R}^{n_0 \times n_1}$ of full rank, that is, $\operatorname{rank}(\mathsf{R}_0) = n_0$, is called a **restriction matrix**. Any matrix $\mathsf{P}_0 \in \mathbb{R}^{n_1 \times n_0}$ of full rank, that is, $\operatorname{rank}(\mathsf{P}_0) = n_0$, is called a **prolongation matrix**.

We say that $\mathsf{R}_0$ and $\mathsf{P}_0$ form a **balanced symmetric pair** if

$$\mathsf{P}_0 = \mathsf{R}_0^\top \in \mathbb{R}^{n_1 \times n_0}.$$

We say that the stiffness matrices satisfy the **Galerkin condition** if

$$\mathsf{A}_0 = \mathsf{R}_0 \mathsf{A}_1 \mathsf{P}_0 \in \mathbb{R}^{n_0 \times n_0}. \tag{3.1.1}$$

■

**Remark 3.1.2** (Balanced pair). We will often assume that $\mathsf{R}_0$ and $\mathsf{P}_0$ form a balanced symmetric pair. In this chapter, this is always assumed. On the other hand, in this and every chapter that follows, we always make clear where and when the Galerkin assumption is being used.

In Chapters 6 and 8, we will describe what happens when this balance between $\mathsf{R}_0$ and $\mathsf{P}_0$ is not satisfied. ◄

The Galerkin condition is graphically depicted in Figure 3.1.

**Proposition 3.1.3** ($\mathsf{A}_0$ is SPD). *If* $\mathsf{A}_1 \in \mathbb{R}^{n_1 \times n_1}$ *is SPD and* $\mathsf{A}_0 \in \mathbb{R}^{n_0 \times n_0}$ *is defined via* (3.1.1), *then* $\mathsf{A}_0$ *is also SPD.*

https://doi.org/10.1515/9783111354880-003

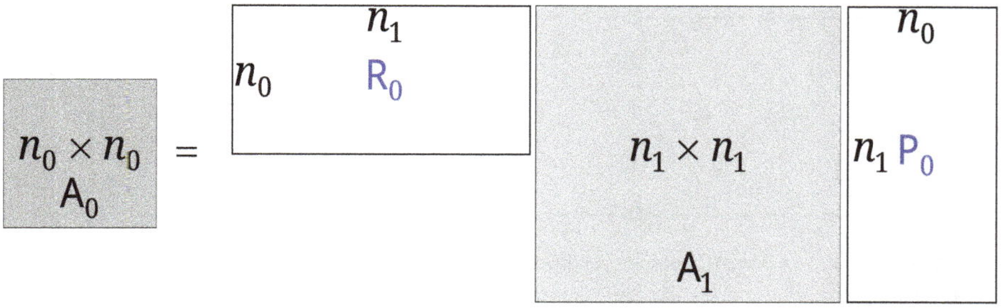


**Figure 3.1:** The creation of the coarse grid matrix via the Galerkin condition.

*Proof.* (Symmetry): Since $A_0 = R_0 A_1 R_0^T$ and $A_1$ is SPD, we have

$$A_0^T = R_0 A_1^T R_0^T = R_0 A_1 R_0^T = A_0.$$

(Positive definiteness): Let $\boldsymbol{w} \in \mathbb{R}^{n_0}$ an arbitrary nonzero vector. Since $R_0$ has full rank, $R_0^T \boldsymbol{w} \neq \boldsymbol{0}$. Moreover, since $A_1$ is SPD,

$$\boldsymbol{w}^T A_0 \boldsymbol{w} = \boldsymbol{w}^T R_0 A_1 R_0^T \boldsymbol{w} = (R_0^T \boldsymbol{w})^T A_1 (R_0^T \boldsymbol{w}) > 0.$$ □

**Remark 3.1.4** (Restriction and prolongation). Observe the actions of $R_0$ and $P_0$:

$$R_0 \in \mathbb{R}^{n_0 \times n_1}, \quad R_0 \boldsymbol{v}_1 \in \mathbb{R}^{n_0}, \quad \forall \boldsymbol{v}_1 \in \mathbb{R}^{n_1},$$

and

$$P_0 \in \mathbb{R}^{n_1 \times n_0}, \quad P_0 \boldsymbol{v}_0 \in \mathbb{R}^{n_1}, \quad \forall \boldsymbol{v}_0 \in \mathbb{R}^{n_0}.$$

In other words, $R_0$ sends a vector from a higher-dimensional space into a lower-dimensional one; $R_0$ destroys information. $P_0$ sends a vector from a lower-dimensional space into a higher-dimensional one; $P_0$ creates information. See Figures 3.2 and 3.3 for the one-dimensional case and Figure 3.4 for the two-dimensional one. ◂

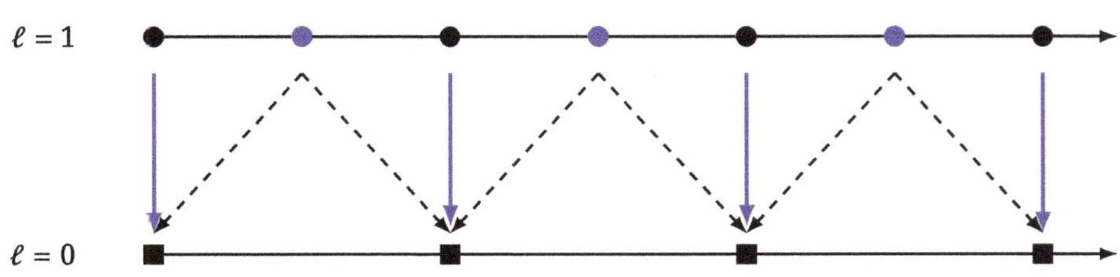


**Figure 3.2:** Restriction from a 1D fine grid (level 1, or $\ell = 1$) to a 1D coarse grid (level 0, or $\ell = 0$). Information is lost through restriction, since there are fewer degrees of freedom left in the coarse grid. Restriction can be thought of as a kind of lossy data compression.

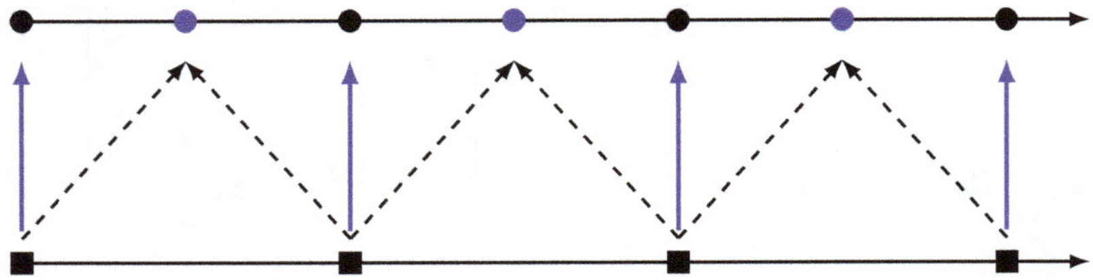


**Figure 3.3:** Prolongation from a 1D coarse grid (level 0, or $\ell = 0$) to a 1D fine grid (level 1, or $\ell = 1$). Information is created through prolongation, usually through some process of interpolation.

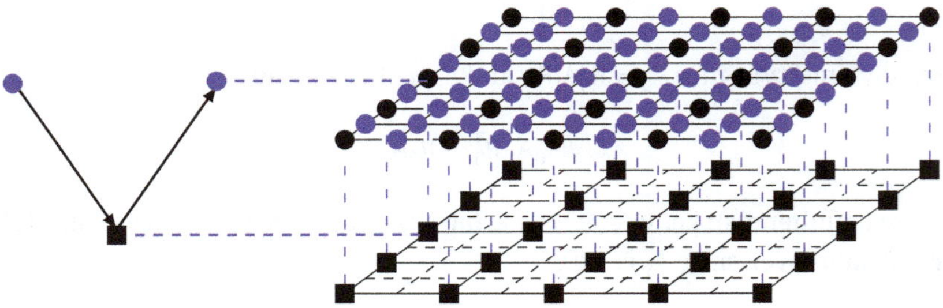


**Figure 3.4:** 2D fine and coarse grids. The restriction matrix moves information from the fine grid (level 1, or $\ell = 1$) to the coarse grid (level 0, or $\ell = 0$). The prolongation goes the other way.

We wish to solve the following problem. Given $\boldsymbol{f}_1 \in \mathbb{R}^{n_1}$ and $\mathsf{A}_1 \in \mathbb{R}^{n_1 \times n_1}$ (SPD), find $\boldsymbol{u}_1^{\mathrm{E}} \in \mathbb{R}^{n_1}$ such that

$$\mathsf{A}_1 \boldsymbol{u}_1^{\mathrm{E}} = \boldsymbol{f}_1.$$

To do this, we define the sequence of approximations

$$\boldsymbol{u}_1^{k+1} := \mathrm{TG}(\boldsymbol{f}_1, \boldsymbol{u}_1^k)$$

using the two-grid operator, which we now define.

**Definition 3.1.5** (Two-grid operator). Suppose that $\boldsymbol{g}_1, \boldsymbol{u}_1^{(0)} \in \mathbb{R}^{n_1}$ are given. The vector $\boldsymbol{u}_1^{(3)} \in \mathbb{R}^{n_1}$ is computed via the **two-grid operator**

$$\boldsymbol{u}_1^{(3)} := \mathrm{TG}(\boldsymbol{g}_1, \boldsymbol{u}_1^{(0)}),$$

which, given $m_1, m_2 \in \mathbb{N}_0$, proceeds as follows:
1. Presmoothing:
   - $\boldsymbol{u}_1^{(1,0)} := \boldsymbol{u}_1^{(0)}$;
   - $\boldsymbol{u}_1^{(1,\sigma+1)} := \boldsymbol{u}_1^{(1,\sigma)} + \mathsf{S}_1(\boldsymbol{g}_1 - \mathsf{A}_1 \boldsymbol{u}_1^{(1,\sigma)}), 0 \le \sigma \le m_1 - 1$;
   - $\boldsymbol{u}_1^{(1)} := \boldsymbol{u}_1^{(1,m_1)}$.

2. Coarse grid correction:
   - $r_1^{(1)} := g_1 - A_1 u_1^{(1)}$;
   - $r_0^{(1)} := R_0 r_1^{(1)}$;
   - $q_0^{(1,E)} := A_0^{-1} r_0^{(1)}$;
   - $q_0^{(1)} := q_0^{(1,E)}$;
   - $q_1^{(1)} := P_0 q_0^{(1)}$;
   - $u_1^{(2)} := u_1^{(1)} + q_1^{(1)}$.
3. Postsmoothing:
   - $u_1^{(3,0)} := u_1^{(2)}$;
   - $u_1^{(3,\sigma+1)} := u_1^{(3,\sigma)} + S_1^\top (g_1 - A_1 u_1^{(3,\sigma)}), \quad 0 \le \sigma \le m_2 - 1$;
   - $u_1^{(3)} := u_1^{(3,m_2)}$.

The element $q_0^{(1)} \in \mathbb{R}^{n_0}$ is called the **two-grid coarse grid correction.** The element $q_0^{(1,E)} \in \mathbb{R}^{n_0}$ is called the **exact coarse grid correction.** The matrix $S_1$ is called the **smoothing operator** or **smoothing matrix.** The parameters $m_1$ and $m_2$ are, respectively, the number of pre- and postsmoothing steps. ■

**Remark 3.1.6** (Equality). In the two-grid setting, $q_0^{(1)} = q_0^{(1,E)}$. This is a very important feature of the two-grid algorithm. In the multigrid algorithm, these two vectors will not be the same, and much of our analysis will focus on estimating their difference. ◂

**Remark 3.1.7** (Coarse grid correction). The element $q_1^{(1)} \in \mathbb{R}^{n_1}$ is also often referred to as the *coarse grid correction*, even though it lives in level 1 (the fine level). Perhaps, $q_1^{(1)}$ would be more appropriately called the *fine grid correction*, but this name has not stuck. In the finite element context, $q_0^{(1)} \in \mathbb{R}^{n_0}$ and $q_1^{(1)} \in \mathbb{R}^{n_1}$ are typically different basis coordinates for the same function, the finite element coarse grid correction function. ◂

**Remark 3.1.8** (Symmetry). Notice that we have presymmetrized the smoothing operations by using the iterator $S_1^\top$ in the postsmoothing step. These smoothing operations are classical GLISs, which we introduced in Chapter 1. Notice also that if $m_1 \neq m_2$, then the pre- and postsmoothing steps are not symmetric. ◂

**Definition 3.1.9** (Two-grid algorithm). Let the numbers of pre- and postsmoothing steps $m_1$ and $m_2$ be given. Suppose that $u_1^k \in \mathbb{R}^{n_1}$ is given. Then the iteration process

$$u_1^{k+1} := \mathrm{TG}(f_1, u_1^k)$$

defines the **generic two-grid algorithm** for solving

$$A_1 u_1^E = f_1.$$

The two-grid algorithm is called **one-sided** if either $m_1 \ge 1$ and $m_2 = 0$ or $m_1 = 0$ and $m_2 \ge 1$. The algorithm is called **symmetric** if $m_1 = m_2$. ■

**Remark 3.1.10** (A special case). Let us examine the coarse-grid correction in a very special case. Suppose that $\mathsf{A}_0$ satisfies the Galerkin condition, $n_0 = n_1$, and

$$\mathsf{R}_0 = \mathrm{I},$$

that is, $\mathsf{R}_0$ is the $n_1 \times n_1$ identity matrix. Then

$$\mathsf{A}_0 = \mathsf{R}_0\mathsf{A}_1\mathsf{P}_0 = \mathsf{R}_0\mathsf{A}_1\mathsf{R}_0^\top = \mathsf{A}_1.$$

Let us compute the coarse grid correction:

$$\begin{aligned}
\boldsymbol{r}_1^{(1)} &= \boldsymbol{f}_1 - \mathsf{A}_1\boldsymbol{u}_1^{(1)};\\
\boldsymbol{r}_0^{(1)} &= \mathsf{R}_0\boldsymbol{r}_1^{(1)} = \boldsymbol{r}_1^{(1)};\\
\boldsymbol{q}_0^{(1)} &= \mathsf{A}_0^{-1}\boldsymbol{r}_0^{(1)} = \mathsf{A}_1^{-1}\boldsymbol{r}_1^{(1)} = \boldsymbol{e}_1^{(1)}.
\end{aligned}$$

Then

$$\begin{aligned}
\boldsymbol{u}_1^{(2)} &= \boldsymbol{u}_1^{(1)} + \mathsf{P}_0\boldsymbol{q}_0^{(1)}\\
&= \boldsymbol{u}_1^{(1)} + \mathsf{P}_0\boldsymbol{e}_1^{(1)}\\
&= \boldsymbol{u}_1^{(1)} + \boldsymbol{e}_1^{(1)}\\
&= \boldsymbol{u}_1^{(1)} + \boldsymbol{u}_1^{\mathrm{E}} - \boldsymbol{u}_1^{(1)}\\
&= \boldsymbol{u}_1^{\mathrm{E}},
\end{aligned}$$

where

$$\mathsf{A}_1\boldsymbol{u}_1^{\mathrm{E}} = \boldsymbol{f}_1.$$

Of course, the two-grid algorithm should terminate at this stage, because we have the exact solution. In general,

$$\boldsymbol{q}_0^{(1)} \neq \boldsymbol{e}_1^{(1)}.$$

However, the approximation

$$\mathsf{P}_0\boldsymbol{q}_0^{(1)} \approx \boldsymbol{e}_1^{(1)}$$

can be quite good in some sense. We will revisit this issue later. ◂

## 3.2 The coarse-grid Ritz projection

Now let us show that the two-grid algorithm is a GLIS. We need only derive the iterator matrix or, equivalently, the error propagation matrix. We first need the following definitions.

**Definition 3.2.1** (Coarse-grid Ritz projection). In the setting of Definition 3.1.1, define the matrix

$$\tilde{\Pi}_1 := \mathsf{R}_0^\top \Pi_0 \in \mathbb{R}^{n_1 \times n_1},$$

where

$$\Pi_0 := \underbrace{\mathsf{A}_0^{-1}}_{n_0 \times n_0} \underbrace{\mathsf{R}_0}_{n_0 \times n_1} \underbrace{\mathsf{A}_1}_{n_1 \times n_1} \in \mathbb{R}^{n_0 \times n_1}.$$

$\tilde{\Pi}_1$ is called the **coarse-grid Ritz projection matrix.** ■

**Proposition 3.2.2** ($\tilde{\Pi}_1$ is a projection). *Assume that* $\mathsf{A}_0$ *satisfies the Galerkin condition. The matrix* $\tilde{\Pi}_1$*, introduced in Definition* 3.2.1, *is a projection, that is,*

$$\tilde{\Pi}_1 \tilde{\Pi}_1 = \tilde{\Pi}_1.$$

*Proof.* Using the Galerkin condition (3.1.1) and the fact that by assumption we have a balanced pair, let us expand

$$\begin{aligned}
\tilde{\Pi}_1 \tilde{\Pi}_1 &= \mathsf{R}_0^\top \Pi_0 \mathsf{R}_0^\top \Pi_0 \\
&= \mathsf{R}_0^\top \mathsf{A}_0^{-1} \mathsf{R}_0 \mathsf{A}_1 \mathsf{R}_0^\top \mathsf{A}_0^{-1} \mathsf{R}_0 \mathsf{A}_1 \\
&\overset{(3.1.1)}{=} \mathsf{R}_0^\top \mathsf{A}_0^{-1} \mathsf{A}_0 \mathsf{A}_0^{-1} \mathsf{R}_0 \mathsf{A}_1 \\
&= \mathsf{R}_0^\top \Pi_0 \\
&= \tilde{\Pi}_1,
\end{aligned}$$

where we also used the definition of $\Pi_0$. □

**Corollary 3.2.3** (Projection). *If the Galerkin condition holds, then* $\mathrm{I}_1 - \tilde{\Pi}_1$ *is a projection matrix, that is,*

$$(\mathrm{I}_1 - \tilde{\Pi}_1)(\mathrm{I}_1 - \tilde{\Pi}_1) = \mathrm{I}_1 - \tilde{\Pi}_1.$$

*Proof.* Recall that for any matrix $\mathsf{P}$, $\mathrm{I} - \mathsf{P}$ is a projection iff $\mathsf{P}$ is a projection as well. □

The smoothing operation is just the application of a GLIS a certain number of times. We define the error transfer for this process in the expected way.

**Definition 3.2.4** (Smoothing error transfer matrix). The **presmoothing error transfer matrix** is defined as

$$\mathsf{K}_1 := \mathrm{I}_1 - \mathsf{S}_1 \mathsf{A}_1.$$

■

**Definition 3.2.5** (Level-1 adjoint). The **level-1 inner product** is

$$(\boldsymbol{u}_1, \boldsymbol{v}_1)_1 := \boldsymbol{v}_1^\top \boldsymbol{u}_1, \quad \forall\, \boldsymbol{u}_1, \boldsymbol{v}_1 \in \mathbb{R}^{n_1}.$$

Let $\mathsf{M} \in \mathbb{R}^{n_1 \times n_1}$. The **level-1 adjoint of** $\mathsf{M}$ is the unique matrix $\mathsf{M}^\top \in \mathbb{R}^{n_1 \times n_1}$ satisfying

$$(\mathsf{M}\boldsymbol{u}_1, \boldsymbol{v}_1)_1 = (\boldsymbol{u}_1, \mathsf{M}^\top \boldsymbol{v}_1)_1, \quad \forall\, \boldsymbol{u}_1, \boldsymbol{v}_1 \in \mathbb{R}^{n_1}.$$

We say that $\mathsf{M}$ is **level-1 self-adjoint** or **level-1 symmetric** if $\mathsf{M} = \mathsf{M}^\top$. When the context is clear, we will drop the **"level-1"** qualifier of the terms above. ■

It is easy to see that in the definition above, the matrix $\mathsf{M}^\top$ is the usual transpose of $\mathsf{M}$.

**Definition 3.2.6** ($\mathsf{A}_1$-adjoint). The $\mathsf{A}_1$**-inner product** is

$$(\boldsymbol{u}_1, \boldsymbol{v}_1)_{\mathsf{A}_1} := (\mathsf{A}_1\boldsymbol{u}_1, \boldsymbol{v}_1)_1, \quad \forall\, \boldsymbol{u}_1, \boldsymbol{v}_1 \in \mathbb{R}^{n_1}.$$

Let $\mathsf{M} \in \mathbb{R}^{n_1 \times n_1}$. The $\mathsf{A}_1$**-adjoint** of $\mathsf{M}$ is the unique matrix $\mathsf{M}^* \in \mathbb{R}^{n_1 \times n_1}$ satisfying

$$(\mathsf{M}\boldsymbol{u}_1, \boldsymbol{v}_1)_{\mathsf{A}_1} = (\boldsymbol{u}_1, \mathsf{M}^* \boldsymbol{v}_1)_{\mathsf{A}_1}, \quad \forall\, \boldsymbol{u}_1, \boldsymbol{v}_1 \in \mathbb{R}^{n_1}.$$

We say that $\mathsf{M}$ is $\mathsf{A}_1$**-self-adjoint** or $\mathsf{A}_1$**-symmetric** if $\mathsf{M} = \mathsf{M}^*$. We say that $\mathsf{M}$ is $\mathsf{A}_1$**-symmetric positive definite** if it is $\mathsf{A}_1$-symmetric and

$$(\mathsf{M}\boldsymbol{u}_1, \boldsymbol{u}_1)_{\mathsf{A}_1} > 0, \quad \forall\, \boldsymbol{u}_1 \in \mathbb{R}^{n_1}_\star.$$ ■

**Proposition 3.2.7** ($\mathsf{A}_1$-adjoint). *Let* $\mathsf{K}_1$ *be the presmoothing error transfer matrix. Then*

$$\mathsf{K}_1^* = \mathsf{I}_1 - \mathsf{S}_1^\top \mathsf{A}_1.$$

*Proof.* To see this, let $\boldsymbol{u}_1, \boldsymbol{v}_1 \in \mathbb{R}^{n_1}$ be arbitrary. Then

$$\begin{aligned}
(\mathsf{K}_1\boldsymbol{u}_1, \boldsymbol{v}_1)_{\mathsf{A}_1} &= ((\mathsf{I}_1 - \mathsf{S}_1\mathsf{A}_1)\boldsymbol{u}_1, \boldsymbol{v}_1)_{\mathsf{A}_1} \\
&= (\mathsf{A}_1(\mathsf{I}_1 - \mathsf{S}_1\mathsf{A}_1)\boldsymbol{u}_1, \boldsymbol{v}_1)_1 \\
&= (\boldsymbol{u}_1, (\mathsf{I}_1 - \mathsf{S}_1\mathsf{A}_1)^\top \mathsf{A}_1^\top \boldsymbol{v}_1)_1 \\
&= (\boldsymbol{u}_1, (\mathsf{I}_1 - \mathsf{A}_1^\top \mathsf{S}_1^\top)\mathsf{A}_1\boldsymbol{v}_1)_1 \\
&= (\boldsymbol{u}_1, (\mathsf{I}_1 - \mathsf{A}_1\mathsf{S}_1^\top)\mathsf{A}_1\boldsymbol{v}_1)_1 \\
&= (\boldsymbol{u}_1, (\mathsf{A}_1 - \mathsf{A}_1\mathsf{S}_1^\top\mathsf{A}_1)\boldsymbol{v}_1)_1 \\
&= (\boldsymbol{u}_1, \mathsf{A}_1(\mathsf{I}_1 - \mathsf{S}_1^\top\mathsf{A}_1)\boldsymbol{v}_1)_1 \\
&= (\mathsf{A}_1\boldsymbol{u}_1, (\mathsf{I}_1 - \mathsf{S}_1^\top\mathsf{A}_1)\boldsymbol{v}_1)_1 \\
&= (\boldsymbol{u}_1, (\mathsf{I}_1 - \mathsf{S}_1^\top\mathsf{A}_1)\boldsymbol{v}_1)_{\mathsf{A}_1}.
\end{aligned}$$ □

## 3.3 The two-grid error transfer matrix

We now have the tools to identify the structure of the error transfer matrix for the two-grid algorithm.

**Theorem 3.3.1** (Two-grid error transfer matrix). *Let* $\boldsymbol{f}_1 \in \mathbb{R}^{n_1}$ *and* $\boldsymbol{u}_1^{\mathrm{E}} = \mathsf{A}_1^{-1}\boldsymbol{f}_1$. *Let the sequence* $\{\boldsymbol{u}_1^k\}_{k=1}^{\infty}$ *be produced by the two-grid algorithm introduced Definition* 3.1.9. *Then the error at the kth step,* $\boldsymbol{e}_1^k := \boldsymbol{u}_1^{\mathrm{E}} - \boldsymbol{u}_1^k$, *satisfies*

$$\boldsymbol{e}_1^{k+1} = \mathsf{E}_1\boldsymbol{e}_1^k,$$

*where the error transfer matrix is*

$$\mathsf{E}_1 = (\mathsf{K}_1^*)^{m_2}(\mathrm{I}_1 - \tilde{\Pi}_1)(\mathsf{K}_1)^{m_1}.$$

*Furthermore, the two-grid algorithm is a GLIS, that is, there is* $\mathsf{B}_1 \in \mathbb{R}^{n_1\times n_1}$ *such that*

$$\mathsf{E}_1 = \mathrm{I}_1 - \mathsf{B}_1\mathsf{A}_1.$$

*Proof.* Let $\boldsymbol{u}_1^k$ be given. Using the notation of Definition 3.1.5, we set

$$\boldsymbol{e}_1^{(i)} := \boldsymbol{u}_1^{\mathrm{E}} - \boldsymbol{u}_1^{(i)}, \quad i \in \{1,2,3\}.$$

The error after the presmoothing step is

$$\boldsymbol{e}_1^{(1)} = (\mathrm{I}_1 - \mathsf{S}_1\mathsf{A}_1)^{m_1}\boldsymbol{e}_1^k = \mathsf{K}_1^{m_1}\boldsymbol{e}_1^k.$$

Next, we compute the error of the coarse-grid correction step:

$$\begin{aligned}\boldsymbol{e}_1^{(2)} &= \boldsymbol{u}_1^{\mathrm{E}} - \boldsymbol{u}_1^{(2)}\\ &= \boldsymbol{u}_1^{\mathrm{E}} - \boldsymbol{u}_1^{(1)} - \mathsf{R}_0^{\mathsf{T}}\boldsymbol{q}_0^{(1)}\\ &= \boldsymbol{e}_1^{(1)} - \mathsf{R}_0^{\mathsf{T}}\mathsf{A}_0^{-1}\boldsymbol{r}_0^{(1)}\\ &= \boldsymbol{e}_1^{(1)} - \mathsf{R}_0^{\mathsf{T}}\mathsf{A}_0^{-1}\mathsf{R}_0\boldsymbol{r}_1^{(1)}\\ &= \boldsymbol{e}_1^{(1)} - \mathsf{R}_0^{\mathsf{T}}\mathsf{A}_0^{-1}\mathsf{R}_0\mathsf{A}_1\boldsymbol{e}_1^{(1)}\\ &= (\mathrm{I}_1 - \tilde{\Pi}_1)\boldsymbol{e}_1^{(1)}.\end{aligned}$$

Consequently,

$$\boldsymbol{e}_1^{(2)} = (\mathrm{I}_1 - \tilde{\Pi}_1)\mathsf{K}_1^{m_1}\boldsymbol{e}_1^k.$$

Finally, after the postsmoothing step,

$$\boldsymbol{e}_1^{k+1} = \boldsymbol{e}_1^{(3)} = (\mathrm{I}_1 - \mathsf{S}_1^{\mathsf{T}}\mathsf{A}_1)^{m_2}\boldsymbol{e}_1^{(2)}.$$

In conclusion,

$$\boldsymbol{e}_1^{k+1} = (\mathsf{K}_1^*)^{m_2}(\mathrm{I}_1 - \tilde{\Pi}_1)\mathsf{K}_1^{m_1}\boldsymbol{e}_1^k,$$

as claimed.

We leave the second part of the proof as an exercise; see Problem 3.2. □

**Remark 3.3.2** (Particular case). In the case $m_1 = m_2 = 1$, we have the following simple form for $\mathsf{B}_1$:

$$\mathsf{B}_1 = \mathsf{S}_1^\top(\mathsf{S}_1^{-\top} + \mathsf{S}_1^{-1} - \mathsf{A}_1)\mathsf{S}_1 + \mathsf{K}_1^*\mathsf{R}_0^\top\mathsf{A}_0^{-1}\mathsf{R}_0\mathsf{K}_1^{*\top},$$

which is, of course, symmetric. The first term represents a symmetric smoothing, whereas the second one is a coarse grid correction. In general, we can show that $\mathsf{B}_1$ is symmetric in the Euclidean sense iff $m_1 = m_2 = m$. See Problem 3.3. ◀

**Theorem 3.3.3** (Symmetry and positivity of $\mathsf{E}_1$). *The error transfer matrix of the two-grid algorithm, $\mathsf{E}_1$, is symmetric with respect to $(\cdot,\cdot)_{\mathsf{A}_1}$ iff $m_1 = m_2 = m$. Furthermore, if $m_1 = m_2 = m$ and the Galerkin condition holds for $\mathsf{A}_0$, then $\mathsf{E}_1$ is symmetric positive semidefinite (SPSD) with respect to $(\cdot,\cdot)_{\mathsf{A}_1}$.*

*Proof.* (Symmetry): Let $\boldsymbol{u}_1, \boldsymbol{v}_1 \in \mathbb{R}^{n_1}$. Then we have

$$\begin{aligned}
(\mathsf{E}_1\boldsymbol{u}_1, \boldsymbol{v}_1)_{\mathsf{A}_1} &= ((\mathsf{K}_1^*)^{m_2}(\mathrm{I}_1 - \tilde{\Pi}_1)\mathsf{K}_1^{m_1}\boldsymbol{u}_1, \boldsymbol{v}_1)_{\mathsf{A}_1} \\
&= ((\mathrm{I}_1 - \tilde{\Pi}_1)\mathsf{K}_1^{m_1}\boldsymbol{u}_1, \mathsf{K}_1^{m_2}\boldsymbol{v}_1)_{\mathsf{A}_1} \\
&= (\mathsf{K}_1^{m_1}\boldsymbol{u}_1, (\mathrm{I}_1 - \tilde{\Pi}_1)^*\mathsf{K}_1^{m_2}\boldsymbol{v}_1)_{\mathsf{A}_1} \\
&= (\boldsymbol{u}_1, (\mathsf{K}_1^*)^{m_1}(\mathrm{I}_1 - \tilde{\Pi}_1)^*\mathsf{K}_1^{m_2}\boldsymbol{v}_1)_{\mathsf{A}_1}.
\end{aligned}$$

Observe that

$$(\mathrm{I}_1 - \tilde{\Pi}_1)^* = (\mathrm{I}_1 - \tilde{\Pi}_1).$$

Indeed,

$$((\mathrm{I}_1 - \tilde{\Pi}_1)\boldsymbol{u}_1, \boldsymbol{v}_1)_{\mathsf{A}_1} = (\boldsymbol{u}_1, \boldsymbol{v}_1)_{\mathsf{A}_1} - (\tilde{\Pi}_1\boldsymbol{u}_1, \boldsymbol{v}_1)_{\mathsf{A}_1},$$

and

$$\begin{aligned}
(\tilde{\Pi}_1\boldsymbol{u}_1, \boldsymbol{v}_1)_{\mathsf{A}_1} &= (\mathsf{A}_1\mathsf{R}_0^\top\mathsf{A}_0^{-1}\mathsf{R}_0\mathsf{A}_1\boldsymbol{u}_1, \boldsymbol{v}_1)_1 \\
&= (\mathsf{R}_0^\top\mathsf{A}_0^{-1}\mathsf{R}_0\mathsf{A}_1\boldsymbol{u}_1, \mathsf{A}_1\boldsymbol{v}_1)_1 \\
&= (\mathsf{A}_1\boldsymbol{u}_1, (\mathsf{R}_0^\top\mathsf{A}_0^{-1}\mathsf{R}_0)^\top\mathsf{A}_1\boldsymbol{v}_1)_1 \\
&= (\mathsf{A}_1\boldsymbol{u}_1, \mathsf{R}_0^\top\mathsf{A}_0^{-1}\mathsf{R}_0\mathsf{A}_1\boldsymbol{v}_1)_1 \\
&= (\mathsf{A}_1\boldsymbol{u}_1, \tilde{\Pi}_1\boldsymbol{v}_1)_1 \\
&= (\boldsymbol{u}_1, \tilde{\Pi}_1\boldsymbol{v}_1)_{\mathsf{A}_1}.
\end{aligned}$$

Therefore

$$\tilde{\Pi}_1^* = \tilde{\Pi}_1, \quad (\mathrm{I}_1 - \tilde{\Pi}_1)^* = (\mathrm{I}_1 - \tilde{\Pi}_1).$$

The symmetry of $\mathsf{E}_1$ then follows iff $m_1 = m_2$.
($\mathsf{E}_1$ is PSD): Let $\boldsymbol{u}_1 \in \mathbb{R}^{n_1}$ and suppose $m_1 = m_2 = m$. Since the Galerkin condition holds, from Corollary 3.2.3 we have

$$(\mathrm{I}_1 - \tilde{\Pi}_1)^2 = (\mathrm{I}_1 - \tilde{\Pi}_1).$$

Now

$$\begin{aligned}
(\mathsf{E}_1\boldsymbol{u}_1, \boldsymbol{u}_1)_{\mathsf{A}_1} &= ((\mathsf{K}_1^*)^m(\mathrm{I}_1 - \tilde{\Pi}_1)\mathsf{K}_1^m\boldsymbol{u}_1, \boldsymbol{u}_1)_{\mathsf{A}_1} \\
&= ((\mathrm{I}_1 - \tilde{\Pi}_1)\mathsf{K}_1^m\boldsymbol{u}_1, \mathsf{K}_1^m\boldsymbol{u}_1)_{\mathsf{A}_1} \\
&= ((\mathrm{I}_1 - \tilde{\Pi}_1)^2\mathsf{K}_1^m\boldsymbol{u}_1, \mathsf{K}_1^m\boldsymbol{u}_1)_{\mathsf{A}_1} \\
&= ((\mathrm{I}_1 - \tilde{\Pi}_1)\mathsf{K}_1^m\boldsymbol{u}_1, (\mathrm{I}_1 - \tilde{\Pi}_1)^*\mathsf{K}_1^m\boldsymbol{u}_1)_{\mathsf{A}_1} \\
&= ((\mathrm{I}_1 - \tilde{\Pi}_1)\mathsf{K}_1^m\boldsymbol{u}_1, (\mathrm{I}_1 - \tilde{\Pi}_1)\mathsf{K}_1^m\boldsymbol{u}_1)_{\mathsf{A}_1} \\
&= \|(\mathrm{I}_1 - \tilde{\Pi}_1)\mathsf{K}_1^m\boldsymbol{u}_1\|_{\mathsf{A}_1}^2 \geq 0.
\end{aligned}$$

□

**Remark 3.3.4** (Galerkin condition). We remark that in the previous result, the Galerkin condition is not required for the symmetry of $\mathsf{E}_1$. All that was used to establish symmetry was the definition

$$\tilde{\Pi}_1 = \mathsf{R}_0^\top \mathsf{A}_0^{-1}\mathsf{R}_0\mathsf{A}_1.$$

On the other hand, we used the Galerkin condition to establish that $\tilde{\Pi}_1$ and $(\mathrm{I}_1 - \tilde{\Pi}_1)$ are projections, and these facts played an important role in showing that $\mathsf{E}_1$ is $\mathsf{A}_1$-SPSD.

The Galerkin condition will fail in some multigrid applications, even if the above definition of $\tilde{\Pi}_1$ remains valid. ◂

## 3.4 Problems

3.1 Let $\mathsf{A} \in \mathbb{R}^{m\times m}$ be symmetric positive definite (SPD). Suppose $\mathsf{P} \in \mathbb{R}^{m\times n}$, $m \geq n$, is of full rank.

(a) Show that $\mathsf{A}_C := \mathsf{P}^\top \mathsf{A}\mathsf{P}$ is invertible. What else can we say about the properties of $\mathsf{A}_C$?

(b) Define $\mathsf{Q}_\mathsf{A} := \mathsf{P}\mathsf{A}_C^{-1}\mathsf{P}^\top \mathsf{A}$. Show that $\mathsf{Q}_\mathsf{A}\boldsymbol{u} \in \operatorname{im}(\mathsf{P})$ is the best approximation of $\boldsymbol{u} \in \mathbb{R}^m$ with respect to the A-norm. In other words, prove that

$$\mathsf{Q}_\mathsf{A}\boldsymbol{u} = \operatorname{argmin}\{\|\boldsymbol{z} - \boldsymbol{u}\|_\mathsf{A}^2 \mid \boldsymbol{z} \in \operatorname{im}(\mathsf{P})\}.$$

(c) Let $\boldsymbol{f} \in \mathbb{R}^m$. Assume that $\boldsymbol{u} \in \mathbb{R}^m$ is the unique solution to

$$\mathsf{A}\boldsymbol{u} = \boldsymbol{f}.$$

Define $\boldsymbol{f}_C := \mathsf{P}^\top \boldsymbol{f}$ and let $\boldsymbol{u}_C \in \mathbb{R}^n$ be the unique solution to

$$\mathsf{A}_C\boldsymbol{u}_C = \boldsymbol{f}_C.$$

Relate $\boldsymbol{u}_C$ to $\boldsymbol{u}$ using $\mathsf{Q}_\mathsf{A}$ and P.

(d) Interpret the result of the last problem in the context of the two-grid concept.

3.2 Complete the proof of Theorem 3.3.1, that is, prove that there is $\mathsf{B}_1 \in \mathbb{R}^{n_1\times n_1}$ such that the two-grid error transfer matrix satisfies

$$\mathsf{E}_1 = \mathsf{I}_1 - \mathsf{B}_1\mathsf{A}_1,$$

or, equivalently, that the two-grid algorithm may be expressed as

$$\boldsymbol{u}_1^{k+1} = \boldsymbol{u}_1^k + \mathsf{B}_1(\boldsymbol{f}_1 - \mathsf{A}_1\boldsymbol{u}_1^k).$$

3.3 Prove that $\mathsf{B}_1 \in \mathbb{R}^{n_1\times n_1}$ from the last problem is (Euclidean) symmetric iff $m_1 = m_2$.

# 4 Fourier analysis of the two-grid algorithm

In this chapter, we consider the model elliptic boundary value problem in 1D (1.7.1). We will examine two kinds of discretization for the problem, and then we will apply the two-grid algorithm to solve the resulting linear systems of equations. We will use a Fourier-type analysis to show that the algorithm converges.

The inspiration for this chapter is a short subsection of Chapter 8 in the classic numerical analysis text by Stoer and Bulirsch [33]. We have greatly expanded the content and placed it in the larger context of the general two-grid and multigrid algorithms. However, the simplicity and brevity of the presentation in [33] is always inspiring and certainly worth reading.

## 4.1 Discretization of the model problem in 1D

### 4.1.1 The finite difference method

As a first method for the discretization of (1.7.1), we consider the finite difference method as described in Section 1.7. We will, however, slightly adapt the notation to suit our discussion. The number of fine grid points will be $n_1 \in \mathbb{N}$, so that the grid spacing is

$$h_1 \coloneqq \frac{1}{n_1 + 1}.$$

The grid is $\mathcal{G}_{h_1} = \{x_{1,i}\}_{i=0}^{n_1+1}$ with points

$$x_{1,i} = i \cdot h_1, \quad i = 0, 1, 2, \ldots, n_1 + 1. \tag{4.1.1}$$

Thus, we look for the grid function $u_1$ such that

$$\begin{cases} u_{1,0} = 0, \\ \dfrac{-u_{1,i-1} + 2u_{1,i} - u_{1,i+1}}{h_1^2} = \tilde{f}_{1,i} \coloneqq f(x_{1,i}), \quad i = 1, \ldots, n_1, \\ u_{1,n_1+1} = 0. \end{cases} \tag{4.1.2}$$

Now, let us set

$$\boldsymbol{u}_1^{\mathrm{FD}} \coloneqq \begin{bmatrix} u_{1,1} \\ u_{1,2} \\ \vdots \\ u_{1,n_1-1} \\ u_{1,n_1} \end{bmatrix}, \quad \boldsymbol{f}_1^{\mathrm{FD}} \coloneqq \begin{bmatrix} h_1\tilde{f}_{1,1} \\ h_1\tilde{f}_{1,2} \\ \vdots \\ h_1\tilde{f}_{1,n_1-1} \\ h_1\tilde{f}_{1,n_1} \end{bmatrix} \in \mathbb{R}^{n_1},$$

https://doi.org/10.1515/9783111354880-004

and

$$\mathsf{A}_1 := \frac{1}{h_1}\begin{bmatrix} 2 & -1 & & & \\ -1 & 2 & -1 & & \\ & \ddots & \ddots & \ddots & \\ & & -1 & 2 & -1 \\ & & & -1 & 2 \end{bmatrix} \in \mathbb{R}^{n_1 \times n_1}.$$

In matrix form, the finite difference approximation is as follows: find $\boldsymbol{u}_1^{\mathrm{FD}} \in \mathbb{R}^{n_1}$ such that

$$\mathsf{A}_1 \boldsymbol{u}_1^{\mathrm{FD}} = \boldsymbol{f}_1^{\mathrm{FD}}. \tag{4.1.3}$$

Here $\mathsf{A}_1$ is our finite difference fine grid stiffness matrix. Notice that it only differs from the finite difference stiffness matrix defined in (1.7.4) by a factor of $\frac{1}{h_1}$. This is a deliberate choice to obtain the same scaling as in the finite element case, which we define next. Before doing so, we recall that this matrix is SPD.

### 4.1.2 The finite element method

Let us now use the piecewise linear finite element method to approximate the solution to the model problem. The finite element method is based on the weak formulation of the model problem: find $u \in H_0^1(0,1)$ such that

$$\left(\frac{\mathrm{d}u}{\mathrm{d}x}, \frac{\mathrm{d}v}{\mathrm{d}x}\right)_{L^2(0,1)} = (f, v)_{L^2(0,1)}, \quad \forall\, v \in H_0^1(0,1).$$

Consider the grid comprised of $n_1 + 1$ equally sized intervals

$$K_{1,i} := (x_{1,i-1}, x_{1,i}), \quad i = 1, \dots, n_1 + 1,$$

where the set of grid points $\mathcal{G}_{h_1}$ is defined in (4.1.1). We define the finite dimensional subspace $V_1$ of $H_0^1(0,1)$ as follows:

$$V_1 := \{v \in S_1 \mid v(0) = v(1) = 0\}, \tag{4.1.4}$$

where

$$S_1 := \{v \in C^0([0,1]) \mid v|_{K_{1,i}} \in \mathbb{P}_1(K_{1,i}),\ 1 \le i \le n_1 + 1\}. \tag{4.1.5}$$

The finite-dimensional space $S_1$ is called the $\mathbb{P}_1$ finite element space. The space $V_1$ is subspace of $S_1$ comprised of all $\mathbb{P}_1$ finite element functions that vanish at the boundary points.

**Definition 4.1.1** (Hat function). For $i = 1, \dots, n_1$, define $\psi_{1,i} \in V_1$ by

$$\psi_{1,i}(x_{1,j}) = \delta_{i,j}, \quad 1 \le j \le n_1.$$

The function $\psi_{1,i}$ is called a **fine-level** or **level-1 hat function**. ∎

The three hat functions for $n_1 = 3$ are shown in Figure 4.1.

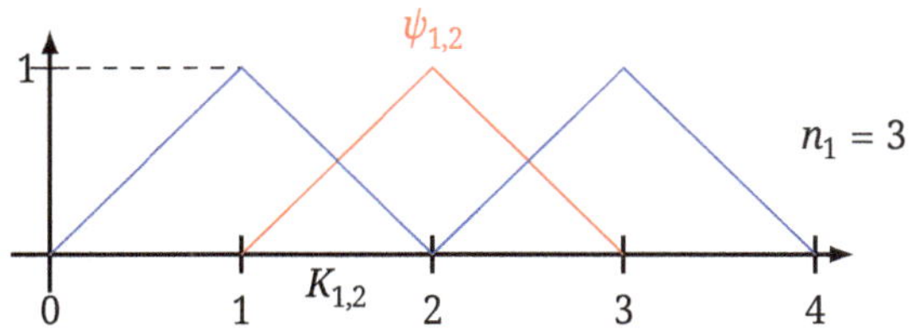


**Figure 4.1:** Piecewise linear basis (hat) functions in 1D.

**Proposition 4.1.2.** *The space $V_1$ defined in* (4.1.4) *is a subspace of $H_0^1(0,1)$, and the set $\mathcal{B}_1 = \{\psi_{1,i}\}_{i=1}^{n_1}$ is a basis for $V_1$. Thus* $\dim V_1 = n_1$.

*Proof.* Exercise; see Problem 4.1. □

The finite element approximation of the model problem is as follows: find $u_1 \in V_1$ such that for all $i = 1, 2, \dots, n_1$,

$$\left( \frac{du_1}{dx}, \frac{d\psi_{1,i}}{dx} \right)_{L^2(0,1)} = (f, \psi_{1,i})_{L^2(0,1)}. \tag{4.1.6}$$

Now set

$$\boldsymbol{f}_1^{\mathrm{FE}} \coloneqq \begin{bmatrix} (f, \psi_{1,1})_{L^2(0,1)} \\ (f, \psi_{1,2})_{L^2(0,1)} \\ \vdots \\ (f, \psi_{1,n_1-1})_{L^2(0,1)} \\ (f, \psi_{1,n_1})_{L^2(0,1)} \end{bmatrix} \in \mathbb{R}^{n_1}.$$

We expand $u_1$ in the basis of hat functions:

$$u_1 = \sum_{j=1}^{n_1} u_{1,j} \psi_{1,j} \in V_1. \tag{4.1.7}$$

Set

$$\boldsymbol{u}_1^{\text{FE}} := \begin{bmatrix} u_{1,1} \\ u_{1,2} \\ \vdots \\ u_{1,n_1-1} \\ u_{1,n_1} \end{bmatrix} \in \mathbb{R}^{n_1}.$$

This represents the coordinate vector of $u_1$ in the basis of hat functions $\mathcal{B}_1$. Now plugging (4.1.7) into (4.1.6), we get

$$\begin{aligned} (f, \psi_{1,i})_{L^2(0,1)} &\overset{(4.1.6)}{=} \left( \frac{\mathrm{d}u_1}{\mathrm{d}x}, \frac{\mathrm{d}\psi_{1,i}}{\mathrm{d}x} \right)_{L^2(0,1)} \\ &\overset{(4.1.7)}{=} \left( \sum_{j=1}^{n_1} u_{1,j} \frac{\mathrm{d}\psi_{1,j}}{\mathrm{d}x}, \frac{\mathrm{d}\psi_{1,i}}{\mathrm{d}x} \right)_{L^2(0,1)} \\ &= \sum_{j=1}^{n_1} \left( \frac{\mathrm{d}\psi_{1,j}}{\mathrm{d}x}, \frac{\mathrm{d}\psi_{1,i}}{\mathrm{d}x} \right)_{L^2(0,1)} u_{1,j}. \end{aligned} \tag{4.1.8}$$

Define the matrix $\mathsf{A}_1 = [a_{1,i,j}] \in \mathbb{R}^{n_1 \times n_1}$ by

$$a_{1,i,j} := \left( \frac{\mathrm{d}\psi_{1,j}}{\mathrm{d}x}, \frac{\mathrm{d}\psi_{1,i}}{\mathrm{d}x} \right)_{L^2(0,1)} = \left( \frac{\mathrm{d}\psi_{1,i}}{\mathrm{d}x}, \frac{\mathrm{d}\psi_{1,j}}{\mathrm{d}x} \right)_{L^2(0,1)}. \tag{4.1.9}$$

Then (4.1.8) can be rewritten as

$$[\mathsf{A}_1 \boldsymbol{u}_1^{\text{FE}}]_i = [\boldsymbol{f}_1^{\text{FE}}]_i.$$

Since we require this to hold for all $i = 1, \ldots, n_1$, this is equivalent to

$$\mathsf{A}_1 \boldsymbol{u}_1^{\text{FE}} = \boldsymbol{f}_1^{\text{FE}}. \tag{4.1.10}$$

Lastly, let us compute the elements of the matrix $\mathsf{A}_1$. Some entries of $\mathsf{A}_1$ are obvious. Notice that if $|i-j| \geq 2$, then the supports of the two hat functions do not overlap at more than a single point and thus do not interact. Consequently,

$$a_{1,i,j} = 0, \quad |i-j| \geq 2.$$

Otherwise, the two hat functions will have interactions. Consider Figure 4.2, and observe that for $2 \leq i \leq n_1$,

$$\begin{aligned} a_{1,i,i-1} &= \left( \frac{\mathrm{d}\psi_{1,i-1}}{\mathrm{d}x}, \frac{\mathrm{d}\psi_{1,i}}{\mathrm{d}x} \right)_{L^2(0,1)} \\ &= \left( \frac{\mathrm{d}\psi_{1,i-1}}{\mathrm{d}x}, \frac{\mathrm{d}\psi_{1,i}}{\mathrm{d}x} \right)_{L^2(K_{1,i})} \end{aligned}$$

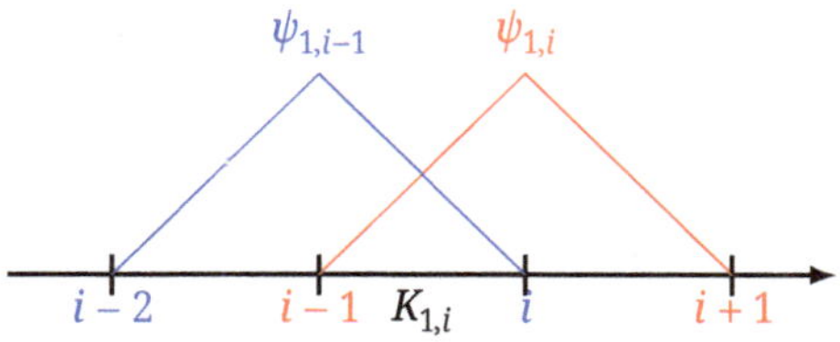


**Figure 4.2:** Interacting hat functions in 1D.

$$= \left(-\frac{1}{h_1}, \frac{1}{h_1}\right)_{L^2(K_{1,i})}$$
$$= -\frac{1}{h_1}.$$

Likewise, for $1 \le i \le n_1 - 1$,

$$a_{1,i,i+1} = -\frac{1}{h_1}.$$

Lastly, for $1 \le i \le n_1$,

$$a_{1,i,i} = \left(\frac{\mathrm{d}\psi_{1,i}}{\mathrm{d}x}, \frac{\mathrm{d}\psi_{1,i}}{\mathrm{d}x}\right)_{L^2(0,1)} = \frac{2}{h_1}.$$

Thus, as in the finite difference case, we have

$$\mathsf{A}_1 = \frac{1}{h_1}\begin{bmatrix} 2 & -1 & & & \\ -1 & 2 & -1 & & \\ & \ddots & \ddots & \ddots & \\ & & -1 & 2 & -1 \\ & & & -1 & 2 \end{bmatrix} \in \mathbb{R}^{n_1 \times n_1}. \tag{4.1.11}$$

**Definition 4.1.3** (Stiffness matrix). The matrix $\mathsf{A}_1 \in \mathbb{R}^{n_1 \times n_1}$ defined in (4.1.11) is called the **fine-level stiffness matrix** or the **level-1 stiffness matrix**. ■

## 4.2 Properties of the stiffness matrix

We have shown that the finite difference approximation to (1.7.1) reduces to a matrix problem: find $\boldsymbol{u}_1^{\mathrm{FD}} \in \mathbb{R}^{n_1}$ such that (4.1.3) holds. Similarly, the finite element approximation can be cast as follows: find $\boldsymbol{u}_1^{\mathrm{FE}} \in \mathbb{R}^{n_1}$ such that (4.1.10) holds. Let us investigate the properties of the stiffness matrix $\mathsf{A}_1$ to establish the existence and uniqueness of solutions to these problems. We will also determine the condition number of the stiffness matrix. The next result shows that $\mathsf{A}_1$ is always invertible.

**Theorem 4.2.1** (Stiffness matrix is SPD). *The level-1 stiffness matrix* $\mathsf{A}_1 \in \mathbb{R}^{n_1 \times n_1}$ *is SPD. Its eigenvalues are*

$$\lambda_1^{(k)} := \frac{4}{h_1}\sin^2\left(\frac{k\pi h_1}{2}\right) = \frac{2}{h_1}(1 - \cos(k\pi h_1)), \tag{4.2.1}$$

$k = 1, \dots, n_1$, *and the corresponding eigenvectors* $\{\boldsymbol{v}_1^{(k)}\}_{k=1}^{n_1} \subset \mathbb{R}^{n_1}$ *are*

$$[\boldsymbol{v}_1^{(k)}]_i = v_{1,i}^{(k)} := \sin(k\pi x_{1,i}), \quad 1 \le i \le n_1, \tag{4.2.2}$$

*where, we recall,* $x_{1,i} = i \cdot h_1$ *for* $i = 0, \dots, n_1 + 1$.

*Proof.* Up to a scaling, the content of this result is given in Problem 1.15a. Nevertheless, let us now provide a direct proof.

Clearly, $\mathsf{A}_1$ is symmetric. We will show that it is positive definite by showing that all of its eigenvalues are strictly positive. For $1 \le i \le n_1$,

$$\begin{aligned}[\mathsf{A}_1\boldsymbol{v}_1^{(k)}]_i &= -\frac{1}{h_1}\sin(k\pi x_{1,i-1}) + \frac{2}{h_1}\sin(k\pi x_{1,i}) - \frac{1}{h_1}\sin(k\pi x_{1,i+1})\\ &= \frac{2}{h_1}[1 - \cos(k\pi h_1)]\sin(k\pi x_{1,i}).\end{aligned}$$

Thus

$$[\mathsf{A}_1\boldsymbol{v}_1^{(k)}]_i = \lambda_1^{(k)} v_{1,i}^{(k)}$$

for $1 \le i \le n_1$. Since the eigenvalues are positive (see Figure 4.3), the matrix $\mathsf{A}_1$ is SPD. □

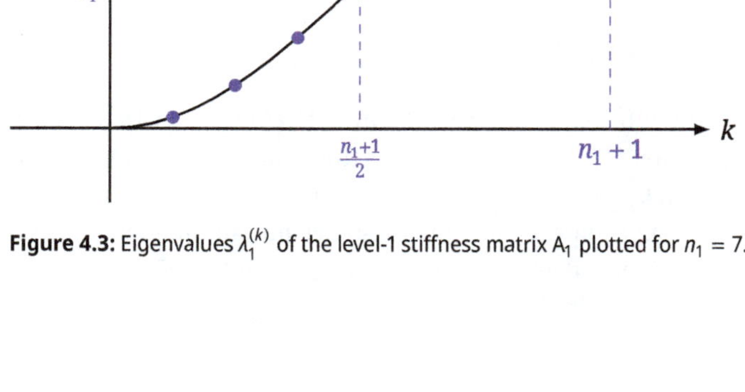


**Figure 4.3:** Eigenvalues $\lambda_1^{(k)}$ of the level-1 stiffness matrix $\mathsf{A}_1$ plotted for $n_1 = 7$.

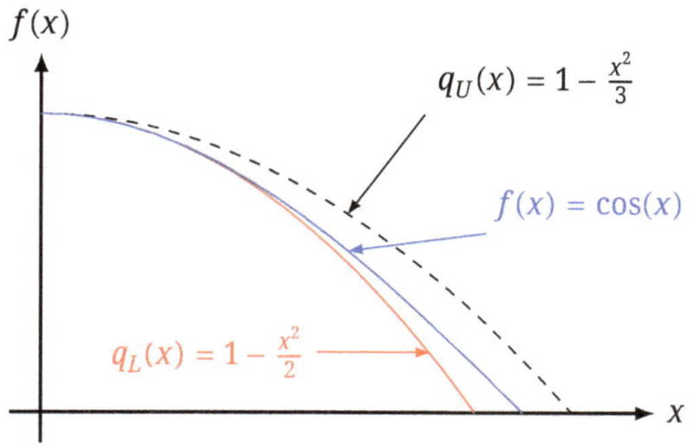


**Figure 4.4:** Upper and lower polynomial bounds for $\cos(x)$, $x \in (0, \frac{\pi}{2}]$.

Before we get into the next result concerned with approximating the condition number, it will be useful to have a technical lemma. Its essence are the polynomial bounds depicted in Figure 4.4.

**Lemma 4.2.2** (Asymptotics for cosine). *We have the following polynomial estimates:*

$$1 - \frac{x^2}{2} < \cos(x), \quad x \in \left(0, \frac{\pi}{2}\right), \tag{4.2.3}$$

*and*

$$\cos(x) < 1 - \frac{x^2}{3}, \quad x \in \left(0, \frac{\pi}{2}\right). \tag{4.2.4}$$

*Proof.* (Lower bound): By Taylor's theorem, for $x \in (0, \frac{\pi}{2})$,

$$\cos(x) = 1 - \frac{x^2}{2} + \frac{x^4}{4!}\cos(\xi),$$

for some $\xi \in (0, \frac{\pi}{2})$. Hence $\cos(\xi) \in (0, 1)$, and it follows that for $x \in (0, \frac{\pi}{2}]$,

$$0 < \frac{x^4}{4!}\cos(\xi) = \cos(x) - 1 + \frac{x^2}{2},$$

which yields the lower bound (4.2.3).

(Upper bound): By Taylor's theorem again, we have, for $x \in (0, \frac{\pi}{2}]$,

$$\cos(x) = 1 - \frac{x^2}{2} + \frac{x^4}{24} - \frac{x^6}{6!}\cos(\xi)$$

for some $\xi \in (0, \frac{\pi}{2})$. Thus

$$\cos(x) - 1 + \frac{x^2}{2} - \frac{x^4}{24} = -\frac{x^6}{6!}\cos(\xi) < 0,$$

and therefore, for $x \in (0, \frac{\pi}{2}]$,

$$\frac{x^2}{2} - \frac{x^4}{24} < 1 - \cos(x).$$

Notice, in addition, that for $x \in (0, 2]$,

$$\frac{x^2}{2} - \frac{x^4}{24} \geq \frac{x^2}{3}.$$

This finally implies that for $x \in (0, \frac{\pi}{2}]$,

$$\frac{x^2}{3} < 1 - \cos(x).$$

which is the upper bound (4.2.4). □

Now we can obtain an estimate of the condition number of $\mathsf{A}_1$.

**Theorem 4.2.3** (Condition number estimate). *The spectral condition number of the level-1 stiffness matrix, $\mathsf{A}_1$, that is,*

$$\kappa(\mathsf{A}_1) = \|\mathsf{A}_1\|\|\mathsf{A}_1^{-1}\| = \frac{\lambda_1^{(n_1)}}{\lambda_1^{(1)}},$$

*satisfies the estimates*

$$C_1 h_1^{-2} \leq \kappa(\mathsf{A}_1) \leq C_2 h_1^{-2}$$

*for some constants $0 < C_1 \leq C_2$.*

*Proof.* Since $\mathsf{A}_1$ is SPD, it follows that

$$\kappa(\mathsf{A}_1) = \frac{\lambda_1^{(n_1)}}{\lambda_1^{(1)}} = \frac{1 - \cos(n_1 \pi h_1)}{1 - \cos(\pi h_1)}.$$

Consider the function $f(x) = 1 - \cos(\pi x)$, as shown in Figure 4.5. Observe that since $n_1 \geq 1$, it follows that

$$0 < h_1 \leq \frac{1}{2} \quad \Longleftrightarrow \quad 0 < \pi h_1 \leq \frac{\pi}{2}.$$

Furthermore,

$$\frac{1}{2} \leq n_1 h_1 < 1 \quad \Longleftrightarrow \quad \frac{\pi}{2} \leq n_1 \pi h_1 < \pi,$$

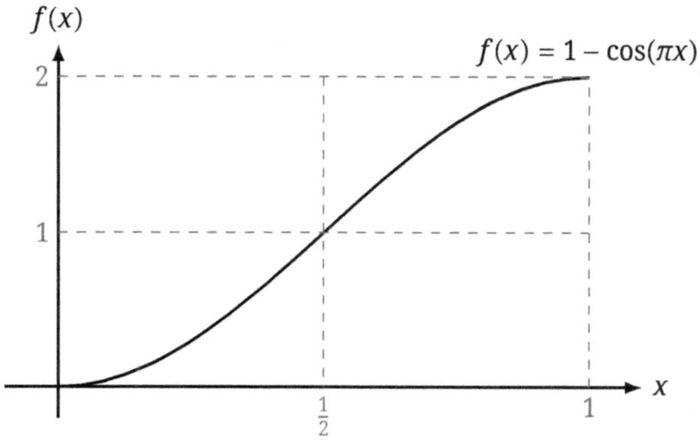


**Figure 4.5:** A plot of the function $f(x) = 1 - \cos(\pi x)$.

and we find, referring to Figure 4.5, that

$$1 \le 1 - \cos(n_1 \pi h_1) < 2.$$

From this, using (4.2.1), it follows that

$$\frac{1}{1 - \cos(\pi h_1)} \le \kappa(\mathsf{A}_1) \le \frac{2}{1 - \cos(\pi h_1)}.$$

Next, using the quadratic estimate (4.2.3), we have

$$1 - \cos(\pi h_1) < \frac{\pi^2}{2} h_1^2, \quad 0 < \pi h_1 \le \frac{\pi}{2},$$

and we conclude the desired lower bound on the condition number:

$$\frac{2}{\pi^2 h_1^2} \le \frac{1}{1 - \cos(\pi h_1)}, \quad 0 < h_1 \le \frac{1}{2}.$$

Similarly, using the quadratic estimate (4.2.4), we have

$$\frac{\pi^2 h_1^2}{3} < 1 - \cos(\pi h_1), \quad 0 < \pi h_1 \le \frac{\pi}{2},$$

which implies

$$\frac{1}{1 - \cos(\pi h_1)} < \frac{3}{\pi^2 h_1^2}, \quad 0 < h_1 \le \frac{1}{2},$$

and in conclusion,

$$\frac{2}{\pi^2 h_1^2} \le \kappa(\mathsf{A}_1) \le \frac{6}{\pi^2 h_1^2}.$$

□

**Definition 4.2.4** (Big theta). Let $f : [0,1] \to \mathbb{R}$ be a continuous function. If there exist constants $0 < C_1 \le C_2$ and numbers $x_0 \in (0,1)$ and $r \in \mathbb{R}$ such that

$$C_1 x^r \le f(x) \le C_2 x^r \quad \text{for all } x \in (0, x_0),$$

then we write

$$f(x) = \Theta(x^r) \quad \text{as} \quad x \to 0,$$

and we say that $f(x)$ **is big theta of** $x^r$ **as** $x \to 0$. ■

**Remark 4.2.5** (Notation). In light of our last definition, Theorem 4.2.3 says that

$$\kappa(\mathsf{A}_1) = \Theta(h_1^{-2}) \quad \text{as } h_1 \to 0.$$

In other words, the spectral condition number blows up as $h_1 \to 0$. This usually means that the performance of any classical GLIS is degraded for small grid sizes, as we will see. Two-grid and multigrid algorithms are designed to overcome this degradation. ◂

## 4.3 The damped Jacobi smoother

Recall that for both the finite difference and finite element methods, we eventually need to solve a linear system of the form:

$$\mathsf{A}_1 \boldsymbol{u}_1^{\square} = \boldsymbol{f}_1^{\square}. \tag{4.3.1}$$

Here the superscript □ can be either "FD" or "FE". We will omit this superscript for simplicity and brevity from this point. To approximate the solution of (4.3.1), let us apply the damped Jacobi method, which is again a splitting method. This requires a splitting (1.4.1) of $\mathsf{A}_1$, that is,

$$\mathsf{A}_1 = \mathsf{D} - \mathsf{U} - \mathsf{L}, \tag{4.3.2}$$

where

$$\mathsf{D} = \frac{2}{h_1} \mathrm{I}_{n_1} \in \mathbb{R}^{n_1 \times n_1}, \tag{4.3.3}$$

$$\mathsf{U} = \frac{1}{h_1} \begin{bmatrix} 0 & 1 & & & \\ & 0 & 1 & & \\ & & \ddots & \ddots & \\ & & & 0 & 1 \\ & & & & 0 \end{bmatrix} \in \mathbb{R}^{n_1 \times n_1}, \tag{4.3.4}$$

and, of course, $\mathsf{L} = \mathsf{U}^\top$.

**Definition 4.3.1** (Damped Jacobi method). Let $\mathsf{A} \in \mathbb{R}^{n \times n}$ be nonsingular, and let $\boldsymbol{f} \in \mathbb{R}^n$. The **damped Jacobi method** is a GLIS to approximate $\boldsymbol{x} = \mathsf{A}^{-1}\boldsymbol{f}$, which, for $\omega \in (0,1]$ and $\boldsymbol{u}^{(0)} \in \mathbb{R}^n$, generates a sequence $\{\boldsymbol{u}^{(\sigma)}\}_{\sigma=1}^{\infty} \subset \mathbb{R}^n$ as follows:

$$\begin{aligned} \boldsymbol{z}^{(\sigma)} &:= \mathsf{D}^{-1}(\mathsf{L} + \mathsf{U})\boldsymbol{u}^{(\sigma)} + \mathsf{D}^{-1}\boldsymbol{f}, \\ \boldsymbol{u}^{(\sigma+1)} &:= \omega \boldsymbol{z}^{(\sigma)} + (1-\omega)\boldsymbol{u}^{(\sigma)}, \end{aligned}$$

where $\mathsf{A} = \mathsf{D} - \mathsf{L} - \mathsf{U}$ is, as in (1.4.1), a standard decomposition of the matrix $\mathsf{A}$. Notice that we may eliminate $\boldsymbol{z}^{(\sigma)}$ to obtain the equivalent version

$$\boldsymbol{u}^{(\sigma+1)} = \boldsymbol{u}^{(\sigma)} + \omega \mathsf{D}^{-1}(\boldsymbol{f} - \mathsf{A}\boldsymbol{u}^{(\sigma)}).$$

■

Let us now apply the damped Jacobi method of Definition 4.3.1 to problem (4.3.1) as a smoother. Then the error transfer matrix can be realized as

$$\mathsf{S}_1 = \omega \mathsf{D}^{-1}$$

and

$$\begin{aligned} \mathsf{K}_1 &= \mathrm{I}_1 - \mathsf{S}_1 \mathsf{A}_1 \\ &= \mathrm{I}_1 - \omega \mathsf{D}^{-1} \mathsf{A}_1 \\ &= \mathrm{I}_1 - \omega \frac{h_1}{2} \mathsf{A}_1. \end{aligned} \tag{4.3.5}$$

Notice that $\mathsf{S}_1 = \mathsf{S}_1^\top$ and $\mathsf{K}_1 = \mathsf{K}_1^*$.

The damped Jacobi method may be written in component form as

$$\begin{aligned} z_{1,i}^{(\sigma)} &= \frac{h_1}{2}\left[\frac{1}{h_1} u_{1,i-1}^{(\sigma)} + \frac{1}{h_1} u_{1,i+1}^{(\sigma)}\right] + \frac{h_1}{2} f_{1,i}, \\ u_{1,i}^{(\sigma+1)} &= \omega z_{1,i}^{(\sigma)} + (1-\omega) u_{1,i}^{(\sigma)}. \end{aligned}$$

**Listing 4.1:** The main damped Jacobi iteration block. The full code is provided in Listing 4.9.

```
% Compute smoothing step.
z(1) = (hl*f(1)+u(2))/2.0;
z(2:nl-1) = (hl*f(2:nl-1)+u(1:nl-2)+u(3:nl))/2.0;
z(nl) = (hl*f(nl)+u(nl-1))/2.0;

% Apply damping.
z(1:nl) = omega*z(1:nl)+(1.0-omega)*u(1:nl);
```

Unlike the Gauss–Seidel method, we do not use componentwise updates immediately after they are generated. Therefore the order in which we pass through the components is irrelevant.

**Theorem 4.3.2** (Eigenpairs of damped Jacobi). *Let* $\mathsf{K}_1 = \mathsf{I}_1 - \omega\mathsf{D}^{-1}\mathsf{A}_1$ *be the error transfer matrix for the damped Jacobi method applied to the model problem* (4.3.1). *The eigenvectors of* $\mathsf{K}_1$ *are the same as those of the level-1 stiffness matrix* $\mathsf{A}_1$, *that is,*

$$[\boldsymbol{v}_1^{(k)}]_i = v_{1,i}^{(k)} = \sin(k\pi x_{1,i}), \quad 1 \le i \le n_1,$$

*for* $k = 1, \dots, n_1$. *The eigenvalues of* $\mathsf{K}_1$ *are, for* $k = 1, \dots, n_1$,

$$\begin{aligned} \mu_1^{(k)}(\omega) &= \omega\cos(k\pi h_1) + 1 - \omega \\ &= 1 - 2\omega\sin^2\left(\frac{k\pi h_1}{2}\right). \end{aligned} \tag{4.3.6}$$

*Proof.* We simply compute

$$\begin{aligned} \mathsf{K}_1\boldsymbol{v}_1^{(k)} &= \boldsymbol{v}_1^{(k)} - \omega\mathsf{D}^{-1}\mathsf{A}_1\boldsymbol{v}_1^{(k)} \\ &= \boldsymbol{v}_1^{(k)} - \omega\frac{h_1}{2}\lambda_1^{(k)}\boldsymbol{v}_1^{(k)} \\ &\overset{(4.2.1)}{=} \left(1 - \omega\frac{h_1}{2}\frac{2}{h_1}(1 - \cos(k\pi h_1))\right)\boldsymbol{v}_1^{(k)} \\ &= (1 - \omega(1 - \cos(k\pi h_1)))\boldsymbol{v}_1^{(k)} \\ &= \left(1 - 2\omega\sin^2\left(\frac{k\pi h_1}{2}\right)\right)\boldsymbol{v}_1^{(k)}. \end{aligned}$$

This shows that

$$\begin{aligned} \mu_1^{(k)}(\omega) &= 1 - \omega + \omega\cos(k\pi h_1) \\ &= 1 - 2\omega\sin^2\left(\frac{k\pi h_1}{2}\right). \end{aligned}$$

□

The next result shows that the performance of the damped Jacobi method, as a standalone solver, is degraded as $h_1 \to 0$.

**Theorem 4.3.3** (Spectral radius of damped Jacobi). *Let* $\mathsf{K}_1 = \mathsf{I}_1 - \omega\mathsf{D}^{-1}\mathsf{A}_1$ *be the error transfer matrix for the damped Jacobi method applied to the model problem* (4.3.1). *Then there is a constant* $C > 0$, *independent of* $h_1$ *and* $\omega$, *such that for all* $\omega \in (0, 1]$ *and* $h_1 \in (0, \frac{1}{2}]$,

$$0 < 1 - C\omega h_1^2 \le \rho(\mathsf{K}_1) < 1.$$

*Proof.* Notice that the function $f(x) = \sin^2(x)$, $x \in (0, \frac{\pi}{2})$, is positive and strictly increasing and $f(x) < \frac{1}{2}$. Therefore, since $\omega \in (0, 1]$,

$$0 < \mu_1^{(n_1)}(\omega) < \cdots < \mu_1^{(1)}(\omega) < 1,$$

which implies

$$\rho(\mathsf{K}_1) = \mu_1^{(1)}(\omega) < 1;$$

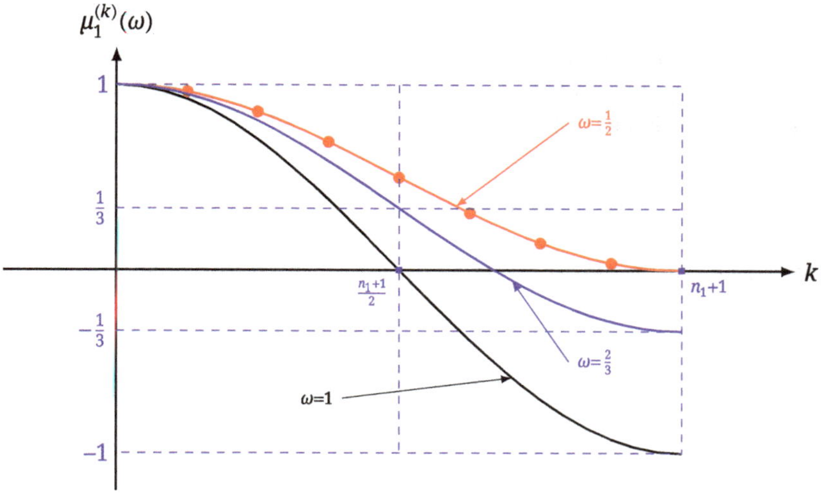


**Figure 4.6:** Plots of the eigenvalues of $K_1$, $\mu_1^{(k)}(\omega) = 1 - 2\omega\sin^2(\frac{k\pi h_1}{2})$, as functions of $k$ for various values of $\omega \in (0,1]$. The discrete nature of the set of eigenvalues is illustrated for the case $\omega = \frac{1}{2}$, where $n_1 = 7$.

see also Figure 4.6. Recall that

$$\mu_1^{(1)}(\omega) = \omega\cos(\pi h_1) + 1 - \omega.$$

We proved earlier estimates (4.2.3) and (4.2.4), which showed that for $x \in (0, \frac{\pi}{2}]$,

$$1 - \frac{x^2}{2} < \cos(x) < 1 - \frac{x^2}{3}.$$

Thus, for all $\omega \in (0,1]$ and $x \in (0, \frac{\pi}{2}]$,

$$\omega\frac{x^2}{2} > \omega(1 - \cos(x)) > \omega\frac{x^2}{3}.$$

Consequently, since $\pi h_1 \in (0, \frac{\pi}{2}]$,

$$\frac{\omega\pi^2}{2}h_1^2 > 1 - \rho(K_1) > \frac{\omega\pi^2}{3}h_1^2.$$

From this it follows that

$$1 > \rho(K_1) > 1 - \frac{\omega\pi^2}{2}h_1^2,$$

which is the claimed result with $C = \frac{\pi^2}{2}$. □

**Remark 4.3.4** (Smoothing effect). In the multigrid setting, we want the Jacobi method (the smoother) to have a "smoothing" effect on the error. In other words, we want to dampen high-frequency modes of the error faster than low-frequency modes. Recall that for the damped Jacobi method,

$$\boldsymbol{e}_1^{(\sigma+1)} = \mathsf{K}_1 \boldsymbol{e}_1^{(\sigma)}, \quad \mathsf{K}_1 = \mathsf{I}_1 - \frac{\omega h_1}{2} \mathsf{A}_1.$$

Let us now expand $\boldsymbol{e}_1^{(\sigma)}$ in the basis of eigenvectors $\{\boldsymbol{v}_1^{(k)}\}_{k=1}^{n_1}$: there exist unique $\{\epsilon_k^{(\sigma)}\}_{k=1}^{n_1} \subset \mathbb{R}$ such that

$$\boldsymbol{e}_1^{(\sigma)} = \sum_{k=1}^{n_1} \epsilon_k^{(\sigma)} \boldsymbol{v}_1^{(k)}.$$

Then

$$\boldsymbol{e}_1^{(\sigma+1)} = \sum_{k=1}^{n_1} \mu_1^{(k)}(\omega) \epsilon_k^{(\sigma)} \boldsymbol{v}_1^{(k)}. \tag{4.3.7}$$

By choosing $\omega$ judiciously, we can bias the smoothing process to favor the dampening of "high-frequency" error components. ◄

## 4.4 The Gauss–Seidel smoother

To approximate the solution of problem (4.3.1), we will next consider the forward Gauss–Seidel method. To define the method, we again split $\mathsf{A}_1$ as in (4.3.2), obtaining $\mathsf{D}$ and $\mathsf{U}$ as in (4.3.3) and (4.3.4), respectively. As before, $\mathsf{L} = \mathsf{U}^\top$. The Gauss–Seidel method can be expressed as

$$\boldsymbol{u}_1^{(\sigma+1)} = (\mathsf{D} - \mathsf{L})^{-1} \mathsf{U} \boldsymbol{u}_1^{(\sigma)} + (\mathsf{D} - \mathsf{L})^{-1} \boldsymbol{f}_1$$

or, equivalently,

$$\boldsymbol{u}_1^{(\sigma+1)} = \boldsymbol{u}_1^{(\sigma)} + (\mathsf{D} - \mathsf{L})^{-1} (\boldsymbol{f}_1 - \mathsf{A}_1 \boldsymbol{u}_1^{(\sigma)}).$$

In our two-grid terminology,

$$\mathsf{S}_1 = (\mathsf{D} - \mathsf{L})^{-1},$$

and

$$\mathsf{K}_1 = \mathsf{I}_1 - \mathsf{S}_1 \mathsf{A}_1 = \mathsf{I}_1 - (\mathsf{D} - \mathsf{L})^{-1} \mathsf{A}_1.$$

Clearly, $\mathsf{S}_1 \neq \mathsf{S}_1^\top$ and $\mathsf{K}_1 \neq \mathsf{K}_1^*$.

For our particular application, the Gauss–Seidel method may be written in component form as

$$u_{1,i}^{(\sigma+1)} = \frac{h_1}{2}\left\{\frac{1}{h_1}u_{1,i-1}^{(\sigma+1)} + \frac{1}{h_1}u_{1,i+1}^{(\sigma)}\right\} + \frac{h_1}{2}f_{1,i},$$

stepping through components from $i = 1$, in order, to $i = n_1$. We use componentwise updates immediately after they are generated. Therefore, the order in which we pass through the components is important.

**Theorem 4.4.1** (Eigenpairs of Gauss–Seidel). *Let* $\mathsf{K}_1 = \mathrm{I}_1 - (\mathsf{D} - \mathsf{L})^{-1}\mathsf{A}_1$ *be the error transfer matrix for the Gauss–Seidel method applied to the model problem* (4.3.1)*. The eigenvectors of* $\mathsf{K}_1$ *are*

$$[\boldsymbol{w}_1^{(k)}]_i = w_{1,i}^{(k)} := [\cos(k\pi h_1)]^i \sin(k\pi x_{1,i}), \quad 1 \le i \le n_1,$$

*for* $k = 1, \ldots, n_1$*. The eigenvalues of* $\mathsf{K}_1$ *are*

$$\nu_1^{(k)} = \cos^2(k\pi h_1), \quad 1 \le k \le n_1. \tag{4.4.1}$$

*Proof.* We begin by writing

$$\mathsf{K}_1 = (\mathsf{D} - \mathsf{L})^{-1}\mathsf{U}.$$

Then

$$\mathsf{K}_1\boldsymbol{w}_1^{(k)} = \nu_1^{(k)}\boldsymbol{w}_1^{(k)}$$

iff

$$\mathsf{U}\boldsymbol{w}_1^{(k)} = \nu_1^{(k)}(\mathsf{D} - \mathsf{L})\boldsymbol{w}_1^{(k)}.$$

The rest of the details are left to the reader; see Problem 4.2. □

**Remark 4.4.2** (Different spectra). Notice that the eigenvectors of the error transfer matrix for the forward Gauss–Seidel method differ from those of the stiffness matrix. This complicates the analysis of the multigrid method when Gauss–Seidel smoothing is used. We will not consider this smoother further until we develop new analysis techniques based on subspace decompositions; see Chapter 12. ◂

## 4.5 Two-level grids and high-frequency smoothing

Why are the first and third stages of the two-grid algorithm referred to as *smoothing* stages? Why did we just refer to the damped Jacobi and Gauss–Seidel methods as *smoothers*? The reason is hinted at in Remark 4.3.4, and we will give a thorough answer to these questions in this section. We have to address the issue that the error grid function, or the finite element error function, will become smoothed out at these stages, and the concept is of fundamental importance in the two-grid and multigrid algorithms. To understand smoothing, we must distinguish between low- and high-frequency modes. First things first, a mode is just a sine wave. Low-frequency modes (sine waves) are essentially those that can be accurately represented on the coarse grid. Let us make explicit now the structure of the coarse grid.

**Definition 4.5.1** (Two-level grid). An object $\mathcal{G}$ of the form

$$\mathcal{G} = \mathcal{G}_{h_0} \cup \mathcal{G}_{h_1}, \quad \mathcal{G}_{h_\ell} = \{x_{\ell,i}\}_{i=0}^{n_\ell+1},$$

is called a **two-level grid** for $\Omega = (0,1)$ if

(i) $n_1 > n_0$,

(ii) for $\ell = 0, 1$, we have

$$0 = x_{\ell,0} < x_{\ell,1} < \cdots < x_{\ell,n_\ell} < x_{\ell,n_\ell+1} = 1,$$

and

(iii) $h_1 < h_0$, where

$$h_\ell = \max_{1 \le i \le n_\ell+1} (x_{\ell,i} - x_{\ell,i-1}).$$ ■

**Definition 4.5.2** (Nested and uniform grids). We say that a two-level grid $\mathcal{G}$ is **uniform and nested** if

(i) $n_1 + 1 \ge 4$, and $n_1 + 1$ is even;

(ii)

$$n_0 := \frac{n_1 + 1}{2} - 1;$$

and

(iii) the grid points satisfy

$$x_{\ell,i} = ih_\ell, \quad 0 \le i \le n_\ell + 1, \quad \ell = 0, 1,$$

where

$$h_\ell = \frac{1}{n_\ell + 1}, \quad \ell = 0, 1.$$ ■

**Example 4.5.3** (Uniform and nested grid). Suppose $n_1 = 3$. A uniform and nested two-level grid is shown in Figure 4.7. A uniform and nested two-level grid for the case $n_1 = 7$ is shown in Figure 4.8. Clearly, once the value of $n_1$ is specified, all of the details of the uniform and nested two-level grid $\mathcal{G}$ for the domain $\Omega = (0,1)$ are uniquely defined, so long as $n_1 + 1$ is an even integer greater than or equal to 4. ▲

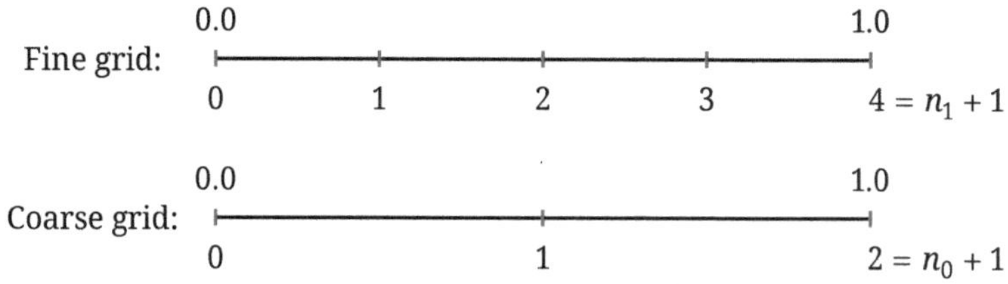


**Figure 4.7:** A uniform and nested two-level grid with $n_1 = 3$.

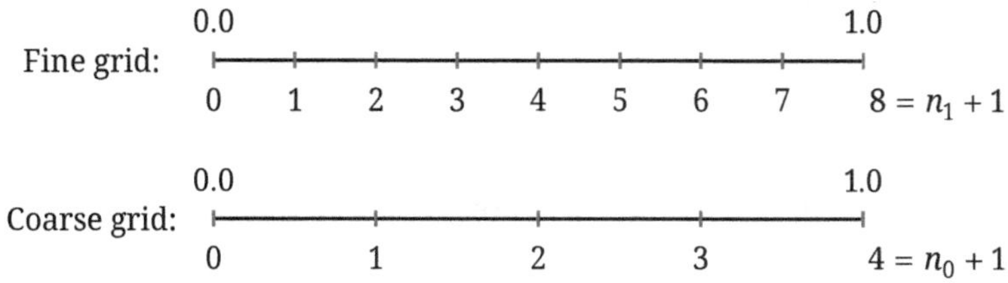


**Figure 4.8:** A uniform and nested two-level grid with $n_1 = 7$.

**Remark 4.5.4** (Convention). Throughout this chapter, we will only be interested in two-level grids for the one-dimensional domain $\Omega = (0,1)$. However, it should be clear how this definition could be generalized along various lines. In any case, to save some writing, we will shorten the assumption "suppose that the two-level grid $\mathcal{G}$ for $\Omega = (0,1)$ is uniform and nested" to "suppose that the two-level grid is uniform and nested." ◂

**Definition 4.5.5** (High- and low-frequency modes). Suppose the two-level grid is uniform and nested. Let

$$\boldsymbol{v}_1 = \sum_{k=1}^{n_1} c_k \boldsymbol{v}_1^{(k)} \in \mathbb{R}^{n_1},$$

with, as in (4.2.2), $\boldsymbol{v}_1^{(k)}$ being the $k$th eigenvector of the level-1 stiffness matrix. This uniquely defines an element $v_1 \in V_1$. For such a function, we say that the $k$th mode $c_k \boldsymbol{v}_1^{(k)}$ is of **high frequency** if

$$n_0 + 1 = \frac{n_1 + 1}{2} \leq k \leq n_1.$$

Otherwise, we say that the mode is of **low frequency**. ■

**Theorem 4.5.6** (Optimal dampening). *Suppose that the two-level grid is uniform and nested and that $\mu_1^{(k)}(\omega)$ is the $k$th eigenvalue of $\mathsf{K}_1$, the error transfer matrix of the damped Jacobi smoother, and $0 < \omega \leq 1$. The quantity*

$$S(\omega) = \max_{\frac{n_1+1}{2} \leq k \leq n_1+1} \left|\mu_1^{(k)}(\omega)\right|$$

*is minimized by*

$$\omega = \omega_0 := \frac{2}{3},$$

*in which case*

$$|\mu_1^{(k)}(\omega_0)| \leq \frac{1}{3}$$

*for the high-frequency range, that is, for all* $\frac{n_1+1}{2} \leq k \leq n_1 + 1$. *More generally, if* $0 < \omega \leq 1$, *then for all* $1 \leq k \leq n_1$,

$$|\mu_1^{(k)}(\omega)| < 1.$$

*Proof.* First of all, recall that the eigenvalues of $\mathsf{K}_1$ are given in (4.3.6). The proof follows by a careful examination of the plots of

$$\mu_1^{(k)}(\omega) = 1 - 2\omega \sin^2\left(\frac{k\pi h_1}{2}\right)$$

for various values of $\omega$; see Figure 4.6. □

**Remark 4.5.7** (Error smoothing). Recall that with $\omega = \omega_0$ we have

$$\boldsymbol{e}_1^{(\sigma+1)} = \sum_{k=1}^{n_1} \mu_1^{(k)}(\omega_0)\epsilon_k^{(\sigma)}\boldsymbol{v}_1^{(k)}$$

for the error after one smoothing iteration by the damped Jacobi method. High-frequency modes will be damped much faster than those of low frequency. In fact the modes $\frac{n_1+1}{2} \leq k \leq n_1$ will be reduced by at least $\frac{1}{3}$ after a single smoothing iteration.

Furthermore, observe that the modes with very low frequency, that is, those close to $k = 1$, are hardly dampened at all. ◀

At this point, we can mention for the first time the principles behind the performance of two-grid or multigrid methods:

**First multigrid principle.** Many classical GLIS have an error smoothing property: high-frequency modes of the error are damped much more rapidly than those of low frequency. However, these converge very slowly, especially as $h_1 \to 0$.

**Second multigrid principle.** We can approximate well low-frequency information on a coarse grid.

Throughout the text, we will see how, in different incarnations, these two principles are used to prove the convergence of our various algorithms.

**Remark 4.5.8** (Error and residual). We observed that a smoothed error is well approximated on a coarse grid. By smoothed we mean that the high-frequency modes are greatly diminished. We can show that if the error is smooth, then the residual may also be smooth. Let us explain, because we need to be very careful. Suppose that

$$\boldsymbol{e}_1^{(\sigma)} = \sum_{k=1}^{n_1} \epsilon_k^{(\sigma)} \boldsymbol{v}_1^{(k)}.$$

Since $\boldsymbol{r}_1^{(\sigma)} = \mathsf{A}_1 \boldsymbol{e}_1^{(\sigma)}$,

$$\begin{aligned} \boldsymbol{r}_1^{(\sigma)} &= \sum_{k=1}^{n_1} \epsilon_k^{(\sigma)} \mathsf{A}_1 \boldsymbol{v}_1^{(k)} \\ &= \sum_{k=1}^{n_1} \epsilon_k^{(\sigma)} \lambda_1^{(k)} \boldsymbol{v}_1^{(k)}. \end{aligned}$$

If the high-frequency modes of the error are totally absent, then, of course, they will be absent from the residual as well. However, if the high-frequency components are not completely diminished in the error, then the sizes of the high-frequency components in the residual will depend upon the products $\epsilon_k^{(\sigma)} \lambda_1^{(k)}$ for $k \geq (n_1 + 1)/2$, which may or may not be small. ◂

Later, in Section 4.9, we will present some computational experiments. There we will clearly see the error smoothing property in action.

## 4.6 Prolongation and restriction operators

We are going to approach the definition of prolongation and restriction matrices in what seems to be the reverse order. We first define the prolongation matrix $\mathsf{P}_0$, and then we set $\mathsf{R}_0 = \mathsf{P}_0^{\mathsf{T}}$.

We will assume that our two-level grid $\mathcal{G}$ for $\Omega = (0, 1)$ is uniform and nested. Our construction comes from the FEM point of view. First of all, we define the coarse-level grid as comprised of $n_0 + 1$ equally sized intervals

$$K_{0,i} := (x_{0,i-1}, x_{0,i}), \quad i = 1, \ldots, n_0 + 1,$$

where the coarse-grid points $\mathcal{G}_{h_0} = \{x_{0,i}\}_{i=0}^{n_0+1}$ are

$$x_{0,i} = i \cdot h_0, \quad h_0 := \frac{1}{n_0 + 1}.$$

With this at hand, we can define the coarse-level finite element spaces

$$\begin{aligned} S_0 &:= \{v \in C^0([0,1]) \mid v|_{K_{0,i}} \in \mathbb{P}_1(K_{0,i}),\ 1 \leq i \leq n_0 + 1\}, \\ V_0 &:= \{v \in S_0 \mid v(0) = v(1) = 0\}. \end{aligned} \tag{4.6.1}$$

**Definition 4.6.1** (Hat function). For $i = 1, \ldots, n_0$, define $\psi_{0,i} \in V_0$ by

$$\psi_{0,i}(x_{0,j}) = \delta_{i,j}, \quad 1 \le j \le n_0.$$

The function $\psi_{0,i}$ is called a **level-0 hat function**. ■

**Example 4.6.2** ($V_0 \subset V_1$). Let $u_0 \in V_0$. See, for example, the function depicted in Figure 4.9, where $n_0 = 3$. Recall that, based on our uniform nested grids assumption,

$$n_0 = \frac{n_1 + 1}{2} - 1 \iff n_1 = 2(n_0 + 1) - 1.$$

Consider next the function in Figure 4.10, where $n_1 = 7$. The exact same function is shown in both Figures 4.9 and 4.10. Since, it turns out, $V_0 \subset V_1$, $u_0 \in V_1$. In other words, any coarse grid function can be trivially represented on the fine grid. ▲

The previous example is a particular case of the following general result.

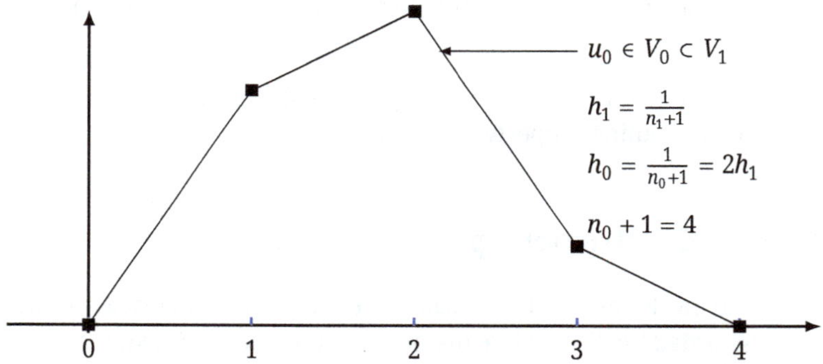


**Figure 4.9:** A piecewise linear function defined on the coarse grid withe $n_0 = 3$.

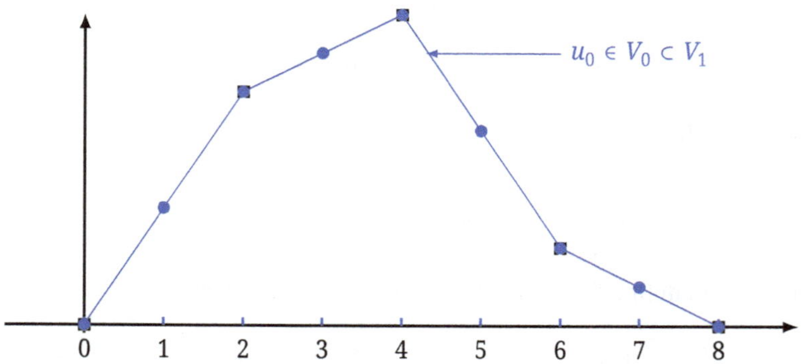


**Figure 4.10:** The coarse grid function from Figure 4.9 as it would be represented on the fine grid.

**Proposition 4.6.3** ($V_0 \subset V_1$). *Assume that the two-level grid is uniform and nested. Then the coarse-level finite element space $V_0$, defined in (4.6.1), is a subspace of the fine-level finite element space $V_1$ defined in (4.1.4). The sets*

$$\mathcal{B}_\ell := \{\psi_{\ell,j}\}_{j=1}^{n_\ell}, \quad \ell = 0, 1,$$

*are bases of $V_\ell$, $\ell = 0, 1$, respectively. Consequently, the dimensions of the spaces are $n_0$ and $n_1$, respectively.*

*Proof.* Exercise; see Problem 4.3. □

The sets $\mathcal{B}_\ell$, for $\ell = 0, 1$, are depicted in Figures 4.11 and 4.12, respectively.

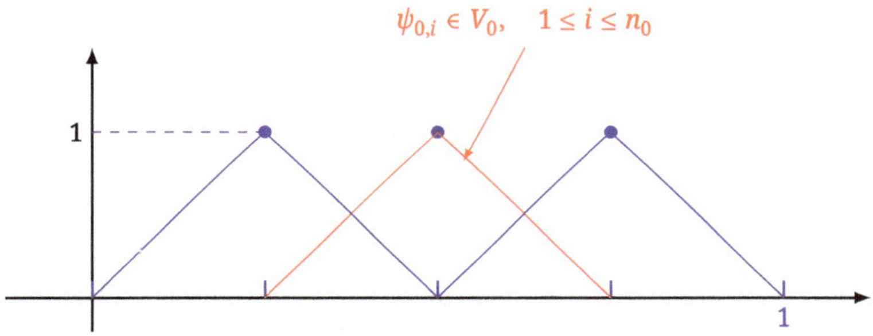


**Figure 4.11:** Level-0 hat basis functions.

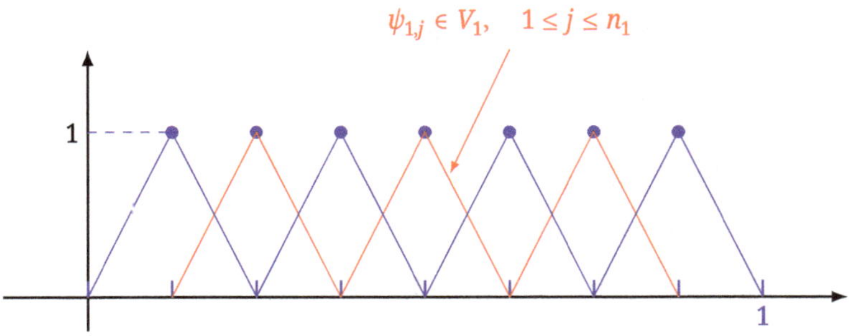


**Figure 4.12:** Level-1 hat basis functions.

It is now time to define the prolongation matrix $\mathsf{P}_0$. The action of the matrix $\mathsf{P}_0 \in \mathbb{R}^{n_1 \times n_0}$ is as follows. Let $u_0 \in V_0$ have $\boldsymbol{u}_0 \in \mathbb{R}^{n_0}$ as a coordinate vector in the level-0 hat function basis of $V_0$. Since $V_0 \subset V_1$, the function $u_0$ can also be represented in the basis of level-1 hat functions. Denote by $\boldsymbol{u}_1 \in \mathbb{R}^{n_1}$ the coordinate vector of this representation. Then

$$\mathsf{P}_0 \boldsymbol{u}_0 = \boldsymbol{u}_1 \in \mathbb{R}^{n_1}.$$

**Example 4.6.4** (Prolongation matrix). Recall that Example 4.6.2, where $n_0 = 3$ and $n_1 = 7$, is a particular instance of the fact that $V_0 \subset V_1$. Our task here will be to find a matrix that maps the 3 coarse grid degrees of freedom (DOFs) that uniquely characterize the coarse grid function into the 7 fine grid DOFs that uniquely characterize that function on the fine grid. We have

$$\mathsf{P}_0 = \begin{bmatrix} \frac{1}{2} & & \\ 1 & & \\ \frac{1}{2} & \frac{1}{2} & \\ & 1 & \\ & \frac{1}{2} & \frac{1}{2} \\ & & 1 \\ & & \frac{1}{2} \end{bmatrix} \in \mathbb{R}^{7\times 3}.$$

Therefore

$$\mathsf{R}_0 = \begin{bmatrix} \frac{1}{2} & 1 & \frac{1}{2} & & & & \\ & & \frac{1}{2} & 1 & \frac{1}{2} & & \\ & & & & \frac{1}{2} & 1 & \frac{1}{2} \end{bmatrix} \in \mathbb{R}^{3\times 7}.$$ ▲

**Definition 4.6.5** (Prolongation and restriction). Assume that the two-level grid is uniform and nested. In other words, the positive integers $n_0$ and $n_1$ must satisfy

$$n_1 = 2(n_0 + 1) - 1.$$

The action of the prolongation matrix, $\mathsf{P}_0 \in \mathbb{R}^{n_1\times n_0}$ on $\boldsymbol{u}_0 \in \mathbb{R}^{n_0}$ is defined as follows:

$$\begin{aligned} [\mathsf{P}_0\boldsymbol{u}_0]_1 &:= \frac{1}{2}u_{0,1}, \\ [\mathsf{P}_0\boldsymbol{u}_0]_{n_1} &:= \frac{1}{2}u_{0,n_0}, \\ [\mathsf{P}_0\boldsymbol{u}_0]_{2i} &:= u_{0,i}, \quad 1 \le i \le n_0, \\ [\mathsf{P}_0\boldsymbol{u}_0]_{2i+1} &:= \frac{1}{2}(u_{0,i} + u_{0,i+1}), \quad 1 \le i \le n_0 - 1. \end{aligned}$$

In addition, the restriction matrix $\mathsf{R}_0 \in \mathbb{R}^{n_0\times n_1}$ is defined as

$$\mathsf{R}_0 = \mathsf{P}_0^\top \in \mathbb{R}^{n_0\times n_1}.$$ ■

**Theorem 4.6.6** (Prolongation components). *Assume that the two-level grid is uniform and nested. In other words, the positive integers $n_0$ and $n_1$ must satisfy*

$$n_1 = 2(n_0 + 1) - 1.$$

*Suppose that the prolongation operator $\mathsf{P}_0 = [p_{0,i,j}] \in \mathbb{R}^{n_1\times n_0}$ is as in Definition 4.6.5. Let*

$$\mathcal{B}_\ell := \{\psi_{\ell,j}\}_{j=1}^{n_\ell}, \quad \ell = 0, 1,$$

*be the hat function bases of $V_\ell$. Then, for $1 \le i \le n_1$ and $1 \le j \le n_0$, the numbers $p_{0,i,j}$ are the unique ones that satisfy*

$$\psi_{0,j} = \sum_{i=1}^{n_1} p_{0,i,j}\psi_{1,i}. \tag{4.6.2}$$

*Thus, if $u_0 \in V_0 \subset V_1$ and*

$$u_0 = \sum_{j=1}^{n_0} u_{0,j}\psi_{0,j},$$

*then the unique representation of $u_0$ in the basis $\mathcal{B}_1$ is given by*

$$\begin{aligned} u_0 &= \sum_{i=1}^{n_1}\sum_{j=1}^{n_0} p_{0,i,j}u_{0,j}\psi_{1,i} \\ &= \sum_{i=1}^{n_1} [\mathsf{P}_0 \boldsymbol{u}_0]_i \psi_{1,i}. \end{aligned}$$

*Proof.* Since $V_0 \subset V_1$, each level-0 hat basis function satisfies $\psi_{0,j} \in V_1$. Therefore, since $\mathcal{B}_1 = \{\psi_{1,i}\}_{i=1}^{n_1}$ is a basis for $V_1$, there exist unique, for $1 \le i \le n_1$ and $1 \le j \le n_0$, numbers, which we conveniently label $p_{0,i,j}$, that satisfy

$$\psi_{0,j} = \sum_{i=1}^{n_1} p_{0,i,j}\psi_{1,i}, \quad \forall j = 1, \dots, n_0.$$

We leave it as an exercise for the reader to prove that these are exactly the elements of the matrix $\mathsf{P}_0$; see Problem 4.9. This proves (4.6.2).

Assume now that $u_0 \in V_0 \subset V_1$ and

$$u_0 = \sum_{j=1}^{n_0} u_{0,j}\psi_{0,j}.$$

The coordinate vector of this function in the basis of coarse-level hat functions $\mathcal{B}_0 = \{\psi_{0,j}\}_{j=1}^{n_0}$ is precisely

$$\boldsymbol{u}_0 = \begin{bmatrix} u_{0,1} & u_{0,2} & \cdots & u_{0,n_0} \end{bmatrix}^\intercal.$$

Using (4.6.2), we have

$$u_0 = \sum_{j=1}^{n_0} u_{0,j}\psi_{0,j}$$

$$
\begin{aligned}
&= \sum_{j=1}^{n_0} u_{0,j} \sum_{i=1}^{n_1} p_{0,i,j} \psi_{1,i} \\
&= \sum_{i=1}^{n_1} \sum_{j=1}^{n_0} p_{0,i,j} u_{0,j} \psi_{1,i} \\
&= \sum_{i=1}^{n_1} [\mathsf{P}_0 \boldsymbol{u}_0]_i \psi_{1,i}.
\end{aligned}
$$

The result is proven. □

**Listing 4.2:** Elementwise prolongation operator. The full code is provided in Listing 4.7.

```
% Linear interpolation at endpoints.
vProl(1) = v(1)/2.0;
vProl(nf) = v(nc)/2.0;

% Linear interpolation at interior points.
vProl(2:2:2*nc) = v(1:nc);
vProl(3:2:2*nc-1) = (v(1:nc-1)+v(2:nc))/2.0;
```

**Listing 4.3:** Elementwise restriction operator. The full code is provided in Listing 4.8.

```
vRest(1:nc) = 0.5*(v(1:2:2*nc-1)+v(3:2:2*nc+1))...
              +v(2:2:2*nc);
```

**Remark 4.6.7** (Implementation of prolongation and restriction). In multigrid, the prolongation and restriction operators are mathematically given by matrix–vector multiplications with the respective restriction and prolongation matrices. However, for practical implementations, these are often performed using elementwise operations to reduce memory usage and improve efficiency. In particular, in MATLAB©, we can implement vectorization to take advantage of its internal parallelization features and further speed up the code. ◂

**Corollary 4.6.8** (Coordinate vectors). *Let $u_0 \in V_0$ and suppose that $\boldsymbol{u}_0 \in \mathbb{R}^{n_0}$ is its coordinate vector with respect to the basis $\mathcal{B}_0$. Then $\mathsf{P}_0 \boldsymbol{u}_0$ is the coordinate vector of $u_0$ with respect to the basis $\mathcal{B}_1$.*

*Proof.* Exercise; see Problem 4.4. □

**Theorem 4.6.9** (Galerkin condition). *Let the two-level grid be uniform and nested. Let $\mathsf{P}_0$ and $\mathsf{R}_0$ be as in Definition* 4.6.5, *with the level-1 stiffness matrix $\mathsf{A}_1$ defined as in* (4.1.9). *The matrix $\mathsf{A}_0 = [a_{0,i,j}] \in \mathbb{R}^{n_0 \times n_0}$ satisfies the Galerkin condition, that is,*

$$\mathsf{A}_0 = \mathsf{R}_0 \mathsf{A}_1 \mathsf{P}_0,$$

*iff*

$$\begin{aligned} a_{0,i,j} &= (\psi'_{0,j}, \psi'_{0,i})_{L^2(0,1)} \\ &= (\psi'_{0,i}, \psi'_{0,j})_{L^2(0,1)}. \end{aligned}$$

*In this case,*

$$\mathsf{A}_0 = \frac{1}{h_0}\begin{bmatrix} 2 & -1 & & & \\ -1 & 2 & -1 & & \\ & \ddots & \ddots & \ddots & \\ & & -1 & 2 & -1 \\ & & & -1 & 2 \end{bmatrix} \in \mathbb{R}^{n_0 \times n_0}.$$

*Proof.* We prove only one direction. The other will be obvious; see Problem 4.5.
($\Longleftarrow$): Recall that by (4.1.9) the level-1 stiffness matrix verifies

$$[\mathsf{A}_1]_{i,j} = (\psi'_{1,j}, \psi'_{1,i})_{L^2(0,1)}.$$

Defining the level-0 stiffness matrix analogously and using (4.6.2), we have

$$\begin{aligned} [\mathsf{A}_0]_{i,j} &= (\psi'_{0,j}, \psi'_{0,i})_{L^2(0,1)} \\ &= (\psi'_{0,i}, \psi'_{0,j})_{L^2(0,1)} \\ &\overset{(4.6.2)}{=} \left( \sum_{k=1}^{n_1} p_{0,k,i}\psi'_{1,k}, \sum_{\ell=1}^{n_1} p_{0,\ell,j}\psi'_{1,\ell} \right)_{L^2(0,1)} \\ &= \sum_{k=1}^{n_1}\sum_{\ell=1}^{n_1} p_{0,k,i}p_{0,\ell,j}(\psi'_{1,k}, \psi'_{1,\ell})_{L^2(0,1)} \\ &= \sum_{k=1}^{n_1}\sum_{\ell=1}^{n_1} [\mathsf{R}_0]_{i,k}[\mathsf{A}_1]_{k,\ell}[\mathsf{P}_0]_{\ell,j} \\ &= [\mathsf{R}_0\mathsf{A}_1\mathsf{P}_0]_{i,j}. \end{aligned}$$

Finally, since

$$\begin{aligned} a_{0,i,j} &= (\psi'_{0,j}, \psi'_{0,i})_{L^2(0,1)} \\ &= (\psi'_{0,i}, \psi'_{0,j})_{L^2(0,1)}, \end{aligned}$$

we can use the same techniques used to construct the level-1 stiffness matrix to show that the matrix $\mathsf{A}_0$ has the stated structure. □

**Corollary 4.6.10** (Spectrum of $\mathsf{A}_0$). *Suppose the two-level grid is uniform and nested and the level-0 stiffness matrix* $\mathsf{A}_0 = [a_{0,i,j}] \in \mathbb{R}^{n_0 \times n_0}$ *satisfies the Galerkin condition. Then* $\mathsf{A}_0$ *is SPD and has the eigenpairs*

$$[\boldsymbol{v}_0^{(k)}]_i := \sin(k\pi x_{0,i}), \quad 1 \le i \le n_0,$$
$$\lambda_0^{(k)} := \frac{2}{h_0}(1 - \cos(k\pi h_0))$$

*for* $k = 1, \ldots, n_0$, *where*

$$x_{0,i} = i \cdot h_0, \quad 0 \le i \le n_0 + 1, \quad h_0 = 2h_1 = \frac{1}{n_0 + 1}.$$

*Proof.* Exercise; see Problem 4.6. □

## 4.7 Other restriction matrices

The definition of the prolongation matrix makes geometric sense. If $\boldsymbol{u}_0 \in \mathbb{R}^{n_0}$ is the unique coordinate vector of the function $u_0 \in V_0$ with respect to the basis $\mathcal{B}_0$, then $\mathsf{P}_0\boldsymbol{u}_0$ is its unique coordinate vector in the basis $\mathcal{B}_1$. This, however, leaves open the interpretation of the restriction matrix.

Consider the situation depicted in Figure 4.13, where $n_1 = 7$. Given the fine grid function $u_1 \in V_1$, suppose its coordinate vector in the basis $\mathcal{B}_1$ is represented as $\boldsymbol{u}_1 \in \mathbb{R}^7$. Let $R_0u_1 \in V_0$ ($\frac{1}{2}R_0u_1 \in V_0$) be the function whose coordinate representation in the basis $\mathcal{B}_0$ is $\mathsf{R}_0\boldsymbol{u}_1$ ($\frac{1}{2}\mathsf{R}_0\boldsymbol{u}_1$); see again Figure 4.13. To understand what is pictured in Figure 4.13, we need to understand what is the action of $\mathsf{R}_0$ on $\boldsymbol{u}_1$. If $\boldsymbol{u}_1 \in \mathbb{R}^{n_1}$ is the coordinate vector

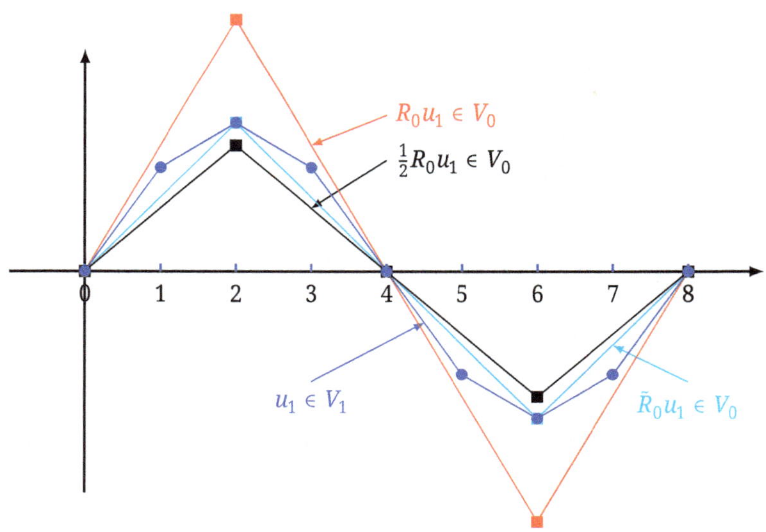


**Figure 4.13:** A fine grid function $u_1 \in V_1$ for $n_1 = 7$ and three possible restrictions $R_0u_1 \in V_0$, $\frac{1}{2}R_0u_1 \in V_0$, and $\tilde{R}_0u_1 \in V_0$. The restriction $\frac{1}{2}R_0u_1$ is called the full weighting restriction.

of $u_1 \in V_1$ with respect to the basis $\mathcal{B}_1$, then, clearly,

$$\mathsf{R}_0 \boldsymbol{u}_1 = \begin{bmatrix} \frac{1}{2} & 1 & \frac{1}{2} & & & & \\ & & \frac{1}{2} & 1 & \frac{1}{2} & & \\ & & & & \frac{1}{2} & 1 & \frac{1}{2} \end{bmatrix} \begin{bmatrix} u_{1,1} \\ u_{1,2} \\ u_{1,3} \\ u_{1,4} \\ u_{1,5} \\ u_{1,6} \\ u_{1,7} \end{bmatrix} = \begin{bmatrix} \frac{1}{2}u_{1,1} + u_{1,2} + \frac{1}{2}u_{1,3} \\ \frac{1}{2}u_{1,3} + u_{1,4} + \frac{1}{2}u_{1,5} \\ \frac{1}{2}u_{1,5} + u_{1,6} + \frac{1}{2}u_{1,7} \end{bmatrix}.$$

The results of this calculation give coordinates that are twice as large as they should be to be geometrically and analytically sensible. Thus dividing by 2, we get a more satisfying result.

There is another possibility that might seem obvious. If $\boldsymbol{u}_1 \in \mathbb{R}^{n_1}$ is the coordinate vector of $u_1 \in V_1$ with respect to the hat function basis $\mathsf{B}_1$, then we could define

$$\tilde{\mathsf{R}}_0 \boldsymbol{u}_1 = \begin{bmatrix} 0 & 1 & 0 & & & & \\ & & 0 & 1 & 0 & & \\ & & & & 0 & 1 & 0 \end{bmatrix} \begin{bmatrix} u_{1,1} \\ u_{1,2} \\ u_{1,3} \\ u_{1,4} \\ u_{1,5} \\ u_{1,6} \\ u_{1,7} \end{bmatrix} = \begin{bmatrix} u_{1,2} \\ u_{1,4} \\ u_{1,6} \end{bmatrix}.$$

In Figure 4.13, $\tilde{R}_0 u_1 \in V_0$ is the function whose coordinate representation in the basis $\mathcal{B}_0$ is $\tilde{\mathsf{R}}_0 \boldsymbol{u}_1$. This also gives a more satisfying result than the standard restriction.

**Definition 4.7.1** (Full weighting restriction). The matrix $\frac{1}{2}\mathsf{R}_0$, where $\mathsf{R}_0$ is as in Definition 4.6.5, is called the **full weighting restriction matrix**. ■

**Remark 4.7.2** (Galerkin condition). The matrices $\frac{1}{2}\mathsf{R}_0$ and $\tilde{\mathsf{R}}_0$ seem to give more geometrically meaningful results than $\mathsf{R}_0$. However, these are not used for one simple reason, the Galerkin condition. If we use either $\frac{1}{2}\mathsf{R}_0$ or $\tilde{\mathsf{R}}_0$, then the Galerkin condition will be broken. As we will see, this condition is too useful to dismiss. ◂

## 4.8 Fourier error analysis

Recall the result of Theorem 3.3.1 regarding the two-grid algorithm:

$$\boldsymbol{e}_1^{k+1} = \mathsf{E}_1 \boldsymbol{e}_1^k,$$

where

$$\mathsf{E}_1 = (\mathsf{K}_1^*)^{m_2} (\mathrm{I}_1 - \tilde{\Pi}_1)(\mathsf{K}_1)^{m_1}.$$

Recall, in addition, that for the present case, we have

$$\mathsf{K}_1 = \mathrm{I}_1 - \omega_0 \frac{h_1}{2} \mathsf{A}_1 = \mathsf{K}_1^*, \quad \omega_0 = \frac{2}{3}.$$

The matrix $\tilde{\Pi}_1$ is the coarse grid Ritz projection matrix

$$\tilde{\Pi}_1 = \mathsf{R}_0^\top \mathsf{A}_0^{-1} \mathsf{R}_0 \mathsf{A}_1.$$

Assuming that the level-0 stiffness matrix $\mathsf{A}_0$ satisfies the Galerkin condition, we have

$$\tilde{\Pi}_1^2 = \tilde{\Pi}_1.$$

Let us introduce the following simplifying notation, which will be used throughout this section:

$$\begin{aligned} \mathsf{C}_m &:= \cos^2\left(\frac{m\pi h_1}{2}\right) \\ &= \frac{1}{2}(1 + \cos(m\pi h_1)), \\ \mathsf{S}_m &:= \sin^2\left(\frac{m\pi h_1}{2}\right) \\ &= \frac{1}{2}(1 - \cos(m\pi h_1)). \end{aligned}$$

We introduce the following definition.

**Definition 4.8.1** (Reciprocal wave numbers). For all $1 \le k \le n_1$, define

$$k' := n_1 + 1 - k,$$

which is called the **reciprocal wave number** of $k$. ■

Clearly, $(k')' = k$. See Table 4.1 for the relationship between $k$ and its reciprocal wave number $k'$.

**Table 4.1:** The values of $k'$ as a function of $k$ for the case $n_1 = 7$, $n_0 = 3$.

| ***k* frequency characterization** | ***k*** | ***k′ = n₁ + 1 − k*** | ***k′* frequency characterization** |
|---|---|---|---|
| low | 1 | $7 = n_1$ | high |
| low | 2 | 6 | high |
| low | 3 | 5 | high |
| high | $n_0 + 1 = 4$ | $4 = n_0 + 1$ | high |
| high | 5 | 3 | low |
| high | 6 | 2 | low |
| high | $n_1 = 7$ | 1 | low |

To analyze the error quantitatively, we need to know what are the actions of our various matrices on the eigenvectors from the coarse and fine grids, which, as we recall, are defined as

$$[\boldsymbol{v}_\ell^{(k)}]_i = \sin(k\pi x_{\ell,i}), \quad 1 \le i, k \le n_\ell, \quad \ell = 0, 1,$$

with the grid points for the coarse and fine grids determined as

$$x_{\ell,i} = i \cdot h_\ell, \quad \ell = 0, 1, \quad 0 \le i \le n_\ell + 1.$$

We begin by listing some aliasing-type results that will be needed in our Fourier analysis.

**Lemma 4.8.2** (Aliasing). *Suppose the two-level grid is uniform and nested. Then for* $1 \le k \le n_1$ *and* $1 \le i \le n_0$,

$$[\boldsymbol{v}_0^{(k')}]_i = \sin(k'\pi x_{0,i}) = -\sin(k\pi x_{0,i}) = -[\boldsymbol{v}_0^{(k)}]_i. \tag{4.8.1}$$

*Furthermore, for* $1 \le k \le n_1$,

$$\cos(k'\pi h_1) = -\cos(k\pi h_1). \tag{4.8.2}$$

*Finally, for* $1 \le k \le n_1$ *and* $1 \le i \le n_1$,

$$\begin{aligned} [\boldsymbol{v}_1^{(k')}]_i &= \sin(k'\pi x_{1,i}) \\ &= (-1)^{i+1}\sin(k\pi x_{1,i}) \\ &= (-1)^{i+1}[\boldsymbol{v}_1^{(k)}]_i. \end{aligned} \tag{4.8.3}$$

*Proof.* Observe, first of all, that for all $1 \le k \le n_1$, we have

$$\begin{aligned} [\boldsymbol{v}_0^{(k')}]_i &= [\boldsymbol{v}_0^{(n_1+1-k)}]_i \\ &= \sin((n_1+1-k)\pi i h_0) \\ &= \sin\left((n_1+1-k)\pi \frac{2i}{n_1+1}\right) \\ &= \sin(2\pi i - k\pi x_{0,i}) \\ &= \sin(-k\pi x_{0,i}) \\ &= -\sin(k\pi x_{0,i}) \\ &= -[\boldsymbol{v}_0^{(k)}]_i. \end{aligned}$$

Similarly,

$$\cos(k'\pi h_1) = \cos((n_1+1-k)\pi h_1)$$

$$= \cos(\pi - k\pi h_1)$$
$$= -\cos(k\pi h_1).$$

For the last statement, consider

$$\begin{aligned}[\boldsymbol{v}_1^{(k')}]_i &= \sin((n_1 + 1 - k)\pi x_{1,i}) \\ &= \sin\left((n_1 + 1 - k)\pi \frac{i}{n_1 + 1}\right) \\ &= \sin(i\pi - k\pi x_{1,i}) \\ &= (-1)^{i+1}[\boldsymbol{v}_1^{(k)}]_i. \end{aligned}$$

□

**Remark 4.8.3** (Aliasing). Identity (4.8.1) quantifies the phenomenon known as *aliasing*, that is, high-frequency modes (sine waves) are distorted on a coarse grid. ◂

**Remark 4.8.4** (Terminology). Strictly speaking, in Lemma 4.8.2, only the first result is an aliasing result. We will abuse terminology and refer to any result where a trigonometric function of an argument containing the reciprocal wave number is related to that of a trigonometric function of an argument containing the original wave number as an *aliasing-type* result. ◂

**Proposition 4.8.5** (Restriction modal analysis). *Suppose that the two-level grid is uniform and nested. Then*

$$\mathsf{R}_0 \boldsymbol{v}_1^{(k)} = \begin{cases} 2\mathsf{C}_k \boldsymbol{v}_0^{(k)}, & 1 \le k \le n_0, \\ -2\mathsf{S}_{k'} \boldsymbol{v}_0^{(k')}, & n_0 + 1 \le k \le n_1. \end{cases} \tag{4.8.4}$$

*Proof.* (Low-frequency components): For $1 \le i, k \le n_0$, consider

$$\begin{aligned}[\mathsf{R}_0 \boldsymbol{v}_1^{(k)}]_i &= \frac{1}{2}[\sin(k\pi x_{1,2i-1}) + 2\sin(k\pi x_{1,2i}) + \sin(k\pi x_{1,2i+1})] \\ &= \frac{1}{2}[\sin(k\pi x_{1,2i} - k\pi h_1) + 2\sin(k\pi x_{1,2i}) + \sin(k\pi x_{1,2i} + k\pi h_1)] \\ &= \frac{1}{2}[\sin(k\pi x_{1,2i})\cos(k\pi h_1) - \cos(k\pi x_{1,2i})\sin(k\pi h_1) \\ &\quad + \sin(k\pi x_{1,2i})\cos(k\pi h_1) + \cos(k\pi x_{1,2i})\sin(k\pi h_1) + 2\sin(k\pi x_{1,2i})] \\ &= \sin(k\pi x_{1,2i})\cos(k\pi h_1) + \sin(k\pi x_{1,2i}) \\ &= \sin(k\pi x_{0,i})(\cos(k\pi h_1) + 1) \\ &= 2\cos^2\left(\frac{k\pi h_1}{2}\right)\sin(k\pi x_{0,i}) \\ &= 2\mathsf{C}_k[\boldsymbol{v}_0^{(k)}]_i. \end{aligned}$$

Here we have used the simple fact that

$$x_{1,2i} = x_{0,i}, \quad 0 \le i \le n_0 + 1.$$

The first part is complete.
(High-frequency components): The first part of the calculation above is still valid for $n_0 + 1 \le k \le n_1$. We now use the aliasing results (4.8.1) and (4.8.2) from the technical lemma to obtain

$$\begin{aligned}
[\mathsf{R}_0 \boldsymbol{v}_1^{(k)}]_i &= \frac{1}{2}[\sin(k\pi x_{1,2i-1}) + 2\sin(k\pi x_{1,2i}) + \sin(k\pi x_{1,2i+1})] \\
&= \sin(k\pi x_{1,2i})(\cos(k\pi h_1) + 1) \\
&= (\cos(k\pi h_1) + 1)[\boldsymbol{v}_0^{(k)}]_i \\
&= -(1 - \cos(k'\pi h_1))[\boldsymbol{v}_0^{(k')}]_i \\
&= -2\sin^2\left(\frac{k'\pi h_1}{2}\right)[\boldsymbol{v}_0^{(k')}]_i \\
&= -2\mathsf{S}_{k'}[\boldsymbol{v}_0^{(k')}]_i.
\end{aligned}$$

The result is proved. □

**Proposition 4.8.6** (Prolongation modal analysis). *Suppose that the two-level grid is uniform and nested. For* $1 \le k \le n_0$,

$$\mathsf{P}_0 \boldsymbol{v}_0^{(k)} = \mathsf{C}_k \boldsymbol{v}_1^{(k)} - \mathsf{S}_k \boldsymbol{v}_1^{(k')}. \tag{4.8.5}$$

*Proof.* For $1 \le i \le n_1$, define

$$j := \begin{cases} \dfrac{i+1}{2}, & i \notin 2\mathbb{N}, \\ \dfrac{i}{2}, & i \in 2\mathbb{N}. \end{cases}$$

Using this and the aliasing result (4.8.3), working with the right-hand side of the desired result, for $1 \le i \le n_1$, and $1 \le k \le n_0$, it follows that

$$\begin{aligned}
I_i^{(k)} &:= \mathsf{C}_k[\boldsymbol{v}_1^{(k)}]_i - \mathsf{S}_k[\boldsymbol{v}_1^{(k')}]_i \\
&\overset{(4.8.3)}{=} \begin{cases} \cos(k\pi h_1)\sin(k\pi x_{1,i}), & i \notin 2\mathbb{N}, \\ \sin(k\pi x_{1,i}), & i \in 2\mathbb{N}, \end{cases} \\
&= \begin{cases} \cos(k\pi h_1)\sin(k\pi x_{1,2j-1}), & i \notin 2\mathbb{N}, \\ \sin(k\pi x_{1,2j}), & i \in 2\mathbb{N}, \end{cases} \\
&= \begin{cases} \frac{1}{2}\sin(k\pi x_{1,2j-1} + k\pi h_1) + \frac{1}{2}\sin(k\pi x_{1,2j-1} - k\pi h_1), & i \notin 2\mathbb{N}, \\ \sin(k\pi x_{1,2j}), & i \in 2\mathbb{N}, \end{cases}
\end{aligned}$$

$$= \begin{cases} \frac{1}{2}\sin(k\pi x_{1,2j}) + \frac{1}{2}\sin(k\pi x_{1,2j-2}), & i \notin 2\mathbb{N}, \\ \sin(k\pi x_{1,2j}), & i \in 2\mathbb{N}, \end{cases}$$

$$= \begin{cases} [\mathsf{P}_0 \boldsymbol{v}_0^{(k)}]_{2j-1}, & i \notin 2\mathbb{N}, \\ [\mathsf{P}_0 \boldsymbol{v}_0^{(k)}]_{2j}, & i \in 2\mathbb{N}. \end{cases}$$

Thus, for $1 \le i \le n_1$, and $1 \le k \le n_0$,

$$I_i^{(k)} = \mathsf{C}_k[\boldsymbol{v}_1^{(k)}]_i - \mathsf{S}_k[\boldsymbol{v}_1^{(k')}]_i = [\mathsf{P}_0 \boldsymbol{v}_0^{(k)}]_i.$$

The reader should provide the missing details; see Problem 4.7. □

We list now a couple more aliasing-type results that will be needed in the analysis. The proofs are straightforward and omitted.

**Lemma 4.8.7** (Aliasing). *Let the two-level grid be uniform and nested. Then*

$$\sin(k'\pi h_1) = \sin(k\pi h_1) \tag{4.8.6}$$

*and*

$$\sin\left(\frac{k'\pi h_1}{2}\right) = \cos\left(\frac{k\pi h_1}{2}\right) \tag{4.8.7}$$

*for all* $1 \le k \le n_1$.

*Proof.* Exercise; see Problem 4.8. □

**Proposition 4.8.8** (Ritz projection modal analysis). *Suppose that the two-level grid is uniform and nested and* $\mathsf{A}_0$ *satisfies the Galerkin condition. Then*

$$(\mathrm{I}_1 - \tilde{\Pi}_1)\boldsymbol{v}_1^{(k)} = \mathsf{S}_k \boldsymbol{v}_1^{(k)} + \mathsf{S}_k \boldsymbol{v}_1^{(k')} \tag{4.8.8}$$

*for* $1 \le k \le n_0$. *Similarly,*

$$(\mathrm{I}_1 - \tilde{\Pi}_1)\boldsymbol{v}_1^{(k')} = \mathsf{C}_k \boldsymbol{v}_1^{(k)} + \mathsf{C}_k \boldsymbol{v}_1^{(k')} \tag{4.8.9}$$

*for* $1 \le k \le n_0 + 1$.

*Proof.* Recall that

$$\tilde{\Pi}_1 = \mathsf{P}_0 \mathsf{A}_0^{-1} \mathsf{R}_0 \mathsf{A}_1 \in \mathbb{R}^{n_1 \times n_1},$$

where, since $\mathsf{A}_0$ satisfies the Galerkin condition,

$$\mathsf{A}_0 = \mathsf{R}_0 \mathsf{A}_1 \mathsf{P}_0 \in \mathbb{R}^{n_0 \times n_0}.$$

Then, for $1 \le k \le n_0$,

$$\begin{aligned}\tilde{\Pi}_1 \boldsymbol{v}_1^{(k)} &= \mathsf{P}_0\mathsf{A}_0^{-1}\mathsf{R}_0\mathsf{A}_1\boldsymbol{v}_1^{(k)} \\ &= \lambda_1^{(k)}\mathsf{P}_0\mathsf{A}_0^{-1}\mathsf{R}_0\boldsymbol{v}_1^{(k)} \\ &\overset{(4.8.4)}{=} 2\mathsf{C}_k\lambda_1^{(k)}\mathsf{P}_0\mathsf{A}_0^{-1}\boldsymbol{v}_0^{(k)} \\ &= \frac{2\mathsf{C}_k\lambda_1^{(k)}}{\lambda_0^{(k)}}\mathsf{P}_0\boldsymbol{v}_0^{(k)} \\ &\overset{(4.8.5)}{=} \frac{2\mathsf{C}_k\lambda_1^{(k)}}{\lambda_0^{(k)}}\left[\mathsf{C}_k\boldsymbol{v}_1^{(k)} - \mathsf{S}_k\boldsymbol{v}_1^{(k')}\right].\end{aligned}$$

Observe that

$$\begin{aligned}\frac{\lambda_1^{(k)}}{\lambda_0^{(k)}} &= \frac{\frac{2}{h_1}(1-\cos(k\pi h_1))}{\frac{2}{h_0}(1-\cos(k\pi h_0))} \\ &= \frac{h_0}{h_1}\frac{\sin^2(\frac{k\pi h_1}{2})}{\sin^2(k\pi h_1)} \\ &= \frac{2\mathsf{S}_k}{\sin^2(k\pi h_1)}.\end{aligned}$$

Thus

$$\tilde{\Pi}_1 \boldsymbol{v}_1^{(k)} = \frac{4\mathsf{C}_k\mathsf{S}_k}{\sin^2(k\pi h_1)}\left[\mathsf{C}_k\boldsymbol{v}_1^{(k)} - \mathsf{S}_k\boldsymbol{v}_1^{(k')}\right].$$

However,

$$\mathsf{C}_k\mathsf{S}_k = \frac{1}{4}\sin^2(k\pi h_1),$$

as the reader can easily verify. So we have

$$\tilde{\Pi}_1 \boldsymbol{v}_1^{(k)} = \mathsf{C}_k\boldsymbol{v}_1^{(k)} - \mathsf{S}_k\boldsymbol{v}_1^{(k')}, \quad 1 \le k \le n_0.$$

Therefore, for $1 \le k \le n_0$,

$$\begin{aligned}(\mathrm{I}_1 - \tilde{\Pi}_1)\boldsymbol{v}_1^{(k)} &= (1-\mathsf{C}_k)\boldsymbol{v}_1^{(k)} + \mathsf{S}_k\boldsymbol{v}_1^{(k')} \\ &= \mathsf{S}_k\left(\boldsymbol{v}_1^{(k)} + \boldsymbol{v}_1^{(k')}\right),\end{aligned}$$

which is (4.8.8).

Equation (4.8.9) is established in an analogous way. We must use the aliasing results (4.8.6) and (4.8.7); see Problem 4.10. □

Using the aliasing results again, we can establish the following equivalent identity.

**Remark 4.8.9** (Ritz projection modal analysis). Suppose that the two-level grid is uniform and nested and $\mathsf{A}_0$ is constructed via the Galerkin condition. Then

$$(\mathrm{I}_1 - \tilde{\Pi}_1)\boldsymbol{v}_1^{(k)} = \mathsf{S}_k \boldsymbol{v}_1^{(k)} + \mathsf{S}_k \boldsymbol{v}_1^{(k')} \tag{4.8.10}$$

for $1 \le k \le n_1$, that is, for all frequencies, high and low. ◂

We now state a convergence result for the one-sided two-grid algorithm, that is, the usual two-grid algorithm with no postsmoothing, $m_2 = 0$. This nonsymmetric version of the algorithm, as we will see, works perfectly for the quantitative Fourier analysis we use. Other forms of analysis will require the symmetry.

**Theorem 4.8.10** (Convergence of the one-sided two-grid algorithm). *Suppose that the two-level grid is uniform and nested and* $\mathsf{A}_0$ *satisfies the Galerkin condition. Assume that* $m_2 = 0$, $\omega_0 = \frac{2}{3}$, *and* $m_1 \ge 1$. *Then, for* $\ell \ge 0$,

$$\|\boldsymbol{e}_1^{\ell+1}\|_1 \le \gamma(m_1)\|\boldsymbol{e}_1^{\ell}\|_1,$$

*where*

$$\gamma(m_1) := \left\{\frac{1}{2} + \left(\frac{1}{3}\right)^{m_1}\right\}^{1/2}. \tag{4.8.11}$$

*Proof.* We begin with the basis expansion

$$\boldsymbol{e}_1^{\ell} = \sum_{k=1}^{n_1} \epsilon_k \boldsymbol{v}_1^{(k)}.$$

The error transfer matrix $\mathsf{E}_1$ for the two-grid scheme, in this case, satisfies

$$\boldsymbol{e}_1^{\ell+1} = \mathsf{E}_1 \boldsymbol{e}_1^{\ell},$$

where

$$\mathsf{E}_1 = (\mathrm{I}_1 - \tilde{\Pi}_1)\mathsf{K}_1^{m_1}.$$

Hence

$$\boldsymbol{e}_1^{\ell+1} = \sum_{k=1}^{n_1} \epsilon_k (\mathrm{I}_1 - \tilde{\Pi}_1)\mathsf{K}_1^{m_1}\boldsymbol{v}_1^{(k)}.$$

By Proposition 4.8.8, for $1 \le k \le n_0 + 1$,

$$(\mathrm{I}_1 - \tilde{\Pi}_1)\mathsf{K}_1^{m_1}\boldsymbol{v}_1^{(k)} = \alpha_k\{\boldsymbol{v}_1^{(k)} + \boldsymbol{v}_1^{(k')}\}, \tag{4.8.12}$$

and, for $1 \le k \le n_0 + 1$,

$$(\mathrm{I}_1 - \tilde{\Pi}_1)\mathrm{K}_1^{m_1}\boldsymbol{v}_1^{(k')} = \beta_k\{\boldsymbol{v}_1^{(k)} + \boldsymbol{v}_1^{(k')}\}, \tag{4.8.13}$$

where, for $1 \le k \le n_0 + 1$,

$$\alpha_k := (\mu_1^{(k)}(\omega_0))^{m_1}\mathrm{s}_k, \quad \beta_k := (\mu_1^{(k')}(\omega_0))^{m_1}\mathrm{c}_k.$$

Observe that we have extended the upper limit of the index $k$ in (4.8.12) by 1 up to $k = n_0 + 1$. The result is still valid in this case. Further, observe that

$$\alpha_{n_0+1} = \beta_{n_0+1},$$

as the reader can easily verify.

By Theorem 4.5.6, for the high frequencies $n_0 + 1 \le k \le n_1$, we have

$$|\mu_1^{(k)}(\omega_0)| \le \frac{1}{3},$$

and, for all frequencies $1 \le k \le n_1$,

$$|\mu_1^{(k)}(\omega_0)| < 1.$$

We can now estimate $\alpha_k$ and $\beta_k$ as follows:

$$|\alpha_k| \le 1^{m_1}\mathrm{s}_k \le 1 \cdot \frac{1}{2}, \quad 1 \le k \le n_0 + 1,$$

and

$$|\beta_k| \le \left(\frac{1}{3}\right)^{m_1}\mathrm{c}_k \le \left(\frac{1}{3}\right)^{m_1} \cdot 1, \quad 1 \le k \le n_0 + 1.$$

We can now represent the error at iteration $\ell$ as

$$\boldsymbol{e}_1^\ell = \sum_{k=1}^{n_0+1} \delta_k(\epsilon_k\boldsymbol{v}_1^{(k)} + \epsilon_{k'}\boldsymbol{v}_1^{(k')}),$$

where

$$\delta_k = \begin{cases} \frac{1}{2}, & k = n_0 + 1, \\ 1, & 1 \le k \le n_0, \end{cases}$$

which reconciles the double counting of the term indexed by $k = n_0 + 1 = k'$. Then

$$\boldsymbol{e}_1^{\ell+1} = \mathrm{E}_1\boldsymbol{e}_1^\ell$$

$$= \sum_{k=1}^{n_0+1} \delta_k (\epsilon_k \alpha_k + \epsilon_{k'} \beta_k)(\boldsymbol{v}_1^{(k)} + \boldsymbol{v}_1^{(k')}).$$

Now notice that the eigenvectors satisfy the orthogonality condition

$$\begin{aligned}(\boldsymbol{v}_1^{(i)}, \boldsymbol{v}_1^{(j)})_1 &= \sum_{m=1}^{n_1} \sin(i\pi x_{1,m}) \sin(j\pi x_{1,m}) \\ &= \delta_{ij} \frac{n_1+1}{2}.\end{aligned}$$

Using this orthogonality, we can prove that

$$\|\boldsymbol{e}_1^\ell\|_1^2 = \frac{n_1+1}{2} \sum_{k=1}^{n_0+1} \delta_k (\epsilon_k^2 + \epsilon_{k'}^2) \tag{4.8.14}$$

and

$$\begin{aligned}\|\boldsymbol{e}_1^{\ell+1}\|_1^2 &= (\boldsymbol{e}_1^{\ell+1}, \boldsymbol{e}_1^{\ell+1})_1 \\ &= (\mathsf{E}_1 \boldsymbol{e}_1^\ell, \mathsf{E}_1 \boldsymbol{e}_1^\ell)_1 \\ &= (n_1+1) \sum_{k=1}^{n_0+1} \delta_k (\epsilon_k^2 \alpha_k^2 + \epsilon_{k'}^2 \beta_k^2 + 2\epsilon_k \epsilon_{k'} \alpha_k \beta_k).\end{aligned} \tag{4.8.15}$$

We leave it as an exercise for the reader to show these results; see Problem 4.11.

Using the arithmetic–geometric mean inequality, we can now estimate

$$\begin{aligned}\|\boldsymbol{e}_1^{\ell+1}\|_1^2 &\le (n_1+1) \sum_{k=1}^{n_0+1} \delta_k [\epsilon_k^2 \alpha_k^2 + \epsilon_{k'}^2 \beta_k^2 + (\epsilon_k^2 + \epsilon_{k'}^2)|\alpha_k||\beta_k|] \\ &\le (n_1+1) \sum_{k=1}^{n_0+1} \delta_k \left[\epsilon_k^2 \left(\frac{1}{2}\right)^2 + \epsilon_{k'}^2 \left(\frac{1}{3}\right)^{2m_1} + (\epsilon_k^2 + \epsilon_{k'}^2)\frac{1}{2}\left(\frac{1}{3}\right)^{m_1}\right] \\ &= \frac{n_1+1}{2} \sum_{k=1}^{n_0+1} \delta_k \left[\epsilon_k^2 \left[\frac{1}{2} + \left(\frac{1}{3}\right)^{m_1}\right] + \epsilon_{k'}^2 \left[2\left(\frac{1}{3}\right)^{2m_1} + \left(\frac{1}{3}\right)^{m_1}\right]\right].\end{aligned}$$

Since

$$\left(\frac{1}{3}\right)^{m_1} \le \frac{1}{2},$$

we have

$$\begin{aligned}\|\boldsymbol{e}_1^{\ell+1}\|_1^2 &\le \frac{n_1+1}{2} \left\{\frac{1}{2} + \left(\frac{1}{3}\right)^{m_1}\right\} \sum_{k=1}^{n_0+1} \delta_k (\epsilon_k^2 + \epsilon_{k'}^2) \\ &= \left\{\frac{1}{2} + \left(\frac{1}{3}\right)^{m_1}\right\} \|\boldsymbol{e}_1^\ell\|_1^2.\end{aligned}$$

Finally,

$$\|e_1^{\ell+1}\|_1 \leq \gamma(m_1)\|e_1^{\ell}\|_1,$$

where $\gamma(m_1)$ was defined in (4.8.11). □

**Remark 4.8.11** (Contraction factor $\gamma(m_1)$). The key observation regarding $\gamma(m_1)$ is that it does not depend on $h_1$; see Table 4.2 for guidance on choosing $m_1$. ◀

**Table 4.2:** Numerical values of $\gamma(m_1)$ defined in (4.8.11) for certain values of the number $m_1$ of presmoothings. As we see, it does not make much sense to use $m_1 > 3$, since the geometric convergence rate does not significantly improve.

| $m_1$ | 1 | 2 | 3 | ⋯ | ∞ |
|---|---|---|---|---|---|
| $\gamma(m_1)$ | 0.91287 | 0.78176 | 0.73283 | ⋯ | 0.70710 |

## 4.9 Some computational experiments

To conclude this chapter, let us perform some computational experiments to confirm the predicted convergence results. In particular, let us use the one-sided two-grid algorithm to approximate the solution of

$$\mathsf{A}_1 \boldsymbol{u}_1^{\mathrm{E}} = \boldsymbol{f}_1,$$

where $\mathsf{A}_1$ is the standard finite element stiffness matrix defined in (4.1.9) for the model problem in (1.7.1). In our experiments, we specify the exact solution

$$[\boldsymbol{u}_1^{\mathrm{E}}]_i = u_{1,i}^{\mathrm{E}} = \exp(\sin(3\pi x_{1,i})) - 1.$$

We manufacture the force vector by setting $\boldsymbol{f}_1 := \mathsf{A}_1 \boldsymbol{u}_1^{\mathrm{E}}$. The initial approximation $\boldsymbol{u}_1^0$ is chosen via pseudorandom number selection. The code implementing the two-grid algorithm is given in Listings 4.4 and 4.5.

### First experiment: reference

In the first experiment, we use the parameters $\omega = \frac{2}{3}$ and $m_1 = 3$. The results are presented in Figures 4.14–4.16. The error is plotted as a function of the iteration number $k$ in Figure 4.14. The computed error contraction factor $\gamma_{\mathrm{comp}}$ is much better than the theoretically predicted value, almost 10 times smaller. This contraction factor is computed using a log-linear fit of the last four error values,

$$\|\boldsymbol{u}_1^{\mathrm{E}} - \boldsymbol{u}_1^{\ell}\|_1, \quad \ell = k_\star - 3, k_\star - 2, k_\star - 1, k_\star,$$

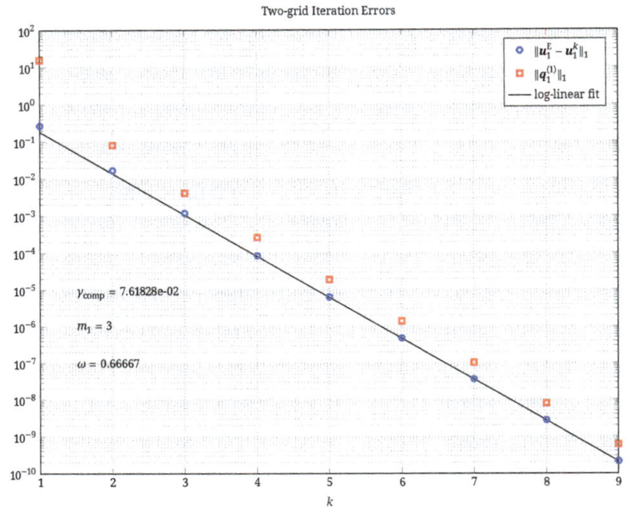


**Figure 4.14:** The error contraction for the one-sided two-grid algorithm, using the parameters $\omega = \frac{2}{3}$, $m_1 = 3$, and $n_1 = 255$. The computed error contraction factor, $\gamma_{comp}$, is much better than the theoretically predicted value, almost 10 times smaller. The factor $\gamma_{comp}$ is computed using a log-linear fit of the last four error values. Note that the norm of the coarse grid correction is a good indicator of the error.

where $k_\star$ is the value of $k$ at which the stopping tolerance is reached. For our experiments, the stopping tolerance is achieved when

$$\|\boldsymbol{u}_1^{\mathrm{E}} - \boldsymbol{u}_1^{k_\star}\|_1 \le 10^{-9}.$$

We expect that the error contraction factor will satisfy

$$\|\boldsymbol{u}_1^{\mathrm{E}} - \boldsymbol{u}_1^{\ell}\|_1 \approx \gamma_{\mathrm{comp}}\|\boldsymbol{u}_1^{\mathrm{E}} - \boldsymbol{u}_1^{\ell-1}\|_1, \quad \ell = k_\star - 2, k_\star - 1, k_\star.$$

Measuring error with the exact error is not practical. However, our computations show that the norm of the coarse grid correction is a good indicator of the error.

In Figures 4.15 and 4.16, we show the error function after presmoothing, $u_1^{\mathrm{E}}(x_{1,i}) - u_1^{(1)}(x_{1,i})$, and after coarse grid correction, $u_1^{\mathrm{E}}(x_{1,i}) - u_1^{(2)}(x_{1,i})$, for iterations from $k = 1$ to $k = 4$. Notice that, as the name suggests, the smoothing operation actually “smooths out” the error. Applying the coarse grid correction then drastically reduces the error, but it reintroduces some high-frequency error. These are later squashed by smoothing, and so on.

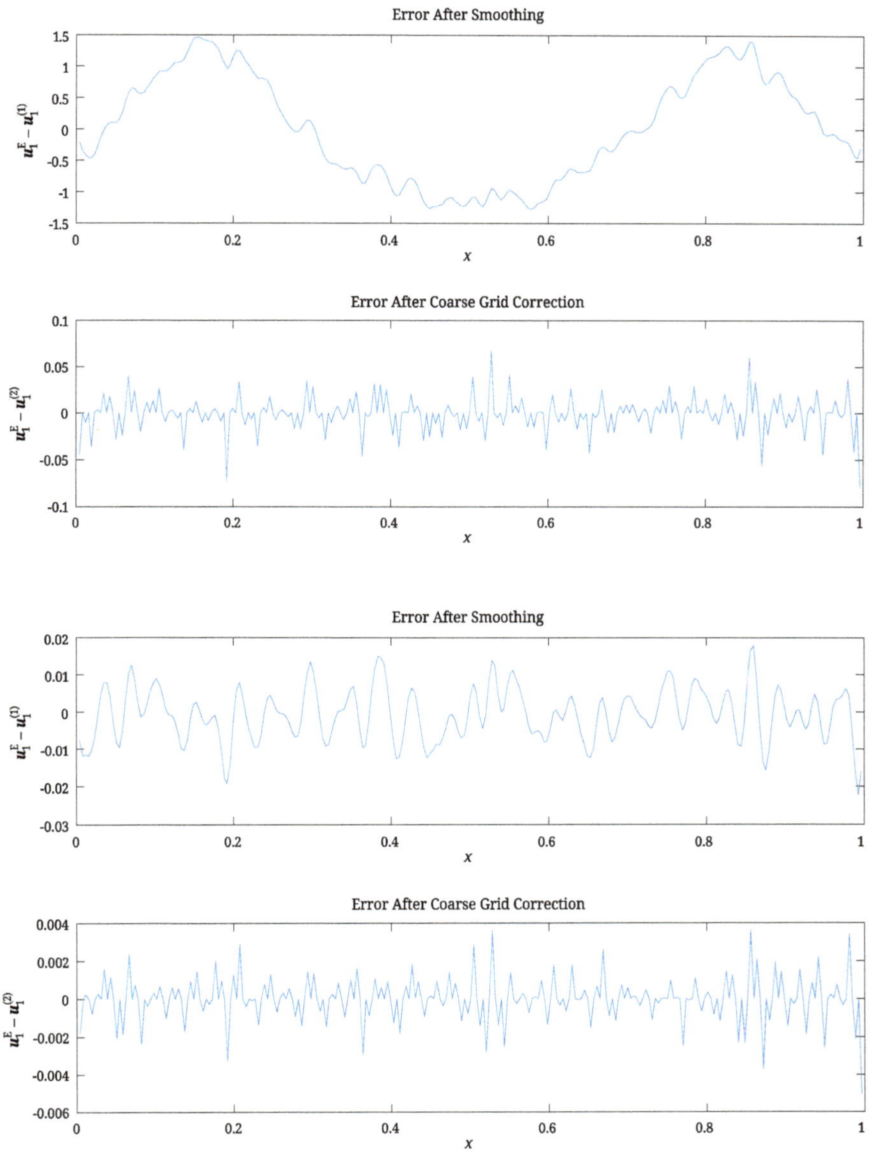


**Figure 4.15:** The errors after presmoothing and after coarse grid correction for the one-sided two-grid algorithm at iterations $k = 1$ (top two plots) and $k = 2$ (bottom two plots), using the parameters $\omega = \frac{2}{3}$, $m_1 = 3$, and $n_1 = 255$; see Figure 4.14.

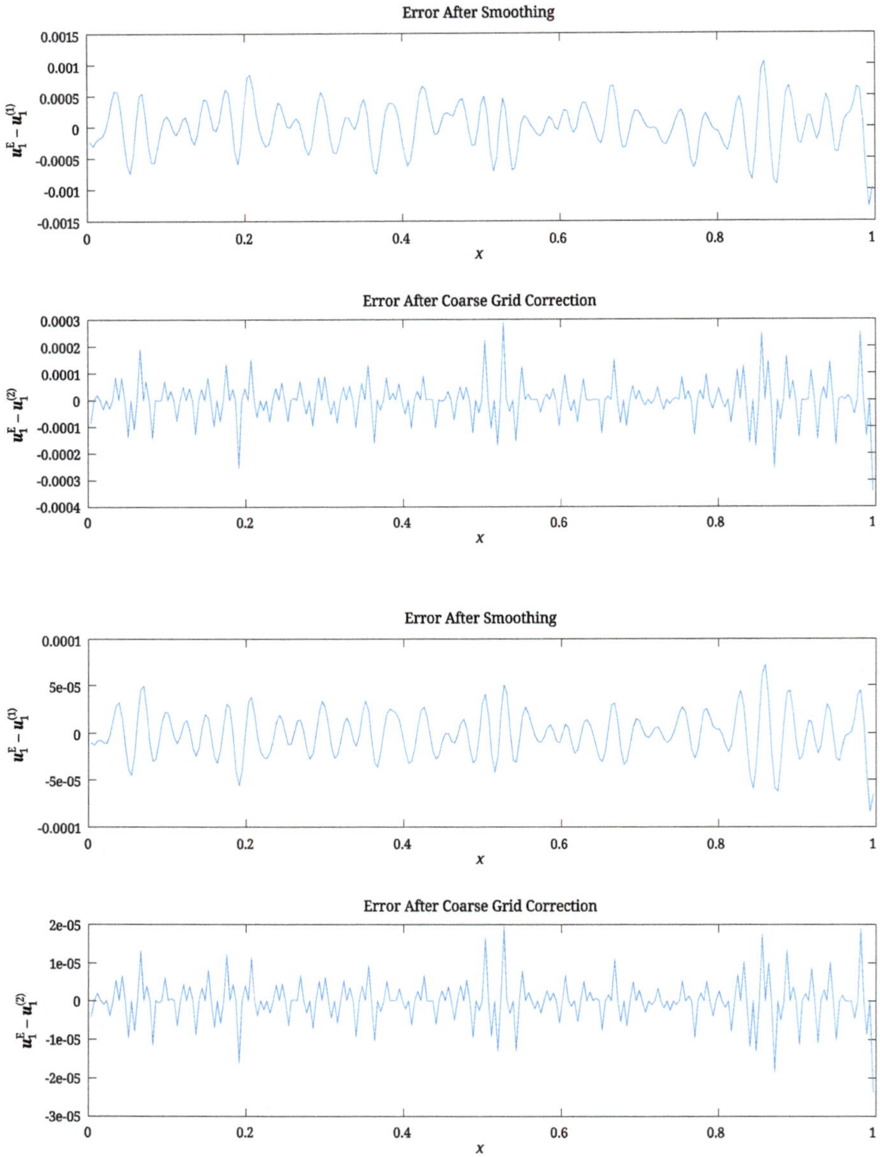


**Figure 4.16:** The errors after presmoothing and after coarse grid correction for the one-sided two-grid algorithm at iterations $k = 3$ (top two plots) and $k = 4$ (bottom two plots) using the parameters $\omega = \frac{2}{3}$, $m_1 = 3$, and $n_1 = 255$; see Figure 4.14.

### Second experiment: large number of smoothing steps

We now use the parameters $\omega = \frac{2}{3}$ and $m_1 = 10$. Figure 4.17 shows the error as a function of the iteration number $k$. The computed error contraction factor $\gamma_{\mathrm{comp}}$ is calculated as in the first experiment, and it turns out to be smaller; see Figure 4.14 for comparison. Doing more smoothing iterations reduces the error contraction factor $\gamma_{\mathrm{comp}}$, but there is a diminishing return with this effort.

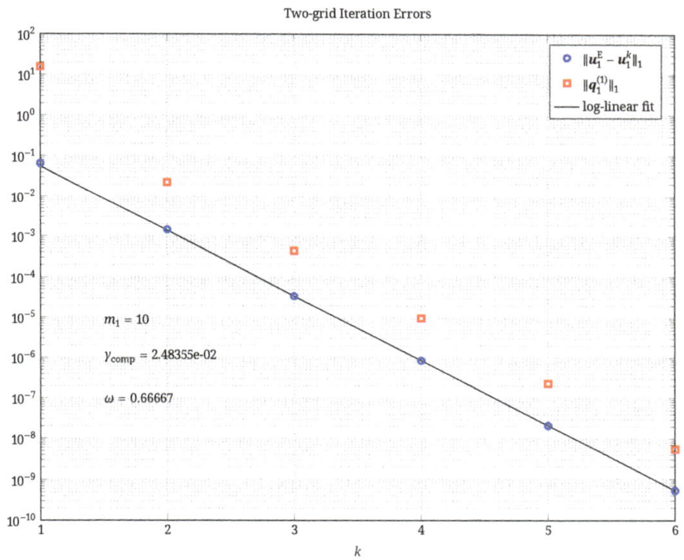


**Figure 4.17:** The error contraction for the one-sided two-grid algorithm using the parameters $\omega = \frac{2}{3}$, $m_1 = 10$, and $n_1 = 255$. The factor $\gamma_{\mathrm{comp}}$ is computed using a log-linear fit of the last four error values. Compare this rate with that from Figure 4.14.

### Third experiment: suboptimal relaxation

For the third experiment, we use the parameters $\omega = \frac{1}{2}$ and $m_1 = 3$. See Figure 4.18. The computed error contraction factor $\gamma_{\mathrm{comp}}$ is larger than that of the first experiment. In the next chapter, we will discuss the use of the parameter $\omega = \frac{1}{2}$ in the damped Jacobi smoother; see Table 4.3 for the computed error contraction factors $\gamma_{\mathrm{comp}}$ for several different values of $\omega$.

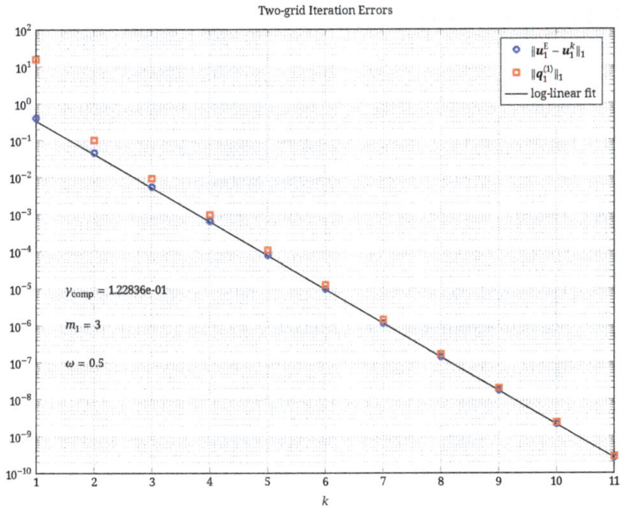


**Figure 4.18:** The error contraction for the one-sided two-grid algorithm using the parameters $\omega = \frac{1}{2}$, $m_1 = 3$, and $n_1 = 255$. The factor $\gamma_{comp}$ is computed using a log-linear fit of the last four error values. Compare this rate with that from Figure 4.14. It appears that this value of $\omega$ is not optimal.

**Table 4.3:** Computed two-grid convergence factors $\gamma_{comp}$ for various parameter choices. The factor $\gamma_{comp}$ is computed using a log-linear fit of the last four error values $\|u_1^E - u_1^k\|_1$.

| $n_1$ | $\omega$ | $m_1$ | $\gamma_{comp}$ |
|---:|---:|---:|---:|
| 63 | 2/3 | 3 | $7.56 \times 10^{-2}$ |
| 127 | 2/3 | 3 | $7.52 \times 10^{-2}$ |
| 255 | 2/3 | 3 | $7.62 \times 10^{-2}$ |
| 511 | 2/3 | 3 | $7.56 \times 10^{-2}$ |
| 1023 | 2/3 | 3 | $7.50 \times 10^{-2}$ |
| 127 | 2/3 | 5 | $4.90 \times 10^{-2}$ |
| 127 | 2/3 | 7 | $3.55 \times 10^{-2}$ |
| 127 | 2/3 | 9 | $2.69 \times 10^{-2}$ |
| 127 | 2/3 | 11 | $2.22 \times 10^{-2}$ |
| 127 | 2/3 | 13 | $1.92 \times 10^{-2}$ |
| 127 | 2/3 | 15 | $1.67 \times 10^{-2}$ |
| 127 | 0.40 | 3 | $2.10 \times 10^{-1}$ |
| 127 | 0.45 | 3 | $1.61 \times 10^{-1}$ |
| 127 | 0.50 | 3 | $1.23 \times 10^{-1}$ |
| 127 | 0.55 | 3 | $9.74 \times 10^{-2}$ |
| 127 | 0.60 | 3 | $8.44 \times 10^{-2}$ |
| 127 | 0.65 | 3 | $7.74 \times 10^{-2}$ |
| 127 | 0.70 | 3 | $6.93 \times 10^{-2}$ |
| 127 | 0.75 | 3 | $1.22 \times 10^{-1}$ |
| 127 | 0.80 | 3 | $2.13 \times 10^{-1}$ |

### Fourth experiment: very large number of smoothings

The fourth experiment uses the parameters $\omega = \frac{2}{3}$ and $m_1 = 20$. The error is plotted as a function of the iteration number $k$ in Figure 4.19. The computed error contraction factor $\gamma_{comp}$ is calculated as in the first experiment, and it turns out to be smaller than that computed for the first two experiments. Note, however, that doing 20 presmoothing steps is not practical. However, this experiment shows the (exaggerated) smoothing effect of the damped Jacobi method with a large number of presmoothing steps; see Figures 4.20 and 4.21.

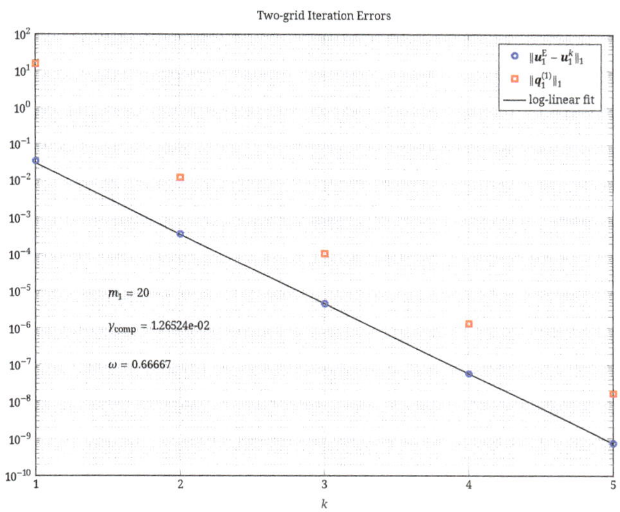


**Figure 4.19:** The error contraction for the one-sided two-grid algorithm using the parameters $\omega = \frac{2}{3}$, $m_1 = 20$, and $n_1 = 255$. The factor $\gamma_{comp}$ is computed using a log-linear fit of the last four error values. Compare this rate with those from Figures 4.14 and 4.18. There is a diminishing return for doing more presmoothing steps. The convergence factor $\gamma_{comp}$ is reduced, but not much; see Table 4.3.

Finally, we comment that doing experiments like those we have presented here, where the solution is known, can help users tune the parameters in their codes for close-to-optimal performance.

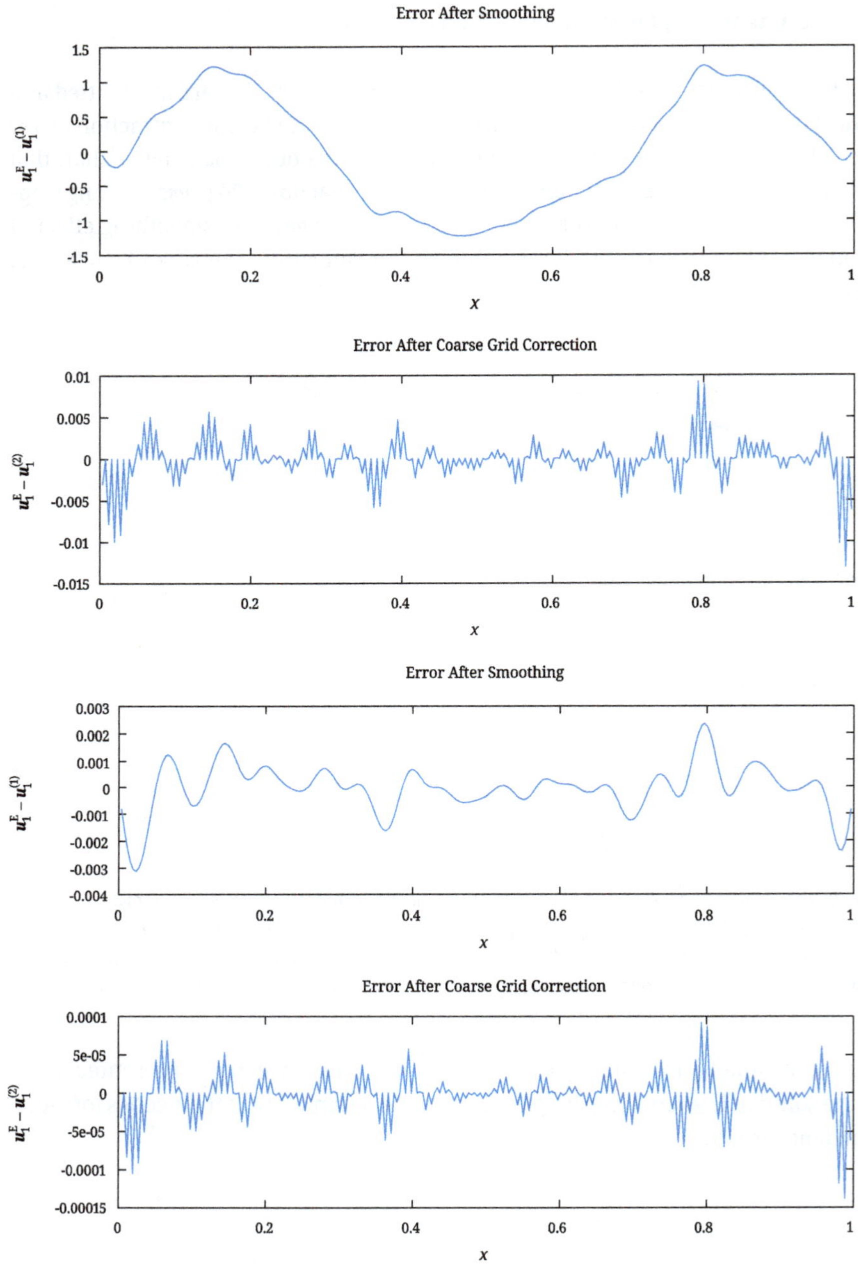


**Figure 4.20:** The errors after presmoothing and after coarse grid correction for the one-sided two-grid algorithm at iterations $k = 1$ (top two plots) and $k = 2$ (bottom two plots) using the parameters $\omega = \frac{2}{3}$, $m_1 = 20$, and $n_1 = 255$. This is an exaggerated number of presmoothing steps. Compare with Figure 4.15.

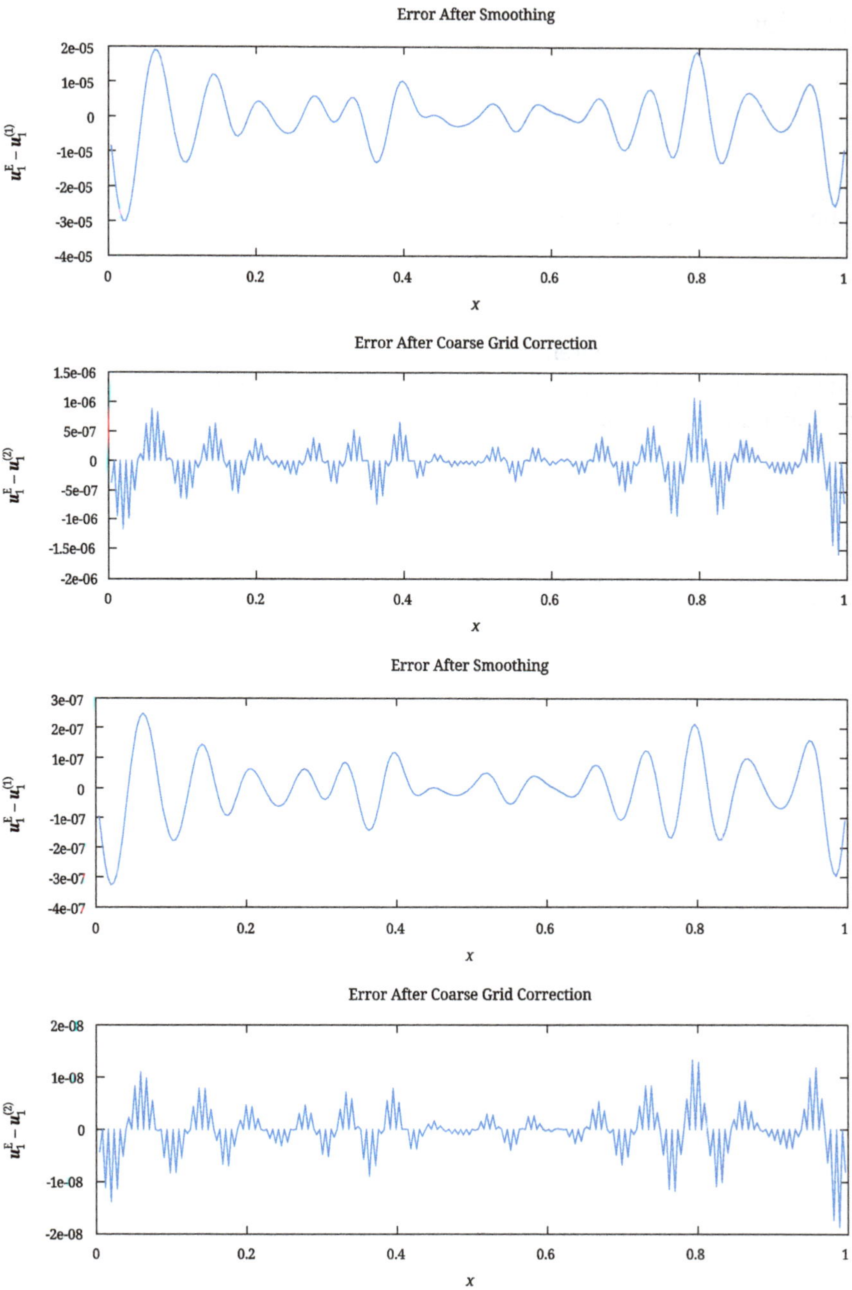


**Figure 4.21:** The errors after presmoothing and after coarse grid correction for the one-sided two-grid algorithm at iterations $k = 3$ (top two plots) and $k = 4$ (bottom two plots) using the parameters $\omega = \frac{2}{3}$, $m_1 = 20$, and $n_1 = 255$. This is an exaggerated number of presmoothing steps. Compare with Figure 4.16. We use this exaggeration to show the error "smoothing" effect of the presmoothing operation.

## 4.10 Problems

4.1 Prove Proposition 4.1.2.

4.2 Complete the proof of Theorem 4.4.1.

4.3 Prove Proposition 4.6.3.

4.4 Prove Corollary 4.6.8.

4.5 Complete the proof of Theorem 4.6.9.

4.6 Prove Corollary 4.6.10.

4.7 Fill in the missing details in the proof of Proposition 4.8.6.

4.8 Prove Lemma 4.8.7.

4.9 Complete the proof of Theorem 4.6.6.

4.10 Complete the proof of Proposition 4.8.8 by establishing the validity of (4.8.9).

4.11 Prove identities (4.8.14) and (4.8.15).

## Listings

```
./1DFEM/
    poisson1dTG.m
    twoGridSolver.m
    coarseGridCorrection.m
    prolongation.m
    restriction.m
    smoothJacobiDamped.m
    smoothGaussSeidel.m
    getResidual.m
```

**Listing 4.4:** `poisson1dTG.m`; Calling script for the two-grid solver.

```
%
% poisson1DTG.m
%
% Two-grid calling script for solving the 1D Poisson
% equation
%
%   -u'' = f
%
% on an interval with homogeneous Dirichlet boundary
% conditions.
%
%% Initialization and Parameter

% Clear command line window and workspace.
clear; clc;

% Tolerance of residual to stop two-grid iterations.
tol = 1.0e-09;

% Number of interior cells at finest level.
n1 = 255;

% Define the grid spacing.
h1 = 1/(n1+1);

% Parameters for the two-grid solver.
m1     = 3;
omega = 2/3;
```

```
kMax   = 100;

%% Setting up the solver

% Seed a random initial guess.
rng(1234);
u = rand(n1,1)-0.5;

% Construct the exact solution and necessary forcing
% vector.
x = linspace(h1,1.0-h1,n1)';
A1 = gallery('tridiag',n1,-1.0/h1,2.0/h1,-1.0/h1);
uExact = exp(sin(3.0*pi*x))-1.0;
f = A1*uExact;

%% Calling the two-grid solver

tic
[u,errVals,kStop] = twoGridSolver(u,f,n1,m1,omega,...
                                  kMax,tol,uExact);
toc

finalErr = norm(uExact-u,2);
fprintf('\nFinal error in L2 sense = %g\n',finalErr);

%% Plotting the numerical solution and error plots

figure(1)
clf
plot(x,u,'k-','LineWidth',1.5)
xlabel('$x$','Interpreter','latex');
ylabel('$u_1(x)$','Interpreter','latex');
title(['Finite Element Approximation of the ',...
       'Poisson Equation']);
axis([0.0,1.0,-1.0,2.0])
printstr = strcat('Solution_n1_',num2str(n1),'.pdf');
exportgraphics(gca,printstr)

figure(2)
clf
semilogy(errVals(1:kStop,3),'bo','LineWidth',1.5)
hold on
```

```
semilogy(errVals(1:kStop,2),'rs','LineWidth',1.5)

% Estimate the rate of contraction.
rate = 1.0;
if kStop >= 4
  kv = kStop-3:kStop;
  le = log(errVals(kStop-3:kStop,3));
  p1 = polyfit(kv,le,1);
  rate = exp(p1(1));
  p1k = polyval(p1,1:kStop);
  semilogy(1:kStop,exp(p1k),'-k')
end

xlabel('$k$','Interpreter','latex');
title('Two-grid Iteration Errors');
legend(...
  '$\left\|{\bf u}_1^{\rm E}-{\bf u}_1^k\right\|_1$',...
  '$\left\|{\bf q_1^{(1)}}\right\|_1$',...
  'log-linear fit','Interpreter','latex');
text(1.5,128*tol,...
  strcat('$\gamma_{\rm comp} =\hspace{.1cm}$',...
  num2str(rate,'%10.5e')),'FontSize',14,...
  'Interpreter','latex')
text(1.5,32*tol,strcat('$m_1 =\hspace{.1cm}$',...
  num2str(m1)),'FontSize',14,'Interpreter','latex')
text(1.5,8*tol,strcat('$\omega =\hspace{.1cm}$',...
  num2str(omega)),'FontSize',14,'Interpreter','latex')
printstr = strcat('Err_n1_',num2str(n1),'_m1_',...
  num2str(m1),'_omega_',num2str(omega),'.pdf');
exportgraphics(gca,printstr)
grid on
grid minor
hold off
```

**Listing 4.5:** twoGridSolver.m; Two-grid solver.

```
function [u,errVals,kStop] = twoGridSolver(u,f,n1,m1,...
                                           omega,kMax,...
                                           tol,uExact)

errVals = zeros(kMax,3);
```

```
res = zeros(n1,1);
cgc = zeros(n1,1);

h1 = 1/(n1+1);
x = linspace(h1,1.0-h1,n1)';

k = 0;
err = tol+1.0;

while ((k < kMax) && (err > tol))

  % Pre-smoothing (using damped Jacobi by default).
  u = smoothJacobiDamped(f,u,m1,omega);

  % Alternative pre-smoothing options:
  % u = smoothGaussSeidel(f,u,m1,'Forward');

  % Compute and apply coarse grid correction.
  res = getResidual(f,u);
  cgc = coarseGridCorrection(res);
  u = u+cgc;

  % Compute errors in discrete 2-norm as the level 1
  % norm.
  errVals(k+1,1) = norm(res,2);
  errVals(k+1,2) = norm(cgc,2);
  errVals(k+1,3) = norm(uExact-u,2);

  % We use the exact error for the stopping criteria.
  % In practice, the exact solution is not available,
  % and one can use either the residual or the coarse
  % grid correction as the measurement for the stopping
  % criteria.
  err = norm(uExact-u,2);

  k = k+1;

  % Print iteration information.
  fprintf('Two-grid iteration %d:\n',k);
  fprintf('    residual   = %.6e\n',errVals(k,1));
  fprintf('    correction= %.6e\n',errVals(k,2));
  fprintf('    true error= %.6e\n',errVals(k,3));
```

```
end

kStop = k;

end
```

**Listing 4.6:** `coarseGridCorrection.m`; Coarse grid correction function.

```
function cgc = coarseGridCorrection(res)

n1 = length(res);
cgc = zeros(n1,1);
n0 = (n1+1)/2-1;
h0 = 1/(n0+1);

% Assemble coarse grid matrix.
A0 = gallery('tridiag',n0,-1.0/h0,2.0/h0,-1.0/h0);

% Solve coarse grid problem and interpolate.
cgc = prolongation(A0\restriction(res));

end
```

**Listing 4.7:** `prolongation.m`; Prolongation function.

```
function vProl = prolongation(v)

nc = length(v);
nf = 2*(nc+1)-1;
vProl = zeros(nf,1);

% Linear interpolation at endpoints.
vProl(1) = v(1)/2.0;
vProl(nf) = v(nc)/2.0;

% Linear interpolation at interior points.
vProl(2:2:2*nc) = v(1:nc);
vProl(3:2:2*nc-1) = (v(1:nc-1)+v(2:nc))/2.0;

end
```

**Listing 4.8:** `restriction.m`; Restriction function.

```
function vRest = restriction(v)

nf = length(v);
nc = (nf+1)/2-1;
vRest = zeros(nc,1);

vRest(1:nc) = 0.5*(v(1:2:2*nc-1)+v(3:2:2*nc+1))...
              +v(2:2:2*nc);

end
```

**Listing 4.9:** `smoothJacobiDamped.m`; Damped Jacobi smoothing function.

```
function u = smoothJacobiDamped(f,u,m,omega)

if m == 0
  return
end

nl = length(f);
hl = 1/(nl+1);
z = zeros(nl,1);

for j = 1:m

  % Compute smoothing step.
  z(1) = (hl*f(1)+u(2))/2.0;
  z(2:nl-1) = (hl*f(2:nl-1)+u(1:nl-2)+u(3:nl))/2.0;
  z(nl) = (hl*f(nl)+u(nl-1))/2.0;

  % Apply damping.
  z(1:nl) = omega*z(1:nl)+(1.0-omega)*u(1:nl);

  u = z;

end

end
```

**Listing 4.10:** `smoothGaussSeidel.m`; Gauss–Seidel smoothing function.

```
function u = smoothGaussSeidel(f,u,m,sweepDirection)

if m == 0
  return
end

nl = length(f);
hl = 1/(nl+1);

for sweep = 1:m
  if strcmp(sweepDirection,'Forward')

    % Update first point.
    u(1) = (hl*f(1)+u(2))/2.0;

    % Update interior points.
    for i = 2:nl-1
      u(i) = (hl*f(i)+u(i-1)+u(i+1))/2.0;
    end

    % Update last point.
    u(nl) = (hl*f(nl)+u(nl-1))/2.0;

  else

    % Update last point.
    u(nl) = (hl*f(nl)+u(nl-1))/2.0;

    % Update interior points.
    for i = nl-1:-1:2
      u(i) = (hl*f(i)+u(i-1)+u(i+1))/2.0;
    end

    % Update first point.
    u(1) = (hl*f(1)+u(2))/2.0;

  end
end

end
```

**Listing 4.11:** `getResidual.m`; Residual calculation.

```
function res = getResidual(f,u)

nl = length(f);
hl = 1.0/(nl+1);
res = zeros(nl,1);

% Compute residual at boundaries.
res(1) = f(1)-(2.0*u(1)-u(2))/hl;
res(nl) = f(nl)-(-u(nl-1)+2.0*u(nl))/hl;

% Compute residual at interior points.
res(2:nl-1) = f(2:nl-1)...
              -(-u(1:nl-2)+2*u(2:nl-1)-u(3:nl))/hl;

end
```

# 5 An energy-norm analysis of the two-grid method

In this chapter, we continue to consider the finite element approximation of the model problem (1.7.1). We will again use a uniform mesh and prove that the two-grid algorithm converges, borrowing many of the results from the previous chapter. The difference here is that the type of analysis we carry out in this chapter will foreshadow the more general, and more modern, approach used in the later chapters. In particular, the convergence theory will depend upon the two pillars of multigrid mentioned in the previous chapter, an *approximation property* and a *smoothing property*. Moreover, convergence will be measured in the more appropriate energy norm.

The reader will quickly realize how much simpler and more powerful is this energy approach to convergence. It is nothing short of remarkable.

## 5.1 The approximation property

The level-1 stiffness matrix $\mathsf{A}_1 = [a_{1,i,j}] \in \mathbb{R}^{n_1 \times n_1}$ was defined in (4.1.9). For convenience, we recall here that for $1 \le i, j \le n_1$, its entries have the form

$$\begin{aligned} a_{1,i,j} &:= (\psi'_{1,j}, \psi'_{1,i})_{L^2(0,1)} \\ &= (\psi'_{1,i}, \psi'_{1,j})_{L^2(0,1)}. \end{aligned}$$

We also recall that this matrix is SPD, and thus it is orthogonally diagonalizable:

$$\mathsf{A}_1 = \mathsf{V}_1 \mathsf{D}_1 \mathsf{V}_1^\top,$$

where $\mathsf{V}_1$ is the orthogonal matrix containing the normalized eigenvectors of $\mathsf{A}_1$ as columns

$$\mathsf{V}_1 := \begin{bmatrix} | & | & & | \\ \tilde{\boldsymbol{v}}_1^{(1)} & \tilde{\boldsymbol{v}}_1^{(2)} & \cdots & \tilde{\boldsymbol{v}}_1^{(n_1)} \\ | & | & & | \end{bmatrix} \in \mathbb{R}^{n_1 \times n_1},$$

and $\mathsf{D}_1$ is the diagonal matrix containing the eigenvalues of $\mathsf{A}_1$,

$$\mathsf{D}_1 := \operatorname{diag}[\lambda_1^{(1)}, \lambda_1^{(2)}, \ldots, \lambda_1^{(n_1)}] \in \mathbb{R}^{n_1 \times n_1}.$$

The eigenvalues $\lambda_1^{(k)}$ and eigenvectors $\boldsymbol{v}_1^{(k)}$ of $\mathsf{A}_1$ were defined in (4.2.1) and (4.2.2), respectively, and

$$\tilde{\boldsymbol{v}}_1^{(k)} := \frac{\boldsymbol{v}_1^{(k)}}{\|\boldsymbol{v}_1^{(k)}\|_1}, \quad 1 \le k \le n_1.$$

Since $\mathsf{A}_1$ is SPD, we can define its square root as

https://doi.org/10.1515/9783111354880-005

$$\mathsf{A}_1^{1/2} := \mathsf{V}_1 \mathsf{D}_1^{1/2} \mathsf{V}_1^\top,$$

where

$$\mathsf{D}_1^{1/2} := \operatorname{diag}\left[\sqrt{\lambda_1^{(1)}}, \sqrt{\lambda_1^{(2)}}, \ldots, \sqrt{\lambda_1^{(n_1)}}\right] \in \mathbb{R}^{n_1 \times n_1}.$$

Of course, $\mathsf{A}_1^{1/2}$ is SPD, and $\mathsf{A}_1^{1/2}\mathsf{A}_1^{1/2} = \mathsf{A}_1$.

Using the square root of $\mathsf{A}_1$, we can relate the energy norm and the $\|\cdot\|_1$-norm as follows. For any $\boldsymbol{v}_1 \in \mathbb{R}^{n_1}$, we have

$$\begin{aligned}
\|\boldsymbol{v}_1\|_{\mathsf{A}_1}^2 &= (\boldsymbol{v}_1, \boldsymbol{v}_1)_{\mathsf{A}_1} \\
&= (\mathsf{A}_1 \boldsymbol{v}_1, \boldsymbol{v}_1)_1 \\
&= (\mathsf{A}_1^{1/2} \boldsymbol{v}_1, \mathsf{A}_1^{1/2} \boldsymbol{v}_1)_1 \\
&= \|\mathsf{A}_1^{1/2} \boldsymbol{v}_1\|_1^2.
\end{aligned}$$

In a similar way, since $\mathsf{A}_1^2$ is SPD, we can define

$$\begin{aligned}
\|\boldsymbol{v}_1\|_{\mathsf{A}_1^2}^2 &:= (\mathsf{A}_1^2 \boldsymbol{v}_1, \boldsymbol{v}_1)_1 \\
&= (\mathsf{A}_1 \boldsymbol{v}_1, \mathsf{A}_1 \boldsymbol{v}_1)_1 \\
&= \|\mathsf{A}_1 \boldsymbol{v}_1\|_1^2.
\end{aligned}$$

**Lemma 5.1.1** (Norm of projection). *Let the two-level grid be uniform and nested, and assume that $\mathsf{A}_0$ satisfies the Galerkin condition. Then*

$$\|(\mathrm{I}_1 - \tilde{\Pi}_1)\boldsymbol{v}_1\|_1 = \sqrt{2\Lambda_1^{-1}} \|(\mathrm{I}_1 - \tilde{\Pi}_1)\boldsymbol{v}_1\|_{\mathsf{A}_1}, \quad \forall\, \boldsymbol{v}_1 \in \mathbb{R}^{n_1}, \tag{5.1.1}$$

*where*

$$\Lambda_1 := \frac{4}{h_1}.$$

*Proof.* Expand $\boldsymbol{v}_1$ in the basis of eigenvectors of $\mathsf{A}_1$:

$$\boldsymbol{v}_1 = \sum_{k=1}^{n_1} v_{1,k} \boldsymbol{v}_1^{(k)}.$$

Recall that, as shown in (4.8.10),

$$(\mathrm{I}_1 - \tilde{\Pi}_1)\boldsymbol{v}_1^{(k)} = \mathsf{S}_k \boldsymbol{v}_1^{(k)} + \mathsf{S}_k \boldsymbol{v}_1^{(k')}$$

for $1 \le k \le n_0 + 1$, and

$$(\mathrm{I}_1 - \tilde{\Pi}_1)\boldsymbol{v}_1^{(k')} = \mathsf{C}_k \boldsymbol{v}_1^{(k)} + \mathsf{C}_k \boldsymbol{v}_1^{(k')}$$

for $1 \le k \le n_0 + 1$. Then, as in (4.8.15),

$$\begin{aligned}\|(\mathrm{I}_1 - \tilde{\Pi}_1)\boldsymbol{v}_1\|_1^2 &= ((\mathrm{I}_1 - \tilde{\Pi}_1)\boldsymbol{v}_1, (\mathrm{I}_1 - \tilde{\Pi}_1)\boldsymbol{v}_1)_1 \\ &= (n_1 + 1)\sum_{k=1}^{n_0+1} \delta_k \gamma_k^2,\end{aligned} \tag{5.1.2}$$

where

$$\delta_k = \begin{cases} \frac{1}{2}, & k = n_0 + 1, \\ 1, & k \neq n_0 + 1, \end{cases}$$

and

$$\gamma_k := v_{1,k}\mathrm{S}_k + v_{1,k'}\mathrm{C}_k.$$

Similarly,

$$\begin{aligned}\|(\mathrm{I}_1 - \tilde{\Pi}_1)\boldsymbol{v}_1\|_{\mathrm{A}_1}^2 &= ((\mathrm{I}_1 - \tilde{\Pi}_1)\boldsymbol{v}_1, (\mathrm{I}_1 - \tilde{\Pi}_1)\boldsymbol{v}_1)_{\mathrm{A}_1} \\ &= (\mathrm{A}_1^{1/2}(\mathrm{I}_1 - \tilde{\Pi}_1)\boldsymbol{v}_1, \mathrm{A}_1^{1/2}(\mathrm{I}_1 - \tilde{\Pi}_1)\boldsymbol{v}_1)_1 \\ &= (n_1 + 1)\sum_{k=1}^{n_0+1} \delta_k \gamma_k^2 \left(\frac{\lambda_1^{(k)} + \lambda_1^{(k')}}{2}\right).\end{aligned} \tag{5.1.3}$$

The proofs of (5.1.2) and (5.1.3) are similar to that used to establish (4.8.15) and are left to the reader as exercises; see Problems 5.1 and 5.2.

Finally, since

$$\frac{\lambda_1^{(k)} + \lambda_1^{(k')}}{2} = \frac{2}{h_1},$$

it follows that

$$\begin{aligned}\|(\mathrm{I}_1 - \tilde{\Pi}_1)\boldsymbol{v}_1\|_{\mathrm{A}_1}^2 &= \frac{2}{h_1}(n_1 + 1)\sum_{k=1}^{n_0+1} \delta_k \gamma_k^2 \\ &= \frac{1}{2}\Lambda_1\|(\mathrm{I}_1 - \tilde{\Pi}_1)\boldsymbol{v}_1\|_1^2.\end{aligned}$$

□

**Remark 5.1.2** (General setting). We will prove a similar result later in the general FEM framework, where, in particular, the meshes need not be uniform. In that setting, we will prove that there is a constant $C_{\mathrm{A}0} > 0$ such that for all $\boldsymbol{v}_1 \in \mathbb{R}^{n_1}$,

$$\|(\mathrm{I}_1 - \tilde{\Pi}_1)\boldsymbol{v}_1\|_1 \le C_{\mathrm{A}0}\sqrt{\Lambda_1^{-1}}\|(\mathrm{I}_1 - \tilde{\Pi}_1)\boldsymbol{v}_1\|_{\mathrm{A}_1}.$$

◂

We are now in position to prove the approximation property.

**Theorem 5.1.3** (Approximation property). *Assume that the two-level grid is uniform and nested and that* $\mathsf{A}_0$ *satisfies the Galerkin condition. Then, for all* $\boldsymbol{v}_1 \in \mathbb{R}^{n_1}$, *we have*

$$\left\|(\mathrm{I}_1 - \tilde{\Pi}_1)\boldsymbol{v}_1\right\|_{\mathsf{A}_1} \le C_{\mathrm{A0}}\sqrt{\Lambda_1^{-1}}\|\boldsymbol{v}_1\|_{\mathsf{A}_1^2}, \tag{5.1.4}$$

*where*

$$\Lambda_1 = \frac{4}{h_1} \qquad C_{\mathrm{A0}} = \sqrt{2}.$$

*Proof.* Let $\boldsymbol{v}_1 \in \mathbb{R}^{n_1}$. Then

$$\begin{aligned}\left\|(\mathrm{I}_1 - \tilde{\Pi}_1)\boldsymbol{v}_1\right\|_{\mathsf{A}_1}^2 &= (\mathsf{A}_1(\mathrm{I}_1 - \tilde{\Pi}_1)\boldsymbol{v}_1, (\mathrm{I}_1 - \tilde{\Pi}_1)\boldsymbol{v}_1)_1 \\ &= ((\mathrm{I}_1 - \tilde{\Pi}_1)\boldsymbol{v}_1, \mathsf{A}_1(\mathrm{I}_1 - \tilde{\Pi}_1)\boldsymbol{v}_1)_1 \\ &= ((\mathrm{I}_1 - \tilde{\Pi}_1)\boldsymbol{v}_1, \mathsf{A}_1\boldsymbol{v}_1)_1 - ((\mathrm{I}_1 - \tilde{\Pi}_1)\boldsymbol{v}_1, \mathsf{A}_1\tilde{\Pi}_1\boldsymbol{v}_1)_1.\end{aligned}$$

The second term on the right-hand side of this last expression is zero, as we now show:

$$\begin{aligned}((\mathrm{I}_1 - \tilde{\Pi}_1)\boldsymbol{v}_1, \tilde{\Pi}_1\boldsymbol{v}_1)_{\mathsf{A}_1} &= ((\mathrm{I}_1 - \tilde{\Pi}_1)\boldsymbol{v}_1, \mathsf{A}_1\tilde{\Pi}_1\boldsymbol{v}_1)_1 \\ &= ((\mathrm{I}_1 - \tilde{\Pi}_1)\boldsymbol{v}_1, \mathsf{A}_1(\mathsf{P}_0\mathsf{A}_0^{-1}\mathsf{R}_0\mathsf{A}_1)\boldsymbol{v}_1)_1 \\ &= ((\mathrm{I}_1 - \tilde{\Pi}_1)\boldsymbol{v}_1, (\mathsf{A}_1\mathsf{P}_0\mathsf{A}_0^{-1}\mathsf{R}_0)\mathsf{A}_1\boldsymbol{v}_1)_1 \\ &= ((\mathsf{A}_1\mathsf{P}_0\mathsf{A}_0^{-1}\mathsf{R}_0)^\top(\mathrm{I}_1 - \tilde{\Pi}_1)\boldsymbol{v}_1, \mathsf{A}_1\boldsymbol{v}_1)_1 \\ &= (\tilde{\Pi}_1(\mathrm{I}_1 - \tilde{\Pi}_1)\boldsymbol{v}_1, \boldsymbol{v}_1)_{\mathsf{A}_1} \\ &= 0,\end{aligned}$$

since $\tilde{\Pi}_1$ is a projection, so that

$$\begin{aligned}\tilde{\Pi}_1(\mathrm{I}_1 - \tilde{\Pi}_1) &= \tilde{\Pi}_1 - \tilde{\Pi}_1^2 \\ &= \tilde{\Pi}_1 - \tilde{\Pi}_1 \\ &= \mathsf{O}_1.\end{aligned}$$

Therefore

$$\begin{aligned}\left\|(\mathrm{I}_1 - \tilde{\Pi}_1)\boldsymbol{v}_1\right\|_{\mathsf{A}_1}^2 &= ((\mathrm{I}_1 - \tilde{\Pi}_1)\boldsymbol{v}_1, \mathsf{A}_1\boldsymbol{v}_1)_1 \\ &\le \left\|(\mathrm{I}_1 - \tilde{\Pi}_1)\boldsymbol{v}_1\right\|_1\|\mathsf{A}_1\boldsymbol{v}_1\|_1 \\ &= \left\|(\mathrm{I}_1 - \tilde{\Pi}_1)\boldsymbol{v}_1\right\|_1\|\boldsymbol{v}_1\|_{\mathsf{A}_1^2} \\ &\overset{(5.1.1)}{=} \sqrt{2\Lambda_1^{-1}}\left\|(\mathrm{I}_1 - \tilde{\Pi}_1)\boldsymbol{v}_1\right\|_{\mathsf{A}_1}\|\boldsymbol{v}_1\|_{\mathsf{A}_1^2}.\end{aligned}$$

The result follows. □

**Corollary 5.1.4** (Orthogonality). *Assume that the two-level grid is uniform and nested and* $\mathsf{A}_0$ *satisfies the Galerkin condition. Then*

$$((\mathrm{I}_1 - \tilde{\Pi}_1)\boldsymbol{v}_1, (\mathrm{I}_1 - \tilde{\Pi}_1)\boldsymbol{v}_1)_{\mathsf{A}_1} = ((\mathrm{I}_1 - \tilde{\Pi}_1)\boldsymbol{v}_1, \boldsymbol{v}_1)_{\mathsf{A}_1},$$

*or, equivalently,*

$$((\mathrm{I}_1 - \tilde{\Pi}_1)\boldsymbol{v}_1, \tilde{\Pi}_1\boldsymbol{v}_1)_{\mathsf{A}_1} = 0.$$

*Proof.* Exercise; see Problem 5.3. □

**Remark 5.1.5** (Approximation property). Estimate (5.1.4) is called an **approximation property** with constant $C_{\mathsf{A}0} > 0$. ◂

## 5.2 Richardson's smoother

In this section, we change the smoothing algorithm from the damped Jacobi to Richardson method. This change is mostly cosmetic, as the underlying structure for the smoother will be essentially the same. Specifically, Richardson's smoother will have the following error transfer matrix:

$$\mathsf{K}_1 = \mathrm{I}_1 - \frac{1}{\Lambda_1}\mathsf{A}_1, \quad \Lambda_1 = \frac{4}{h_1}.$$

Recall that the error transfer matrix of the damped Jacobi smoother is

$$\mathsf{K}_1 = \mathrm{I}_1 - \omega\mathsf{D}^{-1}\mathsf{A}_1 = \mathrm{I}_1 - \omega\frac{h_1}{2}\mathsf{A}_1.$$

Therefore setting $\omega = \frac{1}{2}$ in the damped Jacobi smoother reduces it to the Richardson smoother.

Recall also that the eigenvalues of the level-1 stiffness matrix are

$$\lambda_1^{(k)} = \frac{2}{h_1}(1 - \cos(k\pi h_1)).$$

Thus we have

$$0 < \lambda_1^{(1)} < \lambda_1^{(2)} < \cdots < \lambda_1^{(n_1)} = \rho(\mathsf{A}_1) < \frac{4}{h_1} = \Lambda_1,$$

and $\Lambda_1$ is almost the spectral radius of $\mathsf{A}_1$; the last estimate is asymptotically sharp. The error transfer matrix for Richardson's method has the following eigenvalues:

$$\mu_1^{(k)} = \frac{1}{2}\cos(k\pi h_1) + 1 - \frac{1}{2}$$

$$= 1 - \frac{\lambda_1^{(k)}}{\Lambda_1}.$$

It is not difficult to conclude (see Figure 5.1) that for $n_0 + 1 \le k \le n_1$,

$$\begin{aligned}|\mu_1^{(k)}| = \mu_1^{(k)} &\le \mu_1^{(n_0+1)}\\ &= \frac{1}{2}.\end{aligned}$$

In other words, the smoother reduces all high-frequency modes of the error by at least half.

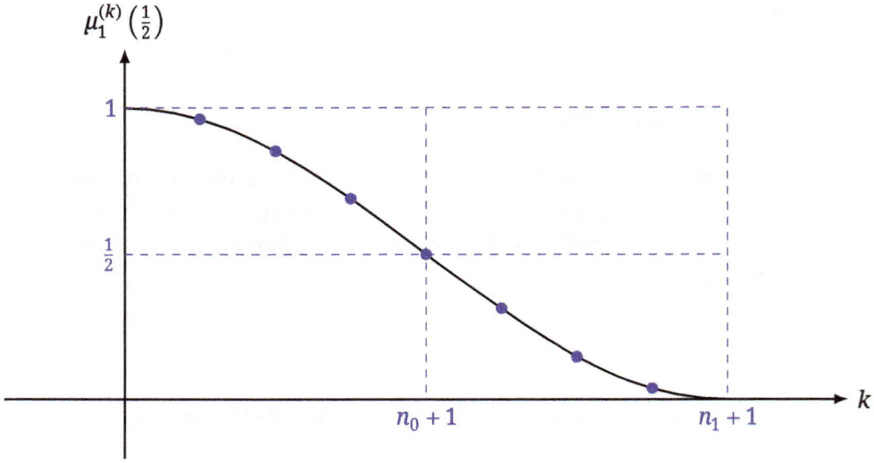


**Figure 5.1:** Eigenvalues of the error transfer matrix for the Richardson smoother plotted for $n_1 = 7$.

## 5.3 The smoothing property

Before we establish the so-called *smoothing property*, we need a technical lemma.

**Lemma 5.3.1** (Uniform bound). *For all $m \in \mathbb{N}$,*

$$\max_{0 \le x \le 1} x(1-x)^{2m} \le \frac{1}{2m}. \tag{5.3.1}$$

*Proof.* Let $m \in \mathbb{N}$ and define

$$f_m(x) = x(1-x)^{2m}, \quad x \in [0,1].$$

Observe that

$$f_m(0) = f_m(1) = 0,$$

and, otherwise,

$$f_m(x) > 0, \quad \forall x \in (0,1).$$

The derivative of $f_m$ is

$$f_m'(x) = (x-1)^{2m} + 2mx(x-1)^{2m-1},$$

and there is a single zero for $f_m'$ in $(0,1)$, which we label $x_0$. It is easy to show that

$$x_0 = \frac{1}{2m+1}.$$

We leave it as an exercise for the reader to show that $f_m$ achieves its maximum value at $x_0$ on the interval $[0,1]$; see Problem 5.4. Evaluating at $x_0$ and estimating, we find

$$f_m(x_0) = \frac{1}{2m+1}\left(\frac{2m}{2m+1}\right)^{2m} \le \frac{1}{2m+1} \le \frac{1}{2m}.$$

The result follows. □

**Theorem 5.3.2** (Smoothing property). *Let the two-level grid be uniform and nested. Assume that smoothing is carried out by the Richardson method or, equivalently, by the damped Jacobi method with* $\omega = \frac{1}{2}$. *Then there is a constant* $C_{S1} > 0$ *such that, for all* $m_1 \in \mathbb{N}$ *and* $\boldsymbol{v}_1 \in \mathbb{R}^{n_1}$,

$$\|\mathsf{K}_1^{m_1}\boldsymbol{v}_1\|_{\mathsf{A}_1^2} \le C_{S1}\sqrt{\frac{\Lambda_1}{m_1}}\|\boldsymbol{v}_1\|_{\mathsf{A}_1}. \tag{5.3.2}$$

*In particular, we can take*

$$C_{S1} = \sqrt{\frac{1}{2}}.$$

*Proof.* Observe, first of all, that

$$\begin{aligned}\|\mathsf{K}_1^{m_1}\boldsymbol{v}_1\|_{\mathsf{A}_1^2}^2 &= \|\mathsf{A}_1\mathsf{K}_1^{m_1}\boldsymbol{v}_1\|_1^2 \\ &= (\mathsf{A}_1\mathsf{K}_1^{m_1}\boldsymbol{v}_1, \mathsf{A}_1\mathsf{K}_1^{m_1}\boldsymbol{v}_1)_1.\end{aligned}$$

Using the eigenvector basis for $\mathsf{A}_1$, let us write

$$\boldsymbol{v}_1 = \sum_{k=1}^{n_1} v_{1,k}\boldsymbol{v}_1^{(k)}.$$

Then

$$
\begin{aligned}
\|\mathsf{K}_1^{m_1}\boldsymbol{v}_1\|_{\mathsf{A}_1^2}^2 &= \frac{n_1+1}{2}\sum_{k=1}^{n_1}(\lambda_1^{(k)}v_{1,k})^2(\mu_1^{(k)})^{2m_1} \\
&= \Lambda_1\left(\frac{n_1+1}{2}\right)\sum_{k=1}^{n_1}\left(\frac{\lambda_1^{(k)}}{\Lambda_1}\right)\left(1-\frac{\lambda_1^{(k)}}{\Lambda_1}\right)^{2m_1}\lambda_1^{(k)}v_{1,k}^2 \\
&\le \Lambda_1 G(m_1)\left(\frac{n_1+1}{2}\right)\sum_{k=1}^{n_1}\lambda_1^{(k)}v_{1,k}^2 \\
&= \Lambda_1 G(m_1)\|\boldsymbol{v}_1\|_{\mathsf{A}_1}^2,
\end{aligned}
$$

where

$$
G(m) := \max_{1\le k\le n_1}\left(\frac{\lambda_1^{(k)}}{\Lambda_1}\right)\left(1-\frac{\lambda_1^{(k)}}{\Lambda_1}\right)^{2m}.
$$

Rescaling and using Lemma 5.3.1, we have

$$
\begin{aligned}
G(m) &\le \max_{0\le x\le 1} x(1-x)^{2m} \\
&\overset{(5.3.1)}{\le} \frac{1}{2m}.
\end{aligned}
$$

Therefore

$$
\|\mathsf{K}_1^{m_1}\boldsymbol{v}_1\|_{\mathsf{A}_1^2} \le \sqrt{\frac{1}{2}}\sqrt{\frac{\Lambda_1}{m_1}}\|\boldsymbol{v}_1\|_{\mathsf{A}_1},
$$

and the result follows with $C_{S1} = \sqrt{\frac{1}{2}}$. □

**Remark 5.3.3** (Smoothing property). Estimate (5.3.2) is called a **smoothing property** with constant $C_{S1} > 0$. ◂

## 5.4 Convergence in the energy norm

Now that we have the powerful smoothing and approximation properties, the energy-norm convergence of the two-grid method is a simple matter.

**Theorem 5.4.1** (Convergence of the one-sided two-grid method). *Assume that the two-level grid is uniform and nested, $\mathsf{A}_0$ satisfies the Galerkin condition, and that smoothing is done by the damped Jacobi method using $\omega = \frac{1}{2}$ (Richardson) with $m_1 > 0$ and $m_2 = 0$. Then the (one-sided) two-grid method converges, provided that $m_1$ is sufficiently large. Moreover, for every $k \in \mathbb{N}_0$, we have the error estimate*

$$
\|\boldsymbol{e}_1^{k+1}\|_{\mathsf{A}_1} \le C_{A0}C_{S1}m_1^{-1/2}\|\boldsymbol{e}_1^k\|_{\mathsf{A}_1},
$$

*where $C_{A0}, C_{S1} > 0$ are as given in Theorem 5.1.3 and Lemma 5.3.2, respectively. Since we can assume that $C_{A0}C_{S1} = 1$, it follows that the method converges if $m_1 \geq 2$.*

*Proof.* Recall that in this case the error transfer matrix is

$$\mathsf{E}_1 = (\mathrm{I}_1 - \tilde{\Pi}_1)\mathsf{K}_1^{m_1}.$$

Thus, for all $\ell \geq 0$, we have

$$\begin{aligned}
\|\boldsymbol{e}_1^{k+1}\|_{\mathsf{A}_1} &= \|(\mathrm{I}_1 - \tilde{\Pi}_1)\mathsf{K}_1^{m_1}\boldsymbol{e}_1^k\|_{\mathsf{A}_1} \\
&\overset{(5.1.4)}{\leq} C_{A0}\sqrt{\Lambda_1^{-1}}\|\mathsf{K}_1^{m_1}\boldsymbol{e}_1^k\|_{\mathsf{A}_1^2} \\
&\overset{(5.3.2)}{\leq} C_{A0}\sqrt{\Lambda_1^{-1}}C_{S1}\sqrt{\Lambda_1}m_1^{-\frac{1}{2}}\|\boldsymbol{e}_1^k\|_{\mathsf{A}_1} \\
&= C_{A0}C_{S1}m_1^{-\frac{1}{2}}\|\boldsymbol{e}_1^k\|_{\mathsf{A}_1}.
\end{aligned}$$

□

**Remark 5.4.2** (General case). In this simplified setting, we know that

$$C_{A0}C_{S1} = 1.$$

In the more general FEM case that we will explore in later chapters — for which the $h_1$-independent constants $C_{A0} > 0$ and $C_{S1} > 0$ may not be known explicitly — we will still be guaranteed that the method converges at a uniform, independent of $h_1$, rate, provided that $m_1 \geq 1$ is large enough so that

$$0 < C_{A0}C_{S1}m_1^{-\frac{1}{2}} < 1.$$

◂

## 5.5 Problems

5.1 Prove the identity in equation (5.1.2).

5.2 Prove the identity in equation (5.1.3).

5.3 Prove Corollary 5.1.4.

5.4 Fill in the missing details of the proof of Lemma 5.3.1.

# 6 Multigrid: algorithms and axiomatic convergence theory

In this chapter, we begin the study of multigrid methods per se. The idea behind multigrid is to replace the exact solution in the coarse-grid correction by a recursive application of the multigrid operator at the coarse level. In this chapter, we learn how to extend the two-grid method to obtain various general multigrid algorithms.

This chapter borrows heavily from the elegant presentations in the books by Brenner and Scott [10, Chap. 06] and Bramble [6]. The books by Braess [4, Chap. V], Hackbusch [23], and Wesseling [38] are also excellent resources and go into significantly greater depth.

We have chosen to use the axiomatic approach of Bramble [6] and Bramble and Zhang [8], which is based on an abstract setting. We break our assumptions into three categories: Galerkin conditions, approximation properties, and smoothing properties. However, our presentation is somewhat unusual, as we use matrices, rather than abstract operators, to define all our objects.

## 6.1 Basic building blocks of multigrid

Suppose that we are given $L \in \mathbb{N}$ and the collection $\{n_\ell\}_{\ell=0}^{L} \subset \mathbb{N}$ with

$$1 \le n_0 < n_1 < \cdots < n_\ell < \cdots < n_L.$$

We refer to $\mathbb{R}^{n_\ell}$ as the level $\ell$ space or, sometimes more informally, as level $\ell$. Symmetry and transposition are understood with respect to the canonical inner product on level $\ell$, $(\cdot,\cdot)_\ell : \mathbb{R}^{n_\ell} \times \mathbb{R}^{n_\ell} \to \mathbb{R}$, defined as

$$\begin{aligned}(\boldsymbol{u}_\ell, \boldsymbol{v}_\ell)_\ell &:= \boldsymbol{v}_\ell^\top \boldsymbol{u}_\ell \\ &= \sum_{j=1}^{n_\ell} u_{\ell,j} v_{\ell,j},\end{aligned}$$

for all $\boldsymbol{u}_\ell, \boldsymbol{v}_\ell \in \mathbb{R}^{n_\ell}$. We will need the following simple property in this and subsequent chapters.

**Proposition 6.1.1** (Coarse-fine transposition). *Let $n_{\ell-1}, n_\ell \in \mathbb{N}$ with $n_{\ell-1} < n_\ell$, and let $\mathsf{B} \in \mathbb{R}^{n_{\ell-1} \times n_\ell}$. Then, for all $\boldsymbol{u}_\ell \in \mathbb{R}^{n_\ell}$ and $\boldsymbol{v}_{\ell-1} \in \mathbb{R}^{n_{\ell-1}}$,*

$$(\mathsf{B}\boldsymbol{u}_\ell, \boldsymbol{v}_{\ell-1})_{\ell-1} = (\boldsymbol{u}_\ell, \mathsf{B}^\top \boldsymbol{v}_{\ell-1})_\ell,$$

*where $\mathsf{B}^\top \in \mathbb{R}^{n_\ell \times n_{\ell-1}}$ is the usual matrix transpose of* $\mathsf{B}$.

*Proof.* Exercise; see Problem 6.1. □

https://doi.org/10.1515/9783111354880-006

We will assume the existence of a family of multigrid prolongation matrices

$$\mathsf{P}_{\ell-1} \in \mathbb{R}^{n_\ell \times n_{\ell-1}}, \quad 1 \le \ell \le L,$$

each member of which is of full rank, that is,

$$\operatorname{rank}(\mathsf{P}_{\ell-1}) = n_{\ell-1}.$$

Similarly, we assume that there is an associated family of multigrid restriction matrices satisfying

$$\mathsf{R}_{\ell-1} = r\mathsf{P}_{\ell-1}^\top \in \mathbb{R}^{n_{\ell-1} \times n_\ell}, \quad 1 \le \ell \le L,$$

for some scale parameter $r > 0$. Clearly, then we also have

$$\operatorname{rank}(\mathsf{R}_{\ell-1}) = n_{\ell-1},$$

and, for all $\boldsymbol{u}_\ell \in \mathbb{R}^{n_\ell}$ and $\boldsymbol{v}_{\ell-1} \in \mathbb{R}^{n_{\ell-1}}$,

$$(\mathsf{R}_{\ell-1}\boldsymbol{u}_\ell, \boldsymbol{v}_{\ell-1})_{\ell-1} = r(\boldsymbol{u}_\ell, \mathsf{P}_{\ell-1}\boldsymbol{v}_{\ell-1})_\ell.$$

**Definition 6.1.2** (Balanced symmetric pair). We say that the restriction and prolongation matrices, respectively,

$$\mathsf{R}_{\ell-1} \in \mathbb{R}^{n_{\ell-1} \times n_\ell} \quad \text{and} \quad \mathsf{P}_{\ell-1} \in \mathbb{R}^{n_\ell \times n_{\ell-1}} \quad 1 \le \ell \le L,$$

form a **balanced symmetric pair** if

$$\mathsf{R}_{\ell-1} = \mathsf{P}_{\ell-1}^\top,$$

that is, if the scale parameter is precisely $r = 1$. ∎

We assume that we have in our possession a family of SPD (stiffness) matrices

$$\mathsf{A}_\ell \in \mathbb{R}^{n_\ell \times n_\ell}, \quad 0 \le \ell \le L.$$

For each level-$\ell$ stiffness matrix, we define the corresponding energy inner product $(\cdot,\cdot)_{\mathsf{A}_\ell}$ on level $\ell$ as

$$(\boldsymbol{u}_\ell, \boldsymbol{v}_\ell)_{\mathsf{A}_\ell} := (\mathsf{A}_\ell\boldsymbol{u}_\ell, \boldsymbol{v}_\ell)_\ell, \quad \forall\, \boldsymbol{u}_\ell, \boldsymbol{v}_\ell \in \mathbb{R}^{n_\ell}.$$

Let us recall a couple of standard concepts. Let $\mathsf{B} \in \mathbb{R}^{n_\ell \times n_\ell}$. The matrix $\mathsf{B}^* \in \mathbb{R}^{n_\ell \times n_\ell}$ that satisfies

$$(\mathsf{B}\boldsymbol{u}_\ell, \boldsymbol{v}_\ell)_{\mathsf{A}_\ell} = (\boldsymbol{u}_\ell, \mathsf{B}^*\boldsymbol{v}_\ell)_{\mathsf{A}_\ell},$$

for all $\boldsymbol{u}_\ell, \boldsymbol{v}_\ell \in \mathbb{R}^{n_\ell}$ is called the $\mathsf{A}_\ell$-adjoint of $\mathsf{B}$ or, equivalently, the adjoint of $\mathsf{B}$ with respect to $(\cdot,\cdot)_{\mathsf{A}_\ell}$. We say that $\mathsf{B}$ is $\mathsf{A}_\ell$-self-adjoint or, equivalently, $\mathsf{B}$ is self-adjoint with respect to $(\cdot,\cdot)_{\mathsf{A}_\ell}$ if

$$\mathsf{B} = \mathsf{B}^*.$$

We say that $\mathsf{B}$ is self-adjoint positive definite (SPD) with respect to $(\cdot,\cdot)_{\mathsf{A}_\ell}$ or, equivalently, $\mathsf{B}$ is $\mathsf{A}_\ell$-SPD if $\mathsf{B}$ is $\mathsf{A}_\ell$-self-adjoint and

$$(\mathsf{B}\boldsymbol{u}_\ell, \boldsymbol{u}_\ell)_{\mathsf{A}_\ell} > 0,$$

for all $\boldsymbol{u}_\ell \in \mathbb{R}^{n_\ell}_\star := \mathbb{R}^{n_\ell} \setminus \{\boldsymbol{0}\}$.

Given $\boldsymbol{f}_L \in \mathbb{R}^{n_L}$, our ultimate goal is to build an efficient iterative solver for the equation

$$\mathsf{A}_L \boldsymbol{u}_L^{\mathrm{E}} = \boldsymbol{f}_L.$$

As usual, we define the error vector as

$$\boldsymbol{e}_L^{\square} := \boldsymbol{u}_L^{\mathrm{E}} - \boldsymbol{u}_L^{\square},$$

where $\boldsymbol{u}_L^{\square} \in \mathbb{R}^{n_L}$ is an approximate solution. Similarly, the residual is defined as

$$\boldsymbol{r}_L^{\square} := \boldsymbol{f}_L - \mathsf{A}_L \boldsymbol{u}_L^{\square} = \mathsf{A}_L \boldsymbol{e}_L^{\square}.$$

**Definition 6.1.3** (Multigrid operator). Let $L \in \mathbb{N}$, and suppose that $\ell \in \{0,1,\dots,L\}$, $\boldsymbol{g}_\ell \in \mathbb{R}^{n_\ell}$, and $\boldsymbol{u}_\ell^{(0)} \in \mathbb{R}^{n_\ell}$. The vector $\boldsymbol{u}_\ell^{(3)} \in \mathbb{R}^{n_\ell}$ is computed via the recursive **multigrid operator**,

$$\boldsymbol{u}_\ell^{(3)} := \mathrm{MG}(\ell, \boldsymbol{g}_\ell, \boldsymbol{u}_\ell^{(0)}),$$

as follows:

- If $\ell = 0$, then

$$\boldsymbol{u}_0^{(3)} := \boldsymbol{u}_0^{(1,\mathrm{E})} := \mathsf{A}_0^{-1} \boldsymbol{g}_0.$$

- Otherwise, if $1 \le \ell \le L$, then
  - presmoothing:
    * $\boldsymbol{u}_\ell^{(1,0)} := \boldsymbol{u}_\ell^{(0)}$;
    * $\boldsymbol{u}_\ell^{(1,\sigma+1)} := \boldsymbol{u}_\ell^{(1,\sigma)} + \mathsf{S}_\ell(\boldsymbol{g}_\ell - \mathsf{A}_\ell \boldsymbol{u}_\ell^{(1,\sigma)}), \quad 0 \le \sigma \le m_1 - 1$;
    * $\boldsymbol{u}_\ell^{(1)} := \boldsymbol{u}_\ell^{(1,m_1)}$.
  - coarse-grid correction:
    * $\boldsymbol{r}_\ell^{(1)} := \boldsymbol{g}_\ell - \mathsf{A}_\ell \boldsymbol{u}_\ell^{(1)}$;
    * $\boldsymbol{r}_{\ell-1}^{(1)} := \mathsf{R}_{\ell-1} \boldsymbol{r}_\ell^{(1)}$;

  * $\boldsymbol{q}_{\ell-1}^{(1,0)} := \mathbf{0}$;
  * $\boldsymbol{q}_{\ell-1}^{(1,\sigma+1)} := \mathrm{MG}(\ell-1, \boldsymbol{r}_{\ell-1}^{(1)}, \boldsymbol{q}_{\ell-1}^{(1,\sigma)}), \quad 0 \le \sigma \le p-1$;
  * $\boldsymbol{q}_{\ell-1}^{(1)} := \boldsymbol{q}_{\ell-1}^{(1,p)}$;
  * $\boldsymbol{q}_{\ell}^{(1)} := \mathsf{P}_{\ell-1}\boldsymbol{q}_{\ell-1}^{(1)}$
  * $\boldsymbol{u}_{\ell}^{(2)} := \boldsymbol{u}_{\ell}^{(1)} + \boldsymbol{q}_{\ell}^{(1)}$.
- postsmoothing:
  * $\boldsymbol{u}_{\ell}^{(3,0)} := \boldsymbol{u}_{\ell}^{(2)}$;
  * $\boldsymbol{u}_{\ell}^{(3,\sigma+1)} := \boldsymbol{u}_{\ell}^{(3,\sigma)} + \mathsf{S}_{\ell}^{\mathsf{T}}(\boldsymbol{g}_{\ell} - \mathsf{A}_{\ell}\boldsymbol{u}_{\ell}^{(3,\sigma)}), \quad 0 \le \sigma \le m_2 - 1$;
  * and, finally, $\boldsymbol{u}_{\ell}^{(3)} := \boldsymbol{u}_{\ell}^{(3,m_2)}$.

The vector $\boldsymbol{q}_{\ell-1}^{(1)} \in \mathbb{R}^{n_{\ell-1}}$ is called the **multigrid coarse-grid correction**, and the vector

$$\boldsymbol{q}_{\ell-1}^{(1,\mathrm{E})} := \mathsf{A}_{\ell-1}^{-1}\boldsymbol{r}_{\ell-1}^{(1)}$$

is the **exact coarse-grid correction**. ■

**Listing 6.1:** Multigrid operator implementation. The full code is provided in Listing 7.5.

```
if level == 0

  % Direct solve at the coarsest level.
  Al = gallery('tridiag',nl,-1.0/hl,2.0/hl,-1.0/hl);
  u = Al\f;
else

  % Pre-smoothing.
  u = smoothJacobiDamped(f,u,MGParam.m1,MGParam.omega);

  % Alternative pre-smoothing options:
  % u = smoothGaussSeidel(f,u,MGParam.m1,'Forward');

  % Get residual and restrict to coarse grid.
  cGr = restriction(getResidual(f,u));
  nlm1 = (nl+1)/2-1;
  cGc = zeros(nlm1,1);

  % Compute the coarse-grid correction with recursive
  % MG call.
  for sig = 1:MGParam.p
    cGc = MGOperator(level-1,cGr,cGc,MGParam);
  end

```

```
    % Correct the solution using coarse-grid correction.
    u = u+prolongation(cGc);

    % Post-smoothing.
    u = smoothJacobiDamped(f,u,MGParam.m2,MGParam.omega);

    % Alternative post-smoothing options:
    % u = smoothGaussSeidel(f,u,MGParam.m2,'Backward');
end
```

**Remark 6.1.4** (Coarse-grid correction). In the two-grid method, we found that $\boldsymbol{q}_{\ell-1}^{(1,\mathrm{E})} = \boldsymbol{q}_{\ell-1}^{(1)}$. In the multigrid method, this is no longer the case, and we can only expect that $\boldsymbol{q}_{\ell-1}^{(1,\mathrm{E})} \approx \boldsymbol{q}_{\ell-1}^{(1)}$. Much of the multigrid analysis revolves around estimating the difference

$$\boldsymbol{q}_{\ell-1}^{(1,\mathrm{E})} - \boldsymbol{q}_{\ell-1}^{(1)}.$$ ◀

**Definition 6.1.5** (Generic multigrid algorithm). Let $m_1$ and $m_2$ be nonnegative integers, and let $p$ be a positive integer. Let $\boldsymbol{u}_L^k \in \mathbb{R}^{n_L}$. Then

$$\boldsymbol{u}_L^{k+1} = \mathrm{MG}(L, \boldsymbol{f}_L, \boldsymbol{u}_L^k)$$

defines the **generic multigrid algorithm** for solving

$$\mathsf{A}_L \boldsymbol{u}_L^{\mathrm{E}} = \boldsymbol{f}_L.$$ ■

**Definition 6.1.6** (Variants of the multigrid method). A multigrid algorithm is called **one-sided** if $m_2 = 0$ and $m_1 \geq 1$. The algorithm is called a **W-cycle** if $p \geq 2$ and is called a **V-cycle** if $p = 1$. The algorithm is called **symmetrized** if $m_1 = m_2 = m$. We say that the algorithm is a **simple symmetric V-cycle** if $p = 1$ and $m_1 = m_2 = 1$; see Figure 6.1. ■

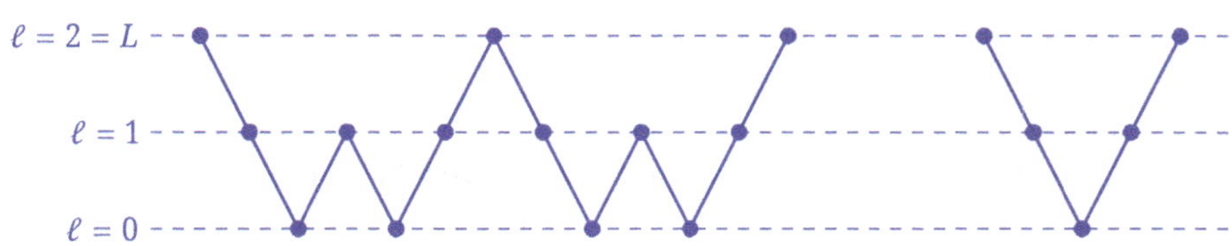


**Figure 6.1:** One full ($p = 2$) W-cycle (left) and one full V-cycle (right) for three levels, $L = 2$.

**Remark 6.1.7** (Reduction to two-grid method). In the case $L = 0$, we just perform a direct solution. When $L = 1$, we recover the two-grid (two-level) method. ◀

We now find the error transfer matrix for the general multigrid algorithm.

**Definition 6.1.8** (Error transfer). Let $\ell \geq 0$. For $\ell = 0$, define

$$\mathsf{E}_0 := \mathsf{O}_{n_0} \in \mathbb{R}^{n_0 \times n_0},$$

where $\mathsf{O}_{n_0} \in \mathbb{R}^{n_0 \times n_0}$ is the zero matrix, and for $\ell \geq 1$, define

$$\mathsf{E}_\ell := (\mathsf{K}_\ell^*)^{m_2}(\mathrm{I}_\ell - \mathsf{P}_{\ell-1}(\mathrm{I}_{\ell-1} - \mathsf{E}_{\ell-1}^p)\Pi_{\ell-1})\mathsf{K}_\ell^{m_1} \in \mathbb{R}^{n_\ell \times n_\ell}, \tag{6.1.1}$$

where

$$\mathsf{K}_\ell := \mathrm{I}_\ell - \mathsf{S}_\ell \mathsf{A}_\ell \in \mathbb{R}^{n_\ell \times n_\ell}, \quad \mathsf{K}_\ell^* = \mathrm{I}_\ell - \mathsf{S}_\ell^\top \mathsf{A}_\ell,$$

and

$$\Pi_{\ell-1} := \mathsf{A}_{\ell-1}^{-1}\mathsf{R}_{\ell-1}\mathsf{A}_\ell \in \mathbb{R}^{n_{\ell-1} \times n_\ell}.$$

For further use, let us also define

$$\tilde{\Pi}_\ell := \mathsf{P}_{\ell-1}\Pi_{\ell-1} \in \mathbb{R}^{n_\ell \times n_\ell}.$$

The matrix $\mathsf{E}_\ell$ is called the **multigrid error matrix**, and $\tilde{\Pi}_\ell$ is called the **coarse-grid Ritz projection matrix.** ■

**Remark 6.1.9** (Adjoint). Observe that, for all $\boldsymbol{u}_\ell, \boldsymbol{v}_\ell \in \mathbb{R}^{n_\ell}$,

$$(\mathsf{K}_\ell \boldsymbol{u}_\ell, \boldsymbol{v}_\ell)_{\mathsf{A}_\ell} = (\boldsymbol{u}_\ell, \mathsf{K}_\ell^* \boldsymbol{v}_\ell)_{\mathsf{A}_\ell}.$$

In other words, $\mathsf{K}_\ell^*$ is the $\mathsf{A}_\ell$-adjoint of $\mathsf{K}_\ell$. ◂

**Theorem 6.1.10** (Multigrid error relation). *Let $\boldsymbol{u}_\ell^{\mathrm{E}}, \boldsymbol{g}_\ell \in \mathbb{R}^{n_\ell}$ satisfy*

$$\mathsf{A}_\ell \boldsymbol{u}_\ell^{\mathrm{E}} = \boldsymbol{g}_\ell.$$

*Then, for all $\boldsymbol{u}_\ell^{(0)} \in \mathbb{R}^{n_\ell}$,*

$$\boldsymbol{u}_\ell^{\mathrm{E}} - \mathrm{MG}(\ell, \boldsymbol{g}_\ell, \boldsymbol{u}_\ell^{(0)}) = \mathsf{E}_\ell(\boldsymbol{u}_\ell^{\mathrm{E}} - \boldsymbol{u}_\ell^{(0)}),$$

*where $\mathsf{E}_\ell$ is the recursively defined matrix from Definition* 6.1.8. *In particular,*

$$\boldsymbol{e}_L^{k+1} = \mathsf{E}_L \boldsymbol{e}_L^k.$$

*Proof.* The proof is by induction. We will use the notation from the definition of the multigrid operator throughout.
(Base cases): Cases $\ell = 0$ and $\ell = 1$ (two-grid method) are clear.
(Induction hypothesis): Assume that the result is true for level $\ell - 1$. In other words, suppose that $\boldsymbol{u}_{\ell-1}^{\mathrm{E}}, \boldsymbol{g}_{\ell-1} \in \mathbb{R}^{n_{\ell-1}}$ satisfy

$$\mathsf{A}_{\ell-1}\boldsymbol{u}_{\ell-1}^{\mathrm{E}} = \boldsymbol{g}_{\ell-1}.$$

Then, for any initial vector $\boldsymbol{u}_{\ell-1}^{(0)} \in \mathbb{R}^{n_{\ell-1}}$,

$$\boldsymbol{u}_{\ell-1}^{\mathrm{E}} - \mathrm{MG}(\ell-1, \boldsymbol{g}_{\ell-1}, \boldsymbol{u}_{\ell-1}^{(0)}) = \mathsf{E}_{\ell-1}(\boldsymbol{u}_{\ell-1}^{\mathrm{E}} - \boldsymbol{u}_{\ell-1}^{(0)}).$$

(Generic case): Suppose that $\boldsymbol{q}_{\ell-1}^{(1,\mathrm{E})}, \boldsymbol{r}_{\ell-1}^{(1)} \in \mathbb{R}^{n_{\ell-1}}$ satisfy

$$\mathsf{A}_{\ell-1}\boldsymbol{q}_{\ell-1}^{(1,\mathrm{E})} = \boldsymbol{r}_{\ell-1}^{(1)}.$$

Then, using the induction hypothesis, we have

$$\begin{aligned}\boldsymbol{q}_{\ell-1}^{(1,\mathrm{E})} - \mathrm{MG}(\ell-1, \boldsymbol{r}_{\ell-1}^{(1)}, \boldsymbol{0}) &= \mathsf{E}_{\ell-1}(\boldsymbol{q}_{\ell-1}^{(1,\mathrm{E})} - \boldsymbol{0}) \\ &= \mathsf{E}_{\ell-1}\boldsymbol{q}_{\ell-1}^{(1,\mathrm{E})}\end{aligned}$$

or, in the notation defining the multigrid operator,

$$\boldsymbol{q}_{\ell-1}^{(1,\mathrm{E})} - \boldsymbol{q}_{\ell-1}^{(1,1)} = \mathsf{E}_{\ell-1}\boldsymbol{q}_{\ell-1}^{(1,\mathrm{E})}.$$

Applying the induction hypothesis again, we have

$$\begin{aligned}\boldsymbol{q}_{\ell-1}^{(1,\mathrm{E})} - \boldsymbol{q}_{\ell-1}^{(1,2)} &= \boldsymbol{q}_{\ell-1}^{(1,\mathrm{E})} - \mathrm{MG}(\ell-1, \boldsymbol{r}_{\ell-1}^{(1)}, \boldsymbol{q}_{\ell-1}^{(1,1)}) \\ &= \mathsf{E}_{\ell-1}(\boldsymbol{q}_{\ell-1}^{(1,\mathrm{E})} - \boldsymbol{q}_{\ell-1}^{(1,1)}) \\ &= \mathsf{E}_{\ell-1}^{2}\boldsymbol{q}_{\ell-1}^{(1,\mathrm{E})}.\end{aligned}$$

Continuing in this way, for all $p \in \mathbb{N}$, we have

$$\boldsymbol{q}_{\ell-1}^{(1,\mathrm{E})} - \boldsymbol{q}_{\ell-1}^{(1,p)} = \mathsf{E}_{\ell-1}^{p}\boldsymbol{q}_{\ell-1}^{(1,\mathrm{E})}.$$

Consequently,

$$\boldsymbol{q}_{\ell-1}^{(1,\mathrm{E})} - \boldsymbol{q}_{\ell-1}^{(1)} = \mathsf{E}_{\ell-1}^{p}\boldsymbol{q}_{\ell-1}^{(1,\mathrm{E})}.$$

Solving for $\boldsymbol{q}_{\ell-1}^{(1)}$, we have

$$\boldsymbol{q}_{\ell-1}^{(1)} = (\mathrm{I}_{\ell-1} - \mathsf{E}_{\ell-1}^{p})\boldsymbol{q}_{\ell-1}^{(1,\mathrm{E})}.$$

Now

$$\begin{aligned}\boldsymbol{q}_{\ell-1}^{(1,\mathrm{E})} &:= \mathsf{A}_{\ell-1}^{-1}\boldsymbol{r}_{\ell-1}^{(1)} \\ &= \mathsf{A}_{\ell-1}^{-1}\mathsf{R}_{\ell-1}(\boldsymbol{g}_{\ell} - \mathsf{A}_{\ell}\boldsymbol{u}_{\ell}^{(1)}) \\ &= \mathsf{A}_{\ell-1}^{-1}\mathsf{R}_{\ell-1}\mathsf{A}_{\ell}(\boldsymbol{u}_{\ell}^{\mathrm{E}} - \boldsymbol{u}_{\ell}^{(1)}) \\ &= \Pi_{\ell-1}(\boldsymbol{u}_{\ell}^{\mathrm{E}} - \boldsymbol{u}_{\ell}^{(1)}).\end{aligned} \tag{6.1.2}$$

Hence

$$\boldsymbol{q}_{\ell-1}^{(1)} = (\mathrm{I}_{\ell-1} - \mathsf{E}_{\ell-1}^{p})\Pi_{\ell-1}(\boldsymbol{u}_\ell^{\mathrm{E}} - \boldsymbol{u}_\ell^{(1)}).$$

For the error after postsmoothing,

$$\boldsymbol{u}_\ell^{\mathrm{E}} - \boldsymbol{u}_\ell^{(3)} = (\mathsf{K}_\ell^{*})^{m_2}(\boldsymbol{u}_\ell^{\mathrm{E}} - \boldsymbol{u}_\ell^{(2)});$$

for the error after coarse-grid correction,

$$\begin{aligned}\boldsymbol{u}_\ell^{\mathrm{E}} - \boldsymbol{u}_\ell^{(2)} &= \boldsymbol{u}_\ell^{\mathrm{E}} - (\boldsymbol{u}_\ell^{(1)} + \mathsf{P}_{\ell-1}\boldsymbol{q}_{\ell-1}^{(1)}) \\ &= \boldsymbol{u}_\ell^{\mathrm{E}} - \boldsymbol{u}_\ell^{(1)} - \mathsf{P}_{\ell-1}(\mathrm{I}_{\ell-1} - \mathsf{E}_{\ell-1}^{p})\Pi_{\ell-1}(\boldsymbol{u}_\ell^{\mathrm{E}} - \boldsymbol{u}_\ell^{(1)}) \\ &= (\mathrm{I}_\ell - \mathsf{P}_{\ell-1}(\mathrm{I}_{\ell-1} - \mathsf{E}_{\ell-1}^{p})\Pi_{\ell-1})(\boldsymbol{u}_\ell^{\mathrm{E}} - \boldsymbol{u}_\ell^{(1)});\end{aligned}$$

and for the error after presmoothing,

$$\boldsymbol{u}_\ell^{\mathrm{E}} - \boldsymbol{u}_\ell^{(1)} = \mathsf{K}_\ell^{m_1}(\boldsymbol{u}_\ell^{\mathrm{E}} - \boldsymbol{u}_\ell^{(0)}).$$

Putting it all together, we have

$$\boldsymbol{u}_\ell^{\mathrm{E}} - \boldsymbol{u}_\ell^{(3)} \overset{(6.1.1)}{=} \mathsf{E}_\ell(\boldsymbol{u}_\ell^{\mathrm{E}} - \boldsymbol{u}_\ell^{(0)}).$$ □

**Corollary 6.1.11** (Multigrid is a GLIS). *For every $0 \le \ell \le L$, there is a (recursively defined) matrix $\mathsf{B}_\ell \in \mathbb{R}^{n_\ell \times n_\ell}$ such that*

$$\mathsf{E}_\ell = \mathrm{I}_\ell - \mathsf{B}_\ell \mathsf{A}_\ell$$

*and*

$$\mathrm{MG}(\ell, \boldsymbol{g}_\ell, \boldsymbol{z}_\ell) = \boldsymbol{z}_\ell + \mathsf{B}_\ell(\boldsymbol{g}_\ell - \mathsf{A}_\ell \boldsymbol{z}_\ell).$$

*In other words, the multigrid method is a GLIS.*

*Proof.* Exercise; see Problem 6.2. □

## 6.2 Galerkin conditions

Let us now begin to present the set of assumptions we will need to analyze the multigrid algorithm. As we mentioned before, this set will be grouped into three classes. The first one is that of Galerkin conditions.

**Assumption 6.2.1** (G0). The multigrid method satisfies the **strong Galerkin condition**, that is, for all $1 \le \ell \le L$, we have

$$\mathsf{R}_{\ell-1} = \mathsf{P}_{\ell-1}^{\top} \tag{G0a}$$

and

$$\mathsf{A}_{\ell-1} = \mathsf{R}_{\ell-1}\mathsf{A}_{\ell}\mathsf{P}_{\ell-1}. \tag{G0b}$$

•

**Assumption 6.2.2** (G1). The multigrid method satisfies the **weak Galerkin condition**, that is, for all $1 \le \ell \le L$, we have

$$\mathsf{R}_{\ell-1} = \mathsf{P}_{\ell-1}^{\top} \tag{G1a}$$

and

$$(\boldsymbol{v}_\ell, \boldsymbol{v}_\ell)_{\mathsf{A}_\ell} \ge (\Pi_{\ell-1}\boldsymbol{v}_\ell, \Pi_{\ell-1}\boldsymbol{v}_\ell)_{\mathsf{A}_{\ell-1}}, \quad \forall\, \boldsymbol{v}_\ell \in \mathbb{R}^{n_\ell}. \tag{G1b}$$

•

In what follows, we will show that Assumption (G0) implies Assumption (G1). To do this, we first need the following result.

**Lemma 6.2.3** (Projection). *For every $1 \le \ell \le L$, we have $\tilde{\Pi}_\ell^* = \tilde{\Pi}_\ell$. In addition, if* (G0) *holds, then*

$$\tilde{\Pi}_\ell = \tilde{\Pi}_\ell^2. \tag{6.2.1}$$

*Proof.* Let $\boldsymbol{u}_\ell, \boldsymbol{v}_\ell \in \mathbb{R}^{n_\ell}$. Using the definition

$$\tilde{\Pi}_\ell = \mathsf{P}_{\ell-1}\mathsf{A}_{\ell-1}^{-1}\mathsf{R}_{\ell-1}\mathsf{A}_\ell$$

and the general relation $\mathsf{R}_{\ell-1}^{\top} = r\mathsf{P}_{\ell-1}$, $r > 0$, we deduce

$$\begin{aligned}
(\tilde{\Pi}_\ell \boldsymbol{u}_\ell, \boldsymbol{v}_\ell)_{\mathsf{A}_\ell} &= (\mathsf{A}_\ell\tilde{\Pi}_\ell \boldsymbol{u}_\ell, \boldsymbol{v}_\ell)_\ell \\
&= (\mathsf{A}_\ell\mathsf{P}_{\ell-1}\mathsf{A}_{\ell-1}^{-1}\mathsf{R}_{\ell-1}\mathsf{A}_\ell\boldsymbol{u}_\ell, \boldsymbol{v}_\ell)_\ell \\
&= ((\mathsf{A}_\ell\mathsf{P}_{\ell-1}\mathsf{A}_{\ell-1}^{-1}\mathsf{R}_{\ell-1})\mathsf{A}_\ell\boldsymbol{u}_\ell, \boldsymbol{v}_\ell)_\ell \\
&= (\mathsf{A}_\ell\boldsymbol{u}_\ell, (\mathsf{A}_\ell\mathsf{P}_{\ell-1}\mathsf{A}_{\ell-1}^{-1}\mathsf{R}_{\ell-1})^{\top}\boldsymbol{v}_\ell)_\ell \\
&= (\mathsf{A}_\ell\boldsymbol{u}_\ell, \tilde{\Pi}_\ell\boldsymbol{v}_\ell)_\ell \\
&= (\boldsymbol{u}_\ell, \tilde{\Pi}_\ell\boldsymbol{v}_\ell)_{\mathsf{A}_\ell}.
\end{aligned}$$

Thus

$$\tilde{\Pi}_\ell^* = \tilde{\Pi}_\ell.$$

Next, recall that if the strong Galerkin condition (G0) holds, then for $1 \le \ell \le L$,

$$\mathsf{A}_{\ell-1} = \mathsf{R}_{\ell-1}\mathsf{A}_\ell\mathsf{P}_{\ell-1}.$$

Consequently,

$$\begin{aligned}\tilde{\Pi}_\ell^2 &= \mathsf{P}_{\ell-1}\mathsf{A}_{\ell-1}^{-1}\mathsf{R}_{\ell-1}\mathsf{A}_\ell\mathsf{P}_{\ell-1}\mathsf{A}_{\ell-1}^{-1}\mathsf{R}_{\ell-1}\mathsf{A}_\ell\\ &= \mathsf{P}_{\ell-1}\mathsf{A}_{\ell-1}^{-1}(\mathsf{R}_{\ell-1}\mathsf{A}_\ell\mathsf{P}_{\ell-1})\mathsf{A}_{\ell-1}^{-1}\mathsf{R}_{\ell-1}\mathsf{A}_\ell\\ &= \mathsf{P}_{\ell-1}\mathsf{A}_{\ell-1}^{-1}\mathsf{A}_{\ell-1}\mathsf{A}_{\ell-1}^{-1}\mathsf{R}_{\ell-1}\mathsf{A}_\ell\\ &= \mathsf{P}_{\ell-1}\mathsf{A}_{\ell-1}^{-1}\mathsf{R}_{\ell-1}\mathsf{A}_\ell\\ &= \tilde{\Pi}_\ell.\end{aligned}$$

□

**Corollary 6.2.4** (Projection). *For every $1 \le \ell \le L$, we have*

$$(\mathrm{I}_\ell - \tilde{\Pi}_\ell)^* = \mathrm{I}_\ell - \tilde{\Pi}_\ell. \tag{6.2.2}$$

*In addition, if Assumption* (G0) *holds, then*

$$(\mathrm{I}_\ell - \tilde{\Pi}_\ell)^2 = \mathrm{I}_\ell - \tilde{\Pi}_\ell. \tag{6.2.3}$$

*Proof.* Exercise; see Problem 6.3. □

**Lemma 6.2.5** ((G0) $\Longrightarrow$ (G1)). *Assumption* (G0) *implies Assumption* (G1).

*Proof.* First, for $\boldsymbol{v}_\ell \in \mathbb{R}^{n_\ell}$, consider

$$\begin{aligned}(\Pi_{\ell-1}\boldsymbol{v}_\ell, \Pi_{\ell-1}\boldsymbol{v}_\ell)_{\mathsf{A}_{\ell-1}} &= (\Pi_{\ell-1}\boldsymbol{v}_\ell, \mathsf{A}_{\ell-1}\Pi_{\ell-1}\boldsymbol{v}_\ell)_{\ell-1}\\ &= (\Pi_{\ell-1}\boldsymbol{v}_\ell, \mathsf{A}_{\ell-1}\mathsf{A}_{\ell-1}^{-1}\mathsf{R}_{\ell-1}\mathsf{A}_\ell\boldsymbol{v}_\ell)_{\ell-1}\\ &= (\Pi_{\ell-1}\boldsymbol{v}_\ell, \mathsf{R}_{\ell-1}\mathsf{A}_\ell\boldsymbol{v}_\ell)_{\ell-1}\\ &= (\mathsf{R}_{\ell-1}^\top\Pi_{\ell-1}\boldsymbol{v}_\ell, \mathsf{A}_\ell\boldsymbol{v}_\ell)_\ell\\ &= (\mathsf{P}_{\ell-1}\Pi_{\ell-1}\boldsymbol{v}_\ell, \mathsf{A}_\ell\boldsymbol{v}_\ell)_\ell\\ &= (\mathsf{P}_{\ell-1}\Pi_{\ell-1}\boldsymbol{v}_\ell, \boldsymbol{v}_\ell)_{\mathsf{A}_\ell}\\ &= (\tilde{\Pi}_\ell\boldsymbol{v}_\ell, \boldsymbol{v}_\ell)_{\mathsf{A}_\ell}.\end{aligned}$$

From this we have

$$\begin{aligned}(\boldsymbol{v}_\ell, \boldsymbol{v}_\ell)_{\mathsf{A}_\ell} - (\Pi_{\ell-1}\boldsymbol{v}_\ell, \Pi_{\ell-1}\boldsymbol{v}_\ell)_{\mathsf{A}_{\ell-1}} &= (\boldsymbol{v}_\ell, \boldsymbol{v}_\ell)_{\mathsf{A}_\ell} - (\tilde{\Pi}_\ell\boldsymbol{v}_\ell, \boldsymbol{v}_\ell)_{\mathsf{A}_\ell}\\ &= ((\mathrm{I}_\ell - \tilde{\Pi}_\ell)\boldsymbol{v}_\ell, \boldsymbol{v}_\ell)_{\mathsf{A}_\ell}\\ &\overset{(6.2.3)}{=} ((\mathrm{I}_\ell - \tilde{\Pi}_\ell)^2\boldsymbol{v}_\ell, \boldsymbol{v}_\ell)_{\mathsf{A}_\ell}\\ &\overset{(6.2.2)}{=} ((\mathrm{I}_\ell - \tilde{\Pi}_\ell)\boldsymbol{v}_\ell, (\mathrm{I}_\ell - \tilde{\Pi}_\ell)\boldsymbol{v}_\ell)_{\mathsf{A}_\ell}\\ &= \|(\mathrm{I}_\ell - \tilde{\Pi}_\ell)\boldsymbol{v}_\ell\|_{\mathsf{A}_\ell}^2\\ &\ge 0.\end{aligned}$$

Thus (G0) implies (G1). □

**Assumption 6.2.6** (G2). We say that **Assumption (G2)** holds iff, for every $1 \le \ell \le L$, we have

$$\mathsf{R}_{\ell-1} = \mathsf{P}_{\ell-1}^{\top} \tag{G2a}$$

and

$$((\mathrm{I}_\ell - \tilde{\Pi}_\ell)\boldsymbol{u}_\ell, \boldsymbol{u}_\ell)_{\mathsf{A}_\ell} \ge 0, \quad \forall\, \boldsymbol{u}_\ell \in \mathbb{R}^{n_\ell}. \tag{G2b}$$

•

**Corollary 6.2.7** (Equivalence). *Assumption* (G1) *is equivalent to Assumption* (G2).

*Proof.* In proving Lemma 6.2.5, we showed, using only the definitions of $\Pi_{\ell-1}$ and $\tilde{\Pi}_\ell$, that

$$(\boldsymbol{u}_\ell, \boldsymbol{u}_\ell)_{\mathsf{A}_\ell} - (\Pi_{\ell-1}\boldsymbol{u}_\ell, \Pi_{\ell-1}\boldsymbol{u}_\ell)_{\mathsf{A}_{\ell-1}} = ((\mathrm{I}_\ell - \tilde{\Pi}_\ell)\boldsymbol{u}_\ell, \boldsymbol{u}_\ell)_{\mathsf{A}_\ell}.$$

Consequently, (G1) holds iff (G2) holds. □

**Theorem 6.2.8** (Error matrix is SPSD). *Assume that* $\mathsf{R}_{\ell-1} = \mathsf{P}_{\ell-1}^{\top}$ *for all* $1 \le \ell \le L$. *If* $m_1 = m_2 = m$, *then*

$$\mathsf{E}_\ell = \mathsf{E}_\ell^*, \quad \forall\, \ell \in \{0, 1, \ldots, L\}.$$

*If, in addition, either* (G1) *or, equivalently,* (G2) *holds, then*

$$(\mathsf{E}_\ell \boldsymbol{u}_\ell, \boldsymbol{u}_\ell)_{\mathsf{A}_\ell} \ge 0, \quad \forall\, \boldsymbol{u}_\ell \in \mathbb{R}^{n_\ell}.$$

*Proof.* The proof proceeds by induction.
(Base cases): The case $\ell = 0$ is trivial. For $\ell = 1$ (the two-grid algorithm), the proof follows the proof of Theorem 3.3.3.
(Induction hypothesis): Assume that, for all $\boldsymbol{u}_{\ell-1}, \boldsymbol{v}_{\ell-1} \in \mathbb{R}^{n_\ell}$,

$$(\mathsf{E}_{\ell-1}\boldsymbol{u}_{\ell-1}, \boldsymbol{v}_{\ell-1})_{\mathsf{A}_{\ell-1}} = (\boldsymbol{u}_{\ell-1}, \mathsf{E}_{\ell-1}\boldsymbol{v}_{\ell-1})_{\mathsf{A}_{\ell-1}}$$

and either (G1) or, equivalently, (G2) holds,

$$(\mathsf{E}_{\ell-1}\boldsymbol{u}_{\ell-1}, \boldsymbol{u}_{\ell-1})_{\mathsf{A}_{\ell-1}} \ge 0.$$

(General case): We will use the definitions of $\Pi_{\ell-1}$ and $\tilde{\Pi}_\ell$ a number of times. Let us recall that

$$\Pi_{\ell-1} = \mathsf{A}_{\ell-1}^{-1}\mathsf{R}_{\ell-1}\mathsf{A}_\ell,$$

which immediately implies

$$\mathsf{A}_{\ell-1}\Pi_{\ell-1} = \mathsf{R}_{\ell-1}\mathsf{A}_\ell.$$

We also have

$$\begin{aligned}\tilde{\Pi}_\ell &= \mathsf{P}_{\ell-1}\Pi_{\ell-1}\\ &= \mathsf{P}_{\ell-1}\mathsf{A}_{\ell-1}^{-1}\mathsf{R}_{\ell-1}\mathsf{A}_\ell.\end{aligned}$$

Let us now evaluate

$$\begin{aligned}(\mathsf{E}_\ell \boldsymbol{u}_\ell, \boldsymbol{v}_\ell)_{\mathsf{A}_\ell} &= ((\mathsf{K}_\ell^*)^m(\mathrm{I}_\ell - \mathsf{P}_{\ell-1}(\mathrm{I}_{\ell-1} - \mathsf{E}_{\ell-1}^p)\Pi_{\ell-1})\mathsf{K}_\ell^m \boldsymbol{u}_\ell, \boldsymbol{v}_\ell)_{\mathsf{A}_\ell}\\
&= ((\mathrm{I}_\ell - \mathsf{P}_{\ell-1}(\mathrm{I}_{\ell-1} - \mathsf{E}_{\ell-1}^p)\Pi_{\ell-1})\mathsf{K}_\ell^m \boldsymbol{u}_\ell, \mathsf{K}_\ell^m \boldsymbol{v}_\ell)_{\mathsf{A}_\ell}\\
&= ((\mathrm{I}_\ell - \mathsf{P}_{\ell-1}\Pi_{\ell-1} + \mathsf{P}_{\ell-1}\mathsf{E}_{\ell-1}^p\Pi_{\ell-1})\mathsf{K}_\ell^m \boldsymbol{u}_\ell, \mathsf{K}_\ell^m \boldsymbol{v}_\ell)_{\mathsf{A}_\ell}\\
&= ((\mathrm{I}_\ell - \mathsf{P}_{\ell-1}\Pi_{\ell-1})\mathsf{K}_\ell^m \boldsymbol{u}_\ell, \mathsf{K}_\ell^m \boldsymbol{v}_\ell)_{\mathsf{A}_\ell}\\
&\quad + (\mathsf{P}_{\ell-1}\mathsf{E}_{\ell-1}^p\Pi_{\ell-1}\mathsf{K}_\ell^m \boldsymbol{u}_\ell, \mathsf{K}_\ell^m \boldsymbol{v}_\ell)_{\mathsf{A}_\ell}\\
&= ((\mathrm{I}_\ell - \tilde{\Pi}_\ell)\mathsf{K}_\ell^m \boldsymbol{u}_\ell, \mathsf{K}_\ell^m \boldsymbol{v}_\ell)_{\mathsf{A}_\ell} + (\mathsf{P}_{\ell-1}\mathsf{E}_{\ell-1}^p\Pi_{\ell-1}\mathsf{K}_\ell^m \boldsymbol{u}_\ell, \mathsf{K}_\ell^m \boldsymbol{v}_\ell)_{\mathsf{A}_\ell}\\
&= (\mathsf{K}_\ell^m \boldsymbol{u}_\ell, (\mathrm{I}_\ell - \tilde{\Pi}_\ell)\mathsf{K}_\ell^m \boldsymbol{v}_\ell)_{\mathsf{A}_\ell} + (\mathsf{P}_{\ell-1}\mathsf{E}_{\ell-1}^p\Pi_{\ell-1}\mathsf{K}_\ell^m \boldsymbol{u}_\ell, \mathsf{A}_\ell\mathsf{K}_\ell^m \boldsymbol{v}_\ell)_\ell\\
&= (\mathsf{K}_\ell^m \boldsymbol{u}_\ell, (\mathrm{I}_\ell - \tilde{\Pi}_\ell)\mathsf{K}_\ell^m \boldsymbol{v}_\ell)_{\mathsf{A}_\ell} + (\mathsf{E}_{\ell-1}^p\Pi_{\ell-1}\mathsf{K}_\ell^m \boldsymbol{u}_\ell, \mathsf{R}_{\ell-1}\mathsf{A}_\ell\mathsf{K}_\ell^m \boldsymbol{v}_\ell)_{\ell-1}\\
&= (\mathsf{K}_\ell^m \boldsymbol{u}_\ell, (\mathrm{I}_\ell - \tilde{\Pi}_\ell)\mathsf{K}_\ell^m \boldsymbol{v}_\ell)_{\mathsf{A}_\ell}\\
&\quad + (\mathsf{E}_{\ell-1}^p\Pi_{\ell-1}\mathsf{K}_\ell^m \boldsymbol{u}_\ell, \mathsf{A}_{\ell-1}\Pi_{\ell-1}\mathsf{K}_\ell^m \boldsymbol{v}_\ell)_{\ell-1}\\
&= (\mathsf{K}_\ell^m \boldsymbol{u}_\ell, (\mathrm{I}_\ell - \tilde{\Pi}_\ell)\mathsf{K}_\ell^m \boldsymbol{v}_\ell)_{\mathsf{A}_\ell} + (\mathsf{E}_{\ell-1}^p\Pi_{\ell-1}\mathsf{K}_\ell^m \boldsymbol{u}_\ell, \Pi_{\ell-1}\mathsf{K}_\ell^m \boldsymbol{v}_\ell)_{\mathsf{A}_{\ell-1}}\\
&= (\mathsf{K}_\ell^m \boldsymbol{u}_\ell, (\mathrm{I}_\ell - \tilde{\Pi}_\ell)\mathsf{K}_\ell^m \boldsymbol{v}_\ell)_{\mathsf{A}_\ell} + (\Pi_{\ell-1}\mathsf{K}_\ell^m \boldsymbol{u}_\ell, \mathsf{E}_{\ell-1}^p\Pi_{\ell-1}\mathsf{K}_\ell^m \boldsymbol{v}_\ell)_{\mathsf{A}_{\ell-1}}\\
&= (\mathsf{K}_\ell^m \boldsymbol{u}_\ell, (\mathrm{I}_\ell - \tilde{\Pi}_\ell)\mathsf{K}_\ell^m \boldsymbol{v}_\ell)_{\mathsf{A}_\ell}\\
&\quad + (\mathsf{A}_{\ell-1}\Pi_{\ell-1}\mathsf{K}_\ell^m \boldsymbol{u}_\ell, \mathsf{E}_{\ell-1}^p\Pi_{\ell-1}\mathsf{K}_\ell^m \boldsymbol{v}_\ell)_{\ell-1}\\
&= (\mathsf{K}_\ell^m \boldsymbol{u}_\ell, (\mathrm{I}_\ell - \tilde{\Pi}_\ell)\mathsf{K}_\ell^m \boldsymbol{v}_\ell)_{\mathsf{A}_\ell} + (\mathsf{R}_{\ell-1}\mathsf{A}_\ell\mathsf{K}_\ell^m \boldsymbol{u}_\ell, \mathsf{E}_{\ell-1}^p\Pi_{\ell-1}\mathsf{K}_\ell^m \boldsymbol{v}_\ell)_{\ell-1}\\
&= (\mathsf{K}_\ell^m \boldsymbol{u}_\ell, (\mathrm{I}_\ell - \tilde{\Pi}_\ell)\mathsf{K}_\ell^m \boldsymbol{v}_\ell)_{\mathsf{A}_\ell} + (\mathsf{A}_\ell\mathsf{K}_\ell^m \boldsymbol{u}_\ell, \mathsf{P}_{\ell-1}\mathsf{E}_{\ell-1}^p\Pi_{\ell-1}\mathsf{K}_\ell^m \boldsymbol{v}_\ell)_\ell\\
&= (\mathsf{K}_\ell^m \boldsymbol{u}_\ell, (\mathrm{I}_\ell - \tilde{\Pi}_\ell)\mathsf{K}_\ell^m \boldsymbol{v}_\ell)_{\mathsf{A}_\ell} + (\mathsf{K}_\ell^m \boldsymbol{u}_\ell, \mathsf{P}_{\ell-1}\mathsf{E}_{\ell-1}^p\Pi_{\ell-1}\mathsf{K}_\ell^m \boldsymbol{v}_\ell)_{\mathsf{A}_\ell}\\
&= (\boldsymbol{u}_\ell, \mathsf{E}_\ell \boldsymbol{v}_\ell)_{\mathsf{A}_\ell}.\end{aligned}$$

Symmetry is thus proven. Notice that in this last computation, we did not assumed that (G2) holds. We only used the definitions and basic properties of $\Pi_{\ell-1}$ and $\tilde{\Pi}_\ell$.

Assume now that (G2) holds. Set $\boldsymbol{v}_\ell = \boldsymbol{u}_\ell$ in the last calculation to obtain

$$(\mathsf{E}_\ell \boldsymbol{u}_\ell, \boldsymbol{u}_\ell)_{\mathsf{A}_\ell} = ((\mathrm{I}_\ell - \tilde{\Pi}_\ell)\mathsf{K}_\ell^m \boldsymbol{u}_\ell, \mathsf{K}_\ell^m \boldsymbol{u}_\ell)_{\mathsf{A}_\ell} + (\mathsf{E}_{\ell-1}^p\Pi_{\ell-1}\mathsf{K}_\ell^m \boldsymbol{u}_\ell, \Pi_{\ell-1}\mathsf{K}_\ell^m \boldsymbol{u}_\ell)_{\mathsf{A}_{\ell-1}}.$$

Now observe that

$$((\mathrm{I}_\ell - \tilde{\Pi}_\ell)\mathsf{K}_\ell^m \boldsymbol{u}_\ell, \mathsf{K}_\ell^m \boldsymbol{u}_\ell)_{\mathsf{A}_\ell} \overset{\text{(G2)}}{\geq} 0$$

$$(\mathsf{E}_{\ell-1}^p \Pi_{\ell-1} \mathsf{K}_\ell^m \boldsymbol{u}_\ell, \Pi_{\ell-1} \mathsf{K}_\ell^m \boldsymbol{u}_\ell)_{\mathsf{A}_{\ell-1}} \geq 0,$$

where the last inequality holds by the induction hypothesis. This implies that

$$(\mathsf{E}_\ell \boldsymbol{u}_\ell, \boldsymbol{u}_\ell)_{\mathsf{A}_\ell} \geq 0,$$

and the result is proven. □

**Remark 6.2.9** (Positive powers). In the proof above, we used the fact that any positive integer power of a self-adjoint positive semidefinite matrix is also positive semidefinite. This is easy to prove and is left as an exercise; see Problem 6.4. ◂

**Corollary 6.2.10** (Symmetric preconditioner). *Let $0 \leq \ell \leq L$, and let $\mathsf{B}_\ell$ be the matrix from Corollary* 6.1.11. *If $m_1 = m_2 = m$, then*

$$\mathsf{B}_\ell = \mathsf{B}_\ell^\top .$$

*Proof.* Exercise; see Problem 6.5. □

## 6.3 Strong and weak approximation properties

The second class of assumptions needed for the analysis of multigrid algorithms is that of approximation properties.

**Assumption 6.3.1** (A0). The multigrid algorithm satisfies the **strong approximation property**: there is a constant $C_{\mathrm{A0}} > 0$ such that, for all $1 \leq \ell \leq L$ and $\boldsymbol{u}_\ell \in \mathbb{R}^{n_\ell}$,

$$\|(\mathrm{I}_\ell - \tilde{\Pi}_\ell)\boldsymbol{u}_\ell\|_\ell^2 \leq \frac{C_{\mathrm{A0}}^2}{\rho_\ell} \|(\mathrm{I}_\ell - \tilde{\Pi}_\ell)\boldsymbol{u}_\ell\|_{\mathsf{A}_\ell}^2, \tag{A0a}$$

where $\rho_\ell := \rho(\mathsf{A}_\ell)$. •

**Assumption 6.3.2** (A1). The multigrid algorithm satisfies the **weak approximation property**: there is $C_{\mathrm{A1}} > 0$ such that, for all $1 \leq \ell \leq L$ and $\boldsymbol{u}_\ell \in \mathbb{R}^{n_\ell}$,

$$((\mathrm{I}_\ell - \tilde{\Pi}_\ell)\boldsymbol{u}_\ell, \boldsymbol{u}_\ell)_{\mathsf{A}_\ell} \leq \frac{C_{\mathrm{A1}}^2}{\rho_\ell} \|\mathsf{A}_\ell \boldsymbol{u}_\ell\|_\ell^2. \tag{A1a}$$

•

**Assumption 6.3.3** (A2). The multigrid algorithm satisfies the $\alpha$**-weak approximation property**: there are constants $\alpha \in (0, 1]$ and $C_{\mathrm{A2}} > 0$ such that, for all $1 \leq \ell \leq L$ and $\boldsymbol{u}_\ell \in \mathbb{R}^{n_\ell}$,

$$|((\mathrm{I}_\ell - \tilde{\Pi}_\ell)\boldsymbol{u}_\ell, \boldsymbol{u}_\ell)_{\mathsf{A}_\ell}| \leq \frac{C_{\mathrm{A2}}^{2\alpha}}{\rho_\ell^\alpha} \|\mathsf{A}_\ell \boldsymbol{u}_\ell\|_\ell^{2\alpha} \|\boldsymbol{u}_\ell\|_{\mathsf{A}_\ell}^{2(1-\alpha)}. \tag{A2a}$$

•

**Theorem 6.3.4** (Strong implies weak). *If the strong Galerkin condition* (G0) *holds, then* (A0) *implies* (A1).

*Proof.* Since (G0) holds,

$$(\mathrm{I}_\ell - \tilde{\Pi}_\ell)^2 \overset{(6.2.3)}{=} \mathrm{I}_\ell - \tilde{\Pi}_\ell.$$

Therefore

$$\begin{aligned}
\|(\mathrm{I}_\ell - \tilde{\Pi}_\ell)\boldsymbol{u}_\ell\|_{\mathsf{A}_\ell}^2 &= ((\mathrm{I}_\ell - \tilde{\Pi}_\ell)\boldsymbol{u}_\ell, (\mathrm{I}_\ell - \tilde{\Pi}_\ell)\boldsymbol{u}_\ell)_{\mathsf{A}_\ell} \\
&= ((\mathrm{I}_\ell - \tilde{\Pi}_\ell)^2\boldsymbol{u}_\ell, \boldsymbol{u}_\ell)_{\mathsf{A}_\ell} \\
&= ((\mathrm{I}_\ell - \tilde{\Pi}_\ell)\boldsymbol{u}_\ell, \boldsymbol{u}_\ell)_{\mathsf{A}_\ell} \\
&= ((\mathrm{I}_\ell - \tilde{\Pi}_\ell)\boldsymbol{u}_\ell, \mathsf{A}_\ell\boldsymbol{u}_\ell)_\ell \\
&\le \|(\mathrm{I}_\ell - \tilde{\Pi}_\ell)\boldsymbol{u}_\ell\|_\ell \|\mathsf{A}_\ell\boldsymbol{u}_\ell\|_\ell \\
&\overset{(\mathrm{A0})}{\le} C_{\mathrm{A0}}\rho_\ell^{-1/2}\|(\mathrm{I}_\ell - \tilde{\Pi}_\ell)\boldsymbol{u}_\ell\|_{\mathsf{A}_\ell}\|\mathsf{A}_\ell\boldsymbol{u}_\ell\|_\ell.
\end{aligned}$$

Thus

$$\|(\mathrm{I}_\ell - \tilde{\Pi}_\ell)\boldsymbol{u}_\ell\|_{\mathsf{A}_\ell} \le C_{\mathrm{A0}}\rho_\ell^{-1/2}\|\mathsf{A}_\ell\boldsymbol{u}_\ell\|_\ell.$$

Squaring, we have

$$((\mathrm{I}_\ell - \tilde{\Pi}_\ell)\boldsymbol{u}_\ell, \boldsymbol{u}_\ell)_{\mathsf{A}_\ell} \le C_{\mathrm{A0}}^2\rho_\ell^{-1}\|\mathsf{A}_\ell\boldsymbol{u}_\ell\|_\ell^2,$$

which is the desired result with $C_{\mathrm{A1}} = C_{\mathrm{A0}}$. □

## 6.4 The first smoothing property

The last collection of assumptions used in the analysis of multigrid algorithms pertains to smoothing properties. In this section, we present the first one.

**Definition 6.4.1** (Richardson's smoother). Let $1 \le \ell \le L$ and assume that $\boldsymbol{u}_\ell^{\mathrm{E}}, \boldsymbol{g}_\ell \in \mathbb{R}^{n_\ell}$ satisfy

$$\mathsf{A}_\ell\boldsymbol{u}_\ell^{\mathrm{E}} = \boldsymbol{g}_\ell.$$

**Richardson's method** for approximating $\boldsymbol{u}_\ell^{\mathrm{E}}$ is the GLIS defined via

$$\boldsymbol{u}_\ell^{(\sigma+1)} := \boldsymbol{u}_\ell^{(\sigma)} + \omega(\boldsymbol{g}_\ell - \mathsf{A}_\ell\boldsymbol{u}_\ell^{(\sigma)}),$$

where $\omega > 0$ is a parameter. Its error transfer matrix is

$$\mathsf{K}_\ell := \mathrm{I}_\ell - \omega\mathsf{A}_\ell = \mathsf{K}_\ell^*.$$

Let $C_R \geq 1$, and let $\Lambda_\ell > 0, 1 \leq \ell \leq L$, be numbers satisfying

$$C_R \rho_\ell \geq \Lambda_\ell \geq \rho_\ell = \rho(\mathsf{A}_\ell). \tag{6.4.1}$$

Choosing

$$\omega := \frac{1}{\Lambda_\ell},$$

we obtain **Richardson's smoother**, and in this case the error transfer matrix becomes

$$\mathsf{K}_\ell = \mathrm{I}_\ell - \Lambda_\ell^{-1}\mathsf{A}_\ell = \mathsf{K}_\ell^*.$$

■

**Assumption 6.4.2** (S1). The multigrid algorithm satisfies the **first smoothing property**: there is $C_{S1} > 0$ such that, for all $1 \leq \ell \leq L$ and $\boldsymbol{u}_\ell \in \mathbb{R}^{n_\ell}$,

$$\|\mathsf{K}_\ell^m \boldsymbol{u}_\ell\|_{\mathsf{A}_\ell^2} \leq C_{S1}\rho_\ell^{1/2} m^{-1/2} \|\boldsymbol{u}_\ell\|_{\mathsf{A}_\ell}. \tag{S1a}$$

•

**Theorem 6.4.3** (Smoothing property). *Richardson's smoother, as described in Definition* 6.4.1, *satisfies the first smoothing property* (S1).

*Proof.* The proof is similar to that of Theorem 5.3.2 and is left as an exercise; see Problem 6.6. □

## 6.5 Convergence of the two-grid method revisited

Let us briefly revisit the convergence of the two-grid method. In fact, the presentation that follows will be our third proof of convergence for the two-grid method. The difference here is that we will use a completely general framework. In other words, there need not be an associated underlying PDE problem to solve. Keep in mind that, as we have mentioned several times before, to obtain the convergence of a "multigrid" method, we always need three components: a Galerkin condition, an approximation property, and a smoothing property.

**Theorem 6.5.1** (Convergence of the two-grid method). *Let $L = 1$ (two-grid) $m_1 = m \geq 1$ and $m_2 = 0$ (one-sided). Assume that the strong Galerkin condition* (G0), *the strong approximation property* (A0), *and the first smoothing property* (S1) *hold. Then*

$$\|\boldsymbol{u}_1^{\mathrm{E}} - \mathrm{TG}(\boldsymbol{f}_1, \boldsymbol{u}_1^{(0)})\|_{\mathsf{A}_1} \leq \frac{C_{A0}C_{S1}}{m^{1/2}}\|\boldsymbol{u}_1^{\mathrm{E}} - \boldsymbol{u}_1^{(0)}\|_{\mathsf{A}_1}, \tag{6.5.1}$$

*where*

$$\mathsf{A}_1\boldsymbol{u}_1^{\mathrm{E}} = \boldsymbol{f}_1.$$

*Written another way,*

$$\|\boldsymbol{e}_1^{k+1}\|_{\mathsf{A}_1} \le C_{\mathrm{A0}} C_{\mathrm{S1}} m^{-1/2} \|\boldsymbol{e}_1^k\|_{\mathsf{A}_1}, \tag{6.5.2}$$

*where* $\boldsymbol{e}_1^k = \boldsymbol{u}_1^{\mathrm{E}} - \boldsymbol{u}_1^{(0)}$ *and* $\boldsymbol{e}_1^{k+1} = \boldsymbol{u}_1^{\mathrm{E}} - \boldsymbol{u}_1^{(3)}$.

*Proof.* Recall that in the present case,

$$\boldsymbol{e}_1^{k+1} = \mathsf{E}_1 \boldsymbol{e}_1^k, \quad \mathsf{E}_1 = (\mathrm{I}_1 - \tilde{\Pi}_1)\mathsf{K}_1^m,$$

or, equivalently,

$$\boldsymbol{u}_1^{\mathrm{E}} - \mathrm{TG}(\boldsymbol{f}_1, \boldsymbol{u}_1^{(0)}) = \mathsf{E}_1(\boldsymbol{u}_1^{\mathrm{E}} - \boldsymbol{u}_1^{(0)}).$$

During the proof of Theorem 6.3.4, we showed that (G0) and (A0) imply

$$\|(\mathrm{I}_\ell - \tilde{\Pi}_\ell)\boldsymbol{u}_\ell\|_{\mathsf{A}_\ell} \le C_{\mathrm{A0}} \rho_\ell^{-1/2} \|\mathsf{A}_\ell \boldsymbol{u}_\ell\|_\ell, \tag{6.5.3}$$

for all $\boldsymbol{u}_\ell \in \mathbb{R}^{n_\ell}$. Applying (6.5.3) (with $\ell = 1$) and using (S1), we have

$$\begin{aligned} \|\boldsymbol{e}_1^{k+1}\|_{\mathsf{A}_1} &= \|(\mathrm{I}_1 - \tilde{\Pi}_1)\mathsf{K}_1^m \boldsymbol{e}_1^k\|_{\mathsf{A}_1} \\ &\overset{(6.5.3)}{\le} C_{\mathrm{A0}} \rho_1^{-1/2} \|\mathsf{A}_1 \mathsf{K}_1^m \boldsymbol{e}_1^k\|_1 \\ &\overset{(\mathrm{S1})}{\le} C_{\mathrm{A0}} \rho_1^{-1/2} C_{\mathrm{S1}} \rho_1^{1/2} m^{-1/2} \|\boldsymbol{e}_1^k\|_{\mathsf{A}_1} \\ &= C_{\mathrm{A0}} C_{\mathrm{S1}} m^{-1/2} \|\boldsymbol{e}_1^k\|_{\mathsf{A}_1}. \end{aligned}$$

□

Next, we will prove that the two-grid method converges even with a significantly weakened approximation property. Before we get that result, we need a technical lemma.

**Lemma 6.5.2** (Stability). *For all* $0 \le \ell \le L$ *and* $\boldsymbol{v}_\ell \in \mathbb{R}^{n_\ell}$, *Richardson's smoother satisfies*

$$\|\mathsf{K}_\ell \boldsymbol{v}_\ell\|_{\mathsf{A}_\ell} \le \|\boldsymbol{v}_\ell\|_{\mathsf{A}_\ell}, \tag{6.5.4}$$

$$(\mathsf{K}_\ell \boldsymbol{v}_\ell, \boldsymbol{v}_\ell)_\ell \le (\boldsymbol{v}_\ell, \boldsymbol{v}_\ell)_\ell. \tag{6.5.5}$$

*Proof.* Let $\ell \ge 0$ and $\boldsymbol{v}_\ell \in \mathbb{R}^{n_\ell}$, and let

$$B_\ell := \{\boldsymbol{w}_\ell^{(1)}, \boldsymbol{w}_\ell^{(2)}, \ldots, \boldsymbol{w}_\ell^{(n_\ell)}\}$$

be a basis of eigenvectors of $\mathsf{A}_\ell$, which is orthonormal with respect to $(\cdot, \cdot)_\ell$. Then there exist unique constants $\alpha_1, \alpha_2, \ldots, \alpha_{n_\ell} \in \mathbb{R}$ such that

$$\boldsymbol{v}_\ell = \sum_{k=1}^{n_\ell} \alpha_k \boldsymbol{w}_\ell^{(k)}.$$

Recall that

$$\mathsf{K}_\ell = \mathrm{I}_\ell - \Lambda_\ell^{-1}\mathsf{A}_\ell$$

with

$$\rho_\ell \le \Lambda_\ell \le C_{\mathrm{R}}\rho_\ell,$$

where $\rho_\ell = \rho(\mathsf{A}_\ell)$, and $C_{\mathrm{R}} \ge 1$ is independent of $\ell$. Then

$$\mathsf{K}_\ell \boldsymbol{w}_\ell^{(k)} = \mu_\ell^{(k)} \boldsymbol{w}_\ell^{(k)},$$

where

$$\mu_\ell^{(k)} := \left(1 - \frac{\lambda_\ell^{(k)}}{\Lambda_\ell}\right).$$

Here $\sigma(\mathsf{A}_\ell) = \{\lambda_\ell^{(k)}\}_{k=1}^{n_\ell}$ and $\sigma(\mathsf{K}_\ell) = \{\mu_\ell^{(k)}\}_{k=1}^{n_\ell}$. Thus

$$\begin{aligned}\|\mathsf{K}_\ell \boldsymbol{v}_\ell\|_{\mathsf{A}_\ell}^2 &= (\mathsf{K}_\ell \boldsymbol{v}_\ell, \mathsf{A}_\ell \mathsf{K}_\ell \boldsymbol{v}_\ell)_\ell \\ &= \sum_{k=1}^{n_\ell} (\mu_\ell^{(k)})^2 \lambda_\ell^{(k)} \alpha_k^2.\end{aligned}$$

Based on (6.4.1),

$$0 \le \mu_\ell^{(k)} = 1 - \frac{\lambda_\ell^{(k)}}{\Lambda_\ell} \le 1, \tag{6.5.6}$$

and we have

$$\|\mathsf{K}_\ell \boldsymbol{v}_\ell\|_{\mathsf{A}_\ell}^2 \le \|\boldsymbol{v}_\ell\|_{\mathsf{A}_\ell}^2.$$

For the second estimate,

$$\begin{aligned}(\mathsf{K}_\ell \boldsymbol{v}_\ell, \boldsymbol{v}_\ell)_\ell &= \left(\sum_{k=1}^{n_\ell} \alpha_k \mathsf{K}_\ell \boldsymbol{w}_\ell^{(k)}, \sum_{j=1}^{n_\ell} \alpha_j \boldsymbol{w}_\ell^{(j)}\right)_\ell \\ &= \left(\sum_{k=1}^{n_\ell} \alpha_k \mu_\ell^{(k)} \boldsymbol{w}_\ell^{(k)}, \sum_{j=1}^{n_\ell} \alpha_j \boldsymbol{w}_\ell^{(j)}\right)_\ell \\ &= \sum_{k=1}^{n_\ell} \alpha_k^2 \mu_\ell^{(k)} \\ &\overset{(6.5.6)}{\le} \sum_{k=1}^{n_\ell} \alpha_k^2 \\ &= (\boldsymbol{v}_\ell, \boldsymbol{v}_\ell)_\ell.\end{aligned}$$

□

**Theorem 6.5.3** (Convergence of the two-grid method). *Suppose that $L = 1$ (two-grid) $m_1 = m \geq 1$ and $m_2 = 0$ (one-sided). Assume that* (G0) *and* (A2) *hold and smoothing is performed with the Richardson method. Then*

$$\|\boldsymbol{u}_1^{\mathrm{E}} - \mathrm{TG}(\boldsymbol{f}_1, \boldsymbol{u}_1^{(0)})\|_{\mathsf{A}_1} \leq \left(\frac{C_{\mathrm{A2}} C_{\mathrm{S1}}}{m^{1/2}}\right)^{\alpha} \|\boldsymbol{u}_1^{\mathrm{E}} - \boldsymbol{u}_1^{(0)}\|_{\mathsf{A}_1}, \tag{6.5.7}$$

*where*

$$\mathsf{A}_1 \boldsymbol{u}_1^{\mathrm{E}} = \boldsymbol{f}_1.$$

*In other words,*

$$\|\boldsymbol{e}_1^{k+1}\|_{\mathsf{A}_1} \leq \left(\frac{C_{\mathrm{A2}} C_{\mathrm{S1}}}{m^{1/2}}\right)^{\alpha} \|\boldsymbol{e}_1^{k}\|_{\mathsf{A}_1}, \tag{6.5.8}$$

*where $\boldsymbol{e}_1^k = \boldsymbol{u}_1^{\mathrm{E}} - \boldsymbol{u}_1^{(0)}$ and $\boldsymbol{e}_1^{k+1} = \boldsymbol{u}_1^{\mathrm{E}} - \boldsymbol{u}_1^{(3)}$.*

*Proof.* Once again, we recall that for the two-grid method,

$$\boldsymbol{e}_1^{k+1} = \mathsf{E}_1 \boldsymbol{e}_1^k, \quad \mathsf{E}_1 = (\mathrm{I}_1 - \tilde{\Pi}_1)\mathsf{K}_1^m.$$

Equivalently,

$$\boldsymbol{u}_1^{\mathrm{E}} - \mathrm{TG}(\boldsymbol{f}_1, \boldsymbol{u}_1^{(0)}) = \mathsf{E}_1(\boldsymbol{u}_1^{\mathrm{E}} - \boldsymbol{u}_1^{(0)}).$$

Since the strong Galerkin condition (G0) holds,

$$(\mathrm{I}_1 - \tilde{\Pi}_1)^2 = \mathrm{I}_1 - \tilde{\Pi}_1,$$

and it is always true that

$$(\mathrm{I}_1 - \tilde{\Pi}_1)^* = \mathrm{I}_1 - \tilde{\Pi}_1.$$

Then the $\alpha$-weak approximation property (A2) can be expressed as follows: for all $\boldsymbol{u}_\ell \in \mathbb{R}^{n_\ell}$,

$$\begin{aligned}
\|(\mathrm{I}_1 - \tilde{\Pi}_1)\boldsymbol{u}_\ell\|_{\mathsf{A}_1}^2 &= ((\mathrm{I}_1 - \tilde{\Pi}_1)\boldsymbol{u}_\ell, \boldsymbol{u}_\ell)_{\mathsf{A}_\ell} \\
&\leq \frac{C_{\mathrm{A2}}^{2\alpha}}{\rho_1^\alpha} \|\mathsf{A}_1 \boldsymbol{u}_\ell\|_1^{2\alpha} \|\boldsymbol{u}_\ell\|_{\mathsf{A}_1}^{2(1-\alpha)},
\end{aligned}$$

for some $C_{\mathrm{A2}} > 0$.

Next, recall that Richardson's method satisfies the smoothing property (S1). Using this fact, the stability of Richardson's method, and the $\alpha$-weak approximation property (A2), we have

$$
\begin{aligned}
\|e_1^{k+1}\|_{A_1} &= \|(I_1 - \tilde{\Pi}_1)K_1^m e_1^k\|_{A_1} \\
&\le \frac{C_{A2}^\alpha}{\rho_1^{\alpha/2}} \|A_1 K_1^m e_1^k\|_1^\alpha \|K_1^m e_1^k\|_{A_1}^{(1-\alpha)} \\
&\overset{(6.5.4)}{\le} \frac{C_{A2}^\alpha}{\rho_1^{\alpha/2}} \|A_1 K_1^m e_1^k\|_1^\alpha \|e_1^k\|_{A_1}^{(1-\alpha)} \\
&\overset{(S1)}{\le} \frac{C_{A2}^\alpha}{\rho_1^{\alpha/2}} C_{S1}^\alpha \rho_1^{\alpha/2} m^{-\alpha/2} \|e_1^k\|_{A_1} \\
&= \left(\frac{C_{A2}C_{S1}}{m^{1/2}}\right)^\alpha \|e_1^k\|_{A_1}.
\end{aligned}
$$

□

## 6.6 Convergence of the W-cycle algorithm

In this section, we prove that the W-cycle converges, provided that we perform enough smoothing iterations per cycle. The proof of convergence of the W-cycle algorithm uses a so-called a perturbation argument. Basically, we will show that the error is equal to the error in the two-grid method plus a perturbation that we can control.

**Theorem 6.6.1** (Convergence of the one-sided W-cycle). *Let $p \ge 2$, $m_1 = m \in \mathbb{N}$, and $m_2 = 0$ (one-sided). Assume that* (G0) *and* (A0) *hold and that the smoothing is done by Richardson's smoother. Then for any $0 < \gamma < 1$, there is a sufficiently large $m$ such that, for all $0 \le \ell \le L$,*

$$
\|u_\ell^E - \mathrm{MG}(\ell, g_\ell, u_\ell^{(0)})\|_{A_\ell} \le \gamma \|u_\ell^E - u_\ell^{(0)}\|_{A_\ell},
$$

*where*

$$
A_\ell u_\ell^E = g_\ell.
$$

*In particular, it suffices to choose $m \in \mathbb{N}$ such that*

$$
m \ge \left(\frac{C_{A0}C_{S1}}{\gamma - \gamma^p}\right)^2. \tag{6.6.1}
$$

*Proof.* Fix $\gamma \in (0,1)$ and choose $m \in \mathbb{N}$ that satisfies (6.6.1). The proof is by induction.
(Base cases): The cases $\ell = 0$ and $\ell = 1$ (Theorem 6.5.1) are clearly true.
(Induction hypothesis): Fix $2 \le \ell \le L$ and assume that

$$
\|E_{\ell-1} w_{\ell-1}\|_{A_{\ell-1}} \le \gamma \|w_{\ell-1}\|_{A_{\ell-1}},
$$

for all $w_{\ell-1} \in \mathbb{R}^{n_{\ell-1}}$.

(General case): Recall that the residual $\boldsymbol{r}_{\ell-1}^{(1)} \in \mathbb{R}^{n_{\ell-1}}$ and the *exact* coarse-grid correction $\boldsymbol{q}_{\ell-1}^{(1,\mathrm{E})} \in \mathbb{R}^{n_{\ell-1}}$ are related by

$$\mathsf{A}_{\ell-1}\boldsymbol{q}_{\ell-1}^{(1,\mathrm{E})} = \boldsymbol{r}_{\ell-1}^{(1)}.$$

Then

$$\begin{aligned}\boldsymbol{u}_\ell^{\mathrm{E}} - \mathrm{MG}(\ell, \boldsymbol{g}_\ell, \boldsymbol{u}_\ell^{(0)}) &= \boldsymbol{u}_\ell^{\mathrm{E}} - \boldsymbol{u}_\ell^{(2)} \\ &= \boldsymbol{u}_\ell^{\mathrm{E}} - [\boldsymbol{u}_\ell^{(1)} + \mathsf{P}_{\ell-1}\boldsymbol{q}_{\ell-1}^{(1)}] \\ &= \boldsymbol{u}_\ell^{\mathrm{E}} - (\boldsymbol{u}_\ell^{(1)} + \mathsf{P}_{\ell-1}\boldsymbol{q}_{\ell-1}^{(1,\mathrm{E})}) + \mathsf{P}_{\ell-1}(\boldsymbol{q}_{\ell-1}^{(1,\mathrm{E})} - \boldsymbol{q}_{\ell-1}^{(1)}) \\ &= \boldsymbol{u}_\ell^{\mathrm{E}} - \mathrm{TG}(\boldsymbol{g}_\ell, \boldsymbol{u}_\ell^{(0)}) + \mathsf{P}_{\ell-1}(\boldsymbol{q}_{\ell-1}^{(1,\mathrm{E})} - \boldsymbol{q}_{\ell-1}^{(1)}).\end{aligned}$$

Observe that by Theorem 6.4.3 Richardson's smoother satisfies (S1). Thus

$$\begin{aligned}\|\boldsymbol{u}_\ell^{\mathrm{E}} - \mathrm{MG}(\ell, \boldsymbol{g}_\ell, \boldsymbol{u}_\ell^{(0)})\|_{\mathsf{A}_\ell} &\le \|\boldsymbol{u}_\ell^{\mathrm{E}} - \mathrm{TG}(\boldsymbol{g}_\ell, \boldsymbol{u}_\ell^{(0)})\|_{\mathsf{A}_\ell} \\ &\quad + \|\mathsf{P}_{\ell-1}(\boldsymbol{q}_{\ell-1}^{(1,\mathrm{E})} - \boldsymbol{q}_{\ell-1}^{(1)})\|_{\mathsf{A}_\ell} \\ &\overset{(6.5.1)}{\le} C_{\mathrm{A0}} C_{\mathrm{S1}} m^{-1/2} \|\boldsymbol{u}_\ell^{\mathrm{E}} - \boldsymbol{u}_\ell^{(0)}\|_{\mathsf{A}_\ell} \\ &\quad + \|\mathsf{P}_{\ell-1}(\boldsymbol{q}_{\ell-1}^{(1,\mathrm{E})} - \boldsymbol{q}_{\ell-1}^{(1)})\|_{\mathsf{A}_\ell}.\end{aligned} \tag{6.6.2}$$

Observe now that, for all $\boldsymbol{w}_{\ell-1} \in \mathbb{R}^{n_{\ell-1}}$,

$$\begin{aligned}\|\mathsf{P}_{\ell-1}\boldsymbol{w}_{\ell-1}\|_{\mathsf{A}_\ell}^2 &= (\mathsf{P}_{\ell-1}\boldsymbol{w}_{\ell-1}, \mathsf{P}_{\ell-1}\boldsymbol{w}_{\ell-1})_{\mathsf{A}_\ell} \\ &= (\mathsf{P}_{\ell-1}\boldsymbol{w}_{\ell-1}, \mathsf{A}_\ell\mathsf{P}_{\ell-1}\boldsymbol{w}_{\ell-1})_\ell \\ &= (\boldsymbol{w}_{\ell-1}, \mathsf{P}_{\ell-1}^{\mathsf{T}}\mathsf{A}_\ell\mathsf{P}_{\ell-1}\boldsymbol{w}_{\ell-1})_{\ell-1} \\ &= (\boldsymbol{w}_{\ell-1}, \mathsf{R}_{\ell-1}\mathsf{A}_\ell\mathsf{P}_{\ell-1}\boldsymbol{w}_{\ell-1})_{\ell-1} \\ &\overset{(\mathrm{G0})}{=} (\boldsymbol{w}_{\ell-1}, \mathsf{A}_{\ell-1}\boldsymbol{w}_{\ell-1})_{\ell-1} \\ &= (\boldsymbol{w}_{\ell-1}, \boldsymbol{w}_{\ell-1})_{\mathsf{A}_{\ell-1}} \\ &= \|\boldsymbol{w}_{\ell-1}\|_{\mathsf{A}_{\ell-1}}^2.\end{aligned}$$

In the proof of Theorem 6.1.10, we showed that

$$\begin{aligned}\boldsymbol{q}_{\ell-1}^{(1,\mathrm{E})} - \boldsymbol{q}_{\ell-1}^{(1)} &= \mathsf{E}_{\ell-1}^{p}\boldsymbol{q}_{\ell-1}^{(1,\mathrm{E})} \\ &\overset{(6.1.2)}{=} \mathsf{E}_{\ell-1}^{p}\Pi_{\ell-1}(\boldsymbol{u}_\ell^{\mathrm{E}} - \boldsymbol{u}_\ell^{(1)}) \\ &= \mathsf{E}_{\ell-1}^{p}\Pi_{\ell-1}\mathsf{K}_\ell^{m}(\boldsymbol{u}_\ell^{\mathrm{E}} - \boldsymbol{u}_\ell^{(0)}).\end{aligned}$$

Using the induction hypothesis, we have

$$\|\mathsf{P}_{\ell-1}(\boldsymbol{q}_{\ell-1}^{(1,\mathrm{E})} - \boldsymbol{q}_{\ell-1}^{(1)})\|_{\mathsf{A}_\ell} = \|\boldsymbol{q}_{\ell-1}^{(1,\mathrm{E})} - \boldsymbol{q}_{\ell-1}^{(1)}\|_{\mathsf{A}_{\ell-1}}$$

$$
\begin{aligned}
&= \|\mathsf{E}_{\ell-1}^{p}\Pi_{\ell-1}\mathsf{K}_\ell^m(\boldsymbol{u}_\ell^{\mathrm{E}} - \boldsymbol{u}_\ell^{(0)})\|_{\mathsf{A}_{\ell-1}} \\
&\le \gamma^p \|\Pi_{\ell-1}\mathsf{K}_\ell^m(\boldsymbol{u}_\ell^{\mathrm{E}} - \boldsymbol{u}_\ell^{(0)})\|_{\mathsf{A}_{\ell-1}}.
\end{aligned}
$$

Since we are assuming that the strong Galerkin condition (G0) holds, it follows that

$$
\|\Pi_{\ell-1}\boldsymbol{w}_\ell\|_{\mathsf{A}_{\ell-1}} = \|\tilde{\Pi}_\ell\boldsymbol{w}_\ell\|_{\mathsf{A}_\ell}.
$$

Furthermore,

$$
\begin{aligned}
\|\tilde{\Pi}_\ell\boldsymbol{w}_\ell\|_{\mathsf{A}_\ell}^2 &= (\tilde{\Pi}_\ell\boldsymbol{w}_\ell, \tilde{\Pi}_\ell\boldsymbol{w}_\ell)_{\mathsf{A}_\ell} \\
&= (\tilde{\Pi}_\ell^2\boldsymbol{w}_\ell, \boldsymbol{w}_\ell)_{\mathsf{A}_\ell} \\
&\overset{(6.2.1)}{=} (\tilde{\Pi}_\ell\boldsymbol{w}_\ell, \boldsymbol{w}_\ell)_{\mathsf{A}_\ell} \\
&\le \|\tilde{\Pi}_\ell\boldsymbol{w}_\ell\|_{\mathsf{A}_\ell}\|\boldsymbol{w}_\ell\|_{\mathsf{A}_\ell},
\end{aligned}
$$

which brings us to the estimate

$$
\|\tilde{\Pi}_\ell\boldsymbol{w}_\ell\|_{\mathsf{A}_\ell} \le \|\boldsymbol{w}_\ell\|_{\mathsf{A}_\ell}. \tag{6.6.3}
$$

Therefore

$$
\begin{aligned}
\|\mathsf{P}_{\ell-1}(\boldsymbol{q}_{\ell-1}^{(1,\mathrm{E})} - \boldsymbol{q}_{\ell-1}^{(1)})\|_{\mathsf{A}_\ell} &\le \gamma^p\|\Pi_{\ell-1}\mathsf{K}_\ell^m(\boldsymbol{u}_\ell^{\mathrm{E}} - \boldsymbol{u}_\ell^{(0)})\|_{\mathsf{A}_{\ell-1}} \\
&= \gamma^p\|\tilde{\Pi}_\ell\mathsf{K}_\ell^m(\boldsymbol{u}_\ell^{\mathrm{E}} - \boldsymbol{u}_\ell^{(0)})\|_{\mathsf{A}_\ell} \\
&\overset{(6.6.3)}{\le} \gamma^p\|\mathsf{K}_\ell^m(\boldsymbol{u}_\ell^{\mathrm{E}} - \boldsymbol{u}_\ell^{(0)})\|_{\mathsf{A}_\ell} \\
&\overset{(6.5.4)}{\le} \gamma^p\|\boldsymbol{u}_\ell^{\mathrm{E}} - \boldsymbol{u}_\ell^{(0)}\|_{\mathsf{A}_\ell}.
\end{aligned}
$$

Combining this with estimate (6.6.2), we have

$$
\begin{aligned}
\|\boldsymbol{u}_\ell^{\mathrm{E}} - \mathrm{MG}(\ell, \boldsymbol{g}_\ell, \boldsymbol{u}_\ell^{(0)})\|_{\mathsf{A}_\ell} &\le C_{\mathrm{A0}}C_{\mathrm{S1}}m^{-1/2}\|\boldsymbol{u}_\ell^{\mathrm{E}} - \boldsymbol{u}_\ell^{(0)}\|_{\mathsf{A}_\ell} \\
&\quad + \|\mathsf{P}_{\ell-1}(\boldsymbol{q}_{\ell-1}^{(1,\mathrm{E})} - \boldsymbol{q}_{\ell-1}^{(1)})\|_{\mathsf{A}_\ell} \\
&\le C_{\mathrm{A0}}C_{\mathrm{S1}}m^{-1/2}\|\boldsymbol{u}_\ell^{\mathrm{E}} - \boldsymbol{u}_\ell^{(0)}\|_{\mathsf{A}_\ell} + \gamma^p\|\boldsymbol{u}_\ell^{\mathrm{E}} - \boldsymbol{u}_\ell^{(0)}\|_{\mathsf{A}_\ell} \\
&\le (C_{\mathrm{A0}}C_{\mathrm{S1}}m^{-1/2} + \gamma^p)\|\boldsymbol{u}_\ell^{\mathrm{E}} - \boldsymbol{u}_\ell^{(0)}\|_{\mathsf{A}_\ell} \\
&= C_\ell\|\boldsymbol{u}_\ell^{\mathrm{E}} - \boldsymbol{u}_\ell^{(0)}\|_{\mathsf{A}_\ell},
\end{aligned}
$$

where

$$
C_\ell \coloneqq C_{\mathrm{A0}}C_{\mathrm{S1}}m^{-1/2} + \gamma^p.
$$

Using (6.6.1), we see that

$$C_\ell = C_{A0}C_{S1}\left(\frac{C_{A0}C_{S1}}{\gamma - \gamma^p}\right)^{-1} + \gamma^p = \gamma,$$

as claimed. □

**Remark 6.6.2** (W-cycle). Notice that we need $p > 1$ for this argument to work. Otherwise, $\gamma - \gamma^p$, which is the denominator in (6.6.1), is zero, and $m$ would need to be infinitely large to get convergence. ◀

We can weaken our approximation property assumptions and still achieve convergence.

**Theorem 6.6.3** (Convergence of the one-sided W-cycle). *Let $p \geq 2$, $m_1 = m \in \mathbb{N}$, and $m_2 = 0$ (one-sided). Assume that* (G0) *and* (A2) *hold and the smoothing is done by Richardson's smoother. Then for any $0 < \gamma < 1$, $m \in \mathbb{N}$ can be chosen large enough so that, for all $\ell \geq 0$,*

$$\|\boldsymbol{u}_\ell^{\mathrm{E}} - \mathrm{MG}(\ell, \boldsymbol{g}_\ell, \boldsymbol{u}_\ell^{(0)})\|_{\mathsf{A}_\ell} \leq \gamma \|\boldsymbol{u}_\ell^{\mathrm{E}} - \boldsymbol{u}_\ell^{(0)}\|_{\mathsf{A}_\ell},$$

*where*

$$\mathsf{A}_\ell \boldsymbol{u}_\ell^{\mathrm{E}} = \boldsymbol{g}_\ell.$$

*In particular, it suffices to choose $m \in \mathbb{N}$ such that*

$$m \geq \left(\frac{C_{A2}C_{S1}}{(\gamma - \gamma^p)^{1/\alpha}}\right)^2.$$

*Proof.* The proof is quite similar to the previous case and is based on a perturbation argument. The main difference is that we need to use the $\alpha$-weak convergence of the two-grid method guaranteed by Theorem 6.5.3; see Problem 6.7. □

## 6.7 Convergence of the simple symmetric V-cycle

In this section, we prove that the simple symmetric V-cycle algorithm ($p = 1$ and $m_1 = m_2 = 1$) converges. First, we need a new useful smoothing assumption.

**Assumption 6.7.1** (S2). The multigrid algorithm satisfies the **second smoothing property**: there is $C_{S2} > 0$ such that, for every $\ell \geq 1$,

$$\|\boldsymbol{v}_\ell\|_\ell^2 \leq \rho_\ell C_{S2}^2 (\overline{\mathsf{K}}_\ell \boldsymbol{v}_\ell, \boldsymbol{v}_\ell)_\ell, \tag{S2a}$$

for all $\boldsymbol{v}_\ell \in \mathbb{R}^{n_\ell}$, where

$$\overline{\mathsf{K}}_\ell := (\mathsf{I}_\ell - \mathsf{K}_\ell^* \mathsf{K}_\ell)\mathsf{A}_\ell^{-1}.$$

•

**Lemma 6.7.2** (Smoothing property). *Richardson's smoother satisfies Assumption* (S2) *with* $\mathsf{S}_\ell = \Lambda_\ell^{-1}\mathrm{I}_\ell = \mathsf{S}_\ell^\top$.

*Proof.* Recall that the choice of $\Lambda_\ell$ guarantees

$$\rho_\ell \le \Lambda_\ell \le C_\mathrm{R}\rho_\ell,$$

for some $C_\mathrm{R} \ge 1$ that is independent of $\ell$. Then

$$\begin{aligned}\overline{\mathsf{K}}_\ell &= (\mathrm{I}_\ell - \mathsf{K}_\ell^*\mathsf{K}_\ell)\mathsf{A}_\ell^{-1}\\ &= (\mathrm{I}_\ell - (\mathrm{I}_\ell - \Lambda_\ell^{-1}\mathsf{A}_\ell)(\mathrm{I}_\ell - \Lambda_\ell^{-1}\mathsf{A}_\ell))\mathsf{A}_\ell^{-1}\\ &= (\mathrm{I}_\ell - (\mathrm{I}_\ell - 2\Lambda_\ell^{-1}\mathsf{A}_\ell + \Lambda_\ell^{-2}\mathsf{A}_\ell^2))\mathsf{A}_\ell^{-1}\\ &= 2\Lambda_\ell^{-1}\mathrm{I}_\ell - \Lambda_\ell^{-2}\mathsf{A}_\ell.\end{aligned}$$

Define

$$\mathsf{J}_\ell := \rho_\ell C_\mathrm{R}\overline{\mathsf{K}}_\ell - \mathrm{I}_\ell,$$

and notice that if we can show that $\mathsf{J}_\ell$ is SPSD with respect to $(\cdot,\cdot)_\ell$, then we get (S2) with $C_{\mathrm{S2}}^2 = C_\mathrm{R}$.

Clearly, $\mathsf{J}_\ell$ is symmetric with respect to $(\cdot,\cdot)_\ell$, that is, $\mathsf{J}_\ell^\top = \mathsf{J}_\ell$. Let now $\{\boldsymbol{v}_\ell^{(1)}, \boldsymbol{v}_\ell^{(2)}, \dots, \boldsymbol{v}_\ell^{(n_\ell)}\}$ be a basis of eigenvectors of $\mathsf{A}_\ell$ that is orthonormal with respect to $(\cdot,\cdot)_\ell$. Then

$$\begin{aligned}\mathsf{J}_\ell\boldsymbol{v}_\ell^{(k)} &= \rho_\ell C_\mathrm{R}\overline{\mathsf{K}}_\ell\boldsymbol{v}_\ell^{(k)} - \boldsymbol{v}_\ell^{(k)}\\ &= \rho_\ell C_\mathrm{R}(2\Lambda_\ell^{-1}\mathrm{I}_\ell - \Lambda_\ell^{-2}\mathsf{A}_\ell)\boldsymbol{v}_\ell^{(k)} - \boldsymbol{v}_\ell^{(k)}\\ &= 2\rho_\ell C_\mathrm{R}\Lambda_\ell^{-1}\boldsymbol{v}_\ell^{(k)} - \rho_\ell C_\mathrm{R}\Lambda_\ell^{-2}\lambda_\ell^{(k)}\boldsymbol{v}_\ell^{(k)} - \boldsymbol{v}_\ell^{(k)}\\ &= (2\rho_\ell C_\mathrm{R}\Lambda_\ell^{-1} - \rho_\ell C_\mathrm{R}\Lambda_\ell^{-2}\lambda_\ell^{(k)} - 1)\boldsymbol{v}_\ell^{(k)}.\end{aligned}$$

Set

$$\eta_\ell^{(k)} := 2\rho_\ell C_\mathrm{R}\Lambda_\ell^{-1} - \rho_\ell C_\mathrm{R}\Lambda_\ell^{-2}\lambda_\ell^{(k)} - 1.$$

We want to show that $\eta_\ell^{(k)} \ge 0$, for all $1 \le k \le n_\ell$. Indeed, since $-\lambda_\ell^{(k)} \ge -\Lambda_\ell$ and $C_\mathrm{R}\rho_\ell \ge \Lambda_\ell$,

$$\begin{aligned}\eta_\ell^{(k)} &= 2C_\mathrm{R}\frac{\rho_\ell}{\Lambda_\ell} - C_\mathrm{R}\frac{\rho_\ell\lambda_\ell^{(k)}}{\Lambda_\ell^2} - 1\\ &\ge 2C_\mathrm{R}\frac{\rho_\ell}{\Lambda_\ell} - C_\mathrm{R}\frac{\rho_\ell}{\Lambda_\ell} - 1\\ &= C_\mathrm{R}\frac{\rho_\ell}{\Lambda_\ell} - 1\\ &\ge 1 - 1\\ &= 0.\end{aligned}$$

Thus the eigenvalues of $\mathsf{J}_\ell$, which we denote $\eta_\ell^{(k)}$, are all nonnegative, and $\mathsf{J}_\ell$ is SPSD. This implies

$$0 \leq (\mathsf{J}_\ell \boldsymbol{v}_\ell, \boldsymbol{v}_\ell)_\ell = \rho_\ell C_{\mathrm{R}}(\overline{\mathsf{K}}_\ell \boldsymbol{v}_\ell, \boldsymbol{v}_\ell)_\ell - (\boldsymbol{v}_\ell, \boldsymbol{v}_\ell)_\ell,$$

and (S2) follows with $C_{\mathrm{S2}}^2 = C_{\mathrm{R}}$. □

Next, we need two more technical lemmas.

**Lemma 6.7.3** (Positive difference). *Let $0 \leq \ell \leq L$. Suppose that $\mathsf{J}_\ell \in \mathbb{R}^{n_\ell \times n_\ell}$ is $\mathsf{A}_\ell$-symmetric, denoted, as usual, by $\mathsf{J}_\ell = \mathsf{J}_\ell^*$, but otherwise arbitrary. Then*

$$(\mathsf{J}_\ell \boldsymbol{v}_\ell, \mathsf{J}_\ell \boldsymbol{v}_\ell)_{\mathsf{A}_\ell} - (\mathsf{J}_\ell^2 \boldsymbol{v}_\ell, \mathsf{J}_\ell^2 \boldsymbol{v}_\ell)_{\mathsf{A}_\ell} \leq (\boldsymbol{v}_\ell, \boldsymbol{v}_\ell)_{\mathsf{A}_\ell} - (\mathsf{J}_\ell \boldsymbol{v}_\ell, \mathsf{J}_\ell \boldsymbol{v}_\ell)_{\mathsf{A}_\ell}, \tag{6.7.1}$$

*for all $\boldsymbol{v}_\ell \in \mathbb{R}^{n_\ell}$.*

*Proof.* Let $\boldsymbol{v}_\ell \in \mathbb{R}^{n_\ell}$. Since $\mathsf{A}_\ell$ is SPD,

$$\begin{aligned}
0 &\leq \|(\mathrm{I}_\ell - \mathsf{J}_\ell^2)\boldsymbol{v}_\ell\|_{\mathsf{A}_\ell}^2 \\
&= ((\mathrm{I}_\ell - \mathsf{J}_\ell^2)\boldsymbol{v}_\ell, (\mathrm{I}_\ell - \mathsf{J}_\ell^2)\boldsymbol{v}_\ell)_{\mathsf{A}_\ell} \\
&= (\mathrm{I}_\ell \boldsymbol{v}_\ell - \mathsf{J}_\ell^2 \boldsymbol{v}_\ell, \mathrm{I}_\ell \boldsymbol{v}_\ell - \mathsf{J}_\ell^2 \boldsymbol{v}_\ell)_{\mathsf{A}_\ell} \\
&= (\boldsymbol{v}_\ell, \boldsymbol{v}_\ell)_{\mathsf{A}_\ell} - (\mathsf{J}_\ell^2 \boldsymbol{v}_\ell, \boldsymbol{v}_\ell)_{\mathsf{A}_\ell} - (\boldsymbol{v}_\ell, \mathsf{J}_\ell^2 \boldsymbol{v}_\ell)_{\mathsf{A}_\ell} + (\mathsf{J}_\ell^2 \boldsymbol{v}_\ell, \mathsf{J}_\ell^2 \boldsymbol{v}_\ell)_{\mathsf{A}_\ell} \\
&\overset{\mathsf{J}_\ell = \mathsf{J}_\ell^*}{=} (\boldsymbol{v}_\ell, \boldsymbol{v}_\ell)_{\mathsf{A}_\ell} - (\mathsf{J}_\ell \boldsymbol{v}_\ell, \mathsf{J}_\ell \boldsymbol{v}_\ell)_{\mathsf{A}_\ell} - (\mathsf{J}_\ell \boldsymbol{v}_\ell, \mathsf{J}_\ell \boldsymbol{v}_\ell)_{\mathsf{A}_\ell} + (\mathsf{J}_\ell^2 \boldsymbol{v}_\ell, \mathsf{J}_\ell^2 \boldsymbol{v}_\ell)_{\mathsf{A}_\ell} \\
&= (\boldsymbol{v}_\ell, \boldsymbol{v}_\ell)_{\mathsf{A}_\ell} - 2(\mathsf{J}_\ell \boldsymbol{v}_\ell, \mathsf{J}_\ell \boldsymbol{v}_\ell)_{\mathsf{A}_\ell} + (\mathsf{J}_\ell^2 \boldsymbol{v}_\ell, \mathsf{J}_\ell^2 \boldsymbol{v}_\ell)_{\mathsf{A}_\ell}.
\end{aligned}$$

Rearranging, we get

$$(\mathsf{J}_\ell \boldsymbol{v}_\ell, \mathsf{J}_\ell \boldsymbol{v}_\ell)_{\mathsf{A}_\ell} - (\mathsf{J}_\ell^2 \boldsymbol{v}_\ell, \mathsf{J}_\ell^2 \boldsymbol{v}_\ell)_{\mathsf{A}_\ell} \leq (\boldsymbol{v}_\ell, \boldsymbol{v}_\ell)_{\mathsf{A}_\ell} - (\mathsf{J}_\ell \boldsymbol{v}_\ell, \mathsf{J}_\ell \boldsymbol{v}_\ell)_{\mathsf{A}_\ell}.$$ □

**Lemma 6.7.4** (Ritz projection). *Assume that the restriction and prolongation matrices form a balanced symmetric pair, that is, for every $1 \leq \ell \leq L$, we have $\mathsf{P}_{\ell-1} = \mathsf{R}_{\ell-1}^\top$. Then, for all $\ell$ and $\boldsymbol{v}_\ell \in \mathbb{R}^{n_\ell}$,*

$$(\Pi_{\ell-1} \boldsymbol{v}_\ell, \Pi_{\ell-1} \boldsymbol{v}_\ell)_{\mathsf{A}_{\ell-1}} = (\boldsymbol{v}_\ell, \boldsymbol{v}_\ell)_{\mathsf{A}_\ell} - ((\mathrm{I}_\ell - \tilde{\Pi}_\ell)\boldsymbol{v}_\ell, \boldsymbol{v}_\ell)_{\mathsf{A}_\ell}. \tag{6.7.2}$$

*Proof.* Recall that we always have

$$\mathsf{R}_{\ell-1}\mathsf{A}_\ell = \mathsf{A}_{\ell-1}\Pi_{\ell-1}$$

and

$$\begin{aligned}
\tilde{\Pi}_\ell &= \mathsf{P}_{\ell-1}\mathsf{A}_{\ell-1}^{-1}\mathsf{R}_{\ell-1}\mathsf{A}_\ell \\
&= \mathsf{P}_{\ell-1}\Pi_{\ell-1}.
\end{aligned}$$

Then

$$\begin{aligned}(\Pi_{\ell-1}\boldsymbol{v}_\ell, \Pi_{\ell-1}\boldsymbol{v}_\ell)_{\mathsf{A}_{\ell-1}} &= (\Pi_{\ell-1}\boldsymbol{v}_\ell, \mathsf{A}_{\ell-1}\Pi_{\ell-1}\boldsymbol{v}_\ell)_{\ell-1} \\ &= (\Pi_{\ell-1}\boldsymbol{v}_\ell, \mathsf{R}_{\ell-1}\mathsf{A}_\ell\boldsymbol{v}_\ell)_{\ell-1} \\ &= (\mathsf{R}_{\ell-1}^\top\Pi_{\ell-1}\boldsymbol{v}_\ell, \mathsf{A}_\ell\boldsymbol{v}_\ell)_\ell \\ &= (\mathsf{P}_{\ell-1}\Pi_{\ell-1}\boldsymbol{v}_\ell, \mathsf{A}_\ell\boldsymbol{v}_\ell)_\ell \\ &= (\tilde{\Pi}_\ell\boldsymbol{v}_\ell, \mathsf{A}_\ell\boldsymbol{v}_\ell)_\ell \\ &= (\tilde{\Pi}_\ell\boldsymbol{v}_\ell, \boldsymbol{v}_\ell)_{\mathsf{A}_\ell} \\ &= (\boldsymbol{v}_\ell, \boldsymbol{v}_\ell)_{\mathsf{A}_\ell} - ((\mathrm{I}_\ell - \tilde{\Pi}_\ell)\boldsymbol{v}_\ell, \boldsymbol{v}_\ell)_{\mathsf{A}_\ell}.\end{aligned}$$ □

The simple symmetric V-cycle method is obtained by setting $p = m_1 = m_2 = 1$. It is somewhat surprising that this method converges, because only one presmoothing and one postsmoothing iteration are performed.

**Theorem 6.7.5** (Contraction). *Let $p = 1$ and $m_1 = m_2 = m = 1$ and assume that* (G1), (A1), *and* (S2) *hold. Assume, in addition, that $\mathsf{S}_\ell = \mathsf{S}_\ell^\top$. Then*

$$0 \le (\mathsf{E}_\ell\boldsymbol{u}_\ell, \boldsymbol{u}_\ell)_{\mathsf{A}_\ell} \le \frac{C_{\mathrm{A1}}^2 C_{\mathrm{S2}}^2}{C_{\mathrm{A1}}^2 C_{\mathrm{S2}}^2 + 1}(\boldsymbol{u}_\ell, \boldsymbol{u}_\ell)_{\mathsf{A}_\ell},$$

*for all $\boldsymbol{u}_\ell \in \mathbb{R}^{n_\ell}$.*

*Proof.* Recall that since $p = m_1 = m_2 = 1$ and $\mathsf{S}_\ell = \mathsf{S}_\ell^\top$,

$$\mathsf{E}_\ell \overset{(6.1.1)}{=} \mathsf{K}_\ell(\mathrm{I}_\ell - \tilde{\Pi}_\ell)\mathsf{K}_\ell + \mathsf{K}_\ell\mathsf{P}_{\ell-1}\mathsf{E}_{\ell-1}\Pi_{\ell-1}\mathsf{K}_\ell.$$

In particular, notice that

$$\mathsf{K}_\ell^* = \mathrm{I}_\ell - \mathsf{S}_\ell^\top\mathsf{A}_\ell = \mathrm{I}_\ell - \mathsf{S}_\ell\mathsf{A}_\ell = \mathsf{K}_\ell.$$

Now set

$$\begin{aligned}\boldsymbol{w}_\ell &:= \mathsf{K}_\ell\boldsymbol{u}_\ell, \\ T_1 &:= ((\mathrm{I}_\ell - \tilde{\Pi}_\ell)\boldsymbol{w}_\ell, \boldsymbol{w}_\ell)_{\mathsf{A}_\ell}, \\ T_2 &:= (\mathsf{P}_{\ell-1}\mathsf{E}_{\ell-1}\Pi_{\ell-1}\boldsymbol{w}_\ell, \boldsymbol{w}_\ell)_{\mathsf{A}_\ell}.\end{aligned}$$

Then

$$(\mathsf{E}_\ell\boldsymbol{u}_\ell, \boldsymbol{u}_\ell)_{\mathsf{A}_\ell} = T_1 + T_2.$$

Let us first consider $T_1$:

$$\begin{aligned}
T_1 &= ((\mathrm{I}_\ell - \tilde{\Pi}_\ell)\boldsymbol{w}_\ell, \boldsymbol{w}_\ell)_{\mathsf{A}_\ell} \\
&\overset{(A1)}{\leq} C_{A1}^2 \rho_\ell^{-1} \|\mathsf{A}_\ell \boldsymbol{w}_\ell\|_\ell^2 \\
&= C_{A1}^2 \rho_\ell^{-1} \|\mathsf{A}_\ell \mathsf{K}_\ell \boldsymbol{u}_\ell\|_\ell^2 \\
&\overset{(S2)}{\leq} C_{A1}^2 \rho_\ell^{-1} C_{S2}^2 \rho_\ell (\overline{\mathsf{K}}_\ell \mathsf{A}_\ell \mathsf{K}_\ell \boldsymbol{u}_\ell, \mathsf{A}_\ell \mathsf{K}_\ell \boldsymbol{u}_\ell)_\ell \\
&= C_{A1}^2 C_{S2}^2 ((\mathrm{I}_\ell - \mathsf{K}_\ell^* \mathsf{K}_\ell)\mathsf{A}_\ell^{-1} \mathsf{A}_\ell \mathsf{K}_\ell \boldsymbol{u}_\ell, \mathsf{A}_\ell \mathsf{K}_\ell \boldsymbol{u}_\ell)_\ell \\
&= C_{A1}^2 C_{S2}^2 ((\mathrm{I}_\ell - \mathsf{K}_\ell^* \mathsf{K}_\ell)\mathsf{K}_\ell \boldsymbol{u}_\ell, \mathsf{A}_\ell \mathsf{K}_\ell \boldsymbol{u}_\ell)_\ell \\
&= C_{A1}^2 C_{S2}^2 ((\mathsf{K}_\ell \boldsymbol{u}_\ell, \mathsf{A}_\ell \mathsf{K}_\ell \boldsymbol{u}_\ell)_\ell - (\mathsf{K}_\ell^* \mathsf{K}_\ell \mathsf{K}_\ell \boldsymbol{u}_\ell, \mathsf{A}_\ell \mathsf{K}_\ell \boldsymbol{u}_\ell)_\ell) \\
&= C_{A1}^2 C_{S2}^2 ((\mathsf{K}_\ell \boldsymbol{u}_\ell, \mathsf{K}_\ell \boldsymbol{u}_\ell)_{\mathsf{A}_\ell} - (\mathsf{K}_\ell^* \mathsf{K}_\ell \mathsf{K}_\ell \boldsymbol{u}_\ell, \mathsf{K}_\ell \boldsymbol{u}_\ell)_{\mathsf{A}_\ell}) \\
&= C_{A1}^2 C_{S2}^2 \{(\mathsf{K}_\ell \boldsymbol{u}_\ell, \mathsf{K}_\ell \boldsymbol{u}_\ell)_{\mathsf{A}_\ell} - (\mathsf{K}_\ell^2 \boldsymbol{u}_\ell, \mathsf{K}_\ell^2 \boldsymbol{u}_\ell)_{\mathsf{A}_\ell}\} \\
&\overset{(6.7.1)}{\leq} C_{A1}^2 C_{S2}^2 \{(\boldsymbol{u}_\ell, \boldsymbol{u}_\ell)_{\mathsf{A}_\ell} - (\mathsf{K}_\ell \boldsymbol{u}_\ell, \mathsf{K}_\ell \boldsymbol{u}_\ell)_{\mathsf{A}_\ell}\}.
\end{aligned} \tag{6.7.3}$$

We are now ready to begin the proof per se. To shorten the notation, we set

$$\gamma := \frac{C_{A1}^2 C_{S2}^2}{C_{A1}^2 C_{S2}^2 + 1}$$

and proceed by induction. The base case is trivial, and so it is skipped. Let now $1 \leq \ell \leq L$. (Induction hypothesis): Assume that, for all $\boldsymbol{w}_{\ell-1} \in \mathbb{R}^{n_{\ell-1}}$,

$$(\mathsf{E}_{\ell-1}\boldsymbol{w}_{\ell-1}, \boldsymbol{w}_{\ell-1})_{\mathsf{A}_{\ell-1}} \leq \gamma(\boldsymbol{w}_{\ell-1}, \boldsymbol{w}_{\ell-1})_{\mathsf{A}_{\ell-1}}.$$

(General case): Now we turn to the bound for $T_2$. First, note that

$$\begin{aligned}
T_2 &= (\mathsf{E}_{\ell-1}\Pi_{\ell-1}\boldsymbol{w}_\ell, \mathsf{R}_{\ell-1}\mathsf{A}_\ell \boldsymbol{w}_\ell)_{\ell-1} \\
&= (\mathsf{E}_{\ell-1}\Pi_{\ell-1}\boldsymbol{w}_\ell, \mathsf{A}_{\ell-1}\Pi_{\ell-1}\boldsymbol{w}_\ell)_{\ell-1} \\
&= (\mathsf{E}_{\ell-1}\Pi_{\ell-1}\boldsymbol{w}_\ell, \Pi_{\ell-1}\boldsymbol{w}_\ell)_{\mathsf{A}_{\ell-1}}.
\end{aligned}$$

Then, by the induction hypothesis,

$$\begin{aligned}
T_2 &\leq \gamma(\Pi_{\ell-1}\boldsymbol{w}_\ell, \Pi_{\ell-1}\boldsymbol{w}_\ell)_{\mathsf{A}_{\ell-1}} \\
&\overset{(6.7.2)}{=} \gamma((\boldsymbol{w}_\ell, \boldsymbol{w}_\ell)_{\mathsf{A}_\ell} - ((\mathrm{I}_\ell - \tilde{\Pi}_\ell)\boldsymbol{w}_\ell, \boldsymbol{w}_\ell)_{\mathsf{A}_\ell}) \\
&= \gamma(\boldsymbol{w}_\ell, \boldsymbol{w}_\ell)_{\mathsf{A}_\ell} - \gamma T_1.
\end{aligned} \tag{6.7.4}$$

To conclude,

$$\begin{aligned}
(\mathsf{E}_\ell \boldsymbol{u}_\ell, \boldsymbol{u}_\ell)_{\mathsf{A}_\ell} &= T_1 + T_2 \\
&\overset{(6.7.4)}{\leq} (1-\gamma)T_1 + \gamma(\boldsymbol{w}_\ell, \boldsymbol{w}_\ell)_{\mathsf{A}_\ell}
\end{aligned}$$

$$
\begin{aligned}
&\overset{(6.7.3)}{\leq} (1-\gamma)C_{\mathrm{A}1}^2C_{\mathrm{S}2}^2\{(\boldsymbol{u}_\ell,\boldsymbol{u}_\ell)_{\mathsf{A}_\ell} - (\boldsymbol{w}_\ell,\boldsymbol{w}_\ell)_{\mathsf{A}_\ell}\} + \gamma(\boldsymbol{w}_\ell,\boldsymbol{w}_\ell)_{\mathsf{A}_\ell} \\
&= \left(1-\frac{C_{\mathrm{A}1}^2C_{\mathrm{S}2}^2}{C_{\mathrm{A}1}^2C_{\mathrm{S}2}^2+1}\right)C_{\mathrm{A}1}^2C_{\mathrm{S}2}^2\{(\boldsymbol{u}_\ell,\boldsymbol{u}_\ell)_{\mathsf{A}_\ell} - (\boldsymbol{w}_\ell,\boldsymbol{w}_\ell)_{\mathsf{A}_\ell}\} \\
&\quad + \frac{C_{\mathrm{A}1}^2C_{\mathrm{S}2}^2}{C_{\mathrm{A}1}^2C_{\mathrm{S}2}^2+1}(\boldsymbol{w}_\ell,\boldsymbol{w}_\ell)_{\mathsf{A}_\ell} \\
&= \frac{C_{\mathrm{A}1}^2C_{\mathrm{S}2}^2}{C_{\mathrm{A}1}^2C_{\mathrm{S}2}^2+1}[(\boldsymbol{u}_\ell,\boldsymbol{u}_\ell)_{\mathsf{A}_\ell} - (\boldsymbol{w}_\ell,\boldsymbol{w}_\ell)_{\mathsf{A}_\ell}] \\
&\quad + \frac{C_{\mathrm{A}1}^2C_{\mathrm{S}2}^2}{C_{\mathrm{A}1}^2C_{\mathrm{S}2}^2+1}(\boldsymbol{w}_\ell,\boldsymbol{w}_\ell)_{\mathsf{A}_\ell} \\
&= \gamma((\boldsymbol{u}_\ell,\boldsymbol{u}_\ell)_{\mathsf{A}_\ell} - (\boldsymbol{w}_\ell,\boldsymbol{w}_\ell)_{\mathsf{A}_\ell}) + \gamma(\boldsymbol{w}_\ell,\boldsymbol{w}_\ell)_{\mathsf{A}_\ell} \\
&= \gamma(\boldsymbol{u}_\ell,\boldsymbol{u}_\ell)_{\mathsf{A}_\ell}. \qquad \square
\end{aligned}
$$

**Corollary 6.7.6** (Convergence of simple symmetric V-cycle). *In the setting of Theorem* 6.7.5, *assume that* $\boldsymbol{u}_\ell^{\mathrm{E}}, \boldsymbol{g}_\ell \in \mathbb{R}^{n_\ell}$ *satisfy*

$$\mathsf{A}_\ell \boldsymbol{u}_\ell^{\mathrm{E}} = \boldsymbol{g}_\ell.$$

*Then, for any* $\boldsymbol{u}_\ell^{(0)} \in \mathbb{R}^{n_\ell}$,

$$\|\boldsymbol{u}_\ell^{\mathrm{E}} - \mathrm{MG}(\ell, \boldsymbol{g}_\ell, \boldsymbol{u}_\ell^{(0)})\|_{\mathsf{A}_\ell} \leq \frac{M}{M+1}\|\boldsymbol{u}_\ell^{\mathrm{E}} - \boldsymbol{u}_\ell^{(0)}\|_{\mathsf{A}_\ell},$$

*where*

$$M := C_{\mathrm{A}1}^2 C_{\mathrm{S}2}^2.$$

*Proof.* We only need to show that, for all $\boldsymbol{v}_\ell \in \mathbb{R}^{n_\ell}$,

$$\|\mathsf{E}_\ell \boldsymbol{v}_\ell\|_{\mathsf{A}_\ell} \leq \frac{M}{M+1}\|\boldsymbol{v}_\ell\|_{\mathsf{A}_\ell}.$$

Since, for $\ell \geq 1$, $\mathsf{E}_\ell$ is SPSD with respect to $(\cdot,\cdot)_{\mathsf{A}_\ell}$, there is a basis of eigenvectors $\{\boldsymbol{w}_\ell^{(1)}, \ldots, \boldsymbol{w}_\ell^{(n_\ell)}\}$ of $\mathsf{E}_\ell$ such that

$$
\begin{aligned}
\mathsf{E}_\ell \boldsymbol{w}_\ell^{(j)} &= \epsilon_\ell^{(j)} \boldsymbol{w}_\ell^{(j)}, \\
(\boldsymbol{w}_\ell^{(i)}, \boldsymbol{w}_\ell^{(j)})_{\mathsf{A}_\ell} &= \delta_{ij},
\end{aligned}
$$

and

$$0 \leq \epsilon_\ell^{(1)} \leq \epsilon_\ell^{(2)} \leq \cdots \leq \epsilon_\ell^{(n_\ell)}.$$

Suppose now that

$$\boldsymbol{v}_\ell = \sum_{k=1}^{n_\ell} c_k \boldsymbol{w}_\ell^{(k)}.$$

Then

$$(\mathsf{E}_\ell \boldsymbol{v}_\ell, \boldsymbol{v}_\ell)_{\mathsf{A}_\ell} = \sum_{k=1}^{n_\ell} c_k^2 \epsilon_\ell^{(k)},$$

and

$$(\boldsymbol{v}_\ell, \boldsymbol{v}_\ell)_{\mathsf{A}_\ell} = \sum_{k=1}^{n_\ell} c_k^2.$$

Theorem 6.7.5 guarantees that

$$\sum_{k=1}^{n_\ell} c_k^2 \epsilon_\ell^{(k)} \le \frac{M}{M+1} \sum_{k=1}^{n_\ell} c_k^2,$$

for all $c_1, \ldots, c_{n_\ell} \in \mathbb{R}$. This, in particular, implies that

$$0 \le \epsilon_\ell^{(k)} \le \frac{M}{M+1}, \quad 1 \le k \le n_\ell.$$

Therefore

$$\begin{aligned} \|\mathsf{E}_\ell \boldsymbol{v}_\ell\|_{\mathsf{A}_\ell}^2 &= (\mathsf{E}_\ell \boldsymbol{v}_\ell, \mathsf{E}_\ell \boldsymbol{v}_\ell)_{\mathsf{A}_\ell} \\ &= \sum_{k=1}^{n_\ell} c_k^2 (\epsilon_\ell^{(k)})^2 \\ &\le \left(\frac{M}{M+1}\right)^2 \sum_{k=1}^{n_\ell} c_k^2 \\ &= \left(\frac{M}{M+1}\right)^2 \|\boldsymbol{v}_\ell\|_{\mathsf{A}_\ell}^2. \end{aligned}$$

□

## 6.8 The Braess–Hackbusch result

We now consider the general symmetric V-cycle. Our goal is to show that the convergence rate can be improved if more smoothing steps are performed. We first need a technical lemma.

**Lemma 6.8.1** (Richardson's smoother). *Suppose that smoothing is done with Richardson's smoother, that is,*

$$\mathsf{S}_\ell = \Lambda_\ell^{-1} \mathsf{I}_\ell,$$

*where*

$$\rho_\ell \le \Lambda_\ell \le C_{\mathrm{R}} \rho_\ell,$$

*for some $C_R \geq 1$ that is independent of $\ell$. Then, for all $m \geq 1$, $\ell \geq 1$, and all $\boldsymbol{v}_\ell \in \mathbb{R}^{n_\ell}$, we have*

$$((\mathsf{I}_\ell - \mathsf{K}_\ell)\mathsf{K}_\ell^{2m}\boldsymbol{v}_\ell, \boldsymbol{v}_\ell)_{\mathsf{A}_\ell} \leq \frac{1}{2m}((\mathsf{I}_\ell - \mathsf{K}_\ell^{2m})\boldsymbol{v}_\ell, \boldsymbol{v}_\ell)_{\mathsf{A}_\ell}. \tag{6.8.1}$$

*Consequently,*

$$\rho_\ell^{-1}\|\mathsf{A}_\ell\mathsf{K}_\ell^m\boldsymbol{v}_\ell\|_\ell^2 \leq \frac{C_R}{2m}(\|\boldsymbol{v}_\ell\|_{\mathsf{A}_\ell}^2 - \|\mathsf{K}_\ell^m\boldsymbol{v}_\ell\|_{\mathsf{A}_\ell}^2). \tag{6.8.2}$$

*Proof.* Let $i, j \in \mathbb{Z}$ with $0 \leq j \leq i$. Then

$$\begin{aligned}((\mathsf{I}_\ell - \mathsf{K}_\ell)\mathsf{K}_\ell^i\boldsymbol{v}_\ell, \boldsymbol{v}_\ell)_{\mathsf{A}_\ell} &= (\mathsf{A}_\ell(\mathsf{I}_\ell - \mathsf{K}_\ell)\mathsf{K}_\ell^i\boldsymbol{v}_\ell, \boldsymbol{v}_\ell)_\ell \\ &= \Lambda_\ell^{-1}(\mathsf{A}_\ell^2\mathsf{K}_\ell^i\boldsymbol{v}_\ell, \boldsymbol{v}_\ell)_\ell \\ &= \Lambda_\ell^{-1}(\mathsf{K}_\ell^i\mathsf{A}_\ell\boldsymbol{v}_\ell, \mathsf{A}_\ell\boldsymbol{v}_\ell)_\ell \\ &\overset{(6.5.5)}{\leq} \Lambda_\ell^{-1}(\mathsf{K}_\ell^j\mathsf{A}_\ell\boldsymbol{v}_\ell, \mathsf{A}_\ell\boldsymbol{v}_\ell)_\ell \\ &= ((\mathsf{I}_\ell - \mathsf{K}_\ell)\mathsf{K}_\ell^j\boldsymbol{v}_\ell, \boldsymbol{v}_\ell)_{\mathsf{A}_\ell}.\end{aligned} \tag{6.8.3}$$

Therefore

$$\begin{aligned}2m((\mathsf{I}_\ell - \mathsf{K}_\ell)\mathsf{K}_\ell^{2m}\boldsymbol{v}_\ell, \boldsymbol{v}_\ell)_{\mathsf{A}_\ell} &= \sum_{j=0}^{2m-1}((\mathsf{I}_\ell - \mathsf{K}_\ell)\mathsf{K}_\ell^{2m}\boldsymbol{v}_\ell, \boldsymbol{v}_\ell)_{\mathsf{A}_\ell} \\ &\overset{(6.8.3)}{\leq} \sum_{j=0}^{2m-1}((\mathsf{I}_\ell - \mathsf{K}_\ell)\mathsf{K}_\ell^{j}\boldsymbol{v}_\ell, \boldsymbol{v}_\ell)_{\mathsf{A}_\ell} \\ &= ((\mathsf{I}_\ell - \mathsf{K}_\ell^{2m})\boldsymbol{v}_\ell, \boldsymbol{v}_\ell)_{\mathsf{A}_\ell},\end{aligned}$$

where the last equality follows since the sum telescopes.

To prove the second estimate, we first use

$$\begin{aligned}\rho_\ell^{-1}\|\mathsf{A}_\ell\mathsf{K}_\ell^m\boldsymbol{v}_\ell\|_\ell^2 &= \rho_\ell^{-1}(\mathsf{A}_\ell\mathsf{K}_\ell^m\boldsymbol{v}_\ell, \mathsf{A}_\ell\mathsf{K}_\ell^m\boldsymbol{v}_\ell)_\ell \\ &= \rho_\ell^{-1}(\mathsf{A}_\ell\mathsf{K}_\ell^m\boldsymbol{v}_\ell, \mathsf{K}_\ell^m\boldsymbol{v}_\ell)_{\mathsf{A}_\ell} \\ &= \rho_\ell^{-1}\Lambda_\ell((\mathsf{I}_\ell - \mathsf{K}_\ell)\mathsf{K}_\ell^m\boldsymbol{u}_\ell, \mathsf{K}_\ell^m\boldsymbol{u}_\ell)_{\mathsf{A}_\ell} \\ &= \rho_\ell^{-1}\Lambda_\ell((\mathsf{I}_\ell - \mathsf{K}_\ell)\mathsf{K}_\ell^{2m}\boldsymbol{v}_\ell, \boldsymbol{v}_\ell)_{\mathsf{A}_\ell} \\ &\overset{(6.8.1)}{\leq} \frac{\Lambda_\ell}{2m\rho_\ell}((\mathsf{I}_\ell - \mathsf{K}_\ell^{2m})\boldsymbol{v}_\ell, \boldsymbol{v}_\ell)_{\mathsf{A}_\ell} \\ &\leq \frac{C_R}{2m}((\mathsf{I}_\ell - \mathsf{K}_\ell^{2m})\boldsymbol{v}_\ell, \boldsymbol{v}_\ell)_{\mathsf{A}_\ell} \\ &= \frac{C_R}{2m}\{(\boldsymbol{v}_\ell, \boldsymbol{v}_\ell)_{\mathsf{A}_\ell} - (\mathsf{K}_\ell^m\boldsymbol{v}_\ell, \mathsf{K}_\ell^m\boldsymbol{v}_\ell)_{\mathsf{A}_\ell}\}.\end{aligned}$$

□

We are now in a position to give the famous result of Braess and Hackbusch [5], which provided the first uniform convergence estimate for the V-cycle algorithm. The proof given here is a simplified, streamlined version of the original based on the presentation in [10].

**Theorem 6.8.2** (Braess–Hackbusch). *Let $p = 1$ and $m_1 = m_2 = m \geq 1$. Assume that* (G1) *and* (A1) *hold. If the smoothing is done with Richardson's smoother, then*

$$0 \leq (\mathsf{E}_\ell \boldsymbol{u}_\ell, \boldsymbol{u}_\ell)_{\mathsf{A}_\ell} \leq \gamma(\boldsymbol{u}_\ell, \boldsymbol{u}_\ell)_{\mathsf{A}_\ell},$$

*for all $\boldsymbol{u}_\ell \in \mathbb{R}^{n_\ell}$, where*

$$\gamma := \frac{M}{M+m}, \quad M := \frac{C_{\mathrm{A1}}^2 C_{\mathrm{R}}}{2}.$$

*Proof.* The proof is similar to that of Theorem 6.7.5. We begin with an expression for the error propagation matrix

$$\mathsf{E}_\ell = \mathsf{K}_\ell^m(\mathrm{I}_\ell - \tilde{\Pi}_\ell)\mathsf{K}_\ell^m + \mathsf{K}_\ell^m \mathsf{P}_{\ell-1}\mathsf{E}_{\ell-1}\Pi_{\ell-1}\mathsf{K}_\ell^m,$$

where

$$\mathsf{K}_\ell = \mathrm{I}_\ell - \Lambda_\ell^{-1}\mathsf{A}_\ell = \mathsf{K}_\ell^*,$$

and, for some $C_{\mathrm{R}} \geq 1$,

$$\rho_\ell \leq \Lambda_\ell \leq C_{\mathrm{R}}\rho_\ell.$$

As before, for $\boldsymbol{u}_\ell \in \mathbb{R}^{n_\ell}$, set

$$\begin{aligned} \boldsymbol{w}_\ell &:= \mathsf{K}_\ell^m \boldsymbol{u}_\ell, \\ T_1 &:= ((\mathrm{I}_\ell - \tilde{\Pi}_\ell)\boldsymbol{w}_\ell, \boldsymbol{w}_\ell)_{\mathsf{A}_\ell}, \\ T_2 &:= (\mathsf{P}_{\ell-1}\mathsf{E}_{\ell-1}\Pi_{\ell-1}\boldsymbol{w}_\ell, \boldsymbol{w}_\ell)_{\mathsf{A}_\ell}. \end{aligned}$$

Then

$$(\mathsf{E}_\ell \boldsymbol{u}_\ell, \boldsymbol{u}_\ell)_{\mathsf{A}_\ell} = T_1 + T_2.$$

We first estimate $T_1$:

$$\begin{aligned} T_1 &= ((\mathrm{I}_\ell - \tilde{\Pi}_\ell)\boldsymbol{w}_\ell, \boldsymbol{w}_\ell)_{\mathsf{A}_\ell} \\ &\overset{\text{(A1)}}{\leq} C_{\mathrm{A1}}^2 \rho_\ell^{-1} \|\mathsf{A}_\ell \boldsymbol{w}_\ell\|_\ell^2 \\ &= C_{\mathrm{A1}}^2 \rho_\ell^{-1} \|\mathsf{A}_\ell \mathsf{K}_\ell^m \boldsymbol{u}_\ell\|_\ell^2 \\ &\overset{(6.8.2)}{\leq} \frac{M}{m}\{(\boldsymbol{u}_\ell, \boldsymbol{u}_\ell)_{\mathsf{A}_\ell} - (\boldsymbol{w}_\ell, \boldsymbol{w}_\ell)_{\mathsf{A}_\ell}\}. \end{aligned} \tag{6.8.4}$$

Exactly as in the proof of Theorem 6.7.5, an induction argument yields

$$T_2 \le \gamma(\boldsymbol{w}_\ell, \boldsymbol{w}_\ell)_{\mathsf{A}_\ell} - \gamma T_1. \tag{6.8.5}$$

Therefore

$$\begin{aligned}(\mathsf{E}_\ell \boldsymbol{u}_\ell, \boldsymbol{u}_\ell)_{\mathsf{A}_\ell} &= T_1 + T_2 \\ &\overset{(6.8.5)}{\le} (1-\gamma)T_1 + \gamma(\boldsymbol{w}_\ell, \boldsymbol{w}_\ell)_{\mathsf{A}_\ell} \\ &\overset{(6.8.4)}{\le} (1-\gamma)\frac{M}{m}\{(\boldsymbol{u}_\ell, \boldsymbol{u}_\ell)_{\mathsf{A}_\ell} - (\boldsymbol{w}_\ell, \boldsymbol{w}_\ell)_{\mathsf{A}_\ell}\} + \gamma(\boldsymbol{w}_\ell, \boldsymbol{w}_\ell)_{\mathsf{A}_\ell} \\ &= \gamma(\boldsymbol{u}_\ell, \boldsymbol{u}_\ell)_{\mathsf{A}_\ell}.\end{aligned}$$

□

**Corollary 6.8.3** (Convergence of general symmetric V-cycle). *In the setting of Theorem 6.8.2, assume that for $0 \le \ell \le L$, $\boldsymbol{u}_\ell^{\mathrm{E}}, \boldsymbol{g}_\ell \in \mathbb{R}^{n_\ell}$ satisfy*

$$\mathsf{A}_\ell \boldsymbol{u}_\ell^{\mathrm{E}} = \boldsymbol{g}_\ell.$$

*Then, for any $\boldsymbol{u}_\ell^{(0)} \in \mathbb{R}^{n_\ell}$,*

$$\|\boldsymbol{u}_\ell^{\mathrm{E}} - \mathrm{MG}(\ell, \boldsymbol{g}_\ell, \boldsymbol{u}_\ell^{(0)})\|_{\mathsf{A}_\ell} \le \frac{M}{M+m}\|\boldsymbol{u}_\ell^{\mathrm{E}} - \boldsymbol{u}_\ell^{(0)}\|_{\mathsf{A}_\ell},$$

*where we recall that $m \ge 1$ and*

$$M = \frac{C_{\mathrm{A1}}^2 C_{\mathrm{R}}}{2}.$$

*Proof.* The proof is similar to that of Corollary 6.7.6 and thus left as an exercise; see Problem 6.8. □

## 6.9 When the standard Galerkin condition fails

In this section, we describe a particular setting in which the "standard" Galerkin conditions (G0)–(G2), fail. We will show that, with a little extra effort, multigrid convergence may still be attained. This general theory, which is based on ideas from [6, 8], will apply to the analysis of multigrid algorithms for the cell-centered finite difference method discussed in Chapter 8.

**Assumption 6.9.1** (G3). The multigrid method satisfies an **imbalanced Galerkin condition**: for all $\ell$,

$$\mathsf{A}_\ell = \mathsf{A}_\ell^\top, \quad \boldsymbol{v}_\ell^\top \mathsf{A}_\ell \boldsymbol{v} > 0,\ \forall\, \boldsymbol{v}_\ell \in \mathbb{R}^{n_\ell}, \quad 0 \le \ell \le L, \tag{G3a}$$

$$\mathrm{rank}(\mathsf{P}_{\ell-1}) = \mathrm{rank}(\mathsf{R}_{\ell-1}) = n_{\ell-1}, \quad 1 \le \ell \le L, \tag{G3b}$$

$$R_{\ell-1} = rP_{\ell-1}^{\top}, \ \exists r \in (0,1), \quad 1 \le \ell \le L, \tag{G3c}$$

$$R_{\ell-1}A_\ell P_{\ell-1} = 2A_{\ell-1}, \quad 1 \le \ell \le L. \tag{G3d}$$

•

**Remark 6.9.2** (Failure of the Galerkin condition). If Assumption (G3) holds, then clearly the strong Galerkin condition (G0) fails. Assumption (G3) implies

$$A_{\ell-1} = \frac{r}{2}P_{\ell-1}^{\top}A_\ell P_{\ell-1},$$

for all $1 \le \ell \le L$. ◀

Let us look at some clear implications of Assumption (G3).

**Proposition 6.9.3** (Not a projection). *Let $1 \le \ell \le L$. Recall that*

$$\tilde{\Pi}_\ell := P_{\ell-1}A_{\ell-1}^{-1}R_{\ell-1}A_\ell.$$

*If assumption* (G3) *holds, then*

$$\tilde{\Pi}_\ell^2 = 2\tilde{\Pi}_\ell.$$

*In other words, the coarse-grid Ritz projection matrix $\tilde{\Pi}_\ell$ is not a projection. However, it is still true that*

$$\tilde{\Pi}_\ell^* = \tilde{\Pi}_\ell,$$

*that is, $\tilde{\Pi}_\ell$ is still self-adjoint in the $A_\ell$ inner product.*

*Proof.* Exercise; see Problem 6.9. □

Since the coarse-grid Ritz projection is not a projection, we introduce a surrogate. For $1 \le \ell \le L$, define

$$\hat{\Pi}_\ell := \frac{1}{2}\tilde{\Pi}_\ell. \tag{6.9.1}$$

It turns out that this rescaling has suitable properties.

**Proposition 6.9.4** (Projection). *For $1 \le \ell \le L$, we have*

$$\hat{\Pi}_\ell^* = \hat{\Pi}_\ell. \tag{6.9.2}$$

*In addition, if the imbalanced Galerkin condition, that is, Assumption* (G3), *holds, then*

$$\hat{\Pi}_\ell^2 = \hat{\Pi}_\ell. \tag{6.9.3}$$

*In other words, if Assumption* (G3) *holds, then $\hat{\Pi}_\ell$ is a projection matrix.*

*Proof.* Exercise; see Problem 6.10. □

From the last result we immediately have the following:

**Corollary 6.9.5** (Projection). *Let, for all* $1 \le \ell \le L$, $\hat{\Pi}_\ell$ *be defined in* (6.9.1). *Then*

$$(\mathrm{I}_\ell - \hat{\Pi}_\ell)^* = \mathrm{I}_\ell - \hat{\Pi}_\ell. \tag{6.9.4}$$

*Furthermore, if Assumption* (G3) *holds, then*

$$(\mathrm{I}_\ell - \hat{\Pi}_\ell)^2 = \mathrm{I}_\ell - \hat{\Pi}_\ell. \tag{6.9.5}$$

*Proof.* Exercise; see Problem 6.11. □

Let us now prove that a version of the two-grid algorithm converges. To do so, we recall that the two-grid error transfer matrix $\mathsf{E}_1$ is defined as

$$\mathsf{E}_1 = (\mathsf{K}_1^*)^{m_2}(\mathrm{I}_1 - \tilde{\Pi}_1)\mathsf{K}_1^{m_1}.$$

**Theorem 6.9.6** ($\alpha$–Convergence of the two-grid Method). *Let* $m_1 = m_2 = m \in \mathbb{N}$. *Assume that* (G3) *and* (A2) *hold, and that the smoothing is performed using Richardson's smoother. Then we have*

$$\left|(\mathsf{E}_1 \boldsymbol{v}_1, \boldsymbol{v}_1)_{\mathsf{A}_1}\right| \le \left(\frac{C_{\mathrm{A2}} C_{\mathrm{S1}}}{m^{1/2}}\right)^{2\alpha} (\boldsymbol{v}_1, \boldsymbol{v}_1)_{\mathsf{A}_1},$$

*for all* $\boldsymbol{v}_1 \in \mathbb{R}^{n_1}$.

*Proof.* The proof is similar to that of Theorem 6.5.3. Let $\boldsymbol{v}_1 \in \mathbb{R}^{n_1}$. The $\alpha$-weak approximation property (A2) implies that

$$\begin{aligned}
\left|(\mathsf{E}_1 \boldsymbol{v}_1, \boldsymbol{v}_1)_{\mathsf{A}_1}\right| &= \left|((\mathrm{I}_1 - \tilde{\Pi}_1)\mathsf{K}_1^m \boldsymbol{v}_1, \mathsf{K}_1^m \boldsymbol{v}_1)_{\mathsf{A}_1}\right| \\
&\le \frac{C_{\mathrm{A2}}^{2\alpha}}{\rho_1^\alpha} \|\mathsf{A}_1 \mathsf{K}_1^m \boldsymbol{v}_1\|_1^{2\alpha} \|\mathsf{K}_1^m \boldsymbol{v}_1\|_{\mathsf{A}_1}^{2(1-\alpha)},
\end{aligned}$$

for some $C_{\mathrm{A2}} > 0$. Next, recall that Richardson's smoother satisfies Assumption (S1). Using this fact, the stability of Richardson's method, and the $\alpha$-weak approximation property (A2), we have

$$\begin{aligned}
\left|(\mathsf{E}_1 \boldsymbol{v}_1, \boldsymbol{v}_1)_{\mathsf{A}_1}\right| &\le \frac{C_{\mathrm{A2}}^{2\alpha}}{\rho_1^\alpha} \|\mathsf{A}_1 \mathsf{K}_1^m \boldsymbol{v}_1\|_1^{2\alpha} \|\mathsf{K}_1^m \boldsymbol{v}_1\|_{\mathsf{A}_1}^{2(1-\alpha)} \\
&\overset{(6.5.4)}{\le} \frac{C_{\mathrm{A2}}^{2\alpha}}{\rho_1^\alpha} \|\mathsf{A}_1 \mathsf{K}_1^m \boldsymbol{v}_1\|_1^{2\alpha} \|\boldsymbol{v}_1\|_{\mathsf{A}_1}^{2(1-\alpha)} \\
&\overset{(\mathrm{S1})}{\le} \frac{C_{\mathrm{A2}}^{2\alpha}}{\rho_1^\alpha} C_{\mathrm{S1}}^{2\alpha} \rho_1^\alpha m^{-\alpha} \|\boldsymbol{v}_1\|_{\mathsf{A}_1}^2 \\
&= \left(\frac{C_{\mathrm{A2}} C_{\mathrm{S1}}}{m^{1/2}}\right)^{2\alpha} (\boldsymbol{v}_1, \boldsymbol{v}_1)_{\mathsf{A}_1}.
\end{aligned}$$

□

**Corollary 6.9.7** ($\alpha$-Convergence of two-grid method). *In the setting of Theorem* 6.9.6, *we have*

$$\|\mathsf{E}_1 \boldsymbol{v}_1\|_{\mathsf{A}_1} \leq \left( \frac{C_{\mathrm{A2}} C_{\mathrm{S1}}}{m^{1/2}} \right)^{2\alpha} \|\boldsymbol{v}_1\|_{\mathsf{A}_1},$$

*for all* $\boldsymbol{v}_1 \in \mathbb{R}^{n_1}$. *In other words, the two-grid method converges, provided that* $m > 0$ *is sufficiently large.*

*Proof.* The proof is similar to that of Corollary 6.7.6. The main difference lies in the fact that while $\mathsf{E}_1$ is still self-adjoint with respect to the $\mathsf{A}_1$-inner product, it is not necessarily positive semidefinite. Therefore some care is needed; see Problem 6.12. □

To conclude this section and, in fact, this chapter, we will prove that a version of the W-cycle algorithm converges. To do so, we need a series of technical lemmas.

**Lemma 6.9.8** (Error transfer). *Let* $m_1 = m_2 = m \in \mathbb{N}$ *and* $p \in \mathbb{N}$, $p > 1$. *Then for* $0 \leq \ell \leq L$, *the error transfer matrix* $\mathsf{E}_\ell$ *for the W-cycle algorithm is self-adjoint with respect to the* $\mathsf{A}_\ell$*-inner product. Furthermore, if there is a number* $\gamma > 0$ *such that*

$$\left|(\mathsf{E}_\ell \boldsymbol{v}_\ell, \boldsymbol{v}_\ell)_{\mathsf{A}_\ell}\right| \leq \gamma (\boldsymbol{v}_\ell, \boldsymbol{v}_\ell)_{\mathsf{A}_\ell}, \quad \forall \boldsymbol{v}_\ell \in \mathbb{R}^{n_\ell},$$

*then*

$$\left|(\mathsf{E}_\ell^p \boldsymbol{v}_\ell, \boldsymbol{v}_\ell)_{\mathsf{A}_\ell}\right| \leq \gamma^p (\boldsymbol{v}_\ell, \boldsymbol{v}_\ell)_{\mathsf{A}_\ell}, \quad \forall \boldsymbol{v}_\ell \in \mathbb{R}^{n_\ell}.$$

*Finally, if p is even, then* $\mathsf{E}_\ell^p$ *is self-adjoint and positive semidefinite with respect to the* $\mathsf{A}_\ell$*-inner product, that is,*

$$0 \leq (\mathsf{E}_\ell^p \boldsymbol{v}_\ell, \boldsymbol{v}_\ell)_{\mathsf{A}_\ell}, \quad \forall \boldsymbol{v}_\ell \in \mathbb{R}^{n_\ell}.$$

*Proof.* Earlier, in the first part of Theorem 6.2.8, using only the general relation $\mathsf{R}_{\ell-1} = r\mathsf{P}_{\ell-1}^\mathsf{T}$, it was established that $\mathsf{E}_\ell$ is self-adjoint with respect to the $\mathsf{A}_\ell$-inner product. Since $\mathsf{E}_\ell$ is self-adjoint with respect to the $\mathsf{A}_\ell$-inner product, there is an orthonormal basis of eigenvectors (with respect to the $\mathsf{A}_\ell$-inner product), and the eigenvalues are all real. Now expand $\boldsymbol{v}_\ell$ in this basis and estimate.

The details of the proof are left to the reader; see Problem 6.13. □

Recall that $\tilde{\Pi}_\ell$, introduced earlier in this section, is a projection. We thus have the following ultimately useful result, which will be used in the convergence proof for the W-cycle.

**Lemma 6.9.9** (Projection estimate). *Assume that the imbalanced Galerkin condition, Assumption* (G3), *holds. Then, for all* $1 \leq \ell \leq L$ *and* $\boldsymbol{v}_\ell \in \mathbb{R}^{n_\ell}$, *we have*

$$-((\mathsf{I}_\ell - \tilde{\Pi}_\ell)\boldsymbol{v}_\ell, \boldsymbol{v}_\ell)_{\mathsf{A}_\ell} \leq (\boldsymbol{v}_\ell, \boldsymbol{v}_\ell)_{\mathsf{A}_\ell}. \tag{6.9.6}$$

*Proof.* Using Corollary 6.9.5, we have

$$\begin{aligned}
0 &\le 2\|(\mathrm{I}_\ell - \hat{\Pi}_\ell)\boldsymbol{v}_\ell\|_{\mathsf{A}_\ell}^2 \\
&= 2((\mathrm{I}_\ell - \hat{\Pi}_\ell)\boldsymbol{v}_\ell, (\mathrm{I}_\ell - \hat{\Pi}_\ell)\boldsymbol{v}_\ell)_{\mathsf{A}_\ell} \\
&\overset{(6.9.5)}{=} 2((\mathrm{I}_\ell - \hat{\Pi}_\ell)\boldsymbol{v}_\ell, \boldsymbol{v}_\ell)_{\mathsf{A}_\ell} \\
&\overset{(6.9.1)}{=} ((2\mathrm{I}_\ell - \tilde{\Pi}_\ell)\boldsymbol{v}_\ell, \boldsymbol{v}_\ell)_{\mathsf{A}_\ell} \\
&= ((\mathrm{I}_\ell - \tilde{\Pi}_\ell)\boldsymbol{v}_\ell, \boldsymbol{v}_\ell)_{\mathsf{A}_\ell} + (\boldsymbol{v}_\ell, \boldsymbol{v}_\ell)_{\mathsf{A}_\ell}.
\end{aligned}$$

□

**Lemma 6.9.10** (Some projection identities). *Assume that* (G3) *holds. Then, for all* $1 \le \ell \le L$ *and* $\boldsymbol{v}_\ell \in \mathbb{R}^{n_\ell}$, *we have*

$$(\Pi_{\ell-1}\boldsymbol{v}_\ell, \Pi_{\ell-1}\boldsymbol{v}_\ell)_{\mathsf{A}_{\ell-1}} = r(\boldsymbol{v}_\ell, \tilde{\Pi}_\ell\boldsymbol{v}_\ell)_{\mathsf{A}_\ell} = r(\tilde{\Pi}_\ell\boldsymbol{v}_\ell, \boldsymbol{v}_\ell)_{\mathsf{A}_\ell}. \tag{6.9.7}$$

*Similarly, for all* $1 \le \ell \le L$, *vectors* $\boldsymbol{v}_\ell \in \mathbb{R}^{n_\ell}$, *and matrices* $\mathsf{B}_{\ell-1} \in \mathbb{R}^{n_{\ell-1}\times n_{\ell-1}}$, *we have*

$$(\mathsf{B}_{\ell-1}\Pi_{\ell-1}\boldsymbol{v}_\ell, \Pi_{\ell-1}\boldsymbol{v}_\ell)_{\mathsf{A}_{\ell-1}} = r(\mathsf{P}_{\ell-1}\mathsf{B}_{\ell-1}\Pi_{\ell-1}\boldsymbol{v}_\ell, \boldsymbol{v}_\ell)_{\mathsf{A}_\ell}. \tag{6.9.8}$$

*Proof.* Suppose $1 \le \ell \le L$ and recall that

$$\mathsf{A}_{\ell-1}\Pi_{\ell-1} = \mathsf{R}_{\ell-1}\mathsf{A}_\ell.$$

Now, for $\boldsymbol{v}_\ell \in \mathbb{R}^{n_\ell}$,

$$\begin{aligned}
(\Pi_{\ell-1}\boldsymbol{v}_\ell, \Pi_{\ell-1}\boldsymbol{v}_\ell)_{\mathsf{A}_{\ell-1}} &= (\mathsf{A}_{\ell-1}\Pi_{\ell-1}\boldsymbol{v}_\ell, \Pi_{\ell-1}\boldsymbol{v}_\ell)_{\ell-1} \\
&= (\mathsf{R}_{\ell-1}\mathsf{A}_\ell\boldsymbol{v}_\ell, \Pi_{\ell-1}\boldsymbol{v}_\ell)_{\ell-1} \\
&= (\mathsf{A}_\ell\boldsymbol{v}_\ell, \mathsf{R}_{\ell-1}^\mathsf{T}\Pi_{\ell-1}\boldsymbol{v}_\ell)_\ell \\
&= r(\mathsf{A}_\ell\boldsymbol{v}_\ell, \mathsf{P}_{\ell-1}\Pi_{\ell-1}\boldsymbol{v}_\ell)_\ell \\
&= r(\mathsf{A}_\ell\boldsymbol{v}_\ell, \tilde{\Pi}_\ell\boldsymbol{v}_\ell)_\ell \\
&= r(\boldsymbol{v}_\ell, \tilde{\Pi}_\ell\boldsymbol{v}_\ell)_{\mathsf{A}_\ell}.
\end{aligned}$$

The second identity is similarly derived. □

We now have all necessary tools to prove the convergence of the symmetric W-cycle algorithm.

**Theorem 6.9.11** ($\alpha$-weak convergence of the W-cycle). *Let* $m_1 = m_2 = m \in \mathbb{N}$, *and let* $p \in \mathbb{N}$ *be even. Assume that* (G3) *and* (A2) *hold and that smoothing is performed using Richardson's smoothing. Then for any* $0 < \gamma < 1$, *there is* $m \in \mathbb{N}$ *such that, for all* $1 \le \ell \le L$, *the W-cycle error matrix* $\mathsf{E}_\ell$ *satisfies*

$$|(\mathsf{E}_\ell\boldsymbol{v}_\ell, \boldsymbol{v}_\ell)_{\mathsf{A}_\ell}| \le \gamma(\boldsymbol{v}_\ell, \boldsymbol{v}_\ell)_{\mathsf{A}_\ell},$$

*for all* $\mathbf{v}_\ell \in \mathbb{R}^{n_\ell}$*. In particular, it suffices to choose*

$$m \geq \left(\frac{1-\gamma^p}{\gamma-\gamma^p}\right)^{1/\alpha} C_{A2}^2 C_{S1}^2. \tag{6.9.9}$$

*Proof.* Fix $1 \leq \ell \leq L$. Recall that

$$\mathsf{E}_\ell = (\mathsf{K}_\ell^*)^m(\mathrm{I}_\ell - \mathsf{P}_{\ell-1}(\mathrm{I}_{\ell-1} - \mathsf{E}_{\ell-1}^p)\Pi_{\ell-1})\mathsf{K}_\ell^m \in \mathbb{R}^{n_\ell \times n_\ell}.$$

We first observe that $\mathsf{E}_\ell$ is not necessarily positive semidefinite. It is, however, self-adjoint with respect to the $\mathsf{A}_\ell$-inner product, as it is guaranteed by Lemma 6.9.8. Set

$$\boldsymbol{w}_\ell = \mathsf{K}_\ell^m \boldsymbol{v}_\ell.$$

Using the definition of $\mathsf{E}_\ell$ and (6.9.8), it follows that

$$(\mathsf{E}_\ell \boldsymbol{v}_\ell, \boldsymbol{v}_\ell)_{\mathsf{A}_\ell} = ((\mathrm{I}_\ell - \tilde{\Pi}_\ell)\boldsymbol{w}_\ell, \boldsymbol{w}_\ell)_{\mathsf{A}_\ell} + \frac{1}{r}(\mathsf{E}_{\ell-1}^p \Pi_{\ell-1}\boldsymbol{w}_\ell, \Pi_{\ell-1}\boldsymbol{w}_\ell)_{\mathsf{A}_{\ell-1}}.$$

We now begin the proof of convergence. We fix $\gamma \in (0,1)$ and proceed by induction on $\ell$. The base case $\ell = 0$ is trivial. The case $\ell = 1$ is true by the related two-grid convergence result. For the induction hypothesis, let us assume that

$$|(\mathsf{E}_{\ell-1}\boldsymbol{v}_{\ell-1}, \boldsymbol{v}_{\ell-1})_{\mathsf{A}_{\ell-1}}| \leq \gamma(\boldsymbol{v}_{\ell-1}, \boldsymbol{v}_{\ell-1})_{\mathsf{A}_{\ell-1}},$$

for all $\boldsymbol{v}_{\ell-1} \in \mathbb{R}^{n_{\ell-1}}$. Using Lemma 6.9.8, it follows that

$$\begin{aligned} 0 &\leq (\mathsf{E}_{\ell-1}^p \boldsymbol{v}_{\ell-1}, \boldsymbol{v}_{\ell-1})_{\mathsf{A}_{\ell-1}} \\ &= |(\mathsf{E}_{\ell-1}^p \boldsymbol{v}_{\ell-1}, \boldsymbol{v}_{\ell-1})_{\mathsf{A}_{\ell-1}}| \\ &\leq \gamma^p(\boldsymbol{v}_{\ell-1}, \boldsymbol{v}_{\ell-1})_{\mathsf{A}_{\ell-1}}. \end{aligned}$$

By this estimate and identity (6.9.7) it follows that

$$\begin{aligned} (\mathsf{E}_\ell \boldsymbol{v}_\ell, \boldsymbol{v}_\ell)_{\mathsf{A}_\ell} &= ((\mathrm{I}_\ell - \tilde{\Pi}_\ell)\boldsymbol{w}_\ell, \boldsymbol{w}_\ell)_{\mathsf{A}_\ell} + \frac{1}{r}(\mathsf{E}_{\ell-1}^p \Pi_{\ell-1}\boldsymbol{w}_\ell, \Pi_{\ell-1}\boldsymbol{w}_\ell)_{\mathsf{A}_{\ell-1}} \\ &\leq ((\mathrm{I}_\ell - \tilde{\Pi}_\ell)\boldsymbol{w}_\ell, \boldsymbol{w}_\ell)_{\mathsf{A}_\ell} + \frac{\gamma^p}{r}(\Pi_{\ell-1}\boldsymbol{w}_\ell, \Pi_{\ell-1}\boldsymbol{w}_\ell)_{\mathsf{A}_{\ell-1}} \\ &\overset{(6.9.7)}{=} ((\mathrm{I}_\ell - \tilde{\Pi}_\ell)\boldsymbol{w}_\ell, \boldsymbol{w}_\ell)_{\mathsf{A}_\ell} + \gamma^p(\tilde{\Pi}_\ell \boldsymbol{w}_\ell, \boldsymbol{w}_\ell)_{\mathsf{A}_\ell} \\ &= (1-\gamma^p)((\mathrm{I}_\ell - \tilde{\Pi}_\ell)\boldsymbol{w}_\ell, \boldsymbol{w}_\ell)_{\mathsf{A}_\ell} + \gamma^p(\boldsymbol{w}_\ell, \boldsymbol{w}_\ell)_{\mathsf{A}_\ell} \\ &\leq (1-\gamma^p)|((\mathrm{I}_\ell - \tilde{\Pi}_\ell)\boldsymbol{w}_\ell, \boldsymbol{w}_\ell)_{\mathsf{A}_\ell}| + \gamma^p(\boldsymbol{w}_\ell, \boldsymbol{w}_\ell)_{\mathsf{A}_\ell}. \end{aligned} \tag{6.9.10}$$

On the other hand, since $\mathsf{E}_\ell^p$ is self-adjoint positive semidefinite, using (6.9.6), we have

$$
\begin{aligned}
-(\mathsf{E}_\ell \boldsymbol{v}_\ell, \boldsymbol{v}_\ell)_{\mathsf{A}_\ell} &= -((\mathrm{I}_\ell - \tilde{\Pi}_\ell)\boldsymbol{w}_\ell, \boldsymbol{w}_\ell)_{\mathsf{A}_\ell} - \frac{1}{r}(\mathsf{E}^p_{\ell-1}\Pi_{\ell-1}\boldsymbol{w}_\ell, \Pi_{\ell-1}\boldsymbol{w}_\ell)_{\mathsf{A}_{\ell-1}} \\
&\le -((\mathrm{I}_\ell - \tilde{\Pi}_\ell)\boldsymbol{w}_\ell, \boldsymbol{w}_\ell)_{\mathsf{A}_\ell} \\
&= -(1-\gamma^p)((\mathrm{I}_\ell - \tilde{\Pi}_\ell)\boldsymbol{w}_\ell, \boldsymbol{w}_\ell)_{\mathsf{A}_\ell} - \gamma^p((\mathrm{I}_\ell - \tilde{\Pi}_\ell)\boldsymbol{w}_\ell, \boldsymbol{w}_\ell)_{\mathsf{A}_\ell} \\
&\overset{(6.9.6)}{\le} -(1-\gamma^p)((\mathrm{I}_\ell - \tilde{\Pi}_\ell)\boldsymbol{w}_\ell, \boldsymbol{w}_\ell)_{\mathsf{A}_\ell} + \gamma^p(\boldsymbol{w}_\ell, \boldsymbol{w}_\ell)_{\mathsf{A}_\ell} \\
&\le (1-\gamma^p)|((\mathrm{I}_\ell - \tilde{\Pi}_\ell)\boldsymbol{w}_\ell, \boldsymbol{w}_\ell)_{\mathsf{A}_\ell}| + \gamma^p(\boldsymbol{w}_\ell, \boldsymbol{w}_\ell)_{\mathsf{A}_\ell}.
\end{aligned}
\tag{6.9.11}
$$

Combining estimates (6.9.10) and (6.9.11), we have

$$
|(\mathsf{E}_\ell \boldsymbol{v}_\ell, \boldsymbol{v}_\ell)_{\mathsf{A}_\ell}| \le (1-\gamma^p)|((\mathrm{I}_\ell - \tilde{\Pi}_\ell)\boldsymbol{w}_\ell, \boldsymbol{w}_\ell)_{\mathsf{A}_\ell}| + \gamma^p(\boldsymbol{w}_\ell, \boldsymbol{w}_\ell)_{\mathsf{A}_\ell}.
$$

To conclude, using the error estimate for the two-grid method, the stability of Richardson's smoother provided in Lemma 6.5.2, and (6.9.9), we obtain

$$
\begin{aligned}
|(\mathsf{E}_\ell \boldsymbol{v}_\ell, \boldsymbol{v}_\ell)_{\mathsf{A}_\ell}| &\le (1-\gamma^p)\left(\frac{C_{\mathrm{A2}}C_{\mathrm{S1}}}{m^{1/2}}\right)^{2\alpha}(\boldsymbol{v}_\ell, \boldsymbol{v}_\ell)_{\mathsf{A}_\ell} + \gamma^p(\boldsymbol{v}_\ell, \boldsymbol{v}_\ell)_{\mathsf{A}_\ell} \\
&= \left[(1-\gamma^p)\left(\frac{C_{\mathrm{A2}}C_{\mathrm{S1}}}{m^{1/2}}\right)^{2\alpha} + \gamma^p\right](\boldsymbol{v}_\ell, \boldsymbol{v}_\ell)_{\mathsf{A}_\ell} \\
&\le \gamma(\boldsymbol{v}_\ell, \boldsymbol{v}_\ell)_{\mathsf{A}_\ell}.
\end{aligned}
$$

The proof is complete. □

**Corollary 6.9.12** (Convergence of the W-cycle). *With the same hypotheses as in Theorem 6.9.11, it follows that, for every $0 < \gamma < 1$, there is $m > 0$ such that*

$$
\|\mathsf{E}_\ell \boldsymbol{v}_\ell\|_{\mathsf{A}_\ell} \le \gamma\|\boldsymbol{v}_\ell\|_{\mathsf{A}_\ell}, \tag{6.9.12}
$$

*for all $\boldsymbol{v}_\ell \in \mathbb{R}^{n_\ell}$. In other words, the W-cycle method converges, provided that $m > 0$ is sufficiently large.*

*Proof.* Exercise; see Problem 6.14. □

Although the previous results already show the convergence of the W-cycle, following [8], we can do better and prove that the W-cycle converges even when $m_1 = m_2 = m = 1$. This requires some auxiliary results. First, we state a special form of Young's inequality.

**Lemma 6.9.13** (Young's inequality). *Let $p \in (1, \infty)$ and define*

$$
q := \frac{p}{p-1}.
$$

*Then, for all $a, b \in \mathbb{R}$, $\tau > 0$, and $\alpha \in (0, 1)$, we have*

$$|ab| \le \tau^{\alpha p} \frac{|a|^p}{p} + \frac{1}{\tau^{\alpha q}} \frac{|b|^q}{q}.$$

*In particular, choosing*

$$p = \frac{1}{\alpha}, \quad q = \frac{1}{1-\alpha},$$

*it follows that, for all $\tau > 0$ and $\alpha \in (0,1)$,*

$$|ab| \le \tau\alpha|a|^{1/\alpha} + \tau^{-\alpha/(1-\alpha)}(1-\alpha)|b|^{1/(1-\alpha)}. \tag{6.9.13}$$

*Proof.* Exercise; see Problem 6.15. □

This next result is highly technical but is key to our theorem that follows.

**Lemma 6.9.14** (Technical estimate). *Let $\alpha \in (0,1)$ and $K > 0$, and set*

$$K_\alpha := (1-\alpha)^{1-\alpha}\alpha^\alpha K^\alpha. \tag{6.9.14}$$

*For $0 \le x \le 1$, $0 < \beta < 1$, and $m \in \mathbb{N}$, define*

$$\Gamma(\beta; x; m) := (1-\beta)\frac{K^\alpha}{m^\alpha}(1-x)^\alpha x^{1-\alpha} + \beta x. \tag{6.9.15}$$

*Assume that $M > 0$ is chosen so that*

$$M \ge \max\left\{\frac{1}{2}(4K_\alpha)^{1/\alpha}, m^\alpha\right\}. \tag{6.9.16}$$

*Then*

$$\Gamma(\delta^2; x; m) \le \delta,$$

*where*

$$\delta = \frac{M}{M + m^\alpha}. \tag{6.9.17}$$

*Proof.* Using (6.9.13), it follows that

$$\Gamma(\beta; x; m) \le L(\beta; x; m), \tag{6.9.18}$$

where

$$L(\beta; x; m) := (1-\beta)\left[\alpha K\tau(1-x) + (1-\alpha)(\tau m)^{-\alpha/(1-\alpha)}x\right] + \beta x, \tag{6.9.19}$$

for any $\tau > 0$. Observe that $x \mapsto L(\beta; x; m)$ is linear, with

$$L(\beta; 0; m) = (1 - \beta)\alpha K\tau > 0$$

and

$$L(\beta; 1; m) = \beta + (1 - \beta)(1 - \alpha)(\tau m)^{-\alpha/(1-\alpha)} > 0.$$

Thus, for $\beta \in (0,1)$, $x \in [0,1]$, and $m \in \mathbb{N}$,

$$0 \le \Gamma(\beta; x; m) \le \max\{L(\beta; 0; m), L(\beta; 1; m)\}.$$

Consequently, for all $m \in \mathbb{N}$,

$$0 \le \Gamma(\delta^2; x; m) \le \max\{L(\delta^2; 0; m), L(\delta^2; 1; m)\}. \tag{6.9.20}$$

Thus, if we show that

$$L(\delta^2; 0; m) \le \delta \tag{6.9.21}$$

and

$$L(\delta^2; 1; m) \le \delta, \tag{6.9.22}$$

then the proof will be complete.

Let us show (6.9.22). We have

$$\begin{aligned}
(1-\delta)^{\alpha}\left(\frac{1+\delta}{\delta}\right) &\overset{(6.9.17)}{=} \left(\frac{m^{\alpha}}{M+m^{\alpha}}\right)^{\alpha}\frac{2M+m^{\alpha}}{M} \\
&\le \left(\frac{m^{\alpha}}{M+m^{\alpha}}\right)^{\alpha}\frac{2(M+m^{\alpha})}{M} \\
&= m^{\alpha^2}\frac{2(M+m^{\alpha})^{1-\alpha}}{M} \\
&\overset{(6.9.16)}{\le} m^{\alpha^2}\frac{2(2M)^{1-\alpha}}{M} \\
&= m^{\alpha^2}\frac{4}{(2M)^{\alpha}} \\
&\overset{(6.9.16)}{\le} \frac{m^{\alpha^2}}{K_{\alpha}} \\
&\le \frac{m^{\alpha}}{K_{\alpha}}.
\end{aligned}$$

Using (6.9.14) and some algebra, the last estimate is equivalent to

$$\frac{(1-\alpha)^{1-\alpha}(1+\delta)^{1-\alpha}}{m^{\alpha}\delta^{1-\alpha}} \le \left(\frac{\delta}{\alpha(1-\delta^2)K}\right)^{\alpha}.$$

Now we can find a number $\tau > 0$ such that

$$\frac{(1-\alpha)^{1-\alpha}(1+\delta)^{1-\alpha}}{m^\alpha \delta^{1-\alpha}} \le \tau^\alpha \le \left(\frac{\delta}{\alpha(1-\delta^2)K}\right)^\alpha. \tag{6.9.23}$$

This will fix the value of $\tau$ in the definition of $L$.

Finally, inequalities (6.9.23) are clearly equivalent to

$$\tau^\alpha \le \left(\frac{\delta}{\alpha(1-\delta^2)K}\right)^\alpha$$

and

$$\frac{(1-\alpha)^{1-\alpha}(1+\delta)^{1-\alpha}}{m^\alpha \delta^{1-\alpha}} \le \tau^\alpha,$$

which, after some algebra, are equivalent to (6.9.21) and (6.9.22), respectively. The proof is complete. □

**Theorem 6.9.15** ($\alpha$-Weak convergence of the W-cycle). *Let $m_1 = m_2 = m \in \mathbb{N}$ and $p = 2$. Assume that* (G3) *and* (A2) *hold and that the smoothing is performed using Richardson's smoother. Define $K_\alpha$ as in* (6.9.14) *with*

$$K = \frac{C_{\mathrm{A2}}^2 C_{\mathrm{R}}}{2}.$$

*Let $M > 0$ be chosen so that* (6.9.16) *holds. Then, for every $1 \le \ell \le L$, the W-cycle error matrix $\mathsf{E}_\ell$ satisfies*

$$|(\mathsf{E}_\ell \boldsymbol{v}_\ell, \boldsymbol{v}_\ell)_{\mathsf{A}_\ell}| \le \frac{M}{M+m^\alpha}(\boldsymbol{v}_\ell, \boldsymbol{v}_\ell)_{\mathsf{A}_\ell}, \tag{6.9.24}$$

*for all $\boldsymbol{v}_\ell \in \mathbb{R}^{n_\ell}$.*

*Proof.* The proof is by induction. Suppose that

$$|(\mathsf{E}_{\ell-1} \boldsymbol{v}_{\ell-1}, \boldsymbol{v}_{\ell-1})_{\mathsf{A}_{\ell-1}}| \le \gamma(\boldsymbol{v}_{\ell-1}, \boldsymbol{v}_{\ell-1})_{\mathsf{A}_{\ell-1}}, \quad \gamma = \frac{M}{M+m^\alpha}.$$

In the proof of Theorem 6.9.11, we observed that for $\boldsymbol{w}_\ell = \mathsf{K}_\ell^m \boldsymbol{v}_\ell$,

$$|(\mathsf{E}_\ell \boldsymbol{v}_\ell, \boldsymbol{v}_\ell)_{\mathsf{A}_\ell}| \le (1-\gamma^2)|((\mathrm{I}_\ell - \tilde{\Pi}_\ell)\boldsymbol{w}_\ell, \boldsymbol{w}_\ell)_{\mathsf{A}_\ell}| + \gamma^2(\boldsymbol{w}_\ell, \boldsymbol{w}_\ell)_{\mathsf{A}_\ell}.$$

Using (A2) and an estimate used in the proof of the Braess–Hackbusch result, Theorem 6.8.2, it follows that

$$T_\ell := |(\mathsf{E}_\ell \boldsymbol{v}_\ell, \boldsymbol{v}_\ell)_{\mathsf{A}_\ell}|$$

$$
\begin{aligned}
&\overset{\text{(A2)}}{\le} (1-\gamma^2)\frac{C_{\text{A2}}^{2\alpha}}{\rho_\ell^\alpha}\|\mathsf{A}_\ell \boldsymbol{w}_\ell\|_\ell^{2\alpha}\|\boldsymbol{w}_\ell\|_{\mathsf{A}_\ell}^{2(1-\alpha)} + \gamma^2(\boldsymbol{w}_\ell, \boldsymbol{w}_\ell)_{\mathsf{A}_\ell} \\
&\overset{(6.8.2)}{\le} (1-\gamma^2)C_{\text{A2}}^{2\alpha}\left(\frac{C_{\text{R}}}{2m}(\|\boldsymbol{v}_\ell\|_{\mathsf{A}_\ell}^2 - \|\boldsymbol{w}_\ell\|_{\mathsf{A}_\ell}^2)\right)^\alpha \|\boldsymbol{w}_\ell\|_{\mathsf{A}_\ell}^{2(1-\alpha)} + \gamma^2\|\boldsymbol{w}_\ell\|_{\mathsf{A}_\ell}^2 \\
&= \Gamma(\gamma^2; x; m)\|\boldsymbol{v}_\ell\|_{\mathsf{A}_\ell}^2,
\end{aligned}
$$

where $\Gamma$ is defined as in (6.9.15), with

$$
x := \frac{\|\boldsymbol{w}_\ell\|_{\mathsf{A}_\ell}^2}{\|\boldsymbol{v}_\ell\|_{\mathsf{A}_\ell}^2}.
$$

Observe that due to the stability of Richardson's smoother, Lemma 6.5.2, it follows that

$$
0 \le x \le 1.
$$

Thus, with condition (6.9.16) imposed on $M$, we can apply Lemma 6.9.14 to conclude that

$$
\Gamma(\gamma^2; x; m) \le \gamma.
$$

The proof is complete. □

**Remark 6.9.16** (Open problem). The avid reader may have noticed that the convergence results presented in this section pertain only to the W-cycle versions of multigrid. To the best of our knowledge, at the time of this writing, the abstract, axiomatic convergence of the standard V-cycle under condition (G3) remains an open question. Bramble and coworkers (see [6] and [8]) have shown the convergence of something called the "variable V-cycle" multigrid method, though we will not consider that here for brevity. ◂

## 6.10 Problems

6.1 Prove Proposition 6.1.1.

6.2 Prove Corollary 6.1.11.

6.3 Prove Corollary 6.2.4.

6.4 Suppose that the square matrix B is self-adjoint positive semidefinite with respect to the energy inner product $(\cdot, \cdot)_A$. Prove that any positive integer power of B is also self-adjoint positive semidefinite with respect to the energy inner product $(\cdot, \cdot)_A$.

6.5 Prove Corollary 6.2.10.

6.6 Prove Theorem 6.4.3.

6.7 Prove Theorem 6.6.3.

6.8 Prove Corollary 6.8.3.

6.9 Prove Proposition 6.9.3.

6.10 Prove Proposition 6.9.4.

6.11 Prove Corollary 6.9.5.

6.12 Prove Corollary 6.9.7.

6.13 Prove Lemma 6.9.8.

6.14 Prove Corollary 6.9.12.

6.15 Prove Lemma 6.9.13.

# 7 Multigrid and the conforming finite element method

In this chapter, we consider the finite element discretization of a model elliptic boundary value problem. We will use the axiomatic theory developed in the last chapter to prove that basic multigrid algorithms will converge when applied to the finite element approximation of this problem. The path to follow is simple. We only need to check that the basic axioms hold. This will help elucidate the meaning of these, at first glance, abstract and obscure axioms.

## 7.1 Description of the problem

Let us begin by describing the model problem. Let $d \in \{1, 2, 3\}$, and let $\Omega \subset \mathbb{R}^d$ be an open polytopal domain with boundary $\partial\Omega$. Often, we will also assume that $\Omega$ is convex. The problem that we are interested in reads as follows: given a suitable $f : \Omega \to \mathbb{R}$, find $u : \overline{\Omega} \to \mathbb{R}$ that solves

$$\begin{cases} -\Delta u = f & \text{in } \Omega, \\ u = 0 & \text{on } \partial\Omega. \end{cases} \tag{7.1.1}$$

We will understand this problem and its discretization in the framework of weak solutions. The first step is defining the bilinear form $\mathcal{A} : H_0^1(\Omega) \times H_0^1(\Omega) \to \mathbb{R}$ by

$$\mathcal{A}(u, v) := (\nabla u, \nabla v)_{L^2(\Omega)}, \tag{7.1.2}$$

which we call the *energy inner product*.

Next, we assume that the right-hand side satisfies $f \in H^{-1}(\Omega) := (H_0^1(\Omega))'$. Observe that, in particular, every $f \in L^2(\Omega)$ gives rise to a linear functional $L_f \in H^{-1}(\Omega)$ in a natural way:

$$\langle L_f, v\rangle := L_f(v) = (f, v)_{L^2(\Omega)}, \quad \forall\, v \in H_0^1(\Omega).$$

For convenience, we use the shorthand notation

$$\langle f, v\rangle = \langle L_f, v\rangle, \quad \forall\, v \in H_0^1(\Omega).$$

With this notation at hand, the weak formulation of the model problem (7.1.1) can be stated as follows: find $u \in H_0^1(\Omega)$ such that

$$\mathcal{A}(u, v) = \langle f, v\rangle, \quad \forall\, v \in H_0^1(\Omega). \tag{7.1.3}$$

To show that this problem is well posed, that is, has a unique solution $u \in H_0^1(\Omega)$, the following inequality is fundamental.

https://doi.org/10.1515/9783111354880-007

**Theorem 7.1.1** (Poincaré inequality). *Let $d \in \{1, 2, 3\}$ and assume that $\Omega \subset \mathbb{R}^d$ is a bounded polytopal domain. There is a constant $C_\mathrm{P} > 0$ that depends only on the domain $\Omega$ such that*

$$C_\mathrm{P}\|v\|^2_{L^2(\Omega)} \le \mathcal{A}(v, v), \quad \forall\, v \in H^1_0(\Omega). \tag{7.1.4}$$

*Proof.* See Problem 7.1 for a particular case. □

With this result at hand, the existence and uniqueness of a weak solution $u \in H^1_0(\Omega)$ can be proved using either the Lax–Milgram theorem or Riesz representation theorem [4, 10]. For the analysis that follows, we need a few more background results, which can be found, for instance, in [4, 10].

**Definition 7.1.2** (Standard regularity condition). We say that problem (7.1.3) satisfies the **standard regularity condition** iff, whenever $f \in L^2(\Omega)$, it follows that $u \in H^1_0(\Omega) \cap H^2(\Omega)$, with

$$|u|_{H^2(\Omega)} \le C\|f\|_{L^2(\Omega)},$$

for some constant $C > 0$ that depends only on the domain $\Omega$. ■

**Theorem 7.1.3** (Convexity implies standard regularity). *For problem* (7.1.3), *if $\Omega$ is a convex polytope, then the standard regularity condition holds.*

**Remark 7.1.4** (Variable coefficients). We have chosen the bilinear form (7.1.2) for simplicity only. With very little modifications, the reader may verify that, under the standard regularity condition, all the theory presented in this chapter is valid if the bilinear form is replaced by

$$\mathcal{A}_\#(u, v) = \int_\Omega [\nabla v^\top D(\boldsymbol{x})\nabla u + c(\boldsymbol{x})uv]\,\mathrm{d}\boldsymbol{x}, \quad \forall\, u, v \in H^1_0(\Omega),$$

where the so-called *reaction coefficient* $c \in L^\infty(\Omega)$ satisfies $c \ge 0$, and the *diffusion coefficient* $D \in L^\infty(\Omega)$ is such that there are constants $0 < D_\mathrm{min} \le D_\mathrm{max}$ for which

$$D_\mathrm{min} \le D(\boldsymbol{x}) \le D_\mathrm{max},$$

for almost every $\boldsymbol{x} \in \Omega$. The reader is invited to verify all the details; see Problem 7.6. ◂

## 7.2 Nested families of finite element spaces

Let $\mathcal{T}_0$ be a conforming and shape regular triangulation of $\Omega$. This, in particular, means that there are no hanging nodes, and we refer to [4, 10, 31] for a full definition. We define $\mathcal{T}_1$ to be the triangulation of $\Omega$ that results from, in a previously defined and systematic way, uniformly refining the original triangulation $\mathcal{T}_0$. In one dimension, the obvious approach is to bisect every segment; see Figure 7.1. In two and more dimensions, there are

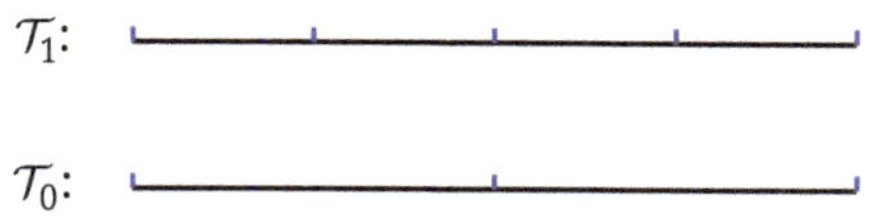


**Figure 7.1:** Bisecting the triangulation $\mathcal{T}_0$ in one dimension.

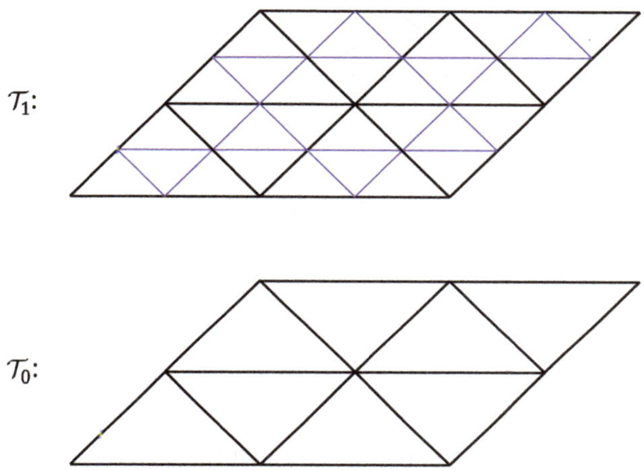


**Figure 7.2:** Refining the triangulation $\mathcal{T}_0$ in two dimensions.

several choices. In Figures 7.2 and 7.3, we present a simple choice that only works in two dimensions. Namely, we join the midpoints of every side of a triangle to obtain four congruent subtriangles. We refer the reader to [21, 24] and [27, Section 4] for a thorough discussion on mesh refinement choices, their complexity, and related topics that go well beyond our scope in this introductory text.

Proceeding in such a way, we can recursively define a nested family of the conforming triangulations, indexed by $\ell = 0, \ldots, L$,

$$\mathcal{T}_0, \mathcal{T}_1, \mathcal{T}_2, \ldots, \mathcal{T}_\ell, \ldots, \mathcal{T}_L.$$

**Definition 7.2.1** (Finite element space). Let $\{\mathcal{T}_\ell\}_{\ell=0}^{L}$ be a nested family of triangulations of $\Omega$. For $0 \le \ell \le L$, define the **grid spacing** $h_\ell$ as

$$h_\ell := \max_{K \in \mathcal{T}_\ell} \operatorname{diam}(K).$$

Subordinate to $\mathcal{T}_\ell$, define the **finite element spaces** of piecewise linear functions $S_\ell$ and $V_\ell$ as

$$S_\ell := \{v_\ell \in C^0(\overline{\Omega}) \mid v_\ell|_K \in \mathbb{P}_1(K),\ \forall K \in \mathcal{T}_\ell\},$$

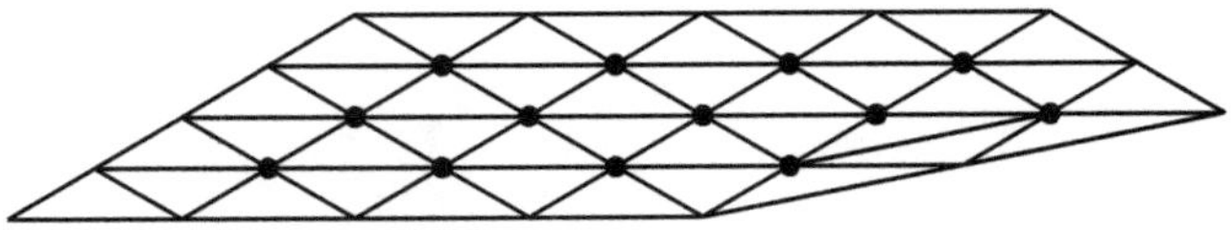

$\mathcal{T}_1 : n_1 = 13.$

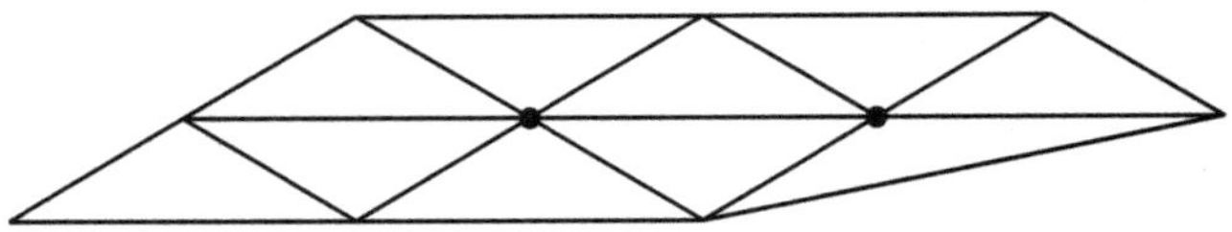

$\mathcal{T}_0 : n_0 = 2$

**Figure 7.3:** Uniform refinement of a two-dimensional triangulation $\mathcal{T}_0$. Interior vertices are marked with a filled dot.

and

$$V_\ell := \{v_\ell \in S_\ell \mid v_\ell|_{\partial\Omega} \equiv 0\}. \qquad \blacksquare$$

In this chapter, we will mostly focus our attention upon $V_\ell$, which is a proper subspace of $S_\ell$. The reason for this choice is that functions in $V_\ell$ satisfy the boundary conditions that our model problem requires.

**Definition 7.2.2** (Lagrange nodal basis). Let $0 \le \ell \le L$, and let $\mathcal{T}_\ell$ and $V_\ell$ be defined as above. Let $\{\boldsymbol{N}_{\ell,j}\}_{j=1}^{n_\ell}$ be the set of interior vertices of the triangulation $\mathcal{T}_\ell$. By $\mathcal{B}_\ell$ we denote the **Lagrange nodal basis** for $V_\ell$, that is,

$$\mathcal{B}_\ell := \{\psi_{\ell,i}\}_{i=1}^{n_\ell}, \tag{7.2.1}$$

where $\psi_{\ell,i} \in V_\ell$ is the unique function such that

$$\psi_{\ell,i}(\boldsymbol{N}_{\ell,j}) = \delta_{i,j}. \qquad \blacksquare$$

The functions $\psi_{\ell,i}$ are sometimes called hat functions. The reason for this name can be seen in Figures 7.4 and 7.5.

**Remark 7.2.3** (Interior vertex). We will always assume that $n_0 > 0$. In other words, the coarsest triangulation always has at least one interior vertex; see Figure 7.3. ◀

The following results describe some properties of $\mathcal{B}_\ell$ and $V_\ell$.

**Lemma 7.2.4** (Subspace and dimension). *For $0 \le \ell \le L$, with $\mathcal{T}_\ell$ and $V_\ell$ defined as above, we have*

$$V_0 \subsetneq V_1 \subsetneq V_2 \subsetneq \cdots \subsetneq V_\ell \subsetneq \cdots \subsetneq V_L \subsetneq H_0^1(\Omega).$$

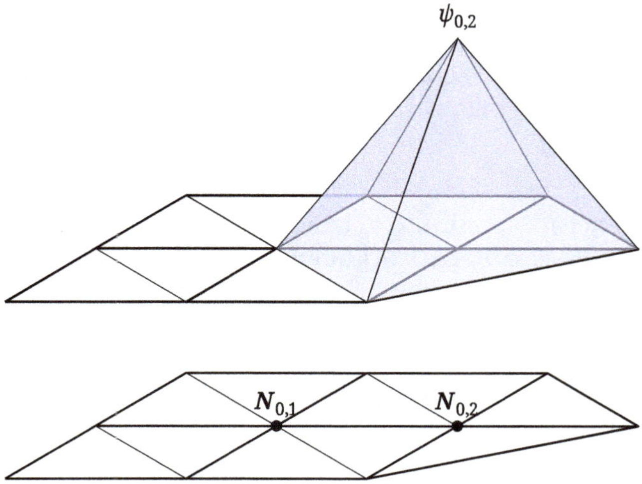


**Figure 7.4:** Interior vertices $\boldsymbol{N}_{0,1}$ and $\boldsymbol{N}_{0,2}$ in the triangulation $\mathcal{T}_0$ and the basis function $\psi_{0,2}$.

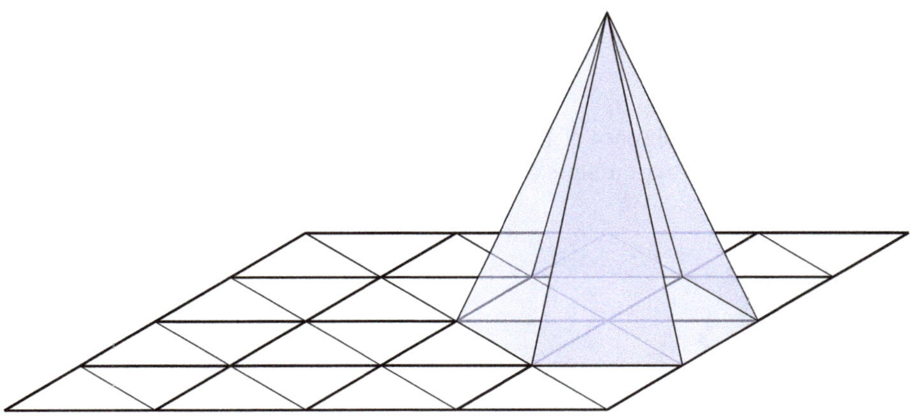

**Figure 7.5:** A uniform mesh in two dimensions and a Lagrange nodal basis (hat) function.

*Each $V_\ell$ is a finite-dimensional vector space with*

$$\dim(V_\ell) = n_\ell.$$

*Furthermore,*

$$0 < n_0 < n_1 < n_2 < \cdots < n_\ell < \cdots < n_L < \infty,$$

*where $n_\ell$ is precisely the number of interior vertices of the triangulation $\mathcal{T}_\ell$. Finally,*

$$h_{\ell-1} = 2h_\ell, \quad 1 \le \ell \le L.$$

*Proof.* Exercise; see Problem 7.2. □

**Proposition 7.2.5** ($\mathcal{B}_\ell$ is a basis). *Let $0 \le \ell \le L$, and assume that $V_\ell$ and $\mathcal{T}_\ell$ are defined as above. Assume that $\mathcal{B}_\ell$ is as in* (7.2.1). *Then $\mathcal{B}_\ell$ is a basis of $V_\ell$.*

*Proof.* Exercise; see Problem 7.3. □

**Remark 7.2.6** (Functions vs. vectors). Suppose $u_\ell \in V_\ell$ is the unique function whose coordinate vector with respect to the basis $\mathcal{B}_\ell$ is $\boldsymbol{u}_\ell$. To express this connection, we write

$$u_\ell \in V_\ell \overset{\mathcal{B}_\ell}{\longleftrightarrow} \boldsymbol{u}_\ell \in \mathbb{R}^{n_\ell}.$$ ◂

**Lemma 7.2.7** (Embedding). *Let $0 \le \ell \le L$, and let $\mathcal{B}_\ell$ be the Lagrange nodal basis for $V_\ell$. For all $1 \le \ell \le L$, there exist unique numbers*

$$p_{\ell-1,j,i} \in \mathbb{R}, \quad 1 \le j \le n_\ell, \quad 1 \le i \le n_{\ell-1},$$

*such that, for each $1 \le i \le n_{\ell-1}$,*

$$\psi_{\ell-1,i} = \sum_{j=1}^{n_\ell} p_{\ell-1,j,i}\psi_{\ell,j}. \tag{7.2.2}$$

*Proof.* Since $V_{\ell-1}$ is a subspace of $V_\ell$ and $\mathcal{B}_\ell$ is a basis for the latter, for every $\psi_{\ell-1,i} \in \mathcal{B}_{\ell-1} \subset V_{\ell-1} \subset V_\ell$, there exist unique coefficients

$$p_{\ell-1,j,i} \in \mathbb{R}, \quad 1 \le j \le n_\ell,$$

such that

$$\psi_{\ell-1,i} = \sum_{j=1}^{n_\ell} p_{\ell-1,j,i}\psi_{\ell,j}.$$

Finally, recall that basis representations are unique. □

### 7.2.1 Finite element approximation

A conforming finite element approximation of problem (7.1.3) is written as follows: for $0 \le \ell \le L$, find $u_\ell \in V_\ell$ such that

$$\mathcal{A}(u_\ell, v_\ell) = \langle f, v_\ell\rangle, \quad \forall\, v_\ell \in V_\ell, \tag{7.2.3}$$

where $\{V_\ell\}_{\ell=0}^{L}$ is the family of nested, conforming finite element subspaces of $H_0^1(\Omega)$ constructed in Definition 7.2.1. It is easy to show that, also, there is a unique finite element approximation $u_\ell \in V_\ell$.

Next, we define the stiffness matrix.

**Definition 7.2.8** (Stiffness matrix). Suppose $0 \le \ell \le L$. Assume that $V_\ell$ and $\mathcal{T}_\ell$ are defined as above. The **level-$\ell$ stiffness matrix** $\mathsf{A}_\ell \in \mathbb{R}^{n_\ell \times n_\ell}$ is defined as

$$[\mathsf{A}_\ell]_{i,j} := \mathcal{A}(\psi_{\ell,j}, \psi_{\ell,i}), \quad 1 \le i,j \le n_\ell,$$

where $\mathcal{A}(\,\cdot\,,\,\cdot\,)$ is the energy inner product (7.1.2). ∎

With the stiffness matrix at hand, it is not difficult to see that the coordinates of the solution may be found by solving the matrix problem

$$\mathsf{A}_\ell \boldsymbol{u}_\ell = \boldsymbol{f}_\ell,$$

where

$$[\boldsymbol{f}_\ell]_i = \langle f, \psi_{\ell,i} \rangle, \quad 1 \le i \le n_\ell.$$

### 7.2.2 Prolongation and restriction

Having defined the finite element discretization, we begin by describing the ingredients of our multigrid method. Here we describe the prolongation and restriction matrices.

**Definition 7.2.9** (Prolongation matrix). For $1 \le \ell \le L$, define the **prolongation matrix**

$$\mathsf{P}_{\ell-1} \in \mathbb{R}^{n_\ell \times n_{\ell-1}}$$

via

$$[\mathsf{P}_{\ell-1}]_{i,j} := p_{\ell-1,i,j}. \tag{7.2.4}$$

∎

**Lemma 7.2.10** (Prolongation operator). *Let $1 \le \ell \le L$, and let $v_{\ell-1} \in V_{\ell-1}$. Suppose $\boldsymbol{v}_{\ell-1}$ is the coordinate vector of $v_{\ell-1}$ in the Lagrange nodal basis $\mathcal{B}_{\ell-1}$ and $\boldsymbol{v}_\ell$ is the coordinate vector of $v_{\ell-1}$ in the basis $\mathcal{B}_\ell$. Then*

$$\boldsymbol{v}_\ell = \mathsf{P}_{\ell-1} \boldsymbol{v}_{\ell-1}.$$

*Proof.* We write, as usual,

$$[\boldsymbol{v}_{\ell-1}]_i = v_{\ell-1,i}, \quad 1 \le i \le n_{\ell-1},$$

and

$$[\boldsymbol{v}_\ell]_i = v_{\ell,i}, \quad 1 \le i \le n_\ell.$$

Thus

$$\begin{aligned} v_{\ell-1} &= \sum_{i=1}^{n_{\ell-1}} v_{\ell-1,i}\psi_{\ell-1,i} \\ &\overset{(7.2.2)}{=} \sum_{i=1}^{n_{\ell-1}} v_{\ell-1,i} \sum_{j=1}^{n_\ell} p_{\ell-1,j,i}\psi_{\ell,j} \\ &= \sum_{j=1}^{n_\ell}\left(\sum_{i=1}^{n_{\ell-1}} p_{\ell-1,j,i}v_{\ell-1,i}\right)\psi_{\ell,j} \\ &= \sum_{j=1}^{n_\ell}[\mathsf{P}_{\ell-1}\boldsymbol{v}_{\ell-1}]_j\psi_{\ell,j}. \end{aligned}$$

However,

$$v_{\ell-1} = \sum_{j=1}^{n_\ell}[\boldsymbol{v}_\ell]_j\psi_{\ell,j}.$$

Since basis representations are unique,

$$\boldsymbol{v}_\ell = \mathsf{P}_{\ell-1}\boldsymbol{v}_{\ell-1}.$$ □

**Definition 7.2.11** (Restriction matrix). For $1 \le \ell \le L$, define the **restriction matrix**

$$\mathsf{R}_{\ell-1} := \mathsf{P}_{\ell-1}^\top \in \mathbb{R}^{n_{\ell-1}\times n_\ell}.$$ ■

**Lemma 7.2.12** (Rank). *Let $1 \le \ell \le L$ and suppose that $\mathsf{P}_{\ell-1}$ and $\mathsf{R}_{\ell-1}$ are as in Definitions* 7.2.9 *and* 7.2.11, *respectively. Then*

$$\operatorname{rank}(\mathsf{P}_{\ell-1}) = \operatorname{rank}(\mathsf{R}_{\ell-1}) = n_{\ell-1}.$$

*Proof.* Suppose $\boldsymbol{v}_{\ell-1} \in \mathbb{R}^{n_{\ell-1}}$ is such that

$$\mathsf{P}_{\ell-1}\boldsymbol{v}_{\ell-1} = \boldsymbol{0} \in \mathbb{R}^{n_\ell}.$$

This represents a linear combination of the $n_{\ell-1}$ columns of $\mathsf{P}_{\ell-1}$. Using the notation from Lemma 7.2.10 and its proof, we have

$$\boldsymbol{v}_\ell = \boldsymbol{0} \in \mathbb{R}^{n_\ell},$$

where $\boldsymbol{v}_{\ell-1}$ and $\boldsymbol{v}_\ell$ are the coordinate vectors of some function $v_{\ell-1} \in V_{\ell-1}$ in the bases $\mathcal{B}_{\ell-1}$ and $\mathcal{B}_\ell$, respectively.

The only way that $\boldsymbol{v}_\ell = \boldsymbol{0}$ is possible is if $v_{\ell-1} \equiv 0$ in $V_{\ell-1}$. Then $\boldsymbol{v}_{\ell-1} = \boldsymbol{0}$. Thus the columns of $\mathsf{P}_{\ell-1}$ are linearly independent, and

$$\operatorname{rank}(\mathsf{P}_{\ell-1}) = n_{\ell-1}.$$ □

## 7.3 Properties of the stiffness matrices

Recall that the level-$\ell$ stiffness matrix was introduced in Definition 7.2.8. Here we study some of its properties. First of all, since $\mathcal{A}$ is symmetric, so are the stiffness matrices. The fact that they are positive definite follows from the Poincaré inequality of Theorem 7.1.1.

**Lemma 7.3.1** ($\mathsf{A}_\ell$ is SPD). *For $0 \le \ell \le L$, the stiffness matrix $\mathsf{A}_\ell \in \mathbb{R}^{n_\ell \times n_\ell}$ is SPD. Moreover, they all satisfy the Galerkin condition* (G0), *that is, for $1 \le \ell \le L$,*

$$\mathsf{A}_{\ell-1} = \mathsf{R}_{\ell-1}\mathsf{A}_\ell\mathsf{P}_{\ell-1}. \tag{7.3.1}$$

*Proof.* (Symmetry):

$$\begin{aligned}[\mathsf{A}_\ell]_{i,j} &= \mathcal{A}(\psi_{\ell,j}, \psi_{\ell,i}) \\ &= \mathcal{A}(\psi_{\ell,i}, \psi_{\ell,j}) \\ &= [\mathsf{A}_\ell]_{j,i}.\end{aligned}$$

(Positivity): Let $\mathbf{0} \neq \boldsymbol{v}_\ell \in \mathbb{R}^{n_\ell}$, and let $\boldsymbol{v}_\ell \overset{\mathcal{B}_\ell}{\longleftrightarrow} v_\ell$. Then using the Poincaré inequality, we have

$$\begin{aligned}0 &< C_{\mathrm{P}}\|v_\ell\|^2_{L^2(\Omega)} \\ &\overset{(7.1.4)}{\le} \mathcal{A}(v_\ell, v_\ell) \\ &= \mathcal{A}\left(\sum_{j=1}^{n_\ell} v_{\ell,j}\psi_{\ell,j}, \sum_{j=1}^{n_\ell} v_{\ell,i}\psi_{\ell,i}\right) \\ &= \sum_{i=1}^{n_\ell}\sum_{j=1}^{n_\ell} v_{\ell,i}\mathcal{A}(\psi_{\ell,j}, \psi_{\ell,i})v_{\ell,j} \\ &= \boldsymbol{v}_\ell^\top \mathsf{A}_\ell \boldsymbol{v}_\ell.\end{aligned}$$

Hence $\mathsf{A}_\ell$ is SPD.
(Galerkin condition): By definition

$$\begin{aligned}[\mathsf{A}_{\ell-1}]_{i,j} &= \mathcal{A}(\psi_{\ell-1,j}, \psi_{\ell-1,i}) \\ &\overset{(7.2.2)}{=} \mathcal{A}\left(\sum_{s=1}^{n_\ell} p_{\ell-1,s,j}\psi_{\ell,s}, \sum_{t=1}^{n_\ell} p_{\ell-1,t,i}\psi_{\ell,t}\right) \\ &= \sum_{s=1}^{n_\ell}\sum_{t=1}^{n_\ell} p_{\ell-1,s,j}\mathcal{A}(\psi_{\ell,s}, \psi_{\ell,t})p_{\ell-1,t,i} \\ &= \sum_{t=1}^{n_\ell}\sum_{s=1}^{n_\ell} p_{\ell-1,t,i}\mathcal{A}(\psi_{\ell,s}, \psi_{\ell,t})p_{\ell-1,s,j}\end{aligned}$$

$$\stackrel{(7.2.4)}{=} \sum_{t=1}^{n_\ell}\sum_{s=1}^{n_\ell}[\mathsf{R}_{\ell-1}]_{i,t}[\mathsf{A}_\ell]_{t,s}[\mathsf{P}_{\ell-1}]_{s,j}.$$

Thus

$$\mathsf{A}_{\ell-1} = \mathsf{R}_{\ell-1}\mathsf{A}_\ell\mathsf{P}_{\ell-1}. \qquad \square$$

**Remark 7.3.2** (Rank of $\mathsf{P}_{\ell-1}$). Observe that this last result also proves that $\mathsf{P}_{\ell-1}$ and $\mathsf{R}_{\ell-1}$ have full rank. Otherwise, $\mathsf{A}_{\ell-1}$ could not be positive definite. ◂

For the level-$\ell$ Euclidean inner product and norm, we use the familiar notation

$$(\boldsymbol{u}_\ell, \boldsymbol{v}_\ell)_\ell := \boldsymbol{v}_\ell^\top \boldsymbol{u}_\ell, \quad \forall\, \boldsymbol{u}_\ell, \boldsymbol{v}_\ell \in \mathbb{R}^{n_\ell},$$

and

$$\|\boldsymbol{u}_\ell\|_\ell := \sqrt{(\boldsymbol{u}_\ell, \boldsymbol{u}_\ell)_\ell}, \quad \forall\, \boldsymbol{u}_\ell \in \mathbb{R}^{n_\ell}.$$

The stiffness matrix-induced inner product and norm are

$$(\boldsymbol{u}_\ell, \boldsymbol{v}_\ell)_{\mathsf{A}_\ell} := (\mathsf{A}_\ell \boldsymbol{u}_\ell, \boldsymbol{v}_\ell)_\ell, \quad \forall\, \boldsymbol{u}_\ell, \boldsymbol{v}_\ell \in \mathbb{R}^{n_\ell},$$

and

$$\|\boldsymbol{u}_\ell\|_{\mathsf{A}_\ell} := \sqrt{(\boldsymbol{u}_\ell, \boldsymbol{u}_\ell)_{\mathsf{A}_\ell}}, \quad \forall\, \boldsymbol{u}_\ell \in \mathbb{R}^{n_\ell}.$$

Let us now estimate the condition number of $\mathsf{A}_\ell$. To do this, we need some results regarding finite element spaces. The proofs can be found in most finite element textbooks; see, for instance, [4, 10].

**Lemma 7.3.3** (Norm equivalence). *Suppose $d \in \{1,2,3\}$. There are constants $0 < C_1 \le C_2$ such that, for all $0 \le \ell \le L$ and $v_\ell \in V_\ell$, with $v_\ell \stackrel{\mathcal{B}_\ell}{\longleftrightarrow} \boldsymbol{v}_\ell \in \mathbb{R}^{n_\ell}$, we have*

$$C_1 h_\ell^d \|\boldsymbol{v}_\ell\|_\ell^2 \le \|v_\ell\|_{L^2(\Omega)}^2 \le C_2 h_\ell^d \|\boldsymbol{v}_\ell\|_\ell^2. \tag{7.3.2}$$

*Proof.* See Problem 7.4 for a particular case. □

**Lemma 7.3.4** (Inverse inequality). *Suppose $d \in \{1,2,3\}$. There is a constant $C_3 > 0$ such that, for all $0 \le \ell \le L$ and $v_\ell \in V_\ell$,*

$$\mathcal{A}(v_\ell, v_\ell) \le C_3 h_\ell^{-2} \|v_\ell\|_{L^2(\Omega)}^2. \tag{7.3.3}$$

*As a consequence of* (7.3.2), *we have*

$$\mathcal{A}(v_\ell, v_\ell) \le C_2 C_3 h_\ell^{d-2} \|\boldsymbol{v}_\ell\|_\ell^2, \tag{7.3.4}$$

*where $v_\ell \stackrel{\mathcal{B}_\ell}{\longleftrightarrow} \boldsymbol{v}_\ell \in \mathbb{R}^{n_\ell}$.*

*Proof.* See Problem 7.5 for a particular case. □

**Remark 7.3.5** (Mesh conditions). These results require some conditions on the underlying family of conforming meshes, such as global quasiuniformity and shape regularity, which hold thanks to our construction of the nested family $\{\mathcal{T}_\ell\}_{\ell=0}^L$. ◂

**Remark 7.3.6** (Norm equivalences). Estimate (7.3.2) is an example of a norm equivalence in finite dimensions. Estimate (7.3.3) is called an *inverse inequality*. This is due to the fact that such estimate is, in general, false. It only holds in finite dimensions, and the dependence on the dimension is made explicit in the equivalence constant. ◂

The following result can be understood as a *localized Poincaré inequality*.

**Lemma 7.3.7** (Localized Poincaré). *Let $0 \le \ell \le L$, $\mathcal{T}_\ell$, and $V_\ell$ be defined as usual. There exist a constant $C_{\mathrm{LP}} > 0$, independent of $\ell$, and a family of nontrivial functions $\tilde{v}_\ell \in V_\ell$ such that, for all $0 \le \ell \le L$,*

$$C_{\mathrm{LP}} h_\ell^{-2} \|\tilde{v}_\ell\|_{L^2(\Omega)}^2 \le \mathcal{A}(\tilde{v}_\ell, \tilde{v}_\ell). \tag{7.3.5}$$

*Proof.* Any nontrivial function $\tilde{v}_\ell \in V_\ell$ with $\operatorname{diam}\operatorname{supp}\tilde{v}_\ell = Ch_\ell$ satisfies this property. For definiteness, we consider a Lagrange nodal basis function. Choose $1 \le i \le n_\ell$ and set $\tilde{v}_\ell = \psi_{\ell,i} \in \mathcal{B}_\ell$. Clearly,

$$\omega_{\ell,i} := \operatorname{supp}\psi_{\ell,i} = \bigcup\{K \in \mathcal{T}_\ell \mid \exists \boldsymbol{x}_0 \in K,\ \psi_{\ell,i}(\boldsymbol{x}_0) \neq 0\}$$

satisfies $\operatorname{diam}\omega_{\ell,i} = Ch_\ell$, where the constant depends only on the shape regularity of the mesh. Moreover, $\psi_{\ell,i} \in H_0^1(\omega_{\ell,i})$. Thus this function satisfies a Poincaré inequality of the form

$$\begin{aligned}
\|\psi_{\ell,i}\|_{L^2(\Omega)}^2 &= \|\psi_{\ell,i}\|_{L^2(\omega_{\ell,i})}^2 \\
&\le C_{\mathrm{LP}'} (\operatorname{diam}\omega_{\ell,i})^2 \|\nabla\psi_{\ell,i}\|_{L^2(\omega_{\ell,i})}^2 \\
&= C_{\mathrm{LP}'} (\operatorname{diam}\omega_{\ell,i})^2 \|\nabla\psi_{\ell,i}\|_{L^2(\Omega)}^2.
\end{aligned}$$

Conclude by using the estimate on the diameter of the support of $\psi_{\ell,i}$. □

**Lemma 7.3.8** (Asymptotic exactness). *There are a constant $C_{\mathrm{P}}^*$ and $\ell_0 \ge 1$ such that if $\ell \ge \ell_0$, then there is a nontrivial function $\hat{v}_\ell \in V_\ell$ such that*

$$\mathcal{A}(\hat{v}_\ell, \hat{v}_\ell) \le C_{\mathrm{P}}^* \|\hat{v}_\ell\|_{L^2(\Omega)}^2. \tag{7.3.6}$$

*Proof.* The Poincaré inequality of Theorem 7.1.1 shows that

$$C_{\mathrm{P},1} = \inf_{v \in H_0^1(\Omega)} \frac{\mathcal{A}(v,v)}{\|v\|_{L^2(\Omega)}^2} \ge C_{\mathrm{P}}.$$

In addition, since $V_\ell \subset H_0^1(\Omega)$, we must have, for every $0 \leq \ell \leq L$,

$$\min_{v_\ell \in V_\ell} \frac{\mathcal{A}(v_\ell, v_\ell)}{\|v_\ell\|^2_{L^2(\Omega)}} \geq C_{\mathrm{P},1},$$

where the infimum is replaced by a minimum due to the fact that $V_\ell$ is finite-dimensional. Next, standard approximation properties of finite elements can be used to state that

$$\lim_{\ell \to \infty} \min_{v_\ell \in V_\ell} \frac{\mathcal{A}(v_\ell, v_\ell)}{\|v_\ell\|^2_{L^2(\Omega)}} = C_{\mathrm{P},1}.$$

Thus, there is $\ell_0 \in \mathbb{N}$ such that, for $\ell \geq \ell_0$,

$$C_{\mathrm{P},1} \leq \min_{v_\ell \in V_\ell} \frac{\mathcal{A}(v_\ell, v_\ell)}{\|v_\ell\|^2_{L^2(\Omega)}} \leq 2C_{\mathrm{P},1}.$$

Let now

$$\hat{v}_\ell \in \operatorname*{argmin}_{v_\ell \in V_\ell} \frac{\mathcal{A}(v_\ell, v_\ell)}{\|v_\ell\|^2_{L^2(\Omega)}}.$$

This function satisfies

$$\mathcal{A}(\hat{v}_\ell, \hat{v}_\ell) \leq C_{\mathrm{P}}^* \|\hat{v}_\ell\|^2_{L^2(\Omega)},$$

with $C_{\mathrm{P}}^* = 2C_{\mathrm{P},1}$, and this implies the result. □

As a next step, we recall some basic facts about the Rayleigh quotient.

**Definition 7.3.9** (Rayleigh quotient). Let $n \in \mathbb{N}$, and let $\mathsf{B} \in \mathbb{R}^{n \times n}$ be SPD. The **Rayleigh quotient** with respect to $\mathsf{B}$ is

$$R_{\mathsf{B}}(\boldsymbol{v}) \coloneqq \frac{\boldsymbol{v}^\top \mathsf{B} \boldsymbol{v}}{\boldsymbol{v}^\top \boldsymbol{v}}, \quad \forall\, \boldsymbol{v} \in \mathbb{R}^n_\star.$$ ■

**Lemma 7.3.10** (Properties of the Rayleigh quotient). *Let $n \in \mathbb{N}$, and let $\mathsf{B} \in \mathbb{R}^{n \times n}$ be SPD with*

$$\mathsf{B}\boldsymbol{v}^{(k)} = \lambda^{(k)} \boldsymbol{v}^{(k)}, \quad 1 \leq k \leq n.$$

*The smallest and largest eigenvalues of $\mathsf{B}$ satisfy, respectively,*

$$\lambda^{(1)} = R_{\mathsf{B}}(\boldsymbol{v}^{(1)}) = \min_{\boldsymbol{v} \in \mathbb{R}^n_\star} R_{\mathsf{B}}(\boldsymbol{v}) > 0 \tag{7.3.7}$$

*and*

$$\lambda^{(n)} = R_{\mathsf{B}}(\boldsymbol{v}^{(n)}) = \max_{\boldsymbol{v} \in \mathbb{R}^n_\star} R_{\mathsf{B}}(\boldsymbol{v}) > 0. \tag{7.3.8}$$

*Proof.* See [31, Chap. 08]. □

**Theorem 7.3.11** (Condition number estimate). *Let $d \in \{1,2,3\}$ and $0 \le \ell \le L$, and let $\mathcal{T}_\ell$ and $V_\ell$ be defined as usual. Recall that the (spectral) condition number of the stiffness matrix $\mathsf{A}_\ell$ is*

$$\kappa(\mathsf{A}_\ell) = \frac{\lambda_\ell^{(n_\ell)}}{\lambda_\ell^{(1)}},$$

*where $\lambda_\ell^{(1)}$ and $\lambda_\ell^{(n_\ell)}$ are, respectively, the smallest and largest eigenvalues of $\mathsf{A}_\ell$. There exist constants $0 < C_4 \le C_5$, independent of $\ell$, such that*

$$C_4 h_\ell^{-2} \le \kappa(\mathsf{A}_\ell) \le C_5 h_\ell^{-2}. \tag{7.3.9}$$

*Consequently, as $\ell \to \infty$,*

$$\kappa(\mathsf{A}_\ell) = \Theta(h_\ell^{-2}).$$

*More specifically, for $i \in \{1, n_\ell\}$, there are constants $C_4^{(i)}, C_5^{(i)} > 0$, such that*

$$C_4^{(n_\ell)} h_\ell^{d-2} \le \lambda_\ell^{(n_\ell)} \le C_5^{(n_\ell)} h_\ell^{d-2}$$

*and*

$$C_5^{(1)} h_\ell^{d} \le \lambda_\ell^{(1)} \le C_4^{(1)} h_\ell^{d}.$$

*Proof.* We prove each of the bounds in (7.3.9) separately.
(Upper bound in (7.3.9)): As usual, we use the correspondence

$$v_\ell \in V_\ell \overset{\mathcal{B}_\ell}{\longleftrightarrow} \boldsymbol{v}_\ell \in \mathbb{R}^{n_\ell}.$$

Then, for arbitrary nonzero $v_\ell \in V_\ell$,

$$\begin{aligned} R_{\mathsf{A}_\ell}(\boldsymbol{v}_\ell) &= \frac{\boldsymbol{v}_\ell^\top \mathsf{A}_\ell \boldsymbol{v}_\ell}{\boldsymbol{v}_\ell^\top \boldsymbol{v}_\ell} \\ &= \frac{\mathcal{A}(v_\ell, v_\ell)}{\|\boldsymbol{v}_\ell\|_\ell^2} \\ &\overset{(7.3.4)}{\le} \frac{C_2 C_3 h_\ell^{d-2} \|\boldsymbol{v}_\ell\|_\ell^2}{\|\boldsymbol{v}_\ell\|_\ell^2} \\ &= C_5^{(n_\ell)} h_\ell^{d-2}. \end{aligned}$$

This implies

$$\lambda_\ell^{(n_\ell)} \le C_5^{(n_\ell)} h_\ell^{d-2}.$$

Similarly,

$$
\begin{aligned}
R_{\mathsf{A}_\ell}(\boldsymbol{v}_\ell) &= \frac{\mathcal{A}(v_\ell, v_\ell)}{\|\boldsymbol{v}_\ell\|_\ell^2} \\
&\overset{(7.1.4)}{\geq} \frac{C_{\mathrm{P}}\|v_\ell\|_{L^2(\Omega)}^2}{\|\boldsymbol{v}_\ell\|_\ell^2} \\
&\overset{(7.3.2)}{\geq} \frac{C_{\mathrm{P}}C_1 h_\ell^d \|\boldsymbol{v}_\ell\|_\ell^2}{\|\boldsymbol{v}_\ell\|_\ell^2} \\
&= C_5^{(1)} h_\ell^d.
\end{aligned}
$$

Therefore

$$\lambda_\ell^{(1)} \geq C_5^{(1)} h_\ell^d.$$

We conclude that

$$\kappa(\mathsf{A}_\ell) = \frac{\lambda_\ell^{(n_\ell)}}{\lambda_\ell^{(1)}} \leq \frac{C_5^{(n_\ell)} h_\ell^{d-2}}{C_5^{(1)} h_\ell^d} = C_5 h_\ell^{-2}.$$

(Lower bound in (7.3.9)): Let $\boldsymbol{v}_\ell^{(n_\ell)} \in \mathbb{R}^{n_\ell}$ be the eigenvector corresponding to $\lambda_\ell^{(n_\ell)}$, and let $\tilde{v}_\ell \in V_\ell$ be one of the functions from Lemma 7.3.7. It follows that

$$
\begin{aligned}
\lambda_\ell^{(n_\ell)} &= R_{\mathsf{A}_\ell}(\boldsymbol{v}_\ell^{(n_\ell)}) \\
&\overset{(7.3.8)}{\geq} R_{\mathsf{A}_\ell}(\tilde{\boldsymbol{v}}_\ell) \\
&= \frac{\mathcal{A}(\tilde{v}_\ell, \tilde{v}_\ell)}{\|\tilde{\boldsymbol{v}}_\ell\|_\ell^2} \\
&\overset{(7.3.5)}{\geq} \frac{C_{\mathrm{LP}} h_\ell^{-2} \|\tilde{v}_\ell\|_{L^2(\Omega)}^2}{\|\tilde{\boldsymbol{v}}_\ell\|_\ell^2} \\
&\overset{(7.3.2)}{\geq} \frac{C_1 C_{\mathrm{LP}} h_\ell^{d-2} \|\tilde{\boldsymbol{v}}_\ell\|_\ell^2}{\|\tilde{\boldsymbol{v}}_\ell\|_\ell^2} \\
&= C_4^{(n_\ell)} h_\ell^{d-2},
\end{aligned}
$$

where we used the correspondence

$$\tilde{v}_\ell \in V_\ell \overset{\mathcal{B}_\ell}{\longleftrightarrow} \tilde{\boldsymbol{v}}_\ell \in \mathbb{R}^{n_\ell}.$$

Finally, let $\lambda_\ell^{(1)}$ be the smallest eigenvalue of $\mathsf{A}_\ell$, let $\boldsymbol{v}_\ell^{(1)}$ be the corresponding eigenvector, and let $\hat{v}_\ell \in V_\ell$ be the function from Lemma 7.3.8. We have

$$\lambda_\ell^{(1)} = R_{\mathsf{A}_\ell}(\boldsymbol{v}_\ell^{(1)})$$

$$
\begin{aligned}
&\overset{(7.3.7)}{\leq} R_{\mathsf{A}_\ell}(\hat{\boldsymbol{v}}_\ell) \\
&= \frac{\mathcal{A}(\hat{v}_\ell, \hat{v}_\ell)}{\|\hat{\boldsymbol{v}}_\ell\|_\ell^2} \\
&\overset{(7.3.6)}{\leq} \frac{C_{\mathrm{P}}^* \|\hat{v}_\ell\|_{L^2(\Omega)}^2}{\|\hat{\boldsymbol{v}}_\ell\|_\ell^2} \\
&\overset{(7.3.2)}{\leq} \frac{C_2 C_{\mathrm{P}}^* h_\ell^d \|\hat{\boldsymbol{v}}_\ell\|_\ell^2}{\|\hat{\boldsymbol{v}}_\ell\|_\ell^2} \\
&= C_4^{(1)} h_\ell^d,
\end{aligned}
$$

where we used the correspondence

$$
\hat{v}_\ell \in V_\ell \overset{\mathcal{B}_\ell}{\longleftrightarrow} \hat{\boldsymbol{v}}_\ell \in \mathbb{R}^{n_\ell}.
$$

We conclude that

$$
\kappa(\mathsf{A}_\ell) \geq \frac{C_4^{(n_\ell)} h_\ell^{d-2}}{C_4^{(1)} h_\ell^d} = C_4 h_\ell^{-2}. \qquad \square
$$

## 7.4 The strong approximation property

Recall that the finite element approximation to (7.1.3) was defined in (7.2.3). Our goal is now to show that the strong approximation property (A0) holds for the present framework. In particular, we will prove that there is a constant $C_{\mathrm{A0}} > 0$, independent of $\ell$, such that, for all $\boldsymbol{u}_\ell \in \mathbb{R}^{n_\ell}$,

$$
\|(\mathrm{I}_\ell - \tilde{\Pi}_\ell)\boldsymbol{u}_\ell\|_\ell^2 \leq C_{\mathrm{A0}}^2 \rho_\ell^{-1} \|(\mathrm{I}_\ell - \tilde{\Pi}_\ell)\boldsymbol{u}_\ell\|_{\mathsf{A}_\ell}^2.
$$

To achieve this, we must use some general facts about Galerkin approximations, which we recall next.

**Theorem 7.4.1** (Galerkin orthogonality and Céa's lemma). *Let $d \in \{1, 2, 3\}$, and let $\Omega \subset \mathbb{R}^d$ be a bounded polyhedral domain. Let $\{\mathcal{T}_h\}_{h>0}$ be a (not necessarily nested) shape-regular family of triangulations of $\Omega$ parameterized by*

$$
h := \max_{K \in \mathcal{T}_h} \operatorname{diam}(K).
$$

*Subordinate to this family, we define the finite element spaces of piecewise linear functions*

$$
S_h := \{v \in C(\overline{\Omega}) \mid v|_K \in \mathbb{P}_1(K),\ \forall K \in \mathcal{T}_h\}
$$

*and*

$$V_h := \{v \in S_h \mid v|_{\partial\Omega} \equiv 0\}.$$

*Suppose that* $f \in H^{-1}(\Omega)$ *and* $u \in H_0^1(\Omega)$ *is the unique solution to* (7.1.3). *Assume that* $u_h \in V_h$ *solves* (7.2.3), *where, formally, the subindex* $\ell$ *is replaced by* $h$. *Then*

$$\mathcal{A}(u - u_h, v_h) = 0, \quad \forall v_h \in V_h. \tag{7.4.1}$$

*Furthermore,*

$$\|u - u_h\|_{H_0^1(\Omega)} = \min_{w_h \in V_h} \|u - w_h\|_{H_0^1(\Omega)}, \tag{7.4.2}$$

*where*

$$|w|_{H^1(\Omega)} := \|w\|_{H_0^1(\Omega)} := \sqrt{\mathcal{A}(w, w)}, \quad \forall w \in H_0^1(\Omega).$$

*Proof.* Since (7.1.3) holds for all $v \in H_0^1(\Omega)$ and $V_h \subset H_0^1(\Omega)$,

$$\mathcal{A}(u, v_h) = \langle f, v_h \rangle, \quad \forall v_h \in V_h. \tag{7.4.3}$$

Subtracting (7.2.3) from (7.4.3), we immediately get (7.4.1).

Next, for all $w_h \in V_h$,

$$\begin{aligned}
\|u - u_h\|_{H_0^1(\Omega)}^2 &= \mathcal{A}(u - u_h, u - u_h) \\
&= \mathcal{A}(u - u_h, u - u_h) + \mathcal{A}(u - u_h, u_h - w_h) \\
&= \mathcal{A}(u - u_h, u - w_h) \\
&\le \|u - u_h\|_{H_0^1(\Omega)} \|u - w\|_{H_0^1(\Omega)}.
\end{aligned}$$

Thus

$$\|u - u_h\|_{H_0^1(\Omega)} \le \|u - w_h\|_{H_0^1(\Omega)},$$

and, since $w \in V_h$ is arbitrary,

$$\|u - u_h\|_{H_0^1(\Omega)} \le \inf_{w_h \in V_h} \|u - w_h\|_{H_0^1(\Omega)}.$$

Consequently, (7.4.2) holds. □

**Definition 7.4.2** (Lagrange nodal interpolation operator). Let $\Omega$, $\mathcal{T}_h$, and $V_h$ be as in Theorem 7.1.3. Set $n_h = \dim V_h$, let $\{\boldsymbol{N}_{h,j}\}_{i=1}^{n_h}$ be the set of interior vertices of $\mathcal{T}_h$, and let $\{\boldsymbol{N}_{h,j}^{+}\}_{i=1}^{n_h^+}$ be the set of all vertices of $\mathcal{T}_h$. The set

$$\mathcal{B}_h := \{\psi_{h,i}\}_{i=1}^{n_h}$$

is the **Lagrange nodal basis** for $V_h$, where the **hat functions** satisfy

$$\psi_{h,i}(\boldsymbol{N}_{h,j}) = \delta_{i,j}, \quad 1 \le i,j \le n_h.$$

Likewise, the set

$$\mathcal{B}_h^+ := \{\psi_{h,i}^+\}_{i=1}^{n_h^+}$$

is the Lagrange nodal basis for the larger space $S_h$, where

$$\psi_{h,i}^+(\boldsymbol{N}_{h,j}^+) = \delta_{i,j}, \quad 1 \le i,j \le n_h^+.$$

The **Lagrange nodal interpolation operator**, denoted $\mathcal{I}_h : C(\overline{\Omega}) \cap H_0^1(\Omega) \to V_h$, is defined as follows: for all $u \in C(\overline{\Omega}) \cap H_0^1(\Omega)$,

$$\mathcal{I}_h u := \sum_{i=1}^{n_h} u(\boldsymbol{N}_{h,i})\psi_{h,i} \in V_h.$$

The **extended Lagrange nodal interpolation operator**, denoted $\mathcal{I}_h^+ : C(\overline{\Omega}) \cap H^1(\Omega) \to S_h$, is defined as follows: for all $u \in C(\overline{\Omega}) \cap H^1(\Omega)$,

$$\mathcal{I}_h^+ u := \sum_{i=1}^{n_h^+} u(\boldsymbol{N}_{h,i}^+)\psi_{h,i}^+ \in S_h.$$ ■

**Remark 7.4.3** (Notation). In the case that the spaces are nested and indexed by $\ell$, we replace the subscripts $h$ by $\ell$. ◂

Next, we need some approximation theory; see, for example, [4, 10].

**Theorem 7.4.4** (Interpolation error estimates). *Let $d \in \{1,2,3\}$, and let $\Omega \subset \mathbb{R}^d$ be a bounded polytopal domain. Suppose $\mathcal{T}_h$, $S_h$, and $V_h$ are defined as above and $\mathcal{I}_h : C(\overline{\Omega}) \cap H_0^1(\Omega) \to V_h$ and $\mathcal{I}_h^+ : C(\overline{\Omega}) \cap H^1(\Omega) \to S_h$ are the Lagrange nodal interpolation operators. Assume that $\{\mathcal{T}_h\}_{h>0}$ is a shape-regular family of triangulations. There exists a constant $C > 0$, independent of $h$, such that, for $s \in \{0,1\}$,*

$$\|u - \mathcal{I}_h u\|_{H^s(\Omega)} \le Ch^{2-s}|u|_{H^2(\Omega)}, \tag{7.4.4}$$

*for all $u \in C(\overline{\Omega}) \cap H_0^1(\Omega) \cap H^2(\Omega)$. Similarly, there exists a constant $C > 0$, independent of $h$, such that for $s \in \{0,1\}$,*

$$\left\|u - \mathcal{I}_h^+ u\right\|_{H^s(\Omega)} \le Ch^{2-s}|u|_{H^2(\Omega)}, \tag{7.4.5}$$

*for all $u \in C(\overline{\Omega}) \cap H^1(\Omega) \cap H^2(\Omega)$.*

**Remark 7.4.5** (Boundary values). This differentiation between $\mathcal{I}_h$ and $\mathcal{I}_h^+$ will become cumbersome to maintain. Thus from this point on we will drop the superscript + from $\mathcal{I}_h^+$. The domain and range of the operator $\mathcal{I}_h$ will be understood from the context. ◂

Combining Céa's lemma and the last result, we immediately obtain the following error estimate.

**Theorem 7.4.6** (Energy norm error estimate). *Let $d \in \{1,2,3\}$, and let $\Omega \subset \mathbb{R}^d$ be a bounded polytopal domain. Suppose $\mathcal{T}_h$ and $V_h$ are as above. Assume that $\{\mathcal{T}_h\}_{h>0}$ is a shape-regular family of triangulations of $\Omega$. Suppose that $f \in H^{-1}(\Omega)$ and $u \in H_0^1(\Omega) \cap H^2(\Omega)$ is the unique solution to* (7.1.3). *Assume that $u_h \in V_h$ is the unique solution to* (7.2.3). *There exists a constant $C > 0$, independent of $h$, such that*

$$\|u - u_h\|_{H_0^1(\Omega)} \leq Ch|u|_{H^2(\Omega)}.$$

To get an optimal error estimate in the $L^2$-norm, we must argue differently. In the literature the following is known as either an *Aubin–Nitsche duality argument* or *Nitsche's trick*.

**Theorem 7.4.7** (Nitsche's trick). *Let $d \in \{1,2,3\}$, and let $\Omega \subset \mathbb{R}^d$ be a bounded polytopal domain. Suppose $\mathcal{T}_h$ and $V_h$ are as above. Assume that $\{\mathcal{T}_h\}_{h>0}$ is a shape-regular family of triangulations. Suppose that for $f \in H^{-1}(\Omega)$, $u \in H_0^1(\Omega)$ is the unique solution to* (7.1.3), *and $u_h \in V_h$ is the unique solution to* (7.2.3). *Then, under the standard regularity condition, there is a constant $C > 0$, independent of $h$, such that*

$$\|u - u_h\|_{L^2(\Omega)} \leq Ch|u - u_h|_{H^1(\Omega)}. \tag{7.4.6}$$

*If, in addition, it is known that $f \in L^2(\Omega)$, so that $u \in H_0^1(\Omega) \cap H^2(\Omega)$, then*

$$|u - u_h|_{H^1(\Omega)} \leq Ch|u|_{H^2(\Omega)}, \tag{7.4.7}$$

*for some $C > 0$. Altogether,*

$$\|u - u_h\|_{L^2(\Omega)} \leq Ch^2|u|_{H^2(\Omega)}. \tag{7.4.8}$$

*Proof.* Set $e = u - u_h \in H_0^1(\Omega) \subset L^2(\Omega)$. Let $z_e \in H_0^1(\Omega)$ be the unique solution of the so-called *dual problem*

$$\mathcal{A}(v, z_e) = (e, v)_{L^2(\Omega)}, \quad \forall\, v \in H_0^1(\Omega).$$

Notice that since $\mathcal{A}(\cdot,\cdot)$ is symmetric, the dual problem is equivalent to the original problem. Since we assume that the standard regularity result holds, we find that $z_e \in H_0^1(\Omega) \cap H^2(\Omega)$ with

$$|z_e|_{H^2(\Omega)} \leq C\|e\|_{L^2(\Omega)}.$$

Let now $z_h \in V_h$ be arbitrary and set $v = e$ in the dual problem. Using Galerkin orthogonality and the Cauchy–Schwarz inequality, we have

$$\begin{aligned}\|e\|_{L^2(\Omega)}^2 &= \mathcal{A}(e, z_e)\\ &= \mathcal{A}(e, z_e - z_h)\\ &\le \|e\|_{H_0^1(\Omega)}\|z_e - z_h\|_{H_0^1(\Omega)}.\end{aligned}$$

Let us choose $z_h = \mathcal{I}_h z_e$, where $\mathcal{I}_h : C(\overline{\Omega}) \cap H_0^1(\Omega) \to V_h$ is the piecewise linear Lagrange nodal interpolation operator. By Theorem 7.4.4

$$\begin{aligned}\|e\|_{L^2(\Omega)}^2 &\le \|e\|_{H_0^1(\Omega)}\|z_e - \mathcal{I}_h z_e\|_{H_0^1(\Omega)}\\ &\le Ch^{2-1}\|e\|_{H_0^1(\Omega)}|z_e|_{H^2(\Omega)}\\ &\le Ch\|e\|_{H_0^1(\Omega)}\|e\|_{L^2(\Omega)}.\end{aligned}$$

Therefore

$$\|e\|_{L^2(\Omega)} \le Ch\|e\|_{H_0^1(\Omega)},$$

and the result follows. □

**Definition 7.4.8** (Ritz projection). Let $\mathcal{T}_h$ and $V_h$ be as above. Let $u \in H_0^1(\Omega)$ be arbitrary. The **Ritz projection** $\mathcal{R}_h : H_0^1(\Omega) \to V_h$ is defined as follows: $\mathcal{R}_h u \in V_h$ is the unique solution to

$$\mathcal{A}(\mathcal{R}_h u, v_h) = \mathcal{A}(u, v_h), \quad \forall\, v_h \in V_h.$$

In the case that $V_h = V_\ell$ and $\mathcal{T}_h = \mathcal{T}_\ell$, we write $\mathcal{R}_\ell \coloneqq \mathcal{R}_h$ and

$$\mathcal{A}(\mathcal{R}_\ell u, v_\ell) = \mathcal{A}(u, v_\ell), \quad \forall\, v_\ell \in V_\ell.$$

■

**Remark 7.4.9** (Approximation). It should be clear that $\mathcal{R}_h u \in V_h$ is just the finite element approximation of $u$ with respect to the $H_0^1$-norm. ◂

**Lemma 7.4.10** (Approximation). *Let $\mathcal{T}_\ell$ and $V_\ell$ be as usual, and let $u_\ell \in V_\ell$. If the standard regularity condition holds, then*

$$\|u_\ell - \mathcal{R}_{\ell-1}u_\ell\|_{L^2(\Omega)} \le Ch_\ell|u_\ell - \mathcal{R}_{\ell-1}u_\ell|_{H^1(\Omega)}, \tag{7.4.9}$$

*for some constant $C > 0$ that is independent of $\ell \ge 1$.*

*Proof.* Observe that $u_\ell \in V_\ell \subset H_0^1(\Omega)$. However, $u_\ell \notin H^2(\Omega)$. For this result, we can think of $u_\ell$ as playing the role of the exact solution to the boundary value problem in

Theorem 7.4.7. The fact that $u_\ell$ is not $H^2$-regular does not matter. Since the standard regularity condition holds, we may still apply (7.4.6) to conclude

$$\|u_\ell - \mathcal{R}_{\ell-1}u_\ell\|_{L^2(\Omega)} \le Ch_{\ell-1}|u_\ell - \mathcal{R}_{\ell-1}u_\ell|_{H^1(\Omega)},$$

for some $C > 0$ that is independent of $\ell$. Now note that

$$h_{\ell-1} = 2h_\ell,$$

and the result follows. □

**Remark 7.4.11** (Shape-regularity). We again point out that for nested triangulations $\mathcal{T}_\ell$, we do not need to assume separately that the family is shape-regular and quasiuniform. These properties hold by construction. ◀

**Theorem 7.4.12** (Strong approximation property). *Let $\mathcal{T}_\ell$ and $V_\ell$ be as usual, and suppose that the strong regularity condition holds. Then the strong approximation property* (A0) *is satisfied. In particular, there is $C_{A0} > 0$, independent of $\ell$, such that*

$$\|\boldsymbol{u}_\ell - \tilde{\Pi}_\ell \boldsymbol{u}_\ell\|_\ell^2 \le C_{A0}^2 \rho_\ell^{-1} \|\boldsymbol{u}_\ell - \tilde{\Pi}_\ell \boldsymbol{u}_\ell\|_{A_\ell}^2, \tag{7.4.10}$$

*for all $\boldsymbol{u}_\ell \in \mathbb{R}^{n_\ell}$.*

*Proof.* Let $\boldsymbol{u}_\ell \in \mathbb{R}^{n_\ell}$, and let

$$u_\ell \in V_\ell \overset{\mathcal{B}_\ell}{\longleftrightarrow} \boldsymbol{u}_\ell \in \mathbb{R}^{n_\ell}.$$

We may rewrite the right-hand side of (7.4.9) as

$$|u_\ell - \mathcal{R}_{\ell-1}u_\ell|_{H^1(\Omega)}^2 = \mathcal{A}(u_\ell - \mathcal{R}_{\ell-1}u_\ell, u_\ell - \mathcal{R}_{\ell-1}u_\ell).$$

Let $\boldsymbol{w}_\ell \in \mathbb{R}^{n_\ell} \overset{\mathcal{B}_\ell}{\longleftrightarrow} u_\ell - \mathcal{R}_{\ell-1}u_\ell \in V_\ell$. We want to show that

$$\boldsymbol{w}_\ell = \boldsymbol{u}_\ell - \tilde{\Pi}_\ell \boldsymbol{u}_\ell = \boldsymbol{u}_\ell - \mathsf{P}_{\ell-1}\mathsf{A}_{\ell-1}^{-1}\mathsf{R}_{\ell-1}\mathsf{A}_\ell \boldsymbol{u}_\ell.$$

We begin with the definition of $\mathcal{R}_{\ell-1}$:

$$\mathcal{A}(\mathcal{R}_{\ell-1}u_\ell, v_{\ell-1}) = \mathcal{A}(u_\ell, v_{\ell-1}), \quad \forall v_{\ell-1} \in V_{\ell-1}.$$

Set $u'_{\ell-1} := \mathcal{R}_{\ell-1}u_\ell \in V_{\ell-1}$ and use the correspondences

$$\boldsymbol{u}'_{\ell-1} \in \mathbb{R}^{n_{\ell-1}} \overset{\mathcal{B}_{\ell-1}}{\longleftrightarrow} u'_{\ell-1} \in V_{\ell-1}$$

and

$$\boldsymbol{v}_{\ell-1} \in \mathbb{R}^{n_{\ell-1}} \overset{\mathcal{B}_{\ell-1}}{\longleftrightarrow} v_{\ell-1} \in V_{\ell-1}.$$

Then

$$\begin{aligned}\mathcal{A}(\mathcal{R}_{\ell-1}u_\ell, v_{\ell-1}) &= (\boldsymbol{u}'_{\ell-1}, \boldsymbol{v}_{\ell-1})_{\mathsf{A}_{\ell-1}}\\ &= (\mathsf{A}_{\ell-1}\boldsymbol{u}'_{\ell-1}, \boldsymbol{v}_{\ell-1})_{\ell-1},\end{aligned}$$

and

$$\begin{aligned}\mathcal{A}(u_\ell, v_{\ell-1}) &= (\boldsymbol{u}_\ell, \mathsf{P}_{\ell-1}\boldsymbol{v}_{\ell-1})_{\mathsf{A}_\ell}\\ &= (\mathsf{A}_\ell\boldsymbol{u}_\ell, \mathsf{P}_{\ell-1}\boldsymbol{v}_{\ell-1})_\ell\\ &= (\mathsf{R}_{\ell-1}\mathsf{A}_\ell\boldsymbol{u}_\ell, \boldsymbol{v}_{\ell-1})_{\ell-1}.\end{aligned}$$

So it follows that

$$\mathsf{A}_{\ell-1}\boldsymbol{u}'_{\ell-1} = \mathsf{R}_{\ell-1}\mathsf{A}_\ell\boldsymbol{u}_\ell$$

and

$$\boldsymbol{u}'_{\ell-1} = \mathsf{A}_{\ell-1}^{-1}\mathsf{R}_{\ell-1}\mathsf{A}_\ell\boldsymbol{u}_\ell = \Pi_{\ell-1}\boldsymbol{u}_\ell.$$

Therefore

$$\begin{aligned}\boldsymbol{w}_\ell &= \boldsymbol{u}_\ell - \mathsf{P}_{\ell-1}\boldsymbol{u}'_{\ell-1}\\ &= \boldsymbol{u}_\ell - \mathsf{P}_{\ell-1}\Pi_{\ell-1}\boldsymbol{u}_\ell\\ &= \boldsymbol{u}_\ell - \tilde{\Pi}_\ell\boldsymbol{u}_\ell.\end{aligned}$$

It follows that

$$\begin{aligned}|u_\ell - \mathcal{R}_{\ell-1}u_\ell|^2_{H^1(\Omega)} &= (\boldsymbol{w}_\ell, \boldsymbol{w}_\ell)_{\mathsf{A}_\ell}\\ &= \|\boldsymbol{w}_\ell\|^2_{\mathsf{A}_\ell}\\ &= \|\boldsymbol{u}_\ell - \tilde{\Pi}_\ell\boldsymbol{u}_\ell\|^2_{\mathsf{A}_\ell}.\end{aligned}$$

Finally, using the norm equivalence in (7.3.2), we have

$$\begin{aligned}C_1 h_\ell^d\|\boldsymbol{u}_\ell - \tilde{\Pi}_\ell\boldsymbol{u}_\ell\|^2_\ell &\overset{(7.3.2)}{\leq} \|u_\ell - \mathcal{R}_{\ell-1}u_\ell\|^2_{L^2(\Omega)}\\ &\overset{(7.4.9)}{\leq} Ch_\ell^2|u_\ell - \mathcal{R}_{\ell-1}u_\ell|^2_{H^1(\Omega)}\\ &= Ch_\ell^2\|\boldsymbol{u}_\ell - \tilde{\Pi}_\ell\boldsymbol{u}_\ell\|^2_{\mathsf{A}_\ell}.\end{aligned}$$

The proof of Theorem 7.3.11 showed that

$$C_6^{(n_\ell)}h_\ell^{d-2} \leq \rho_\ell \leq C_7^{(n_\ell)}h_\ell^{d-2}.$$

Combining this with the last estimate gives the desired result (7.4.10). □

**Corollary 7.4.13** (Weak approximation property). *Let $\mathcal{T}_\ell$ and $V_\ell$ be defined as usual, and let $\mathsf{A}_\ell$ be the standard stiffness matrix for problem* (7.1.3). *Then the weak approximation property* (A1) *holds: there exists a constant $C_{\mathrm{A1}} > 0$, independent of $\ell$, such that*

$$((\mathrm{I}_\ell - \tilde{\Pi}_\ell)\boldsymbol{u}_\ell, \boldsymbol{u}_\ell)_{\mathsf{A}_\ell} \leq C_{\mathrm{A1}}^2 \rho_\ell^{-1} \|\mathsf{A}_\ell \boldsymbol{u}_\ell\|_\ell^2, \tag{7.4.11}$$

*for all $\boldsymbol{u}_\ell \in \mathbb{R}^{n_\ell}$.*

*Proof.* This follows immediately from Theorem 6.3.4, since the Galerkin condition (G0) and the strong approximation property (A0) hold. □

**Remark 7.4.14** (Convergence). Since assumptions (G0) and (A0) hold, using Richardson's smoother, the W-cycle and V-cycle algorithms defined in Chapter 6 converge. There is nothing more to do! The proofs are merely an application of Theorem 6.6.1 and Corollaries 6.7.6 and 6.8.3. ◂

## 7.5 The full multigrid algorithm

Many of the readers of this text may have heard it said that multigrid is an optimal-order method. By this it is meant that a good enough approximation to the finite element solution can be found by some multigrid algorithm in $\mathcal{O}(n_L)$ operations, where $n_L$ is the number of unknowns (degrees of freedom) in our finite element solution. By contrast, if one were to use, say, Gaussian elimination to find the solution, then $\mathcal{O}(n_L^3)$ operations would be required. However, we need to be precise about which multigrid algorithm the discussion is about. In particular, we need another multigrid operator, which we now define.

**Definition 7.5.1** (Full multigrid). Recall that the multigrid operator MG(·) was introduced in Definition 6.1.3. Let $r \in \mathbb{N}$ and $1 \leq s \leq L$. Assume that $f \in L^2(\Omega)$, and define

$$\boldsymbol{f}_\ell := \begin{bmatrix} (f, \psi_{\ell,1})_{L^2(\Omega)} \\ (f, \psi_{\ell,2})_{L^2(\Omega)} \\ \vdots \\ (f, \psi_{\ell,n_\ell})_{L^2(\Omega)} \end{bmatrix} \in \mathbb{R}^{n_\ell}, \quad 0 \leq \ell \leq L.$$

The **full multigrid operator**

$$\hat{\boldsymbol{u}}_s := \mathrm{FMG}(s)$$

is defined as follows:

–
$$\hat{\boldsymbol{u}}_0 := \mathsf{A}_0^{-1} \boldsymbol{f}_0.$$

– For $\ell = 1, 2, \ldots, s$,

–
$$\boldsymbol{u}_\ell^{(0)} := \mathsf{P}_{\ell-1}\hat{\boldsymbol{u}}_{\ell-1};$$

–
$$\boldsymbol{u}_\ell^{(\sigma+1)} := \mathrm{MG}(\ell, \boldsymbol{f}_\ell, \boldsymbol{u}_\ell^{(\sigma)}), \quad 0 \le \sigma \le r-1;$$

–
$$\hat{\boldsymbol{u}}_\ell := \boldsymbol{u}_\ell^{(r)}.$$
■

The shape of the flow of information on the full multigrid algorithm is depicted in Figure 7.6.

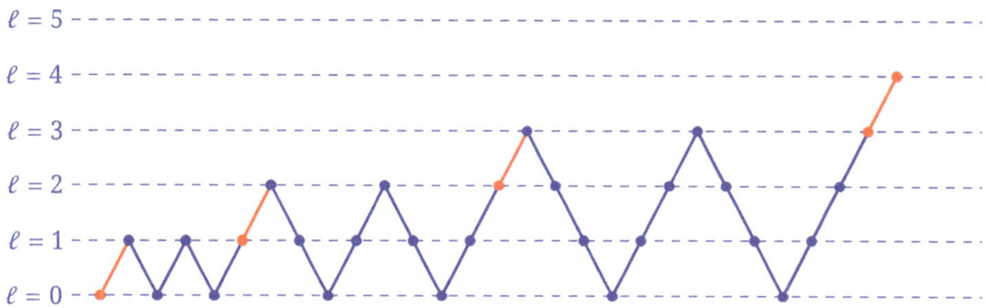


**Figure 7.6:** The shape of the full multigrid algorithm for $r = 2$ and $p = 1$.

**Theorem 7.5.2** (Full multigrid error estimate). *Suppose that there is* $0 < \gamma < 1$ *such that, for all* $0 \le \ell \le L$ *and* $\boldsymbol{u}_\ell^{(0)} \in \mathbb{R}^{n_\ell}$,

$$\|\boldsymbol{u}_\ell^{\mathrm{E}} - \mathrm{MG}(\ell, \boldsymbol{g}_\ell, \boldsymbol{u}_\ell^{(0)})\|_{\mathsf{A}_\ell} \le \gamma \|\boldsymbol{u}_\ell^{\mathrm{E}} - \boldsymbol{u}_\ell^{(0)}\|_{\mathsf{A}_\ell}, \tag{7.5.1}$$

*where*

$$\boldsymbol{u}_\ell^{\mathrm{E}} := \mathsf{A}_\ell^{-1}\boldsymbol{g}_\ell.$$

*Assume that* $f \in L^2(\Omega)$ *and* $r \in \mathbb{N}$, *in the full multigrid algorithm of Definition* 7.5.1, *is large enough so that*

$$\gamma^r < \frac{1}{2}.$$

*Suppose that*

$$\hat{u}_\ell \in V_\ell \overset{\mathcal{B}_\ell}{\longleftrightarrow} \hat{\boldsymbol{u}}_\ell := \mathrm{FMG}(\ell) \in \mathbb{R}^{n_\ell}.$$

*Then, there exists a constant* $C > 0$, *independent of* $\ell$, *such that*

$$|u_\ell - \hat{u}_\ell|_{H^1(\Omega)} = \|u_\ell - \hat{u}_\ell\|_{H_0^1(\Omega)} \le Ch_\ell |u|_{H^2(\Omega)},$$

*where* $u_\ell \in V_\ell$ *is the solution to* (7.2.3), *and* $u \in H_0^1(\Omega) \cap H^2(\Omega)$ *solves* (7.1.3).

*Proof.* Define

$$\hat{e}_\ell := u_\ell - \hat{u}_\ell \in V_\ell.$$

This is the error in computing the finite element approximation. Clearly, $\hat{e}_0 \equiv 0$. In general,

$$|\hat{e}_\ell|^2_{H^1(\Omega)} = \mathcal{A}(\hat{e}_\ell, \hat{e}_\ell) = \|\hat{\boldsymbol{e}}_\ell\|^2_{\mathsf{A}_\ell},$$

where

$$\hat{\boldsymbol{e}}_\ell = \boldsymbol{u}_\ell - \hat{\boldsymbol{u}}_\ell \in \mathbb{R}^{n_\ell} \overset{\mathcal{B}_\ell}{\longleftrightarrow} \hat{e}_\ell = u_\ell - \hat{u}_\ell \in V_\ell.$$

Then

$$\begin{aligned}
|\hat{e}_\ell|_{H^1(\Omega)} &= \|\boldsymbol{u}_\ell - \hat{\boldsymbol{u}}_\ell\|_{\mathsf{A}_\ell} \\
&\overset{(7.5.1)}{\leq} \gamma^r \|\boldsymbol{u}_\ell - \mathsf{P}_{\ell-1}\hat{\boldsymbol{u}}_{\ell-1}\|_{\mathsf{A}_\ell} \\
&= \gamma^r |u_\ell - \hat{u}_{\ell-1}|_{H^1(\Omega)} \\
&\leq \gamma^r \left(|u_\ell - u|_{H^1(\Omega)} + |u - u_{\ell-1}|_{H^1(\Omega)} + |u_{\ell-1} - \hat{u}_{\ell-1}|_{H^1(\Omega)}\right) \\
&\overset{(7.4.7)}{\leq} \gamma^r \left(Ch_\ell |u|_{H^2(\Omega)} + 2Ch_\ell |u|_{H^2(\Omega)} + |\hat{e}_{\ell-1}|_{H^1(\Omega)}\right) \\
&= C\gamma^r h_\ell |u|_{H^2(\Omega)} + \gamma^r |\hat{e}_{\ell-1}|_{H^1(\Omega)}.
\end{aligned} \tag{7.5.2}$$

By the same reasoning,

$$|\hat{e}_{\ell-1}|_{H^1(\Omega)} \leq C\gamma^r h_{\ell-1} |u|_{H^2(\Omega)} + \gamma^r |\hat{e}_{\ell-2}|_{H^1(\Omega)}. \tag{7.5.3}$$

Combining (7.5.2) and (7.5.3), we have

$$|\hat{e}_\ell|_{H^1(\Omega)} \leq C\gamma^r h_\ell |u|_{H^2(\Omega)} + C\gamma^{2r} h_{\ell-1} |u|_{H^2(\Omega)} + \gamma^{2r} |\hat{e}_{\ell-2}|_{H^1(\Omega)}.$$

Continuing in this fashion and using $\hat{e}_0 \equiv 0$, we have

$$\begin{aligned}
|\hat{e}_\ell|_{H^1(\Omega)} &\leq \{Ch_\ell\gamma^r + Ch_{\ell-1}\gamma^{2r} + Ch_{\ell-2}\gamma^{3r} + \cdots + Ch_1\gamma^{\ell r}\}|u|_{H^2(\Omega)} \\
&= \{Ch_\ell\gamma^r + Ch_\ell 2\gamma^{2r} + Ch_\ell 2^2\gamma^{3r} + \cdots + Ch_\ell 2^{\ell-1}\gamma^{\ell r}\}|u|_{H^2(\Omega)} \\
&= \frac{Ch_\ell}{2}\{2\gamma^r + 2^2\gamma^{2r} + 2^3\gamma^{3r} + \cdots + 2^\ell\gamma^{\ell r}\}|u|_{H^2(\Omega)} \\
&\leq \frac{C\gamma^r}{1 - 2\gamma^r} h_\ell |u|_{H^2(\Omega)}.
\end{aligned}$$

The theorem is proven. □

**Remark 7.5.3** (Approximate solution). Let us think about what the last result tells us. Using the triangle inequality, we have

$$\begin{aligned}\|u - \hat{u}_\ell\|_{H_0^1(\Omega)} &\le \|u - u_\ell\|_{H_0^1(\Omega)} + \|u_\ell - \hat{u}_\ell\|_{H_0^1(\Omega)}\\ &\le Ch_\ell|u|_{H^2(\Omega)} + Ch_\ell|u|_{H^2(\Omega)}\\ &= Ch_\ell|u|_{H^2(\Omega)}.\end{aligned}$$

In other words, the solution that we compute using the full multigrid operator, provided that $r$ is sufficiently large, is asymptotically as good as the finite element approximation. Thus it does not make sense to go any further. The next result shows that the computational cost of the full multigrid operator is optimal. ◂

**Proposition 7.5.4** (Work estimate for full multigrid). *Let* $0 \le \ell \le L$ *and* $n_\ell = \dim(V_\ell)$. *Assume that there are constants* $0 < C_1 \le C_2$, *that are independent of* $\ell$, *for which*

$$C_1 2^{d\ell} \le n_\ell \le C_2 2^{d\ell}, \quad 0 \le \ell \le L.$$

*Recall that* $d \in \{1, 2, 3\}$ *is the space dimension. If*

$$p < 2^d,$$

*then the amount of work* $W_s$ *for the full multigrid operator* FMG$(s)$ *satisfies*

$$W_s \le Cn_s,$$

*where* $C > 0$ *is a constant that is independent of* $s$.

*Proof.* Let, for $1 \le \ell \le L$, $w_\ell$ be the work required to compute the output of the multigrid operator, MG$(\ell, \cdot, \cdot)$. Then, assuming that the work is dominated by smoothing,

$$w_\ell \le C(m_1 + m_2)n_\ell + pw_{\ell-1},$$

where $C > 0$ is independent of $\ell$. Similarly,

$$w_{\ell-1} \le C(m_1 + m_2)n_{\ell-1} + pw_{\ell-2}.$$

Combining the last two inequalities gives

$$w_\ell \le C(m_1 + m_2)n_\ell + pC(m_1 + m_2)n_{\ell-1} + p^2 w_{\ell-2}.$$

Continuing in this fashion, we obtain

$$\begin{aligned}w_\ell &\le C(m_1 + m_2)\{n_\ell + pn_{\ell-1} + p^2 n_{\ell-2} + \cdots + p^\ell n_0\}\\ &\le CC_2(m_1 + m_2)2^{d\ell}\left\{1 + \frac{p}{2^d} + \left(\frac{p}{2^d}\right)^2 + \cdots + \left(\frac{p}{2^d}\right)^\ell\right\}\\ &\le \frac{CC_2(m_1 + m_2)2^{d\ell}}{1 - \frac{p}{2^d}}\end{aligned}$$

$$\begin{aligned}
&\leq \frac{CC_2(m_1+m_2)}{C_1(1-\frac{p}{2^d})} n_\ell \\
&= Cn_\ell.
\end{aligned}$$

Finally, neglecting the cost of the prolongation step, we have

$$W_s = W_{s-1} + rw_s \leq W_{s-1} + rCn_s.$$

Likewise, at level $s-1$,

$$W_{s-1} \leq W_{s-2} + rCn_{s-1}.$$

Consequently,

$$W_s \leq rCn_s + rCn_{s-1} + W_{s-2}.$$

Continuing in this fashion,

$$\begin{aligned}
W_s &\leq rCn_s + rCn_{s-1} + \cdots + rCn_0 \\
&\leq rCC_2(2^{ds} + 2^{d(s-1)} + \cdots + 1) \\
&= rCC_2 2^{ds}\left(1 + \frac{1}{2^d} + \left(\frac{1}{2^d}\right)^2 + \cdots + \left(\frac{1}{2^d}\right)^s\right) \\
&\leq \frac{rCC_2}{1-\frac{1}{2^d}} 2^{ds} \\
&\leq \frac{rCC_2}{C_1(1-\frac{1}{2^d})} n_s \\
&\leq Cn_s.
\end{aligned}$$

□

## 7.6 Some computational experiments

To conclude this chapter, as a complement to Section 4.9, let us perform some computational experiments to illustrate the convergence results of our various multigrid algorithms. In particular, let us use the algorithms to approximate the solution of

$$\mathsf{A}_L \boldsymbol{u}_L^{\mathrm{E}} = \boldsymbol{f}_L,$$

where $\mathsf{A}_L$ is the standard finite element stiffness matrix.

### 7.6.1 Numerical experiments in one dimension

In one dimension, we consider the model problem (1.7.1). In our experiments, we specify the exact solution

$$[\boldsymbol{u}_L^{\mathrm{E}}]_i = u_{L,i}^{\mathrm{E}} = \exp(\sin(3\pi x_{L,i})) - 1.0.$$

As in Section 4.9, we use a uniform mesh in one space dimension. The stiffness matrix is defined in (4.1.9). The force vector is manufactured by setting $\boldsymbol{f}_L := \mathsf{A}_L \boldsymbol{u}_L^{\mathrm{E}}$. We report the results of several computational experiments in Table 7.1. The initial approximation $\boldsymbol{u}_L^{(0)}$ is chosen via a pseudorandom number selection. The main MATLAB© codes implementing the multigrid algorithm are given in Listings 7.3, 7.4, and 7.5. The error reduction for the multigrid V-Cycle algorithm ($p = 1$) using the parameters $\omega = \frac{2}{3}$, $m_1 = m_2 = 3$, $n_L = 255$, and $L = 7$ is shown in Figure 7.7.

**Table 7.1:** Computed multigrid convergence factors $\gamma_{\mathrm{comp}}$ for various parameter choices. The factor $\gamma_{\mathrm{comp}}$ is computed using a log-linear fit of the last four error values $\|\boldsymbol{u}_L^{\mathrm{E}} - \boldsymbol{u}_L^k\|_L$.

| $n_L$ | $L$ | $\omega$ | $m_1$ | $m_2$ | $p$ | $\gamma_{\mathrm{comp}}$ |
|---|---|---|---|---|---|---|
| 63 | 5 | 2/3 | 3 | 3 | 1 | $5.93 \times 10^{-02}$ |
| 127 | 6 | 2/3 | 3 | 3 | 1 | $5.99 \times 10^{-02}$ |
| 255 | 7 | 2/3 | 3 | 3 | 1 | $6.01 \times 10^{-02}$ |
| 511 | 8 | 2/3 | 3 | 3 | 1 | $6.01 \times 10^{-02}$ |
| 1023 | 9 | 2/3 | 3 | 3 | 1 | $6.01 \times 10^{-02}$ |
| 127 | 6 | 2/3 | 4 | 4 | 1 | $4.65 \times 10^{-02}$ |
| 127 | 6 | 2/3 | 5 | 5 | 1 | $3.81 \times 10^{-02}$ |
| 127 | 6 | 2/3 | 6 | 6 | 1 | $3.21 \times 10^{-02}$ |
| 127 | 6 | 0.50 | 3 | 3 | 1 | $8.06 \times 10^{-02}$ |
| 127 | 6 | 0.55 | 3 | 3 | 1 | $7.27 \times 10^{-02}$ |
| 127 | 6 | 0.60 | 3 | 3 | 1 | $6.66 \times 10^{-02}$ |
| 127 | 6 | 0.65 | 3 | 3 | 1 | $6.14 \times 10^{-02}$ |
| 127 | 6 | 0.70 | 3 | 3 | 1 | $5.71 \times 10^{-02}$ |
| 127 | 6 | 0.75 | 3 | 3 | 1 | $5.39 \times 10^{-02}$ |
| 127 | 6 | 0.80 | 3 | 3 | 1 | $5.91 \times 10^{-02}$ |
| 127 | 6 | 0.50 | 3 | 3 | 2 | $5.44 \times 10^{-02}$ |
| 127 | 6 | 0.50 | 3 | 3 | 3 | $5.42 \times 10^{-02}$ |

### 7.6.2 The `iFEM` package in MATLAB©

In this section, we introduce `iFEM` [1], which is a MATLAB© package for the approximation of solutions to PDE via the finite element method. `iFEM` is capable of mesh generation, mesh processing (refinement, coarsening, et cetera), assembling of stiffness ma-

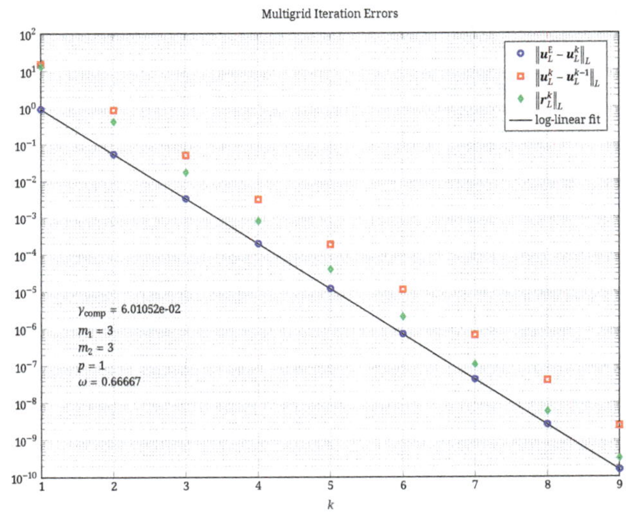


**Figure 7.7:** The error reduction for the multigrid V-cycle algorithm using the parameters $\omega = \frac{2}{3}$, $m_1 = m_2 = 3$, $n_L = 255$, and $L = 7$. The factor $\gamma_{\text{comp}}$ is computed using a log-linear fit of the last four error values. Note that both the Cauchy difference $\|\boldsymbol{u}_L^k - \boldsymbol{u}_L^{k-1}\|_L$ and the residual $\|\boldsymbol{r}_L^k\|_L$ are good indicators of the error.

trices and forcing vectors, and solving the resulting linear systems with multigrid and preconditioned conjugate gradient methods.

For mesh generation, we can manually specify the coordinates of the nodes and the elements by specifying the connectivity. Any polygonal domain, such as an L-shaped domain, can be easily created with such a method. This gives the users the highest degree of autonomy in designing the domain mesh, at the cost of time and manual effort. For some common domains like rectangles and disks, iFEM has built-in commands to create them by specifying the location and mesh size. For even more complicated or specific domains, the users can import their own mesh created by a third-party software. Figure 7.8 shows example meshes, like those described above.

**Listing 7.1:** Creating a mesh by manually specifying the coordinates of the nodes and the connectivity of the elements. The full code is provided in Listing 7.6.

```
% Specifying the coordinates.
nodeL = [0,-1;1,-1;1,0;1,1;0,1;-1,1;-1,0;0,0];

% Specifying the connectivity.
elemL = [1,2,8;3,8,2;8,3,5;4,5,3;7,8,6;5,6,8];
```

To improve the accuracy of the numerical solution, we can automatically refine the inputted conforming mesh. To automate this process, iFEM has internal tools for

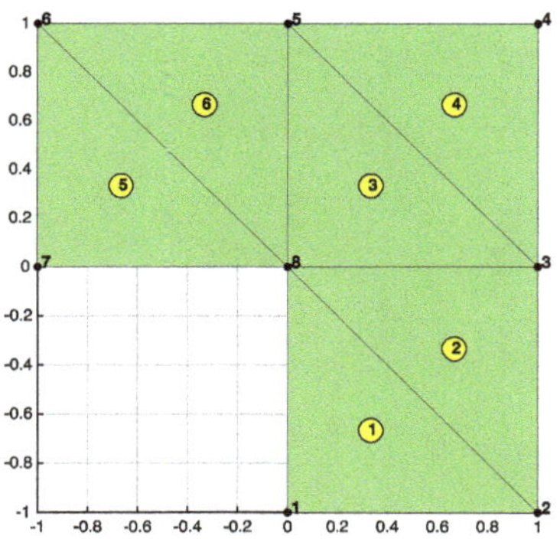

(a) Mesh created by specifying the nodes and elements manually.

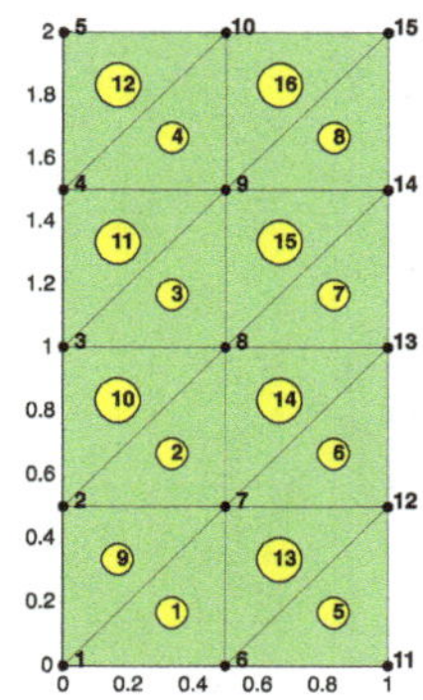

(b) Mesh created by the command `squaremesh`.

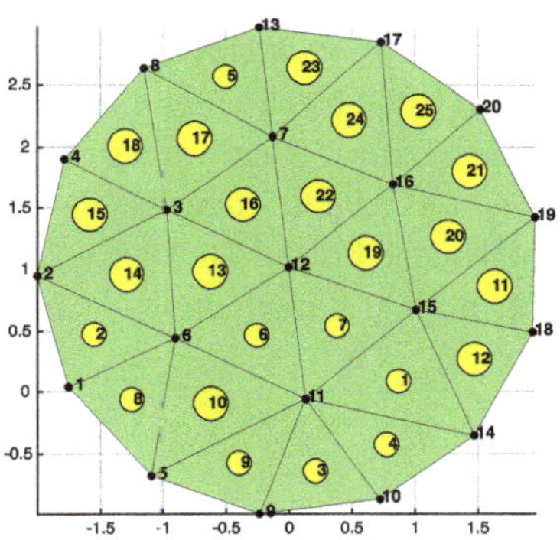

(c) Mesh created by the command `circlemesh`.

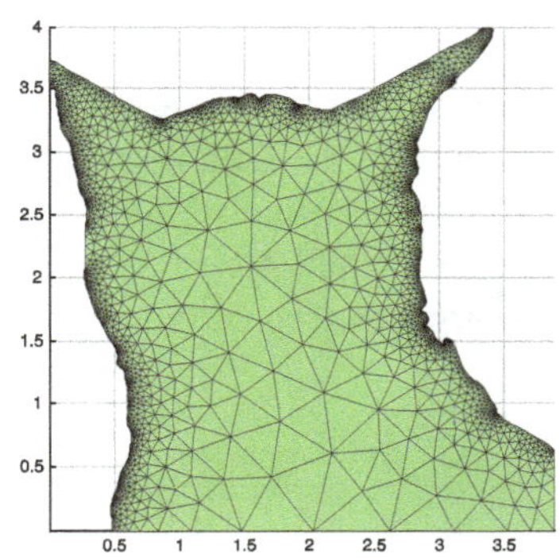

(d) Mesh imported from an external file.

**Figure 7.8:** Meshes visualized with the command `showmesh`.

standard uniform refinement, specifically, bisection for one-dimensional meshes and quadrisection (see Figures 7.2 and 7.9) for two-dimensional meshes. The following code snippet shows the usage of the command `uniformrefine` for the uniform refinement of an L-shaped domain.

**Listing 7.2:** Refining meshes with the `iFEM` `uniformrefine` command. The full code is provided in Listing 7.6.

```
% Refine mesh uniformly by quadrisecting
[nodeRefinedL,elemRefinedL] = uniformrefine(nodeL,elemL);
```

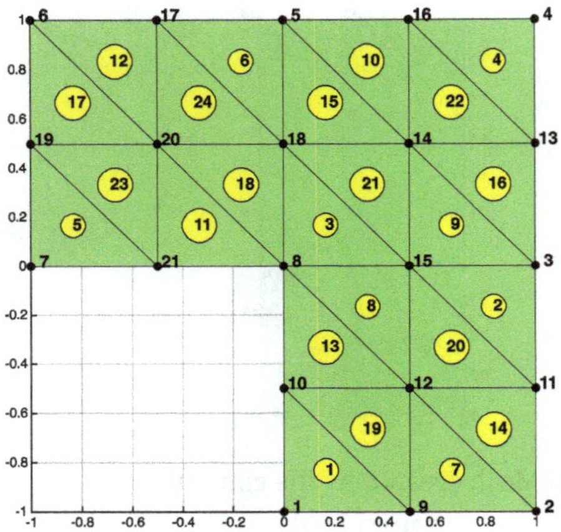

**Figure 7.9:** An L-shaped mesh obtained by refining an initially coarser mesh.

Having constructed the mesh, the next step is to specify the boundary conditions and assemble the stiffness matrix and the forcing vector. For common equations, like the Poisson equation, `iFEM` has built-in scripts to assemble the stiffness matrix and forcing vector. This is accomplished via built-in quadrature rules to efficiently compute and/or approximate the integrals. After this, solvers can be called; see Listing 7.6. Figure 7.10 shows the convergence graph and contraction rate of the multigrid method.

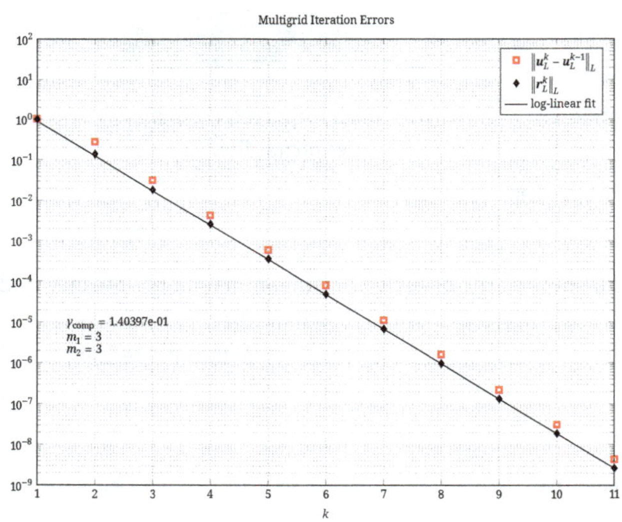


**Figure 7.10:** The error reduction for the multigrid V-Cycle algorithm using `iFEM`. Due to the configuration of the `iFEM` package, the exact error cannot be measured at each iteration.

## 7.7 Problems

7.1 Prove Theorem 7.1.1 for functions in the space $C^1(\overline{\Omega}) \cap H_0^1(\Omega)$ for $d = 1$ and $\Omega = (0, 1)$.

7.2 Prove Lemma 7.2.4.

7.3 Prove Proposition 7.2.5.

7.4 Prove Lemma 7.3.3 for the particular case where $d = 1$ (one space dimension) and the mesh is uniform, as in Chapter 4.

7.5 Prove Lemma 7.3.4 for the particular case where $d = 1$ (one space dimension) and the mesh is uniform, as in Chapter 4.

7.6 Consider the bilinear form $\mathcal{A}_{\#}$ introduced in Remark 7.1.4. Verify that, under the standard regularity condition of Definition 7.1.2, all the theory developed in this chapter follows.

# Listings

## One-dimensional finite element code

```
./1DFEM/
    poisson1dMG.m
    multiGridSolver.m
    MGOperator.m
    prolongation.m (see Listing 4.7)
    restriction.m (see Listing 4.8)
    smoothJacobiDamped.m (see Listing 4.9)
    smoothGaussSeidel.m (see Listing 4.10)
    getResidual.m (see Listing 4.11)
```

**Listing 7.3:** poisson1dMG.m: Calling script for the 1D finite element multigrid solver.

```
%
% poisson1DMG.m
%
% Multigrid calling script for solving the 1D Poisson
% equation
%
%   -u'' = f
%
% on an interval with homogeneous Dirichlet boundary
% conditions.
%
%% Initialization and Parameter

% Clear command line window and workspace.
clear; clc;

% Tolerance of residual to stop multigrid iterations.
tol = 1.0e-09;

% Number of interior cells at finest level.
nL = 255;

% Number of multigrid levels.
L = 7;

% Define the grid spacing at finest level.
```

```
hL = 1/(nL+1);

% Parameters for the multigrid solver.
p      = 1;
m1     = 3;
m2     = 3;
omega  = 2/3;
kMax   = 100;

%% Setting up the multigrid solver

% Check mesh coarsening.
nl = nL;
for k = L-1:-1:0
  if mod(nl+1,2) ~= 0
    error('Coarsening error: L is too large.');
  end
  nl = (nl+1)/2-1;
end

% Give a random initial guess.
rng(1234);
u = rand(nL,1)-0.5;

% Construct the exact solution and necessary forcing
% vector.
x = linspace(hL,1.0-hL,nL)';
AL = gallery('tridiag',nL,-1.0/hL,2.0/hL,-1.0/hL);
uExact = exp(sin(3.0*pi*x))-1.0;
f = AL*uExact;

% Store parameters in a struct.
MGParam.nL = nL;
MGParam.L = L;
MGParam.p = p;
MGParam.m1 = m1;
MGParam.m2 = m2;
MGParam.omega = omega;
MGParam.tol = tol;
MGParam.kMax = kMax;

%% Calling the multigrid solver
```

```
tic
[u,errVals,kStop] = multiGridSolver(u,f,MGParam,uExact);
toc

finalErr = norm(uExact-u,2);
fprintf('\nFinal error in L2 sense = %g\n',finalErr);

%% Plotting the numerical solution and error plots

figure(1)
clf
plot(x,u,'k-','LineWidth',1.5)
xlabel('$x$','Interpreter','latex');
ylabel('$u_L(x)$','Interpreter','latex');
title(['Finite Element Approximation of the ',...
       'Poisson Equation']);
axis([0.0,1.0,-1.0,2.0])
printstr = strcat('Solution_nL_',num2str(nL),'.pdf');
exportgraphics(gca,printstr)

figure(2)
clf

semilogy(errVals(1:kStop,3),'bo','LineWidth',1.5)
hold on
semilogy(errVals(1:kStop,2),'rs','LineWidth',1.5)
semilogy(errVals(1:kStop,1),'gd','LineWidth',1.5)


% Estimate the rate of contraction.
rate = 1.0;
if kStop >= 4
  kv = kStop-3:kStop;
  le = log(errVals(kStop-3:kStop,3));
  p1 = polyfit(kv,le,1);
  rate = exp(p1(1));
  p1k = polyval(p1,1:kStop);
  semilogy(1:kStop,exp(p1k),'-k')
end

xlabel('$k$','Interpreter','latex');
```

```
title('Multigrid Iteration Errors',...
      'Interpreter','latex');
legend(...
  '$\left\|{\bf u}_L^{\rm E}-{\bf u}_L^k\right\|_L$',...
  '$\left\|{\bf u}_L^k-{\bf u}_L^{k-1} \right\|_L$',...
  '$\left\|{\bf r}_L^k \right\|_L$',...
  'log-linear fit','Interpreter','latex');
text(1.5,128*tol,...
  strcat('$\gamma_{\rm comp} =\hspace{.1cm}$',...
  num2str(rate,'%10.5e')),'FontSize',14,...
  'Interpreter','latex')
text(1.5,32*tol,strcat('$m_1 =\hspace{.1cm}$',...
  num2str(MGParam.m1)),'FontSize',14,...
  'Interpreter','latex')
text(1.5,8*tol,strcat('$m_2 =\hspace{.1cm}$',...
  num2str(MGParam.m2)),'FontSize',14,...
  'Interpreter','latex')
text(1.5,2*tol,strcat('$p =\hspace{.1cm}$',...
  num2str(MGParam.p)),'FontSize',14,...
  'Interpreter','latex')
text(1.5,0.5*tol,strcat('$\omega =\hspace{.1cm}$',...
  num2str(MGParam.omega)),'FontSize',14,...
  'Interpreter','latex')
printstr = strcat('Err_nL_',num2str(nL),'_m1_',...
  num2str(MGParam.m1),'.pdf');
exportgraphics(gca,printstr)
grid on
grid minor
hold off
```

**Listing 7.4:** `multiGridSolver.m`: 1D finite element multigrid solver.

```
function [u,errVals,kStop] = multiGridSolver(u,f,...
                                             MGParam,...
                                             uExact)

L = MGParam.L;
nL = MGParam.nL;
kMax = MGParam.kMax;
tol = MGParam.tol;
errVals = zeros(kMax,3);
```

```

hL = 1/(nL+1);
x = linspace(hL,1.0-hL,nL)';

k = 0;
err = tol+1.0;

while ((k < kMax) && (err > tol))
  uo = u;
  u = MGOperator(L,f,u,MGParam);

  errVals(k+1,1) = norm(getResidual(f,u),2);
  errVals(k+1,2) = norm(u-uo,2);
  errVals(k+1,3) = norm(uExact-u,2);

  % We use the exact error for the stopping criteria.
  % In practice, the exact solution is not available,
  % and one can use either the residual or the coarse
  % grid correction as the measurement for the stopping
  % criteria.
  err = norm(uExact-u,2);

  k = k+1;

  % Print iteration information.
  fprintf('Multigrid iteration %d:\n',k);
  fprintf('   residual   = %.6e\n',errVals(k,1));
  fprintf('   correction= %.6e\n',errVals(k,2));
  fprintf('   true error= %.6e\n',errVals(k,3));
end

kStop = k;

end
```

**Listing 7.5:** MGOperator.m: 1D finite element multigrid operator.

```
function u = MGOperator(level,f,u,MGParam)

nl = length(f);
hl = 1/(nl+1);
```

```

if level == 0

  % Direct solve at the coarsest level.
  Al = gallery('tridiag',nl,-1.0/hl,2.0/hl,-1.0/hl);
  u = Al\f;
else

  % Pre-smoothing.
  u = smoothJacobiDamped(f,u,MGParam.m1,MGParam.omega);

  % Alternative pre-smoothing options:
  % u = smoothGaussSeidel(f,u,MGParam.m1,'Forward');

  % Get residual and restrict to coarse grid.
  cGr = restriction(getResidual(f,u));
  nlm1 = (nl+1)/2-1;
  cGc = zeros(nlm1,1);

  % Compute the coarse-grid correction with recursive
  % MG call.
  for sig = 1:MGParam.p
    cGc = MGOperator(level-1,cGr,cGc,MGParam);
  end

  % Correct the solution using coarse-grid correction.
  u = u+prolongation(cGc);

  % Post-smoothing.
  u = smoothJacobiDamped(f,u,MGParam.m2,MGParam.omega);

  % Alternative post-smoothing options:
  % u = smoothGaussSeidel(f,u,MGParam.m2,'Backward');
end

end
```

### Two-dimensional iFEM© finite element code

```
./2DFEM/
    meshDemo.m
    poisson2DMG.m
```

**Listing 7.6:** `meshDemo.m`: Demo of mesh creation and refinement with iFEM©.

```
% Mesh generation, refinement and visualization

hold off; close all; clear all;

%% Create a L-shaped domain

% Specifying the coordinates.
nodeL = [0,-1;1,-1;1,0;1,1;0,1;-1,1;-1,0;0,0];

% Specifying the connectivity.
elemL = [1,2,8;3,8,2;8,3,5;4,5,3;7,8,6;5,6,8];
figure; hold off
showmesh(nodeL,elemL);
axis on;

% Plotting the indices of all triangles.
findelem(nodeL,elemL);

% Plotting the indices of all vertices.
findnode(nodeL);
title('Mesh of L-shaped domain',...
      'Units','normalized','Position',[0.5,1.05,0]);


%% Create a rectangular domain with mesh size specified

% Rectangle [0,1]*[0,2], mesh size 0.5
[nodeRectangular,elemRectangular] =...
                                squaremesh([0,1,0,2],0.5);
figure; hold off
showmesh(nodeRectangular,elemRectangular);
axis on;
findelem(nodeRectangular,elemRectangular);
findnode(nodeRectangular);
```

```
title('Mesh of rectangular domain',...
      'Units','normalized','Position',[0.5,1.05,0]);

%% Create a circular domain with mesh size specified

% Circular mesh with center (0,1) and radius 2,
% mesh size 0.5
[nodeRectangular,elemRectangular] =...
                                   circlemesh(0,1,2,0.5);
figure; hold off
showmesh(nodeRectangular,elemRectangular);
axis on;
findelem(nodeRectangular,elemRectangular);
findnode(nodeRectangular);
title('Mesh of circular domain',...
      'Units','normalized','Position',[0.5,1.05,0]);

%% Create a customized domain

% Cat domain created with Im2Mesh package
% Jiexian Ma (2025). Im2mesh (2D image to triangular
% meshes)
% (https://www.mathworks.com/matlabcentral/...
% fileexchange/71772-im2mesh-2d-image-to-triangular...
% -meshes),
% MATLAB Central File Exchange.
% Retrieved January 30, 2025.

load("cat.mat")
figure; hold off
showmesh(nodeCat,elemCat);
axis on;
title('Mesh of cat domain',...
      'Units','normalized','Position',[0.5,1.05,0]);

%% Uniform refinement of meshes

% Refine mesh uniformly by quadrisecting
[nodeRefinedL,elemRefinedL] = uniformrefine(nodeL,elemL);

figure; hold off
showmesh(nodeRefinedL,elemRefinedL);
```

```
axis on;
findelem(nodeRefinedL,elemRefinedL);
findnode(nodeRefinedL);
title('Mesh of refined L-shaped domain',...
      'Units','normalized','Position',[0.5,1.05,0]);
hold off
```

**Listing 7.7:** `poisson2DMG.m`: Main calling script for solving Poisson equation on the unit square with iFEM©.

```
%
% poisson2DMG.m
%
% Calling script for solving the 2D Poisson equation
%
% -\Delta u = f
%
% on a square domain with homogeneous Dirichlet boundary
% conditions using the finite element method
% (iFEM package) with multigrid solver.
%
%% Initialization and Parameter Setting

% Clear command line window and workspace.
clear; clc; close all;

% Tolerance of residual to stop multigrid iterations.
tol = 1.0e-09;

% Number of multigrid levels.
L = 5;

% Number of interior cells in x,y at finest level.
nL = [2^L-1,2^L-1];

% Define the domain of the equation.
xLower = [0.0,0.0];
xUpper = [1.0,1.0];

% Define the exact solution and source term.
syms x y;
```

```
px = [xUpper(1)-xLower(1),xUpper(2)-xLower(2)];
uExact = exp(sin(4.0*pi*x/px(1))*sin(4.0*pi*y/px(2)))-1;
f = -diff(diff(uExact,x),x)-diff(diff(uExact,y),y);

% Convert to MATLAB functions.
fFunc = matlabFunction(f,'Vars',{[x,y]});
uExactFunc = matlabFunction(uExact,'Vars',{[x,y]});

%% Setting up the Mesh and Boundary Conditions

% Initial coarse mesh.
[node,elem] = squaremesh([xLower(1), xUpper(1),...
                          xLower(2), xUpper(2)],...
                         px(1));

% Refine the mesh.
for i = 1:L+1
  [node,elem] = uniformrefine(node,elem);
end

% Set Dirichlet boundary conditions on all boundaries.
bdFlag = setboundary(node,elem,'Dirichlet','all');

%% Setting up the Problem and Solver Parameters

% Problem definition.

% Homogeneous Dirichlet Condition.
pde.g_D = 0;

% Forcing function.
pde.f = fFunc;

% Exact solution for error calculations.
pde.exactu = uExactFunc;

% Multigrid solver parameters.
option.solver                   = 'mg';
option.mgoption.solver          = 'VCYCLE';
option.mgoption.smoothingstep = 3;
option.mgoption.printlevel      = 2;
option.tol                      = tol;
```

```
option.maxIt                        = 100;

%% Calling the iFEM Solver

[soln,eqn,info] = Poisson(node,elem,bdFlag,pde,option);

%% Plotting the Numerical Solution

figure(1)
showresult(node,elem,soln.u);
title('Numerical Solution of the Poisson Equation');
xlabel('x');
ylabel('y');
zlabel('u');
colorbar;

%% Error Analysis and Convergence Plots

% Extract error values.
errVals = info.error;
kStop = size(errVals,1);

figure(2)
clf

% Plot available error quantities.
semilogy(errVals(1:kStop,1),'rs','LineWidth',1.5)
hold on
semilogy(errVals(1:kStop,2),'kd','LineWidth',1.5)

% Log-linear fit for convergence rate.
rate = 1.0;
if kStop>=4
  kv = kStop-3:kStop;
  le = log(errVals(kStop-3:kStop,2));
  p1 = polyfit(kv,le,1);
  rate = exp(p1(1));
  p1k = polyval(p1,1:kStop);
  semilogy(1:kStop,exp(p1k),'-k')
end

% Labels and title.
```

```
xlabel('$k$','Interpreter','latex');
title(...
  'Multigrid Iteration Errors','Interpreter','latex');
legend(...
  '$\left\|{\bf u}_L^k-{\bf u}_L^{k-1} \right\|_L$',...
  '$\left\|{\bf r}_L^k\right\|_L$',...
  'log-linear fit','Interpreter','latex');

% Add annotations for parameters.
text(1.5,2048*option.tol,strcat(...
  '$\gamma_{\rm comp} =\hspace{.1cm}$',...
  num2str(rate,'%10.5e')),'FontSize',14,...
  'Interpreter','latex');
text(1.5,512*option.tol,strcat('$m_1 =\hspace{.1cm}$',...
  num2str(option.mgoption.smoothingstep)),...
  'FontSize',14,'Interpreter','latex');
text(1.5,128*option.tol,strcat('$m_2 =\hspace{.1cm}$',...
  num2str(option.mgoption.smoothingstep)),...
  'FontSize',14,'Interpreter','latex');

% Save figure as PDF.
printstr = strcat('Err_nL_',num2str(nL(1)),'_m1_',...
  num2str(option.mgoption.smoothingstep),'.pdf');
exportgraphics(gca,printstr)
grid on
grid minor
hold off

%% L2 Error Computation

% Note about error types.
fprintf(['Note: All the errors listed above are ',...
  'residual errors from the iterative solver.\n']);

L2_error = getL2error(node,elem,pde.exactu,soln.u);
fprintf('\nFinal error in L2 sense = %g\n',L2_error);
```

# 8 Cell-centered finite difference methods and multigrid

In this chapter, we develop the cell-centered finite difference method to approximate the solution to the two-point boundary value problem (1.7.1) and its two-dimensional analogue (7.1.1) with $\Omega = (0,1)^2$.

We will describe two-grid and multigrid methods to efficiently find approximations. The reader will observe that the techniques for the analysis of the resulting iterative methods will be similar to those for the finite element approximations. The details, however, are different and require some careful attention.

## 8.1 Discretization in one dimension

We begin by creating multilevel grids that cover the interval $\Omega = (0,1)$. The construction, however, is different from that in Chapter 4. For $\ell = 0,1,\dots,L$, we define the number of grid points and the uniform grid spacing by

$$n_\ell := q \cdot 2^\ell, \quad h_\ell := \frac{1}{n_\ell},$$

where $q \in \mathbb{N}$. The first notable difference from Chapter 4 is that for $1 \le \ell \le L$, the grids we construct here will have an even number of degrees of freedom $n_\ell$. The level-0 grid has size $n_0 = q$, which may be even or odd, as desired by the user.

**Definition 8.1.1** (Cell-centered grid). Let $0 \le \ell \le L$. The **cell-centered grid points** are defined as

$$x_{\ell,i} := \left(i - \frac{1}{2}\right)h_\ell, \quad i = 0,1,2,\dots,n_\ell, n_\ell + 1. \tag{8.1.1}$$

The **set of cell-centered points** is defined as

$$C_\ell := \{x_{\ell,i} \mid 0 \le i \le n_\ell + 1\}.$$

The grid points $x_{\ell,0} < 0$ and $x_{\ell,n_\ell+1} > 1$ are called **ghost points**. The **set of interior cell-centered points** is defined as

$$C_\ell^\circ := \{x_{\ell,i} \mid 1 \le i \le n_\ell\}.$$

The **vertex points** are defined as

$$x_{\ell,i+1/2} := i \cdot h_\ell, \quad i = 0,1,\dots,n_\ell. \tag{8.1.2}$$

https://doi.org/10.1515/9783111354880-008

The disjoint **grid cells** are

$$K_{\ell,i} := (x_{\ell,i-1/2}, x_{\ell,i+1/2}], \quad i = 1, 2, \dots, n_\ell. \tag{8.1.3}$$

■

**Remark 8.1.2** (Nested grid). We refer to the cell-centered multilevel grid structure as being nested and uniform, even though, as we will see, the cell-centered grid points of level $(\ell - 1)$ never appear in level-$\ell$ grid; see Figures 8.1 and 8.2. ◂

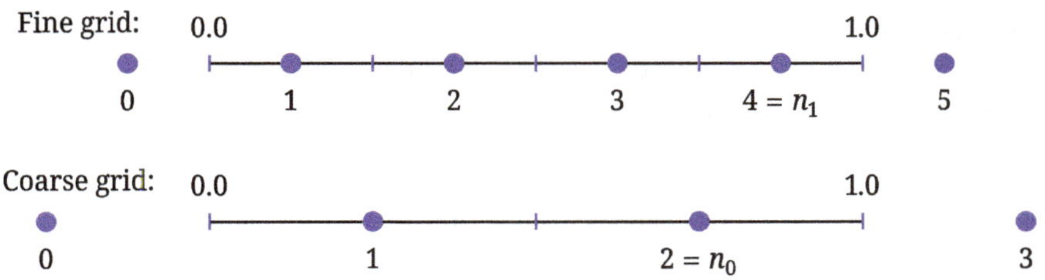


**Figure 8.1:** A uniform and nested two-level cell-centered grid with $q = 2$. Notice that for the cell-centered finite difference case, the coarse and fine grid points (blue dots) are not collocated.

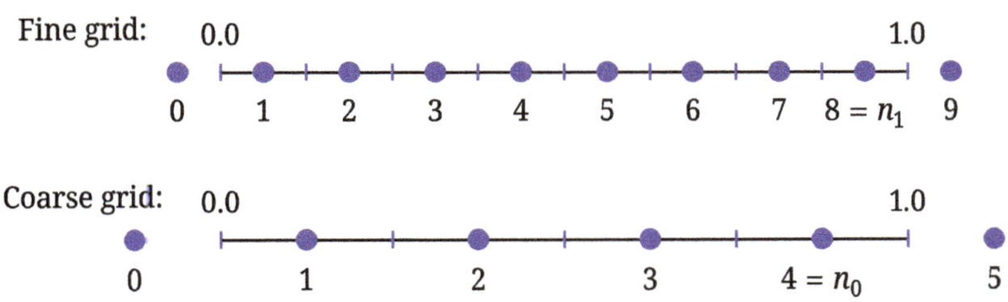


**Figure 8.2:** A uniform and nested two-level cell-centered grid with $q = 4$. Notice that for the cell-centered finite difference case, the coarse and fine grid points (blue dots) are not collocated.

Observe that, for each $0 \leq \ell \leq L$,

$$\bigcup_{i=1}^{n_\ell} K_{\ell,i} = (0, 1].$$

There are two equivalent ways of viewing functions built upon cell-centered grids that will be useful for us.

**Definition 8.1.3** (Grid functions). For every $0 \leq \ell \leq L$, we define the space of **piecewise constant functions** as

$$V_\ell := \{v : (0, 1] \to \mathbb{R} \mid v|_{K_{\ell,i}} \in \mathbb{P}_0, \ 1 \leq i \leq n_\ell\}.$$

The spaces of **cell-centered grid functions** and **interior cell-centered grid functions** are defined, respectively, as

$$\mathcal{C}_\ell := \{v_\ell : C_\ell \to \mathbb{R}\} \quad \text{and} \quad \mathcal{C}_\ell^\circ := \{v_\ell : C_\ell^\circ \to \mathbb{R}\}.$$

The subspace of **homogeneous Dirichlet cell-centered grid functions** is defined as

$$\mathcal{C}_{\ell,0} := \{v_\ell \in \mathcal{C}_\ell \mid v_\ell(x_{\ell,0}) = -v_\ell(x_{\ell,1}),\ v_\ell(x_{\ell,n_\ell+1}) = -v_\ell(x_{\ell,n_\ell})\}.$$ ■

**Remark 8.1.4** (Space identification). In an obvious way, we can identify the spaces $\mathcal{C}_{\ell,0}$ and $\mathcal{C}_\ell^\circ$. For any function $f \in \mathcal{C}_{\ell,0}$, we can find its unique counterpart in $\mathcal{C}_\ell^\circ$, and vice versa. Nevertheless, it will be useful to distinguish the two and retain different notation for them. ◂

For grid functions, we will use the convenient and standard notation

$$v_{\ell,i} := v_\ell(x_{\ell,i}), \quad 0 \le i \le n_\ell + 1.$$

In other words, we prefer to use subscripts when referring to the values, or entries, of grid functions.

Next, we recall that, for a set $A \subset \mathbb{R}^n$, it is standard to denote by $\chi_A$ its characteristic function,

$$\chi_A(\boldsymbol{x}) := \begin{cases} 1, & \boldsymbol{x} \in A, \\ 0, & \boldsymbol{x} \in \mathbb{R}^n \setminus A. \end{cases}$$

**Definition 8.1.5** (Basis functions). For every $0 \le \ell \le L$, define the **basis of piecewise constant functions** as

$$\mathcal{B}_\ell := \{\psi_{\ell,j}\}_{j=1}^{n_\ell} \subset V_\ell,$$

where

$$\psi_{\ell,j} := \chi_{K_{\ell,j}}, \quad 1 \le j \le n_\ell.$$

Define the **canonical grid function basis**

$$\mathcal{G}_\ell := \{g_{\ell,j}\}_{j=1}^{n_\ell} \subset \mathcal{C}_{\ell,0},$$

where $g_{\ell,j} \in \mathcal{C}_{\ell,0}$ satisfies

$$g_{\ell,j}(x_{\ell,i}) = \delta_{i,j}, \quad 1 \le i,j \le n_\ell.$$ ■

**Proposition 8.1.6** (Bases). *For every $0 \leq \ell \leq L$, $\mathcal{B}_\ell$ is indeed a basis of $V_\ell$, and $\mathcal{G}_\ell$ is a basis for $\mathcal{C}_{\ell,0}$. Thus*

$$\dim(V_\ell) = n_\ell = \dim(\mathcal{C}_{\ell,0}).$$

*Proof.* Exercise; see Problem 8.1. □

In light of the previous result, there is a one-to-one correspondence between vectors in $\mathbb{R}^{n_\ell}$ and functions in $V_\ell$:

$$v_\ell \in V_\ell \overset{\mathcal{B}_\ell}{\longleftrightarrow} \boldsymbol{v}_\ell \in \mathbb{R}^{n_\ell}.$$

This identification connects the coordinate vector $\boldsymbol{v}_\ell \in \mathbb{R}^{n_\ell}$ with its unique piecewise constant function $v_\ell \in V_\ell$ through the basis $\mathcal{B}_\ell$ so that

$$v_\ell = \sum_{j=1}^{n_\ell} v_{\ell,j} \psi_{\ell,j}.$$

Likewise, there is a one-to-one correspondence between vectors in $\mathbb{R}^{n_\ell}$ and grid functions in $\mathcal{C}_{\ell,0}$, which also has dimension $n_\ell$. This correspondence is denoted by

$$v_\ell \in \mathcal{C}_{\ell,0} \overset{\mathcal{G}_\ell}{\longleftrightarrow} \boldsymbol{v}_\ell \in \mathbb{R}^{n_\ell}.$$

Once again, this identification connects the coordinate vector $\boldsymbol{v}_\ell \in \mathbb{R}^{n_\ell}$ with its unique grid function $v_\ell \in \mathcal{C}_{\ell,0}$ via

$$v_\ell = \sum_{j=1}^{n_\ell} v_{\ell,j} g_{\ell,j}.$$

This identification is illustrated in Figure 8.3.

Assume, for now, that we are given $f_{\ell,i}$, $1 \leq i \leq n_\ell$, which will be approximations of $f(x_{\ell,i})$ to be specified later. The finite difference approximation of the one-dimensional Poisson problem (1.7.1) can be expressed as follows: find the grid function

$$u_\ell = (u_{\ell,0}, u_{\ell,1}, u_{\ell,2}, \ldots, u_{\ell,n_\ell}, u_{\ell,n_\ell+1}) \in \mathcal{C}_\ell$$

such that

$$\begin{cases} \dfrac{u_{\ell,0} + u_{\ell,1}}{2} = 0, \\ \dfrac{-u_{\ell,i-1} + 2u_{\ell,i} - u_{\ell,i+1}}{h_\ell^2} = f_{\ell,i}, \quad 1 \leq i \leq n_\ell, \\ \dfrac{u_{\ell,n_\ell} + u_{n_\ell+1}}{2} = 0. \end{cases} \tag{8.1.4}$$

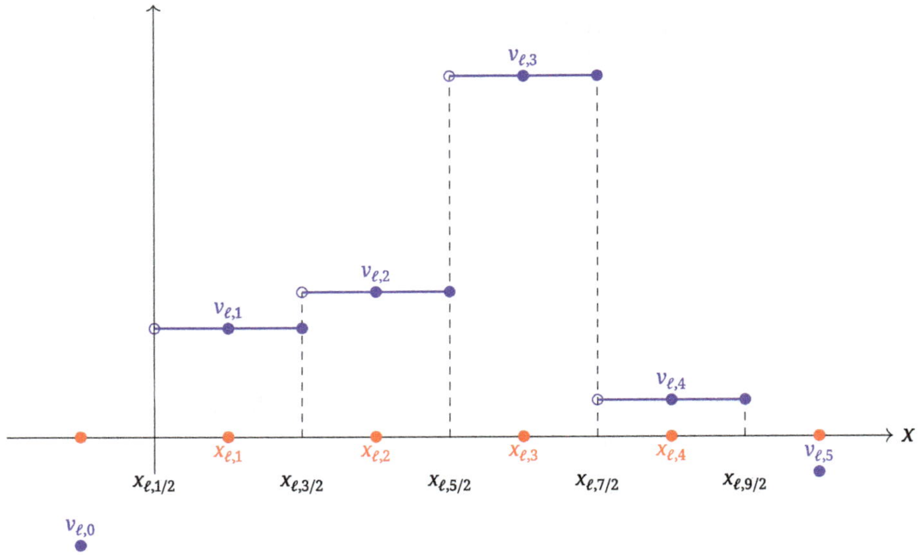


**Figure 8.3:** A piecewise constant function $v_\ell \in V_\ell$. We use the same symbol for the corresponding grid function $v_\ell \in \mathcal{C}_{\ell,0}$, which has values only at the cell-centered points and has homogeneous Dirichlet boundary values.

Equivalently, we can express the problem as follows: find the grid function

$$u_\ell = (u_{\ell,0}, u_{\ell,1}, u_{\ell,2}, \dots, u_{\ell,n_\ell}, u_{\ell,n_\ell+1}) \in \mathcal{C}_{\ell,0}$$

such that

$$\frac{-u_{\ell,i-1} + 2u_{\ell,i} - u_{\ell,i+1}}{h_\ell^2} = f_{\ell,i}, \quad 1 \le i \le n_\ell, \tag{8.1.5}$$

where the boundary conditions are built into the space of solutions.

In the cell-centered finite difference method,

$$u_{\ell,i} \approx u(x_{\ell,i}) = u\left(\left(i - \frac{1}{2}\right)h_\ell\right),$$

where the function $u$ is the solution to the boundary value problem (1.7.1). In fact, we can show that, under certain reasonable assumptions,

$$\sqrt{h_\ell \sum_{i=1}^{n_\ell} |u(x_{\ell,i}) - u_{\ell,i}|^2} \le Ch_\ell^2,$$

where $C > 0$ is a constant independent of $h_\ell$; see [31, Chap. 24].

Let us now set

$$\boldsymbol{u}_\ell = \begin{bmatrix} u_{\ell,1} \\ u_{\ell,2} \\ \vdots \\ u_{\ell,n_\ell-1} \\ u_{\ell,n_\ell} \end{bmatrix} \in \mathbb{R}^{n_\ell}, \quad \boldsymbol{f}_\ell = \begin{bmatrix} f_{\ell,1} \\ f_{\ell,2} \\ \vdots \\ f_{\ell,n_\ell-1} \\ f_{\ell,n_\ell} \end{bmatrix} \in \mathbb{R}^{n_\ell},$$

and

$$\mathsf{A}_\ell := \frac{1}{h_\ell^2} \begin{bmatrix} 3 & -1 & & & \\ -1 & 2 & -1 & & \\ & \ddots & \ddots & \ddots & \\ & & -1 & 2 & -1 \\ & & & -1 & 3 \end{bmatrix} \in \mathbb{R}^{n_\ell \times n_\ell}.$$

The matrix $\mathsf{A}_\ell \in \mathbb{R}^{n_\ell \times n_\ell}$ is called the cell-centered stiffness matrix. Note how it differs from the stiffness matrices we encountered in Chapter 4. In matrix form, the finite difference approximation is as follows: find $\boldsymbol{u}_\ell \in \mathbb{R}^{n_\ell}$ such that

$$\mathsf{A}_\ell \boldsymbol{u}_\ell = \boldsymbol{f}_\ell. \tag{8.1.6}$$

Let us briefly discuss implementation issues. As an example, suppose that we wish to compute the residual

$$\boldsymbol{r}_\ell = \boldsymbol{f}_\ell - \mathsf{A}_\ell \boldsymbol{v}_\ell.$$

To complete this, seemingly simple, calculation, we need to obtain the vector $\mathsf{A}_\ell \boldsymbol{v}_\ell$. To compute this matrix–vector product efficiently, with a minimal amount of memory usage, we perform elementwise operations on the vector $\boldsymbol{v}_\ell$, as if it were a grid function, using the finite difference operator. This is an efficient alternative to storing all the entries, including zeros, of $\mathsf{A}_\ell$ and computing the matrix–vector product explicitly, which is very inefficient.

To get the calculation right, we need to implement boundary conditions. For this, we have two choices. Either we can build them into the definition of a modified finite difference operator, or we can apply boundary conditions and use the standard finite difference operator.

It turns out that the latter approach is much more efficient. This requires that we extend the vector $\boldsymbol{v}_\ell$ so that it includes storage locations associated with the ghost points. Thus, if we first apply the homogeneous Dirichlet conditions, then this sets the ghost points appropriately. After this is done, we can apply the unmodified finite difference operator. This completes the calculation of $\mathsf{A}_\ell \boldsymbol{v}_\ell$. See the code in Listings 8.8 and 8.13.

**Definition 8.1.7** (Level-$\ell$ stiffness matrix). For $0 \le \ell \le L$, the **level-$\ell$ stiffness matrix** is given by

$$\mathsf{A}_\ell := \frac{1}{h_\ell^2}\begin{bmatrix} 3 & -1 & & & \\ -1 & 2 & -1 & & \\ & \ddots & \ddots & \ddots & \\ & & -1 & 2 & -1 \\ & & & -1 & 3 \end{bmatrix} \in \mathbb{R}^{n_\ell \times n_\ell}. \tag{8.1.7}$$

■

**Theorem 8.1.8** (Stiffness matrix is SPD). *The level-$\ell$ stiffness matrix $\mathsf{A}_\ell \in \mathbb{R}^{n_\ell \times n_\ell}$ is SPD. Its eigenvalues are, for $k = 1, 2, \ldots, n_\ell$,*

$$\lambda_\ell^{(k)} = \frac{4}{h_\ell^2}\sin^2\left(\frac{k\pi h_\ell}{2}\right) = \frac{2}{h_\ell^2}(1 - \cos(k\pi h_\ell)), \tag{8.1.8}$$

*and the corresponding eigenvectors are*

$$[\boldsymbol{v}_\ell^{(k)}]_i = v_{\ell,i}^{(k)} = \sin(k\pi x_{\ell,i}), \quad 1 \le i \le n_\ell. \tag{8.1.9}$$

*Proof.* The matrix $\mathsf{A}_\ell$ is clearly symmetric. We will show that it is positive definite by showing that all its eigenvalues are positive. To see this, we first of all observe that, for all $k = 1, \ldots, n_\ell$,

$$\begin{aligned} \sin(k\pi x_{\ell,0}) &= -\sin(k\pi x_{\ell,1}), \\ \sin(k\pi x_{\ell,n_\ell+1}) &= -\sin(k\pi x_{\ell,n_\ell}). \end{aligned}$$

Thus, for $1 \le i \le n_\ell$,

$$\begin{aligned} h_\ell^2[\mathsf{A}_\ell \boldsymbol{v}_\ell^{(k)}]_i &= -\sin(k\pi x_{\ell,i-1}) + 2\sin(k\pi x_{\ell,i}) - \sin(k\pi x_{\ell,i+1}) \\ &= 2[1 - \cos(k\pi h_\ell)]\sin(k\pi x_{\ell,i}). \end{aligned}$$

This shows that, for $1 \le i \le n_\ell$,

$$[\mathsf{A}_\ell \boldsymbol{v}_\ell^{(k)}]_i = \lambda_\ell^{(k)} v_{\ell,i}^{(k)}.$$

Since the eigenvalues are positive, $\mathsf{A}_\ell$ is SPD; see Figure 8.4. □

**Remark 8.1.9** (Scaling). Observe that for the one-dimensional cell-centered finite difference stiffness matrix, the largest eigenvalue always satisfies the *equality*

$$\frac{\lambda_\ell^{(n_\ell)}}{n_\ell^2} = 4\sin^2\left(\frac{n_\ell \pi}{2n_\ell}\right) = 4.$$

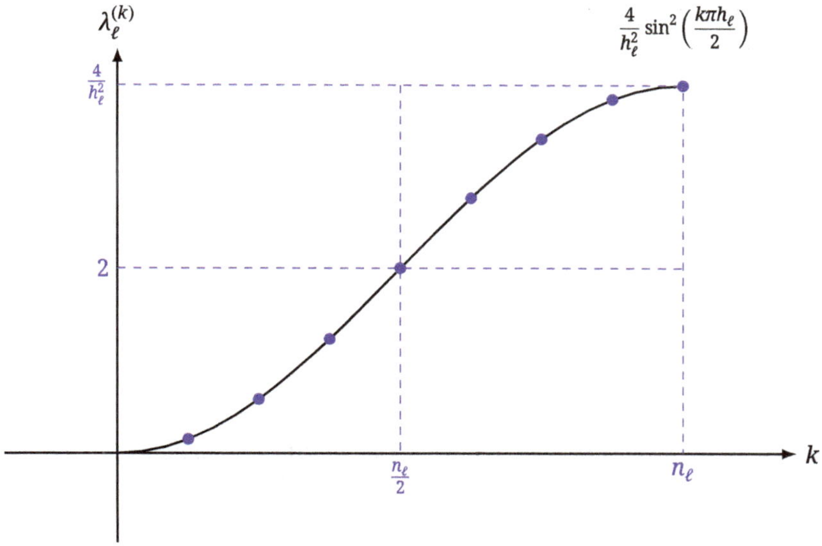


**Figure 8.4:** Eigenvalues of the level-$\ell$ stiffness matrix $A_\ell$. The values of $\lambda_\ell^{(k)}$ are plotted for $n_\ell = 8$.

For the finite element discretization described in Chapter 4, we observed that the largest eigenvalue satisfied the *strict inequality*

$$\frac{\lambda_\ell^{(n_\ell)}}{n_\ell + 1} = 4\sin^2\left(\frac{n_\ell \pi}{2(n_\ell + 1)}\right) < 4$$

with

$$\lim_{n_\ell \uparrow \infty} \frac{\lambda_\ell^{(n_\ell)}}{n_\ell + 1} = 4.$$

Recall that the scalings we used for the respective stiffness matrices and therefore their respective eigenvalues were different. ◂

**Proposition 8.1.10** (Orthogonality). *Assume that $\ell \geq 1$, so that $n_\ell$ is even. The eigenvectors of the level-$\ell$ stiffness matrix are pairwise orthogonal and satisfy the normalization conditions*

$$(\boldsymbol{v}_\ell^{(k)}, \boldsymbol{v}_\ell^{(k)})_\ell = (\boldsymbol{v}_\ell^{(k)})^\top \boldsymbol{v}_\ell^{(k)} = \beta_{\ell,k} := \begin{cases} \frac{n_\ell}{2}, & 1 \leq k \leq n_\ell - 1, \\ n_\ell, & k = n_\ell. \end{cases} \tag{8.1.10}$$

*Proof.* Exercise; see Problem 8.2. □

**Theorem 8.1.11** (Condition number estimate). *There are positive constants $C_1 \leq C_2$ such that, for every $\ell$, the spectral condition number of the level-$\ell$ stiffness matrix $A_\ell$, that is,*

$$\kappa(\mathsf{A}_\ell) := \|\mathsf{A}_\ell\|\|\mathsf{A}_\ell^{-1}\| = \frac{\lambda_\ell^{(n_\ell)}}{\lambda_\ell^{(1)}},$$

*satisfies the estimates*

$$C_1 h_\ell^{-2} \le \kappa(\mathsf{A}_\ell) \le C_2 h_\ell^{-2}.$$

*Proof.* The proof is similar to that of Theorem 4.2.3 and is left as an exercise; see Problem 8.3. □

This last result states that

$$\kappa(\mathsf{A}_\ell) = \Theta(h_\ell^{-2}) \quad \text{as } h_\ell \to 0.$$

In other words, the spectral condition number blows up as $h_\ell \to 0$. As we have seen, this usually means that the performance of any classical GLIS is degraded for small grid sizes.

## 8.2 The damped Jacobi and Richardson smoothers

After discretization, at every level $0 \le \ell \le L$, we have a family of linear systems to consider:

$$\mathsf{A}_\ell \boldsymbol{u}_\ell = \boldsymbol{f}_\ell \in \mathbb{R}^{n_\ell}. \tag{8.2.1}$$

To approximate the solution of (8.2.1) at a given level $\ell$, we could consider a GLIS from Chapter 1.

### 8.2.1 The damped Jacobi smoother

First, let us recall the damped Jacobi method. This method requires the canonical splitting of $\mathsf{A}_\ell$,

$$\mathsf{A}_\ell = \mathsf{D}_\ell - \mathsf{U}_\ell - \mathsf{L}_\ell \in \mathbb{R}^{n_\ell \times n_\ell},$$

where

$$\mathsf{D}_\ell = \frac{1}{h_\ell^2}\begin{bmatrix} 3 & & & & \\ & 2 & & & \\ & & \ddots & & \\ & & & 2 & \\ & & & & 3 \end{bmatrix}, \quad \mathsf{U}_\ell = \frac{1}{h_\ell^2}\begin{bmatrix} 0 & 1 & & & \\ & 0 & 1 & & \\ & & \ddots & \ddots & \\ & & & 0 & 1 \\ & & & & 0 \end{bmatrix},$$

and, of course, $\mathsf{L}_\ell = \mathsf{U}_\ell^\top$. The damped Jacobi method reads

$$\begin{aligned} \boldsymbol{z}_\ell^{\sigma+1} &:= \mathsf{D}_\ell^{-1}(\mathsf{U}_\ell + \mathsf{U}_\ell^\top)\boldsymbol{u}_\ell^\sigma + \mathsf{D}_\ell^{-1}\boldsymbol{f}_\ell, \\ \boldsymbol{u}_\ell^{\sigma+1} &:= \omega \boldsymbol{z}_\ell^{\sigma+1} + (1-\omega)\boldsymbol{u}_\ell^\sigma, \end{aligned}$$

where $0 < \omega \le 1$. Eliminating $\boldsymbol{z}_\ell^{\sigma+1}$, we have the equivalent version

$$\boldsymbol{u}_\ell^{\sigma+1} = \boldsymbol{u}_\ell^\sigma + \omega \mathsf{D}_\ell^{-1}(\boldsymbol{f}_\ell - \mathsf{A}_\ell \boldsymbol{u}_\ell^\sigma). \tag{8.2.2}$$

In our multigrid terminology, the damped Jacobi smoother is selected by using the iterator matrix

$$\mathsf{S}_\ell = \omega \mathsf{D}_\ell^{-1},$$

and the associated error transfer is

$$\begin{aligned} \mathsf{K}_\ell &= \mathrm{I}_\ell - \mathsf{S}_\ell \mathsf{A}_\ell \\ &= \mathrm{I}_\ell - \omega \mathsf{D}_\ell^{-1}\mathsf{A}_\ell. \end{aligned}$$

The difficulty for our analysis with the damped Jacobi method is that the diagonal elements are no longer constant. This means that the eigenvectors of $\mathsf{K}_\ell$ no longer coincide with those of $\mathsf{A}_\ell$. *Consequently, we will abandon it, at least for theoretical pursuits!*

### 8.2.2 The generic Richardson and Λ-Richardson smoothers

Next, let us consider Richardson's method as a smoother, which, as we will see, has more desirable properties. Unlike what we encountered in Chapter 5, the difference between Richardson's smoother and the damped Jacobi smoother is not merely cosmetic. Richardson's smoother can be expressed generically as

$$\boldsymbol{u}_\ell^{\sigma+1} = \boldsymbol{u}_\ell^\sigma + \tilde{\omega}(\boldsymbol{f}_\ell - \mathsf{A}_\ell \boldsymbol{u}_\ell^\sigma),$$

where $\tilde{\omega} > 0$. We will typically choose

$$\tilde{\omega}^{-1} = \Lambda_\ell = \rho_\ell = \lambda_\ell^{(n_\ell)} = \frac{4}{h_\ell^2},$$

where $\Lambda_\ell$ is precisely the spectral radius of $\mathsf{A}_\ell$. We refer to this particular choice as the Λ-Richardson smoother.

For now, we want to let $\tilde{\omega} > 0$ remain generic, but we will rescale it for reasons that will become clear later. In particular, let us choose

$$\tilde{\omega} := \omega \frac{h_\ell^2}{2}, \quad \omega > 0.$$

Thus the generic form of Richardson's smoother we will consider is

$$\boldsymbol{u}_\ell^{\sigma+1} = \boldsymbol{u}_\ell^\sigma + \omega \frac{h_\ell^2}{2} (\boldsymbol{f}_\ell - \mathsf{A}_\ell \boldsymbol{u}_\ell^\sigma), \quad \omega > 0, \tag{8.2.3}$$

and the error transfer matrix for the generic Richardson smoother is

$$\mathsf{K}_\ell = \mathsf{K}_\ell(\omega) = \mathrm{I}_\ell - \frac{\omega h_\ell^2}{2} \mathsf{A}_\ell. \tag{8.2.4}$$

If we set $\omega = \frac{1}{2}$, then we obtain the $\Lambda$-Richardson smoother, which is typically used in multigrid applications.

**Remark 8.2.1** (Scaling). This choice of scaling, which we call the generic Richardson smoother (8.2.3), is not arbitrary. It puts the eigenvalues of the error transfer matrix of the damped Jacobi smoother from Chapter 4 (see Equation (4.3.5)) into correspondence with those of the error transfer matrix in (8.2.4). This can be seen in Theorem 8.3.1. ◂

**Listing 8.1:** A single iteration of Richardson's smoother from the code `smoothRichardson.m`. The full code can is provided in Listing 8.10.

```
    r(1:nf) = getResidual(u,f,DeW,rC,hf);
    u(2:nf+1) = u(2:nf+1)+omega*(hf2/2.0)*r(1:nf);
    u = applyBCs(u);
```

### 8.2.3 The damped quasi-Jacobi method

Let us look at one final method for smoothing, the damped quasi-Jacobi method, which, as we will see, is equivalent to the generic Richardson smoother.

Let $u_\ell^\sigma \in \mathcal{C}_\ell^\circ$. There is, of course, a unique function $u_\ell^\sigma \in \mathcal{C}_{\ell,0}$ — for which we use the same symbol — that agrees with the first one at every interior cell-centered grid point and has discrete homogeneous Dirichlet boundary conditions, that is,

$$u_{\ell,0}^\sigma = -u_{\ell,1}^\sigma \quad \text{and} \quad u_{\ell,n_\ell+1}^\sigma = -u_{\ell,n_\ell}^\sigma.$$

Define the grid function $z_\ell \in \mathcal{C}_\ell^\circ$ as follows: for $1 \le i \le n_\ell$,

$$f_{\ell,i} = \frac{-u_{\ell,i+1}^\sigma + 2z_{\ell,i} - u_{\ell,i-1}^\sigma}{h_\ell^2}.$$

We can explicitly solve for $z_{\ell,i}$ at each index $i$, obtaining

$$z_{\ell,i} = \frac{h_\ell^2 f_{\ell,i} + u_{\ell,i+1}^\sigma + u_{\ell,i-1}^\sigma}{2}.$$

It does not matter in what order we compute the entries of the grid function $z_\ell$. Once all the entries of $z_\ell$ are found, we compute the update via

$$u_\ell^{\sigma+1} = \omega z_\ell + (1-\omega)u_\ell^\sigma \in \mathcal{C}_\ell^\circ,$$

where $\omega \in (0,1]$ is a damping parameter.

This method is equivalent to the generic Richardson method. To see this, let us eliminate the grid function $z_\ell$ in the last equation to get

$$\begin{aligned} u_{\ell,i}^{\sigma+1} &= \omega \frac{h_\ell^2 f_{\ell,i} + u_{\ell,i+1}^\sigma + u_{\ell,i-1}^\sigma}{2} + u_{\ell,i}^\sigma - \omega u_{\ell,i}^\sigma \\ &= u_{\ell,i}^\sigma + \omega \frac{h_\ell^2 f_{\ell,i} + u_{\ell,i+1}^\sigma - 2u_{\ell,i}^\sigma + u_{\ell,i-1}^\sigma}{2} \\ &= u_{\ell,i}^\sigma + \omega \frac{h_\ell^2}{2}\left(f_{\ell,i} - \frac{-u_{\ell,i+1}^\sigma + 2u_{\ell,i}^\sigma - u_{\ell,i-1}^\sigma}{h_\ell^2}\right), \end{aligned}$$

for all $1 \le i \le n_\ell$. Since the discrete homogeneous Dirichlet boundary conditions hold for the grid function $u_\ell^\sigma$, the last equation can be reexpressed in vector form as

$$\boldsymbol{u}_\ell^{\sigma+1} = \boldsymbol{u}_\ell^\sigma + \omega \frac{h_\ell^2}{2}(\boldsymbol{f}_\ell - \mathsf{A}_\ell \boldsymbol{u}_\ell^\sigma), \quad \omega > 0,$$

which is exactly (8.2.3).

The present method is not the same as the damped Jacobi smoother described in (8.2.2). This has to do with the fact that the diagonal entries of the cell-centered stiffness matrix are not all the same. However, the present damped quasi-Jacobi method differs from (8.2.2) only in the way the $i = 1$ and $i = n_\ell$ components of the update are computed. The details are left to the reader to explore.

## 8.3 Smoothing properties

**Theorem 8.3.1** (Eigenpairs of generic Richardson). *Assume that the error transfer matrix $\mathsf{K}_\ell$ for the generic Richardson smoother applied to the model problem* (8.2.1) *is given by* (8.2.4). *The eigenvectors of $\mathsf{K}_\ell$ are the same as those for the level-$\ell$ stiffness matrix $\mathsf{A}_\ell$, that is,*

$$[\boldsymbol{v}_\ell^{(k)}]_i = v_{\ell,i}^{(k)} = \sin(k\pi x_{\ell,i}), \quad 1 \le i \le n_\ell,$$

*for $k = 1, \ldots, n_\ell$. The eigenvalues of* $\mathsf{K}_\ell$ *are*

$$\mu_\ell^{(k)}(\omega) = 1 - 2\omega \sin^2\left(\frac{k\pi h_\ell}{2}\right), \quad 1 \le k \le n_\ell. \tag{8.3.1}$$

*Proof.* Exercise; see Problem 8.4. □

**Remark 8.3.2** (Exactness). Observe that, for the cell-centered finite difference case, we have the *equality*

$$\mu_\ell^{(n_\ell)}(\omega) = 1 - 2\omega \sin^2\left(\frac{n_\ell \pi}{2n_\ell}\right) = 1 - 2\omega.$$

On the other hand, for the one-dimensional finite element case, we have the *inequality*

$$\mu_\ell^{(n_\ell)}(\omega) = 1 - 2\omega \sin^2\left(\frac{n_\ell \pi}{2(n_\ell + 1)}\right) > 1 - 2\omega.$$

The reader should compare Figures 4.6 and 8.5, which depict the finite element and cell-centered finite difference cases, respectively. ◂

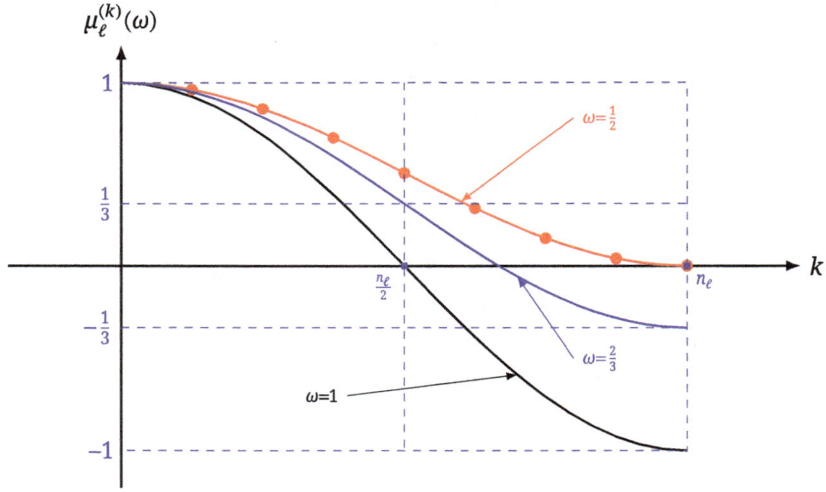


**Figure 8.5:** Plots of the eigenvalues of $\mathsf{K}_\ell$ as functions of $k$ for various values of $\omega \in (0, 1]$. See Theorem 8.3.1. The discrete nature of the set of eigenvalues is shown for the $\omega = \frac{1}{2}$ case, where $n_\ell = 8$ for illustration.

The next result shows why the performance of the Richardson method, as a standalone solver, is degraded as $h_\ell \to 0$.

**Theorem 8.3.3** (Spectral radius of generic Richardson). *Let $1 \le \ell \le L$ and assume that the error transfer matrix for the generic Richardson smoother $\mathsf{K}_\ell$, when applied to the model problem* (8.2.1), *is given by* (8.2.4). *Then, as $h_\ell \to 0$,*

$$\rho(\mathsf{K}_\ell) = \mu_1^{(1)}(\omega) = 1 - \omega\Theta(h_\ell^2),$$

*for all $0 < \omega < 1$.*

*Proof.* Exercise; see Problem 8.5. □

The Λ-Richardson smoother has the following eigenvalues:

$$\mu_\ell^{(k)}\left(\frac{1}{2}\right) = \frac{1}{2}\cos(k\pi h_\ell) + 1 - \frac{1}{2} = 1 - \frac{\lambda_\ell^{(k)}}{\Lambda_\ell}.$$

By looking at Figure 8.6 it clearly follows that, for $\ell \ge 1$,

$$\left|\mu_\ell^{(k)}\left(\frac{1}{2}\right)\right| = \mu_\ell^{(k)}\left(\frac{1}{2}\right) \le \mu_\ell^{(n_{\ell-1})}\left(\frac{1}{2}\right) = \frac{1}{2}, \quad n_{\ell-1} = \frac{n_\ell}{2} \le k \le n_\ell.$$

In other words, the Λ-Richardson smoother reduces all high-frequency modes of the error by at least half. Here we define, as we would naturally expect, high-frequency modes as those modes whose wave numbers $k$ satisfy

$$\frac{n_\ell}{2} \le k \le n_\ell.$$

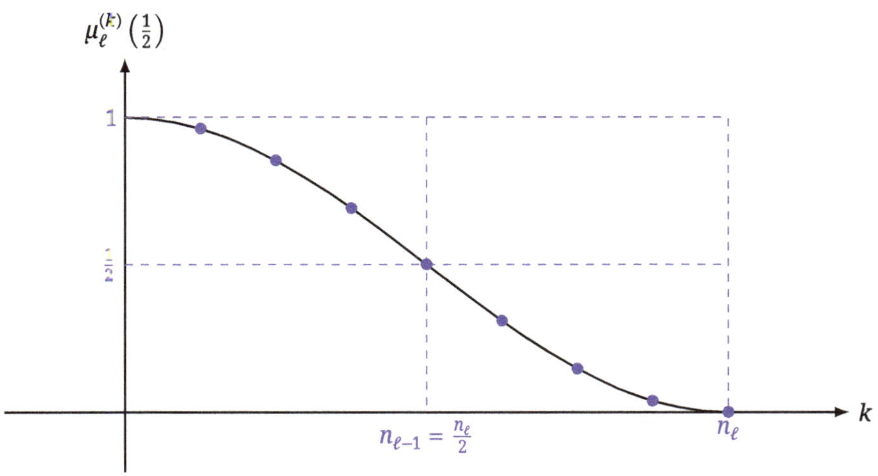


**Figure 8.6:** Eigenvalues $\mu_\ell^{(k)}(\frac{1}{2})$ of the error transfer matrix for the Λ-Richardson smoother plotted for $n_\ell = 8$.

The one-dimensional cell-centered finite difference (level-$\ell$) stiffness matrix was introduced in Definition 8.1.7. Recall that this matrix is SPD, and it is therefore orthogonally diagonalizable, that is,

$$\mathsf{A}_\ell = \mathsf{V}_\ell \mathsf{D}_\ell \mathsf{V}_\ell^\top,$$

where $\mathsf{V}_\ell$ is the orthogonal matrix containing the normalized eigenvectors of $\mathsf{A}_\ell$,

$$\mathsf{V}_\ell = \begin{bmatrix} | & | & & | \\ \tilde{\boldsymbol{v}}_\ell^{(1)} & \tilde{\boldsymbol{v}}_\ell^{(2)} & \cdots & \tilde{\boldsymbol{v}}_\ell^{(n_\ell)} \\ | & | & & | \end{bmatrix} \in \mathbb{R}^{n_\ell \times n_\ell},$$

and $\mathsf{D}_\ell$ is the diagonal matrix containing the positive eigenvalues of $\mathsf{A}_\ell$,

$$\mathsf{D}_\ell = \operatorname{diag}[\lambda_\ell^{(1)}, \lambda_\ell^{(2)}, \ldots, \lambda_\ell^{(n_\ell)}] \in \mathbb{R}^{n_\ell \times n_\ell}.$$

The eigenvalues $\lambda_\ell^{(k)}$ of $\mathsf{A}_\ell$ are defined in (8.1.8). Since $\mathsf{A}_\ell$ is SPD, we can define its square root as

$$\mathsf{A}_\ell^{1/2} = \mathsf{V}_\ell \mathsf{D}_\ell^{1/2} \mathsf{V}_\ell^\top,$$

where

$$\mathsf{D}_\ell^{1/2} = \operatorname{diag}\Big[\sqrt{\lambda_\ell^{(1)}}, \sqrt{\lambda_\ell^{(2)}}, \ldots, \sqrt{\lambda_\ell^{(n_\ell)}}\Big] \in \mathbb{R}^{n_\ell \times n_\ell}.$$

Of course, $\mathsf{A}_\ell^{1/2}$ is SPD, and $\mathsf{A}_\ell^{1/2}\mathsf{A}_\ell^{1/2} = \mathsf{A}_\ell$.

Observe now that, for $\boldsymbol{v}_\ell \in \mathbb{R}^{n_\ell}$,

$$\begin{aligned} \|\boldsymbol{v}_\ell\|_{\mathsf{A}_\ell} &= \sqrt{(\boldsymbol{v}_\ell, \boldsymbol{v}_\ell)_{\mathsf{A}_\ell}} \\ &= \sqrt{(\mathsf{A}_\ell \boldsymbol{v}_\ell, \boldsymbol{v}_\ell)_\ell} \\ &= \sqrt{(\mathsf{A}_\ell^{1/2} \boldsymbol{v}_\ell, \mathsf{A}_\ell^{1/2} \boldsymbol{v}_\ell)_\ell} \\ &= \|\mathsf{A}_\ell^{1/2} \boldsymbol{v}_\ell\|_\ell. \end{aligned}$$

In a similar way,

$$\begin{aligned} \|\boldsymbol{v}_\ell\|_{\mathsf{A}_\ell^2} &:= \sqrt{(\mathsf{A}_\ell^2 \boldsymbol{v}_\ell, \boldsymbol{v}_\ell)_\ell} \\ &= \sqrt{(\mathsf{A}_\ell \boldsymbol{v}_\ell, \mathsf{A}_\ell \boldsymbol{v}_\ell)_\ell} \\ &= \|\mathsf{A}_\ell \boldsymbol{v}_\ell\|_\ell. \end{aligned}$$

Let us now establish a smoothing property for the cell-centered finite difference case in one space dimension.

**Theorem 8.3.4** (First smoothing property). *Let $1 \le \ell \le L$. Assume that smoothing is carried out by the $\Lambda$-Richardson smoother, that is, $\omega = \frac{1}{2}$. Then, there is a constant $C > 0$ such that, for all $m_1 \in \mathbb{N}$ and $\boldsymbol{w}_\ell \in \mathbb{R}^{n_\ell}$,*

$$\|\mathsf{K}_\ell^{m_1} \boldsymbol{w}_\ell\|_{\mathsf{A}_\ell^2} \le C \sqrt{\frac{\Lambda_\ell}{m_1}} \|\boldsymbol{w}_\ell\|_{\mathsf{A}_\ell}. \tag{8.3.2}$$

*In particular, we can take*

$$C = 2^{-1/2}.$$

*In other words, the $\Lambda$-Richardson smoother satisfies the first smoothing property* (S1).

*Proof.* First, observe that

$$\begin{aligned}\|\mathsf{K}_\ell^{m_1} \boldsymbol{w}_\ell\|_{\mathsf{A}_\ell^2}^2 &= \|\mathsf{A}_\ell \mathsf{K}_\ell^{m_1} \boldsymbol{w}_\ell\|_\ell^2 \\ &= (\mathsf{A}_\ell \mathsf{K}_\ell^{m_1} \boldsymbol{w}_\ell, \mathsf{A}_\ell \mathsf{K}_\ell^{m_1} \boldsymbol{w}_\ell)_\ell.\end{aligned}$$

Using the sinusoidal eigenvector basis of $\mathsf{A}_\ell$, we write

$$\boldsymbol{w}_\ell = \sum_{k=1}^{n_\ell} w_{\ell,k} \boldsymbol{v}_\ell^{(k)}.$$

Then, using the identity from Proposition 8.1.10, we have

$$\begin{aligned}\|\mathsf{K}_\ell^{m_1} \boldsymbol{w}_\ell\|_{\mathsf{A}_\ell^2}^2 &= \sum_{k=1}^{n_\ell} \beta_{\ell,k} (\lambda_\ell^{(k)} w_{\ell,k})^2 \left(\mu_\ell^{(k)}\left(\frac{1}{2}\right)\right)^{2m_1} \\ &= \Lambda_\ell \sum_{k=1}^{n_\ell} \beta_{\ell,k} \left(\frac{\lambda_\ell^{(k)}}{\Lambda_\ell}\right)\left(1 - \frac{\lambda_\ell^{(k)}}{\Lambda_\ell}\right)^{2m_1} \lambda_\ell^{(k)} w_{\ell,k}^2 \\ &\le \Lambda_\ell G(m_1) \sum_{k=1}^{n_\ell} \beta_{\ell,k} \lambda_\ell^{(k)} w_{\ell,k}^2 \\ &= \Lambda_\ell G(m_1) \|\boldsymbol{w}_\ell\|_{\mathsf{A}_\ell}^2,\end{aligned}$$

where

$$G(m) := \max_{1 \le k \le n_\ell} \left(\frac{\lambda_\ell^{(k)}}{\Lambda_\ell}\right)\left(1 - \frac{\lambda_\ell^{(k)}}{\Lambda_\ell}\right)^{2m}.$$

Rescaling and using Lemma 5.3.1, we have

$$\begin{aligned}G(m) &\le \max_{0 \le x \le 1} x(1-x)^{2m} \\ &\overset{(5.3.1)}{\le} \frac{1}{2m}.\end{aligned}$$

Therefore

$$\|\mathsf{K}_\ell^{m_1}\boldsymbol{w}_\ell\|_{\mathsf{A}_\ell^2} \leq \sqrt{\frac{1}{2}}\sqrt{\frac{\Lambda_\ell}{m_1}}\|\boldsymbol{w}_\ell\|_{\mathsf{A}_\ell},$$

and the result follows with $C = 2^{-1/2}$. □

## 8.4 Prolongation, restriction, and the imbalanced Galerkin condition

As in the finite element setting, prolongation and restriction matrices communicate information between a coarse-level, say, the level-$(\ell-1)$ grid, and a fine-level, the level-$\ell$ grid.

**Definition 8.4.1** (Prolongation and restriction). Suppose that the multilevel cell-centered grid is uniform and nested and $1 \leq \ell \leq L$. Recall that, by construction $n_\ell = 2n_{\ell-1}$, $1 \leq \ell \leq L$. The **cell-centered prolongation matrix** $\mathsf{P}_{\ell-1} \in \mathbb{R}^{n_\ell \times n_{\ell-1}}$ has the form

$$\mathsf{P}_{\ell-1} = \begin{bmatrix} 1 & 0 & & 0 & 0 \\ 1 & 0 & & 0 & 0 \\ 0 & 1 & & 0 & 0 \\ 0 & 1 & & 0 & 0 \\ 0 & 0 & & 0 & 0 \\ 0 & 0 & & 0 & 0 \\ \vdots & \vdots & \cdots & \vdots & \vdots \\ 0 & 0 & & 0 & 0 \\ 0 & 0 & & 0 & 0 \\ 0 & 0 & & 1 & 0 \\ 0 & 0 & & 1 & 0 \\ 0 & 0 & & 0 & 1 \\ 0 & 0 & & 0 & 1 \end{bmatrix}.$$

Equivalently, the action of $\mathsf{P}_{\ell-1}$ on an arbitrary vector $\boldsymbol{v}_{\ell-1} \in \mathbb{R}^{n_{\ell-1}}$ is

$$[\mathsf{P}_{\ell-1}\boldsymbol{v}_{\ell-1}]_{2i-1} = v_{\ell-1,i}, \quad 1 \leq i \leq n_{\ell-1},$$
$$[\mathsf{P}_{\ell-1}\boldsymbol{v}_{\ell-1}]_{2i} = v_{\ell-1,i}, \quad 1 \leq i \leq n_{\ell-1}.$$

The **cell-centered restriction matrix** $\mathsf{R}_{\ell-1} \in \mathbb{R}^{n_{\ell-1} \times n_\ell}$ is defined as

$$\mathsf{R}_{\ell-1} = \frac{1}{2}\mathsf{P}_{\ell-1}^\top. \tag{8.4.1}$$

■

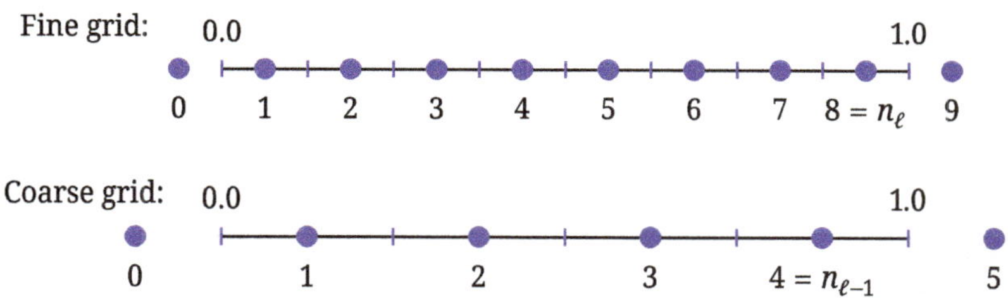


**Figure 8.7:** A uniform and nested two-level cell-centered grid. The prolongation operator moves information from the coarse level (level-($\ell - 1$)) grid to the fine level (level-$\ell$) grid.

Listings 8.5 and 8.6 show vectorized implementations of these two operations in the one-dimensional setting.

**Example 8.4.2** (Prolongation and restriction). Let $n_\ell = 8$ and $n_{\ell-1} = 4$. The associated uniform and nested two-level grid is shown in Figure 8.7. The prolongation operator moves information from the level-($\ell - 1$) (coarse) grid to the level-$\ell$ (fine) grid. The restriction operator moves information in the opposite direction. Since the values at the ghost cells will be set using boundary conditions, only the interior points are involved in the prolongation and restriction processes. To compute the prolongation of the 4 interior degrees of freedom in the coarse grid in this example, we need a prolongation matrix having 8 rows and 4 columns. We employ a restriction matrix having 4 rows and 8 columns. Precisely,

$$\mathsf{P}_{\ell-1} = \begin{bmatrix} 1 & 0 & 0 & 0 \\ 1 & 0 & 0 & 0 \\ 0 & 1 & 0 & 0 \\ 0 & 1 & 0 & 0 \\ 0 & 0 & 1 & 0 \\ 0 & 0 & 1 & 0 \\ 0 & 0 & 0 & 1 \\ 0 & 0 & 0 & 1 \end{bmatrix} \quad \text{and} \quad \mathsf{R}_{\ell-1} = \begin{bmatrix} \frac{1}{2} & \frac{1}{2} & 0 & 0 & 0 & 0 & 0 & 0 \\ 0 & 0 & \frac{1}{2} & \frac{1}{2} & 0 & 0 & 0 & 0 \\ 0 & 0 & 0 & 0 & \frac{1}{2} & \frac{1}{2} & 0 & 0 \\ 0 & 0 & 0 & 0 & 0 & 0 & \frac{1}{2} & \frac{1}{2} \end{bmatrix}. \qquad \blacktriangle$$

**Proposition 8.4.3** (Left inverse). *Suppose that the multilevel cell-centered grid is uniform and nested and* $1 \le \ell \le L$. *Then*

$$\mathsf{R}_{\ell-1}\mathsf{P}_{\ell-1} = \mathrm{I}_{n_{\ell-1}}.$$

*Proof.* Exercise; see Problem 8.6. □

**Proposition 8.4.4** (Imbalanced Galerkin condition). *Suppose the multilevel cell-centered grid is uniform and nested. The strong Galerkin condition* (G0) *fails to hold in this setting. However, for* $1 \le \ell \le L$,

$$\mathsf{R}_{\ell-1}\mathsf{A}_\ell\mathsf{P}_{\ell-1} = 2\mathsf{A}_{\ell-1}, \tag{8.4.2}$$

*so that the imbalanced Galerkin condition* (G3) *holds with* $r = \frac{1}{2}$.

*Proof.* Exercise; see Problem 8.7. □

Proposition 8.4.4 establishes that the imbalanced Galerkin condition, that is, Assumption (G3), holds. Therefore we are in the setting of Section 6.9, which will be our axiomatic template to follow to establish convergence of multigrid methods.

The fact that the standard Galerkin conditions fail has important implications for the analysis of the cell-centered multigrid method. For instance, the coarse-grid Ritz projection matrix is no longer a projection, as the following result states.

**Proposition 8.4.5** (Not a projection). *Let* $1 \le \ell \le L$. *Recall that the coarse-grid Ritz "projection" matrix* $\tilde{\Pi}_\ell$ *is generally defined as*

$$\tilde{\Pi}_\ell = \mathsf{P}_{\ell-1}\mathsf{A}_{\ell-1}^{-1}\mathsf{R}_{\ell-1}\mathsf{A}_\ell.$$

*In the present setting,* $\tilde{\Pi}_\ell$ *is not a projection matrix. In particular,*

$$\tilde{\Pi}_\ell^2 = 2\tilde{\Pi}_\ell.$$

*Furthermore,*

$$(\mathrm{I}_\ell - \tilde{\Pi}_\ell)^2 = \mathrm{I}_\ell.$$

*On the other hand, we always have*

$$\tilde{\Pi}_\ell^* = \tilde{\Pi}_\ell, \quad (\mathrm{I}_\ell - \tilde{\Pi}_\ell)^* = \mathrm{I}_\ell - \tilde{\Pi}_\ell.$$

*Proof.* Using the definitions and Proposition 8.4.4, we have

$$\begin{aligned}
\tilde{\Pi}_\ell^2 &= \mathsf{P}_{\ell-1}\mathsf{A}_{\ell-1}^{-1}\mathsf{R}_{\ell-1}\mathsf{A}_\ell\mathsf{P}_{\ell-1}\mathsf{A}_{\ell-1}^{-1}\mathsf{R}_{\ell-1}\mathsf{A}_\ell \\
&\overset{(8.4.2)}{=} \mathsf{P}_{\ell-1}\mathsf{A}_{\ell-1}^{-1}(2\mathsf{A}_{\ell-1})\mathsf{A}_{\ell-1}^{-1}\mathsf{R}_{\ell-1}\mathsf{A}_\ell \\
&= 2\mathsf{P}_{\ell-1}\mathsf{A}_{\ell-1}^{-1}\mathsf{R}_{\ell-1}\mathsf{A}_\ell \\
&= 2\tilde{\Pi}_\ell.
\end{aligned}$$

The other calculations are simple and skipped for brevity; see Problem 8.8. □

We have the following simple corollary.

**Corollary 8.4.6** (Coarse-grid Ritz error). *Let* $1 \le \ell \le L$. *Recall that in the present setting, the coarse-level stiffness matrix satisfies*

$$\mathsf{A}_{\ell-1} = \frac{1}{2}\mathsf{R}_{\ell-1}\mathsf{A}_\ell\mathsf{P}_{\ell-1}.$$

*Therefore it follows that*

$$\|(\mathsf{I}_\ell - \tilde{\Pi}_\ell)\boldsymbol{v}_\ell\|_{\mathsf{A}_\ell} = \|\boldsymbol{v}_\ell\|_{\mathsf{A}_\ell}, \quad \forall\, \boldsymbol{v}_\ell \in \mathbb{R}^{n_\ell}.$$

*Consequently,*

$$\|\mathsf{I}_\ell - \tilde{\Pi}_\ell\|_{\mathsf{A}_\ell} = 1.$$

*Proof.* Exercise; see Problem 8.9. □

## 8.5 Some additional multilevel tools

We have all of the pieces in place to define the multigrid method. In this section, we add some additional machinery that will aid in the analysis of the cell-centered multigrid method in one space dimension. In particular, we need some tools for establishing a suitable approximation property, which is the subject of the next section.

**Definition 8.5.1** (Level-$\ell$ discrete operators). Let $v_\ell \in V_\ell$ be arbitrary and suppose $\boldsymbol{v}_\ell \in \mathbb{R}^{n_\ell}$ is its coordinate vector with respect to the basis $\mathcal{B}_\ell$, that is,

$$v_\ell \in V_\ell \overset{\mathcal{B}_\ell}{\longleftrightarrow} \boldsymbol{v}_\ell \in \mathbb{R}^{n_\ell}.$$

Suppose that $\mathsf{A}_\ell$ is the cell-centered stiffness matrix defined in (8.1.7). The **level-$\ell$ discrete minus Laplacian operator** on $V_\ell$ is the operator $-\Delta_\ell^\circ : V_\ell \to V_\ell$ defined by

$$-\Delta_\ell^\circ v_\ell := \sum_{i=1}^{n_\ell} [\mathsf{A}_\ell \boldsymbol{v}_\ell]_i \psi_{\ell,i}.$$

The **energy form** $\mathcal{A}_\ell(\cdot,\cdot) : V_\ell \times V_\ell \to \mathbb{R}$ is defined by

$$\mathcal{A}_\ell(u_\ell, v_\ell) := (-\Delta_\ell^\circ u_\ell, v_\ell)_{L^2(0,1)}, \tag{8.5.1}$$

for all $u_\ell, v_\ell \in V_\ell$. ■

**Proposition 8.5.2** (Symmetric and bilinear). *For each $0 \le \ell \le L$, the form $\mathcal{A}_\ell(\cdot,\cdot)$ is symmetric, that is,*

$$\mathcal{A}_\ell(u_\ell, v_\ell) = \mathcal{A}_\ell(v_\ell, u_\ell), \quad \forall\, u_\ell, v_\ell \in V_\ell.$$

*It is also bilinear, that is, linear in each argument. Furthermore, for all $1 \le i,j \le n_\ell$,*

$$\mathcal{A}_\ell(\psi_{\ell,i}, \psi_{\ell,j}) = h_\ell [\mathsf{A}_\ell]_{j,i}. \tag{8.5.2}$$

*Proof.* Exercise; see Problem 8.10. □

**Definition 8.5.3** ($L^2$-projection onto $V_\ell$). For each $0 \le \ell \le L$, we denote the $L^2$**-projection onto** $V_\ell$ by $Q_\ell : L^2(0,1) \to V_\ell$. This operator is defined as follows: for each $v \in L^2(0,1)$, $Q_\ell v \in V_\ell$ is the unique function satisfying

$$(Q_\ell v, w_\ell)_{L^2(0,1)} = (v, w_\ell)_{L^2(0,1)}, \quad \forall\, w_\ell \in V_\ell. \qquad \blacksquare$$

**Proposition 8.5.4** (Coordinate vectors). *Let $0 \le \ell \le L$ and $v \in L^2(0,1)$. If $\mathbf{v}_\ell \in \mathbb{R}^{n_\ell}$ is the coordinate vector of $v_\ell = Q_\ell v \in V_\ell$ with respect to the basis $\mathcal{B}_\ell$, then*

$$v_{\ell,i} = [\mathbf{v}_\ell]_i = \frac{1}{h_\ell} \int_{K_{\ell,i}} v(x)\,dx, \quad i = 1, \dots, n_\ell. \tag{8.5.3}$$

*Proof.* By definition,

$$v_\ell = Q_\ell v = \sum_{i=1}^{n_\ell} v_{\ell,i} \psi_{\ell,i},$$

and

$$\left( \sum_{i=1}^{n_\ell} v_{\ell,i} \psi_{\ell,i}, \psi_{\ell,j} \right)_{L^2(0,1)} = (v, \psi_{\ell,j})_{L^2(0,1)},$$

for $j = 1, \dots, n_\ell$. Using linearity and the fact that

$$(\psi_{\ell,i}, \psi_{\ell,j})_{L^2(0,1)} = h_\ell \delta_{i,j},$$

we have

$$\begin{aligned}
(v, \psi_{\ell,j})_{L^2(0,1)} &= \sum_{i=1}^{n_\ell} v_{\ell,i} (\psi_{\ell,i}, \psi_{\ell,j})_{L^2(0,1)} \\
&= \sum_{i=1}^{n_\ell} v_{\ell,i} h_\ell \delta_{i,j} \\
&= h_\ell v_{\ell,j}.
\end{aligned}$$

Hence we have

$$v_{\ell,j} = \frac{1}{h_\ell} (v, \psi_{\ell,j})_{L^2(0,1)} = \frac{1}{h_\ell} \int_{K_{\ell,j}} v(x)\,dx. \qquad \square$$

**Proposition 8.5.5** (Representation). *Let $1 \le \ell \le L$ and $v_\ell \in V_\ell$. Set $v_{\ell-1} = Q_{\ell-1} v_\ell \in V_{\ell-1}$ and assume that $\mathbf{v}_{\ell-1} \in \mathbb{R}^{n_{\ell-1}}$ is the coordinate vector of $v_{\ell-1}$ with respect to the basis $\mathcal{B}_{\ell-1}$, that is,*

$$v_{\ell-1} \in V_{\ell-1} \xleftrightarrow{\mathcal{B}_{\ell-1}} \boldsymbol{v}_{\ell-1} \in \mathbb{R}^{n_{\ell-1}}.$$

*Then, for each $i \le i \le n_{\ell-1}$, we have*

$$v_{\ell-1,i} = \frac{1}{2}(v_{\ell,2i-1} + v_{\ell,2i}),$$

*where*

$$v_\ell \in V_\ell \xleftrightarrow{\mathcal{B}_\ell} \boldsymbol{v}_\ell \in \mathbb{R}^{n_\ell}.$$

*In other words,*

$$\boldsymbol{v}_{\ell-1} = \mathsf{R}_{\ell-1}\boldsymbol{v}_\ell.$$

*Proof.* Suppose

$$v_\ell = \sum_{i=1}^{n_\ell} v_{\ell,i}\psi_{\ell,i} \in V_\ell \xleftrightarrow{\mathcal{B}_\ell} \boldsymbol{v}_\ell \in \mathbb{R}^{n_\ell}.$$

Then, since

$$K_{\ell-1,i} = K_{\ell,2i-1} \cup K_{\ell,2i},$$

it follows that, for $1 \le i \le n_{\ell-1}$,

$$\begin{aligned}
v_{\ell-1,i} &\overset{(8.5.3)}{=} \frac{1}{h_{\ell-1}} \int_{K_{\ell-1,i}} v_\ell(x)\,\mathrm{d}x \\
&= \frac{1}{h_{\ell-1}} \int_{K_{\ell,2i-1}} v_\ell(x)\,\mathrm{d}x + \frac{1}{h_{\ell-1}} \int_{K_{\ell,2i}} v_\ell(x)\,\mathrm{d}x \\
&= \frac{h_\ell}{h_{\ell-1}} v_{\ell,2i-1} + \frac{h_\ell}{h_{\ell-1}} v_{\ell,2i} \\
&= \frac{1}{2}(v_{\ell,2i-1} + v_{\ell,2i}).
\end{aligned}$$

□

**Proposition 8.5.6** (Prolongation action). *Let $1 \le \ell \le L$ and $v_{\ell-1} \in V_{\ell-1}$ and assume that*

$$v_{\ell-1} \in V_{\ell-1} \xleftrightarrow{\mathcal{B}_{\ell-1}} \boldsymbol{v}_{\ell-1} \in \mathbb{R}^{n_{\ell-1}}.$$

*Then $v_{\ell-1} \in V_\ell$, and if $\boldsymbol{v}_\ell$ is its coordinate vector with respect to the basis $\mathcal{B}_\ell$, that is,*

$$v_{\ell-1} \in V_\ell \xleftrightarrow{\mathcal{B}_\ell} \boldsymbol{v}_\ell \in \mathbb{R}^{n_\ell},$$

*then*

$$\boldsymbol{v}_\ell = \mathsf{P}_{\ell-1}\boldsymbol{v}_{\ell-1}.$$

*Proof.* Exercise; see Problem 8.12. □

One important property of the $L^2$-projection is its $H^1$-stability.

**Lemma 8.5.7** ($H^1$-stability). *For every $0 \le \ell \le L$, the operator $Q_\ell$ is $H^1$-stable in the sense that there is a constant $C_2 > 0$, independent of $\ell$, such that, for all $v \in H_0^1(0,1)$,*

$$\mathcal{A}_\ell(Q_\ell v, Q_\ell v) \le C_2^2 \int_0^1 |v'(x)|^2 \, dx.$$

*Proof.* Exercise; see Problem 8.11. □

From this stability it is not difficult to prove that the $L^2$-projection possesses suitable approximation properties.

**Proposition 8.5.8** (Projection error estimate). *There exist a constant $C > 0$, independent of $\ell$, such that, for all $0 \le \ell \le L$ and $v \in H^1(0,1)$, we have*

$$\|v - Q_\ell v\|_{L^2(0,1)} \le Ch_\ell \|v'\|_{L^2(0,1)}. \tag{8.5.4}$$

*Proof.* Exercise; see Problem 8.13. □

## 8.6 The $\alpha$-weak approximation property

Recall that we have assumed that smoothing will be carried out via the Λ-Richardson smoother and the smoothing property has already been established. All the elements of multigrid are now completely defined. We now have only to prove the convergence of the various algorithms by establishing an appropriate approximation property.

In this section, we demonstrate that the cell-centered multigrid method satisfies the $\alpha$-weak approximation property with precisely $\alpha = \frac{1}{2}$. Our present analysis will cover the one-dimensional case with constant diffusivity, but it can straightforwardly be extended to higher dimensions and possibly even to cases of nonconstant diffusivity.

Let $f \in L^2(0,1)$. The cell-centered finite difference problem is formulated as follows: find $u_\ell \in V_\ell$ such that

$$\mathcal{A}_\ell(u_\ell, v_\ell) = (f, v_\ell)_{L^2(0,1)}, \quad \forall\, v_\ell \in V_\ell. \tag{8.6.1}$$

Set

$$f_\ell := Q_\ell f \in V_\ell \xleftrightarrow{\mathcal{B}_\ell} \boldsymbol{f}_\ell \in \mathbb{R}^{n_\ell}$$

and

$$u_\ell \in V_\ell \overset{\mathcal{B}_\ell}{\longleftrightarrow} \boldsymbol{u}_\ell \in \mathbb{R}^{n_\ell}.$$

It is easy to see that

$$\mathsf{A}_\ell \boldsymbol{u}_\ell = \boldsymbol{f}_\ell. \tag{8.6.2}$$

Of course, we are primarily interested in the approximation on some fine level $\ell = L$.

**Remark 8.6.1** (Right-hand side). Notice that, in the cell centered finite difference problem, the way $\boldsymbol{f}_\ell$ is defined is different from that in the finite difference approximation of previous chapters. Indeed, previously, $f_\ell$ is defined by point evaluation of the right-hand side, whereas here it is defined by local averaging. This is done because here we only assume that $f \in L^2(0,1)$, and so it may not have well-defined point values. However, if it turns out that $f \in C([0,1]) \subset L^2(0,1)$, then these two approaches give very similar approximations; see Problem 8.14. ◂

**Theorem 8.6.2** (Error estimate). *Let $f \in L^2(0,1)$. Assume that $u \in H_0^1(0,1) \cap H^2(0,1)$ satisfies*

$$(u', v')_{L^2(0,1)} = (f, v)_{L^2(0,1)}, \quad \forall v \in H_0^1(0,1).$$

*Assume that $u_\ell \in V_\ell$ is the unique solution to* (8.6.1). *Then there is a constant $C > 0$, independent of $\ell$, such that*

$$\sqrt{\mathcal{A}_\ell(u_\ell - Q_\ell u, u_\ell - Q_\ell u)} \le C h_\ell \|f\|_{L^2(0,1)}. \tag{8.6.3}$$

*Proof.* See [19]. □

**Definition 8.6.3** (Discrete cell-centered Ritz projection). For each $1 \le \ell \le L$, the **discrete cell-centered Ritz projection** $\mathcal{R}_{\ell-1} : V_\ell \to V_{\ell-1}$ is defined as follows: for $v_\ell \in V_\ell$, $\mathcal{R}_{\ell-1} v_\ell \in V_{\ell-1}$ satisfies

$$\mathcal{A}_{\ell-1}(\mathcal{R}_{\ell-1} v_\ell, w_{\ell-1}) = \mathcal{A}_\ell(v_\ell, w_{\ell-1}), \quad \forall w_{\ell-1} \in V_{\ell-1}.$$ ■

**Proposition 8.6.4** (Ritz projection representation). *Let $1 \le \ell \le L$. Let $v_\ell \in V_\ell$, and set $v_{\ell-1} = \mathcal{R}_{\ell-1} v_\ell \in V_{\ell-1}$. Under the usual coordinate identification, that is,*

$$v_{\ell-1} = \mathcal{R}_{\ell-1} v_\ell \in V_{\ell-1} \overset{\mathcal{B}_{\ell-1}}{\longleftrightarrow} \boldsymbol{v}_{\ell-1} \in \mathbb{R}^{n_{\ell-1}},$$
$$v_\ell \in V_\ell \overset{\mathcal{B}_\ell}{\longleftrightarrow} \boldsymbol{v}_\ell \in \mathbb{R}^{n_\ell},$$

*we have*

$$\boldsymbol{v}_{\ell-1} = \Pi_{\ell-1}\boldsymbol{v}_\ell,$$

*where, as before, the matrix* $\Pi_{\ell-1}$ *is defined as*

$$\Pi_{\ell-1} = \mathsf{A}_{\ell-1}^{-1}\mathsf{R}_{\ell-1}\mathsf{A}_\ell \in \mathbb{R}^{n_{\ell-1}\times n_\ell}.$$

*Proof.* Clearly, for all $j = 1, \dots, n_{\ell-1}$,

$$\mathcal{A}_{\ell-1}\left(\sum_{i=1}^{n_{\ell-1}} v_{\ell-1,i}\psi_{\ell-1,i}, \psi_{\ell-1,j}\right) = \mathcal{A}_\ell\left(\sum_{k=1}^{n_\ell} v_{\ell,k}\psi_{\ell,k}, \psi_{\ell-1,j}\right). \tag{8.6.4}$$

In addition, the left-hand side of (8.6.4) is

$$\begin{aligned}\mathcal{A}_{\ell-1}\left(\sum_{i=1}^{n_{\ell-1}} v_{\ell-1,i}\psi_{\ell-1,i}, \psi_{\ell-1,j}\right) &= \sum_{i=1}^{n_{\ell-1}} v_{\ell-1,i}\mathcal{A}_{\ell-1}(\psi_{\ell-1,i}, \psi_{\ell-1,j}) \\ &\overset{(8.5.2)}{=} h_{\ell-1}\sum_{i=1}^{n_{\ell-1}} [\mathsf{A}_{\ell-1}]_{i,j} v_{\ell-1,i} \\ &= h_{\ell-1}[\mathsf{A}_{\ell-1}\boldsymbol{v}_{\ell-1}]_j.\end{aligned}$$

Next, observe that

$$\psi_{\ell-1,j} = \sum_{s=1}^{n_\ell} p_{\ell-1,s,j}\psi_{\ell,s},$$

where

$$[\mathsf{P}_{\ell-1}]_{s,j} = p_{\ell-1,s,j}, \quad 1 \le s \le n_\ell, \quad 1 \le j \le n_{\ell-1}.$$

Thus the right-hand side of (8.6.4) satisfies

$$\begin{aligned}\mathcal{A}_\ell\left(\sum_{k=1}^{n_\ell} v_{\ell,k}\psi_{\ell,k}, \sum_{s=1}^{n_\ell} p_{\ell-1,s,j}\psi_{\ell,s}\right) &= \sum_{k,s=1}^{n_\ell} v_{\ell,k}p_{\ell-1,s,j}\mathcal{A}_\ell(\psi_{\ell,k}, \psi_{\ell,s}) \\ &\overset{(8.5.2)}{=} h_\ell\sum_{k,s=1}^{n_\ell} v_{\ell,k}p_{\ell-1,s,j}[\mathsf{A}_\ell]_{k,s} \\ &= h_\ell[\mathsf{P}_{\ell-1}^\top\mathsf{A}_\ell\boldsymbol{v}_\ell]_j.\end{aligned}$$

Thus, for every $\boldsymbol{v}_\ell \in \mathbb{R}^{n_\ell}$, (8.6.4) is equivalent to

$$\begin{aligned}\mathsf{A}_{\ell-1}\boldsymbol{v}_{\ell-1} &= \frac{h_\ell}{h_{\ell-1}}\mathsf{P}_{\ell-1}^\top\mathsf{A}_\ell\boldsymbol{v}_\ell \\ &= \frac{1}{2}\mathsf{P}_{\ell-1}^\top\mathsf{A}_\ell\boldsymbol{v}_\ell\end{aligned}$$

$$\stackrel{(8.4.1)}{=} \mathsf{R}_{\ell-1}\mathsf{A}_\ell \boldsymbol{v}_\ell.$$

This can be rewritten as

$$\boldsymbol{v}_{\ell-1} = \mathsf{A}_{\ell-1}^{-1}\mathsf{R}_{\ell-1}\mathsf{A}_\ell \boldsymbol{v}_\ell = \Pi_{\ell-1}\boldsymbol{v}_\ell. \qquad \square$$

**Definition 8.6.5** (Lifting operator). Let $0 \le \ell \le L$. The **lifting operator** $\mathcal{L} : V_\ell \to H_0^1(0,1)$ is defined as follows. For $v_\ell \in V_\ell$, $\mathcal{L}v_\ell \in H_0^1(0,1)$ is the solution of the variational problem

$$((\mathcal{L}v_\ell)', w')_{L^2(0,1)} = (-\Delta_\ell^\circ v_\ell, w)_{L^2(0,1)}, \quad \forall\, w \in H_0^1(0,1). \tag{8.6.5}$$

■

It is not difficult to see that the lifting operator $\mathcal{L}$ is well defined. Indeed, (8.6.5) has a unique solution since $-\Delta_\ell^\circ v_\ell \in V_\ell \subset L^2(0,1)$. Furthermore, $v_\ell \in V_\ell$ is the cell-centered approximation of its lifting $v = \mathcal{L}v_\ell$. The following result shows a strange kind of stability for the lifting operator.

**Proposition 8.6.6** (Reverse stability). *For all* $0 \le \ell \le L$ *and* $v_\ell \in V_\ell$, *there is a constant* $C > 0$, *independent of* $\ell$, *such that*

$$\|\mathcal{L}v_\ell\|_{H^1(0,1)}^2 \le C\mathcal{A}_\ell(v_\ell, v_\ell). \tag{8.6.6}$$

*Proof.* By the Poincaré inequality, Theorem 7.1.1, there is a constant $C_1 > 0$ such that

$$\|\mathcal{L}v_\ell\|_{H^1(0,1)}^2 \le C_1((\mathcal{L}v_\ell)', (\mathcal{L}v_\ell)')_{L^2(0,1)}.$$

Now using the definition of the lifting operator (8.6.5) and the $L^2$-projection operator of Definition 8.5.3, we have

$$\begin{aligned}
((\mathcal{L}v_\ell)', (\mathcal{L}v_\ell)')_{L^2(0,1)} &= -(\Delta_\ell^\circ v_\ell, \mathcal{L}v_\ell)_{L^2(0,1)} \\
&= -(\Delta_\ell^\circ v_\ell, Q_\ell\mathcal{L}v_\ell)_{L^2(0,1)} \\
&\stackrel{(8.5.1)}{=} \mathcal{A}_\ell(v_\ell, Q_\ell\mathcal{L}v_\ell) \\
&\le \sqrt{\mathcal{A}_\ell(v_\ell, v_\ell)}\sqrt{\mathcal{A}_\ell(Q_\ell\mathcal{L}v_\ell, Q_\ell\mathcal{L}v_\ell)},
\end{aligned}$$

where the last step follows from the Cauchy–Schwarz inequality.

We may now combine this estimate with the stability of the $L^2$-projection proved in Lemma 8.5.7 to obtain

$$\begin{aligned}
\|\mathcal{L}v_\ell\|_{H^1(0,1)}^2 &\le C_1\sqrt{\mathcal{A}_\ell(v_\ell, v_\ell)}\sqrt{\mathcal{A}_\ell(Q_\ell\mathcal{L}v_\ell, Q_\ell\mathcal{L}v_\ell)} \\
&\le C_1C_2\sqrt{\mathcal{A}_\ell(v_\ell, v_\ell)}\sqrt{((\mathcal{L}v_\ell)', (\mathcal{L}v_\ell)')_{L^2(0,1)}} \\
&\le C_1C_2\sqrt{\mathcal{A}_\ell(v_\ell, v_\ell)}\|\mathcal{L}v_\ell\|_{H^1(0,1)}.
\end{aligned}$$

Setting $C = (C_1C_2)^2$, the proof is complete. □

The following result can be understood as a version of the $\alpha$-weak approximation property but stated for piecewise constant functions instead of vectors.

**Theorem 8.6.7** (Approximation). *For all $1 \le \ell \le L$ and $v_\ell \in V_\ell$, there exists a constant $C > 0$, independent of $\ell$, such that*

$$|\mathcal{A}_\ell(v_\ell - \mathcal{R}_{\ell-1}v_\ell, v_\ell)| \le Ch_\ell \|\Delta_\ell^\circ v_\ell\|_{L^2(0,1)} \sqrt{\mathcal{A}_\ell(v_\ell, v_\ell)}.$$

*Proof.* For $v_\ell \in V_\ell$, by the triangle inequality we have

$$\begin{aligned} |\mathcal{A}_\ell(v_\ell - \mathcal{R}_{\ell-1}v_\ell, v_\ell)| &\le |\mathcal{A}_\ell(v_\ell - Q_\ell \mathcal{L} v_\ell, v_\ell)| \\ &\quad + |\mathcal{A}_\ell(Q_\ell \mathcal{L} v_\ell - Q_{\ell-1}\mathcal{L} v_\ell, v_\ell)| \\ &\quad + |\mathcal{A}_\ell(Q_{\ell-1}\mathcal{L} v_\ell - \mathcal{R}_{\ell-1}v_\ell, v_\ell)| \\ &= T_1 + T_2 + T_3, \end{aligned} \tag{8.6.7}$$

where the three terms $T_i$, $i = 1, 2, 3$, are defined in order of appearence on the right hand side. We now examine each term separately.

We first deal with the term $T_1$. Since $v_\ell \in V_\ell$ is the cell-centered finite difference approximation of $\mathcal{L}v_\ell$, the error estimate (8.6.3) guarantees that there exists $C > 0$ such that

$$\begin{aligned} T_1 &\le \sqrt{\mathcal{A}_\ell(v_\ell - Q_\ell \mathcal{L} v_\ell, v_\ell - Q_\ell \mathcal{L} v_\ell)} \sqrt{\mathcal{A}_\ell(v_\ell, v_\ell)} \\ &\overset{(8.6.3)}{\le} Ch_\ell \|\Delta_\ell^\circ v_\ell\|_{L^2(0,1)} \sqrt{\mathcal{A}_\ell(v_\ell, v_\ell)}. \end{aligned}$$

To handle the last term, $T_3$, we begin with a preliminary observation. Let $v_{\ell-1} \in V_{\ell-1}$ be the cell-centered finite difference approximation of $\mathcal{L}v_\ell$ in the space $V_{\ell-1}$. In other words, $v_{\ell-1}$ is the unique function that satisfies

$$\mathcal{A}_{\ell-1}(v_{\ell-1}, w_{\ell-1}) = -(\Delta_\ell^\circ v_\ell, w_{\ell-1})_{L^2(0,1)}, \quad \forall\, w_{\ell-1} \in V_{\ell-1}.$$

On the other hand, since $V_{\ell-1} \subset V_\ell$, definition (8.5.1) of the energy form shows that

$$-(\Delta_\ell^\circ v_\ell, w_{\ell-1})_{L^2(0,1)} = \mathcal{A}_\ell(v_\ell, w_{\ell-1}), \quad \forall\, w_{\ell-1} \in V_{\ell-1}.$$

Finally, recall that by Definition 8.6.3

$$\mathcal{A}_\ell(v_\ell, w_{\ell-1}) = \mathcal{A}_{\ell-1}(\mathcal{R}_{\ell-1}v_\ell, w_{\ell-1}), \quad \forall\, w_{\ell-1} \in V_{\ell-1}.$$

Let us combine these three identities to obtain that

$$\mathcal{A}_{\ell-1}(v_{\ell-1}, w_{\ell-1}) = \mathcal{A}_{\ell-1}(\mathcal{R}_{\ell-1}v_\ell, w_{\ell-1}), \quad \forall\, w_{\ell-1} \in V_{\ell-1}.$$

In other words, $v_{\ell-1} = \mathcal{R}_{\ell-1}v_\ell$. In summary, $\mathcal{R}_{\ell-1}v_\ell \in V_{\ell-1}$ is the cell-centered finite difference approximation of the function $\mathcal{L}v_\ell \in H_0^1(0,1)$. The error estimate (8.6.3) then guarantees

$$\sqrt{\mathcal{A}_{\ell-1}(\mathcal{R}_{\ell-1}v_\ell - \mathcal{Q}_{\ell-1}\mathcal{L}v_\ell, \mathcal{R}_{\ell-1}v_\ell - \mathcal{Q}_{\ell-1}\mathcal{L}v_\ell)} \le Ch_{\ell-1}\|\Delta_\ell^\circ v_\ell\|_{L^2(0,1)}.$$

Next, Proposition 8.4.4 shows that for all $w_{\ell-1} \in V_{\ell-1}$,

$$\mathcal{A}_\ell(w_{\ell-1}, w_{\ell-1}) = 2\mathcal{A}_{\ell-1}(w_{\ell-1}, w_{\ell-1}).$$

These last two facts together yield

$$\begin{aligned} T_3 &\le \sqrt{\mathcal{A}_\ell(\mathcal{Q}_{\ell-1}\mathcal{L}v_\ell - \mathcal{R}_{\ell-1}v_\ell, \mathcal{Q}_{\ell-1}\mathcal{L}v_\ell - \mathcal{R}_{\ell-1}v_\ell)}\sqrt{\mathcal{A}_\ell(v_\ell, v_\ell)} \\ &\le \sqrt{2}Ch_{\ell-1}\|\Delta_\ell^\circ v_\ell\|_{L^2(0,1)}\sqrt{\mathcal{A}_\ell(v_\ell, v_\ell)} \\ &= 2\sqrt{2}Ch_\ell\|\Delta_\ell^\circ v_\ell\|_{L^2(0,1)}\sqrt{\mathcal{A}_\ell(v_\ell, v_\ell)}. \end{aligned}$$

Finally, we consider the middle term $T_2$ of (8.6.7). Note that

$$\begin{aligned} T_2 &= |(\mathcal{Q}_\ell\mathcal{L}v_\ell - \mathcal{Q}_{\ell-1}\mathcal{L}v_\ell, \Delta_\ell^\circ v_\ell)_{L^2(0,1)}| \\ &\le \|\mathcal{Q}_\ell\mathcal{L}v_\ell - \mathcal{Q}_{\ell-1}\mathcal{L}v_\ell\|_{L^2(0,1)}\|\Delta_\ell^\circ v_\ell\|_{L^2(0,1)} \\ &\le (\|\mathcal{Q}_\ell\mathcal{L}v_\ell - \mathcal{L}v_\ell\|_{L^2(0,1)} + \|\mathcal{L}v_\ell - \mathcal{Q}_{\ell-1}\mathcal{L}v_\ell\|_{L^2(0,1)})\|\Delta_\ell^\circ v_\ell\|_{L^2(0,1)} \\ &\overset{(8.5.4)}{\le} C(h_\ell + h_{\ell-1})\|\mathcal{L}v_\ell\|_{H^1(0,1)}\|\Delta_\ell^\circ v_\ell\|_{L^2(0,1)} \\ &\overset{(8.6.6)}{\le} Ch_\ell\|\Delta_\ell^\circ v_\ell\|_{L^2(0,1)}\sqrt{\mathcal{A}_\ell(v_\ell, v_\ell)}. \end{aligned}$$

It remains to combine the previously obtained estimates of each of the terms, $T_i$, $i = 1, 2, 3$, which form (8.6.7). □

We have reached the point where we can show that the $\alpha$-weak approximation property (A2) holds with

$$\alpha = \frac{1}{2}.$$

**Theorem 8.6.8** ($\alpha$-weak approximation property). *There is a constant $C_{\mathrm{A2}} > 0$ such that for all $1 \le \ell \le L$, we have*

$$|((\mathsf{I}_\ell - \tilde{\Pi}_\ell)\boldsymbol{v}_\ell, \boldsymbol{v}_\ell)_{\mathsf{A}_\ell}| \le C_{\mathrm{A2}}\frac{\|\mathsf{A}_\ell\boldsymbol{v}_\ell\|_\ell}{\sqrt{\rho(\mathsf{A}_\ell)}}\|\boldsymbol{v}_\ell\|_{\mathsf{A}_\ell}, \quad \forall\, \boldsymbol{v}_\ell \in \mathbb{R}^{n_\ell}.$$

*Proof.* Exercise; see Problem 8.15. □

**Remark 8.6.9** (Convergence). We have shown that the smoothing property holds for Richardson's smoother. In addition, Proposition 8.4.4 shows that (G3) holds. Finally, we have just verified assumption (A2) with $\alpha = \frac{1}{2}$. Thus, we can apply the axiomatic theory of Section 6.9 to conclude that the two-grid method (see Corollary 6.9.7), the symmetric W-cycle with a large enough number of smoothing steps (see Corollary 6.9.12), and even the generic symmetric W-cycle (Theorem 6.9.15) converge. There is nothing more to do! The proofs are merely an application of the aforementioned results. ◂

**Remark 8.6.10** (Open question). Recall that (see Remark 6.9.16) convergence of the V-cycle in this setting is an open question. ◂

## 8.7 The variable-diffusivity case

In this section, we develop a cell-centered finite difference method and a multigrid solver for a one-dimensional problem with a nonconstant diffusion coefficient. To be specific, let $f \in C(0,1)$ and $D \in C^1(0,1) \cap C([0,1])$. We assume that there are constants $0 < D_{\min} \leq D_{\max}$ for which

$$D_{\min} \leq D(x) \leq D_{\max}, \quad \forall x \in [0,1].$$

The problem we will be concerned with is to find $u : [0,1] \to \mathbb{R}$ that solves

$$\begin{cases} u(0) = 0, \\ -\dfrac{\mathrm{d}}{\mathrm{d}x}\left(D\dfrac{\mathrm{d}u}{\mathrm{d}x}\right) = f \quad \text{in } (0,1), \\ u(1) = 0. \end{cases} \tag{8.7.1}$$

We will operate in the framework of classical solutions. In other words, a solution to (8.7.1) is understood as a function $u \in C^2(0,1) \cap C([0,1])$ that satisfies (8.7.1) for every point in $[0,1]$.

We now introduce the cell-centered finite difference discretization of problem (8.7.1). We first define the operator $-\Delta_\ell^D : \mathcal{C}_{\ell,0} \to \mathcal{C}_\ell^\circ$: for $v_\ell \in \mathcal{C}_{\ell,0}$ and $1 \leq i \leq n_\ell$,

$$\begin{aligned} (-\Delta_\ell^D v_\ell)_i &:= -\Delta_\ell^D v_{\ell,i} \\ &:= \frac{-D_{\ell,i+1/2}(v_{\ell,i+1} - v_{\ell,i}) + D_{\ell,i-1/2}(v_{\ell,i} - v_{\ell,i-1})}{h_\ell^2}, \end{aligned} \tag{8.7.2}$$

where

$$D_{\ell,i+1/2} := D(i \cdot h_\ell), \quad 0 \leq i \leq n_\ell.$$

The cell-centered finite difference approximation of (8.7.1) is to find $u_\ell \in \mathcal{C}_{\ell,0}$ that satisfies, for $1 \leq i \leq n_\ell$,

$$-\Delta_\ell^D u_\ell = f_\ell, \tag{8.7.3}$$

where $f_\ell \in \mathcal{C}_\ell^\circ$ is defined for $1 \le i \le n_\ell$ as $f_{\ell,i} := f(x_{\ell,i})$. The reader may verify (see Problem 8.16) that this is a formally second-order scheme.

The stiffness matrix can be computed as usual, that is, by considering the $n_\ell$ equations given in (8.7.3) and the fact that $u_\ell \in \mathcal{C}_{\ell,0}$. To obtain the first equation, we set $i = 1$ in (8.7.3) to see that

$$\frac{1}{h_\ell^2}\left[-D_{\ell,1/2}u_{\ell,0} + (D_{\ell,1/2} + D_{\ell,3/2})u_{\ell,1} - D_{\ell,3/2}u_{\ell,2}\right] = f_{\ell,1}.$$

In addition, since $u_\ell \in \mathcal{C}_{\ell,0}$, we must have

$$u_{\ell,0} = -u_{\ell,1}.$$

Combining these last two identities, the first equation is

$$\frac{1}{h_\ell^2}\left[(2D_{\ell,1/2} + D_{\ell,3/2})u_{\ell,1} - D_{\ell,3/2}u_{\ell,2}\right] = f_{\ell,1}.$$

For $2 \le i \le n_\ell - 1$, the equation is (8.7.3). Using the right-hand boundary condition, we have

$$u_{\ell,n_\ell+1} = -u_{\ell,n_\ell}.$$

The last equation is

$$\frac{1}{h_\ell^2}\left[-D_{\ell,n_\ell-1/2}u_{\ell,n_\ell-1} + (D_{\ell,n_\ell-1/2} + 2D_{\ell,n_\ell+1/2})u_{\ell,n_\ell}\right] = f_{\ell,n_\ell}.$$

Thus the cell-centered finite difference approximation for the nonconstant diffusion coefficient can be represented as

$$\mathsf{A}_\ell \boldsymbol{u}_\ell = \boldsymbol{f}_\ell,$$

where the level-$\ell$ stiffness matrix is the following tridiagonal matrix:

$$\mathsf{A}_\ell = \frac{1}{h_\ell^2}\begin{bmatrix} a_1 & b_1 & & & \\ b_1 & a_2 & b_2 & & \\ & b_2 & a_3 & \ddots & \\ & & \ddots & \ddots & b_{n_\ell-1} \\ & & & b_{n_\ell-1} & a_{n_\ell} \end{bmatrix} \in \mathbb{R}^{n_\ell \times n_\ell}, \tag{8.7.4}$$

where

$$a_1 = 2D_{\ell,1/2} + D_{\ell,3/2},$$
$$a_i = D_{\ell,i-1/2} + D_{\ell,i+1/2}, \quad 2 \le i \le n_\ell - 1,$$
$$a_{n_\ell} = D_{\ell,n_\ell-1/2} + 2D_{\ell,n_\ell+1/2},$$

and

$$b_i = -D_{\ell,i+1/2}, \quad 1 \le i \le n_\ell - 1.$$

**Remark 8.7.1** (Operator notation). In what follows, we denote by $-\Delta_\ell^D : \mathcal{C}_{\ell,0} \to \mathcal{C}_\ell^\circ$ the operator representation of the matrix $\mathsf{A}_\ell$, and *vice versa.* In this representation, the boundary conditions are built into the space. We could equally well use the notation $-\Delta_\ell^{D,\circ} : \mathcal{C}_\ell^\circ \to \mathcal{C}_\ell^\circ$ as the operator representation of $\mathsf{A}_\ell$ where it is understood that the boundary conditions are built directly into the definition of the operator $-\Delta_\ell^{D,\circ}$. ◂

As in the constant coefficient case, there is no hope to obtain any of the Galerkin conditions (G0)–(G2). The following result shows that, nevertheless, condition (G3) holds.

**Proposition 8.7.2** (Condition (G3)). *Let $1 \le \ell \le L$. Assume that the stiffness matrices are defined as in* (8.7.4) *and the prolongation and restriction matrices are as in Definition* 8.4.1. *Then the imbalanced Galerkin condition, that is, Assumption* (G3), *holds:*

$$\mathsf{R}_{\ell-1}\mathsf{A}_\ell\mathsf{P}_{\ell-1} = 2\mathsf{A}_{\ell-1}.$$

*Proof.* To simplify the arguments, we introduce some notation. For $\boldsymbol{v}_{\ell-1} \in \mathbb{R}^{n_{\ell-1}}$, define $\boldsymbol{v}_\ell := \mathsf{P}_{\ell-1}\boldsymbol{v}_{\ell-1}$. Using the canonical identifications, we have the grid functions

$$\boldsymbol{v}_{\ell-1} \overset{\mathcal{G}_{\ell-1}}{\longleftrightarrow} v_{\ell-1} \in \mathcal{C}_{\ell-1}^\circ,$$
$$\boldsymbol{v}_\ell \overset{\mathcal{G}_\ell}{\longleftrightarrow} v_\ell \in \mathcal{C}_\ell^\circ.$$

The index on the level-$(\ell-1)$ grid is the symbol $I$, so that $1 \le I \le n_{\ell-1}$, and the index on the level-$\ell$ grid is the symbol $i$, so that $1 \le i \le n_\ell$. See Figure 8.8.

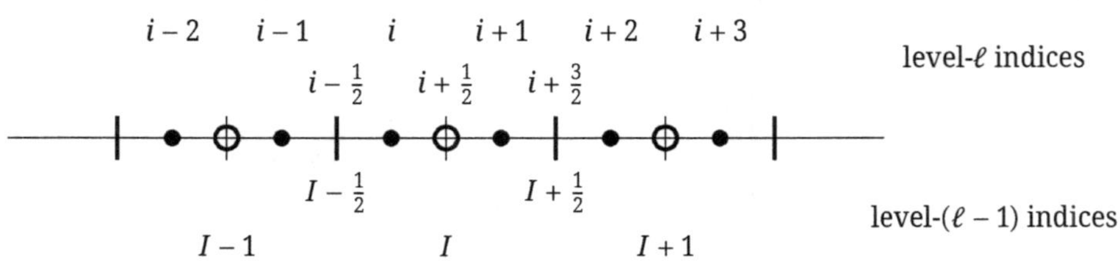


**Figure 8.8:** Setup and notation for the proof of Proposition 8.7.2. The unfilled circles denote the cell-centered grid points for the level-$(\ell-1)$ mesh, whereas filled circles are the level-$\ell$ cell-centered grid points.

Define

$$a_I := [\mathsf{R}_{\ell-1}\mathsf{A}_\ell\mathsf{P}_{\ell-1}\boldsymbol{v}_{\ell-1}]_I.$$

Then

$$\begin{aligned}
a_I &= \frac{1}{2h_\ell^2}[-D_{\ell,i+1/2}(v_{\ell,i+1} - v_{\ell,i}) + D_{\ell,i-1/2}(v_{\ell,i} - v_{\ell,i-1})] \\
&\quad + \frac{1}{2h_\ell^2}[-D_{\ell,i+3/2}(v_{\ell,i+2} - v_{\ell,i+1}) + D_{\ell,i+1/2}(v_{\ell,i+1} - v_{\ell,i})] \\
&= \frac{1}{2h_\ell^2}[-D_{\ell,i+1/2}(v_{\ell-1,I} - v_{\ell-1,I}) + D_{\ell,i-1/2}(v_{\ell-1,I} - v_{\ell-1,I-1})] \\
&\quad + \frac{1}{2h_\ell^2}[-D_{\ell,i+3/2}(v_{\ell-1,I+1} - v_{\ell-1,I}) + D_{\ell,i+1/2}(v_{\ell-1,I} - v_{\ell-1,I})] \\
&= \frac{1}{2}\left[\frac{-D_{\ell,i+3/2}(v_{\ell-1,I+1} - v_{\ell-1,I})}{(\frac{1}{2}h_{\ell-1})^2}\right] + \frac{1}{2}\left[\frac{D_{\ell,i-1/2}(v_{\ell-1,I} - v_{\ell-1,I-1})}{(\frac{1}{2}h_{\ell-1})^2}\right] \\
&= 2\frac{-D_{\ell,i+3/2}(v_{\ell-1,I+1} - v_{\ell-1,I}) + D_{\ell,i-1/2}(v_{\ell-1,I} - v_{\ell,I-1})}{h_{\ell-1}^2} \\
&= 2\frac{-D_{\ell-1,I+1/2}(v_{\ell-1,I+1} - v_{\ell-1,I}) + D_{\ell-1,I-1/2}(v_{\ell-1,I} - v_{\ell,I-1})}{h_{\ell-1}^2} \\
&= 2[\mathsf{A}_{\ell-1}\boldsymbol{v}_{\ell-1}]_I.
\end{aligned}$$

The argument must change slightly when $I = 1$ and $I = n_{\ell-1}$. We leave it to the reader to check those cases separately. After those details are confirmed, we can conclude that $\mathsf{R}_{\ell-1}\mathsf{A}_\ell\mathsf{P}_{\ell-1} = 2\mathsf{A}_{\ell-1}$, as claimed. □

**Remark 8.7.3** (Smoothing and approximation). We have shown that the imbalanced Galerkin condition, Assumption (G3), holds. There are two missing pieces that would be required to establish the convergence of our multigrid method. We need, in particular, a smoothing property using, perhaps, a Richardson-type smoother and an approximation property. Neither of these two pieces is trivial in the variable-diffusivity context. ◂

## 8.8 Cell-centered multigrid in two dimensions

In this section, we describe the cell-centered multigrid method in two space dimensions. For brevity, however, we do not provide a detailed account of the theory, since, at least for the constant-diffusivity case, this only involves minor changes from the one-dimensional case.

Let $\Omega = (0,1)^2, f \in C(\Omega)$, and $D \in C(\overline{\Omega}) \cap C^1(\Omega)$. Assume that there are constants $0 < D_{\min} \leq D_{\max}$ for which

$$D_{\min} \leq D(\boldsymbol{x}) \leq D_{\max}, \quad \forall \boldsymbol{x} \in \overline{\Omega}.$$

We will refer to the function $D$ as the diffusion coefficient, or diffusivity, for short. Presently, let us operate in the framework of classical solutions. In other words, we look for a solution $u \in C^2(\Omega) \cap C(\overline{\Omega})$ that in $\overline{\Omega}$ satisfies

$$\begin{cases} -\nabla \cdot (D\nabla u) = f & \text{in } \Omega, \\ u = 0 & \text{on } \partial\Omega. \end{cases} \tag{8.8.1}$$

The approximation of $u$ will be carried out via the cell-centered finite difference method. We begin by defining a two-dimensional grid of cell-centered points covering the domain $\Omega$. Similarly to the one-dimensional case, for each level $\ell = 0, 1, \ldots, L$, we set the number of grid points in each coordinate direction and the grid size to be, respectively,

$$m_\ell := q \cdot 2^\ell \quad \text{and} \quad h_\ell := \frac{1}{m_\ell},$$

where $q \in \mathbb{N}$ is fixed.

**Definition 8.8.1** (Cell-centered grid). Let $0 \le \ell \le L$. The **cell-centered grid points** are defined as

$$\boldsymbol{x}_{\ell,i,j} := (\zeta_{\ell,i}, \zeta_{\ell,j}), \quad i,j = 0, \ldots, m_\ell + 1,$$

where

$$\zeta_{\ell,k} := \left(k - \frac{1}{2}\right) h_\ell, \quad k = 0, \ldots, m_\ell + 1.$$

The **set of cell-centered grid points** is

$$C_\ell := \{\boldsymbol{x}_{\ell,i,j} \mid 0 \le i,j \le m_\ell + 1\} \setminus X_\ell,$$

where $X_\ell$ is the set of excluded corners,

$$X_\ell := \{\boldsymbol{x}_{\ell,0,0}, \boldsymbol{x}_{\ell,0,m_\ell+1}, \boldsymbol{x}_{\ell,m_\ell+1,0}, \boldsymbol{x}_{\ell,m_\ell+1,m_\ell+1}\}.$$

The grid points that do not belong to $\Omega$ are called **ghost points.** The **set of interior cell-centered grid points** is

$$C_\ell^\circ := \{\boldsymbol{x}_{\ell,i,j} \mid 1 \le i,j \le m_\ell\} = C_\ell \cap \Omega.$$

Finally, the **grid cells** are

$$K_{\ell,i,j} := ((i-1)h_\ell, ih_\ell] \times ((j-1)h_\ell, jh_\ell], \quad i,j = 1, \ldots, m_\ell.$$

■

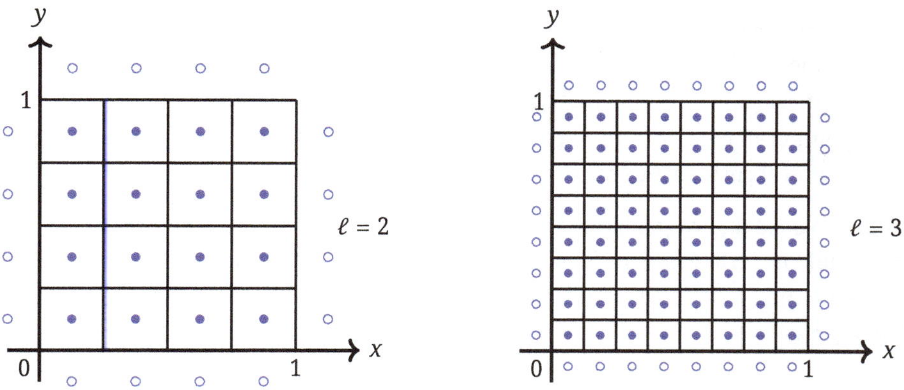


**Figure 8.9:** Example of two-dimensional cell-centered grids and grid points for $\ell = 2$ and $\ell = 3$. The ghost points are indicated by unfilled circles.

We refer to Figure 8.9 for an example of a two dimensional grid, along with its interior and ghost points.

Similarly to the one-dimensional case, we can define the space of piecewise constant functions $V_\ell$ and various spaces of cell-centered grid functions.

**Definition 8.8.2** (Grid functions). For every $0 \le \ell \le L$, we define the space of **piecewise constant functions** as

$$V_\ell \coloneqq \{v : (0,1]^2 \to \mathbb{R} \mid v|_{K_{\ell,i,j}} \in \mathbb{P}_0,\ 1 \le i,j \le m_\ell\}.$$

The spaces of **cell-centered grid functions** and **interior cell-centered grid functions** are defined, respectively, as

$$\mathcal{C}_\ell \coloneqq \{v_\ell : C_\ell \to \mathbb{R}\} \quad \text{and} \quad \mathcal{C}_\ell^\circ \coloneqq \{v_\ell : C_\ell^\circ \to \mathbb{R}\}.$$

The subspace of **homogeneous Dirichlet cell-centered grid functions** is denoted

$$\mathcal{C}_{\ell,0} \coloneqq \{u_\ell \in \mathcal{C}_\ell \mid u_\ell = 0 \text{ on } \partial\Omega\},$$

where $u_\ell = 0$ on $\partial\Omega$ means that

$$\begin{aligned} \frac{u_{\ell,0,j} + u_{\ell,1,j}}{2} &= 0, \quad 1 \le j \le m_\ell, \\ \frac{u_{\ell,n_\ell,j} + u_{\ell,n_\ell+1,j}}{2} &= 0, \quad 1 \le j \le m_\ell, \\ \frac{u_{\ell,i,0} + u_{\ell,i,1}}{2} &= 0, \quad 1 \le i \le m_\ell, \\ \frac{u_{\ell,i,n_\ell} + u_{\ell,i,n_\ell+1}}{2} &= 0, \quad 1 \le i \le m_\ell. \end{aligned} \tag{8.8.2}$$

We say that the **discrete homogeneous Dirichlet boundary conditions** hold for $u_\ell \in \mathcal{C}_\ell$, and we write $u_\ell = 0$ on $\partial\Omega$, if (8.8.2) holds. ■

The canonical bases of these spaces can be constructed as in Definition 8.1.5. We suppress the details for brevity.

As in the one-dimensional case, there is a standard way to relate functions in $V_\ell, \mathcal{C}_{\ell,0}$, and $\mathcal{C}^\circ_\ell$ with vectors in $\mathbb{R}^{m_\ell^2}$. In particular, this can be accomplished using the so-called lexicographical correspondence.

**Definition 8.8.3** (Lexicographic ordering). Suppose that $1 \le i,j \le m$. We say that the indices $i$ and $j$ are **traversed in forward lexicographic order** if the indices are put into one-to-one correspondence with the integers $\{1, 2, \dots, m^2\}$, with counted indices, via the mapping

$$(i,j) \mapsto F(i,j) := i + (j-1)m. \tag{8.8.3}$$

The function $F$ is called the **forward enumeration function.** We say that the indices $i$ and $j$ are **traversed in backward lexicographic order** if the indices are counted via the mapping

$$(i,j) \mapsto B(i,j) := m - i + 1 + (m-j)m. \tag{8.8.4}$$

The function $B$ is called the **backward enumeration function.** ■

An example of the forward and backward lexicographic enumerations of a cell-centered 4-by-4 grid is illustrated in Figure 8.10. The general lexicographic correspondence uses the forward lexicographic enumeration and is defined as follows:

$$v_\ell \in \mathcal{C}^\circ_\ell \overset{\text{lex}}{\longleftrightarrow} \mathbf{v}_\ell \in \mathbb{R}^{m_\ell^2}, \tag{8.8.5}$$

| | | | |
|---|---|---|---|
| 13 | 14 | 15 | 16 |
| 9 | 10 | 11 | 12 |
| 5 | 6 | 7 | 8 |
| 1 | 2 | 3 | 4 |

(a) Forward lexicographic

| | | | |
|---|---|---|---|
| 4 | 3 | 2 | 1 |
| 8 | 7 | 6 | 5 |
| 12 | 11 | 10 | 9 |
| 16 | 15 | 14 | 13 |

(b) Backward lexicographic

**Figure 8.10:** (a) Forward lexicographic and (b) backward lexicographic enumeration of the cells of a $4 \times 4$ grid.

where

$$[\boldsymbol{v}_\ell]_{F(i,j)} = v_{\ell,i,j}, \quad 1 \le i,j \le m_\ell.$$

This identification clearly shows that $\dim(\mathcal{C}_\ell^\circ) = m_\ell^2$. To keep our notation standardized, we define

$$n_\ell := m_\ell^2.$$

Thus, because of the mutual identifications between $V_\ell, \mathcal{C}_{\ell,0}, \mathcal{C}_\ell^\circ$, and $\mathbb{R}^{m_\ell^2}$, we are justified in writing

$$\dim(\mathcal{C}_\ell^\circ) = \dim(\mathcal{C}_{\ell,0}) = \dim(V_\ell) = n_\ell, \quad 0 \le \ell \le L.$$

### 8.8.1 The constant-diffusivity case

Let us first consider the case

$$D \equiv 1 \quad \text{on} \quad \overline{\Omega},$$

the so-called constant-diffusivity case. Thus we need an approximation of the minus Laplacian operator, $-\Delta = -\nabla \cdot (1\nabla)$, on $\Omega$.

For this purpose, for $0 \le \ell \le L$, we construct the cell-centered finite difference minus Laplacian, which is an operator denoted by $-\Delta_\ell : \mathcal{C}_\ell \to \mathcal{C}_\ell^\circ$ and defined as follows: for any grid function $v_\ell \in \mathcal{C}_\ell$ and $1 \le i,j \le m_\ell$,

$$(-\Delta_\ell v_\ell)_{i,j} := -\Delta_\ell v_{\ell,i,j} := \frac{-v_{\ell,i+1,j} - v_{\ell,i-1,j} - v_{\ell,i,j+1} - v_{\ell,i,j-1} + 4v_{\ell,i,j}}{h_\ell^2}.$$

The grid operator $-\Delta_\ell$ can be expressed as a stencil in the usual way. For $1 \le i,j \le m_\ell$,

$$[-\Delta_\ell]_{i,j} = \frac{1}{h_\ell^2} \begin{bmatrix} & -1 & \\ -1 & 4 & -1 \\ & -1 & \end{bmatrix} \begin{matrix} j+1 \\ j \\ j-1 \end{matrix}$$

$$\qquad\qquad i-1 \quad i \quad i+1$$

We are now ready to describe our finite difference scheme. First, we discretize the right-hand side $f$. For each $\ell \ge 0$, define the grid function $f_\ell \in \mathcal{C}_\ell^\circ$ via

$$f_{\ell,i,j} := f(\boldsymbol{x}_{\ell,i,j}), \quad 1 \le i,j \le m_\ell. \tag{8.8.6}$$

We seek the grid function $u_\ell \in \mathcal{C}_{\ell,0}$ such that

$$-\Delta_\ell u_{\ell,i,j} = f_{\ell,i,j}, \quad 1 \leq i,j \leq m_\ell. \tag{8.8.7}$$

Observe that discrete homogeneous Dirichlet boundary conditions (8.8.2) are built into the space of solutions $\mathcal{C}_{\ell,0}$.

We can incorporate the boundary conditions into the definition of the minus discrete Laplacian operator by modifying the stencil of the grid operator $-\Delta_\ell$ near the boundaries. This gives rise to the so-called interior minus discrete Laplacian $-\Delta_\ell^\circ : \mathcal{C}_\ell^\circ \to \mathcal{C}_\ell^\circ$. For $2 \leq i,j \leq m_\ell - 1$, the stencils of $-\Delta_\ell$ and $-\Delta_\ell^\circ$ exactly coincide.

At the left boundary, the stencil of $-\Delta_\ell^\circ$ is modified as follows: for $2 \leq j \leq m_\ell - 1$,

$$[-\Delta_\ell^\circ]_{1,j} = \frac{1}{h_\ell^2}$$
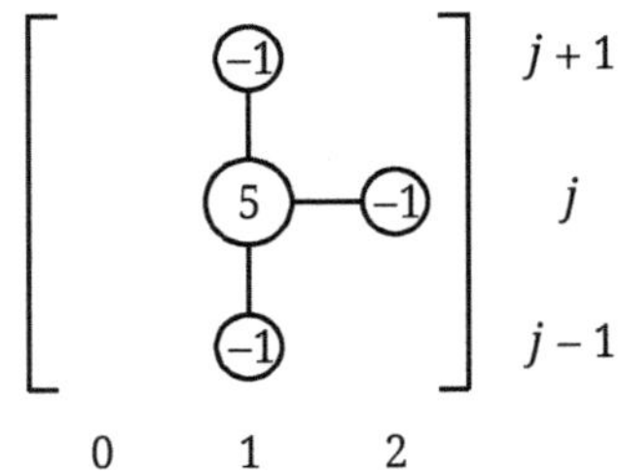


Similarly, at the bottom boundary, the stencil of $-\Delta_\ell^\circ$ is, for $2 \leq i \leq m_\ell - 1$,

$$[-\Delta_\ell^\circ]_{i,1} = \frac{1}{h_\ell^2}$$
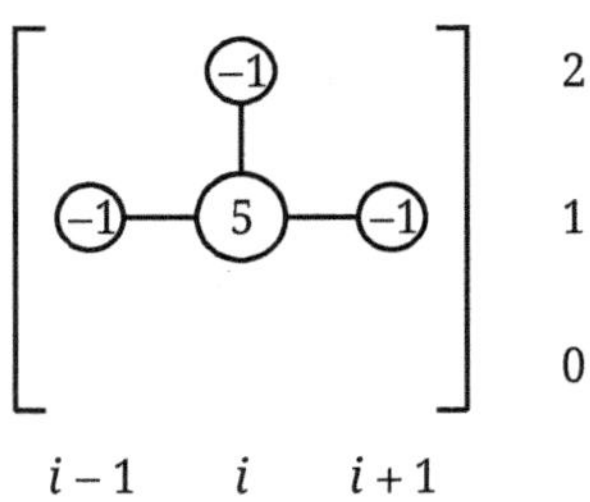


Lastly, at the bottom left corner, we have the stencil

$$[-\Delta_\ell^\circ]_{1,1} = \frac{1}{h_\ell^2}$$
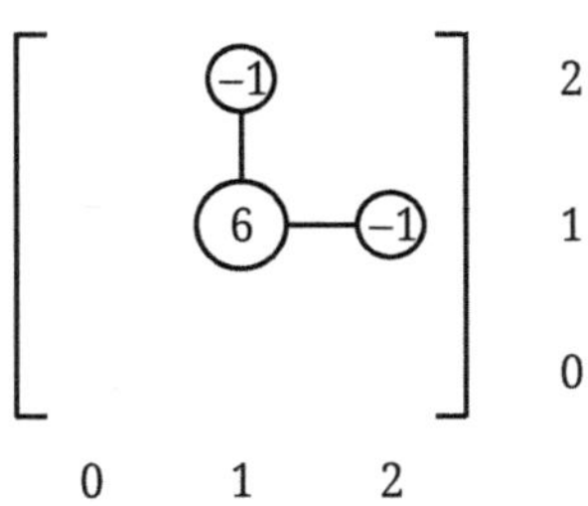


The remaining edge/corner cases can be easily inferred from those presented.

It should be clear that the following operator representations are equivalent:

$$-\Delta_\ell^\circ : \mathcal{C}_\ell^\circ \to \mathcal{C}_\ell^\circ,$$
$$-\Delta_\ell : \mathcal{C}_{\ell,0} \to \mathcal{C}_\ell^\circ.$$

As a result, the following problem statements are equivalent:
($P^\circ$) Given $f_\ell \in \mathcal{C}_\ell^\circ$, find $u_\ell \in \mathcal{C}_\ell^\circ$ satisfying

$$-\Delta_\ell^\circ u_\ell = f_\ell.$$

($P_0$) Given $f_\ell \in \mathcal{C}_\ell^\circ$, find $u_\ell \in \mathcal{C}_{\ell,0}$ satisfying

$$-\Delta_\ell u_\ell = f_\ell.$$

($P_M$) Given

$$\boldsymbol{f}_\ell \in \mathbb{R}^{n_\ell} \overset{\text{lex}}{\longleftrightarrow} f_\ell \in \mathcal{C}_\ell^\circ,$$

find $\boldsymbol{u}_\ell \in \mathbb{R}^{n_\ell}$ such that

$$\mathsf{A}_\ell \boldsymbol{u}_\ell = \boldsymbol{f}_\ell,$$

where the stiffness matrix $\mathsf{A}_\ell$ is defined as

$$\mathsf{A}_\ell = \frac{1}{h_\ell^2}\begin{bmatrix} \mathsf{B}_\ell & -\mathrm{I}_\ell & & & \\ -\mathrm{I}_\ell & \mathsf{C}_\ell & -\mathrm{I}_\ell & & \\ & -\mathrm{I}_\ell & \ddots & \ddots & \\ & & \ddots & \mathsf{C}_\ell & -\mathrm{I}_\ell \\ & & & -\mathrm{I}_\ell & \mathsf{B}_\ell \end{bmatrix} \in \mathbb{R}^{n_\ell \times n_\ell}. \tag{8.8.8}$$

Here $\mathrm{I}_\ell \in \mathbb{R}^{m_\ell \times m_\ell}$ is the identity matrix,

$$\mathsf{B}_\ell = \begin{bmatrix} 6 & -1 & & & & \\ -1 & 5 & -1 & & & \\ & -1 & 5 & \ddots & & \\ & & \ddots & \ddots & -1 & \\ & & & -1 & 5 & -1 \\ & & & & -1 & 6 \end{bmatrix} \in \mathbb{R}^{m_\ell \times m_\ell},$$

and

$$\mathsf{C}_\ell = \begin{bmatrix} 5 & -1 & & & & \\ -1 & 4 & -1 & & & \\ & -1 & 4 & \ddots & & \\ & & \ddots & \ddots & -1 & \\ & & & -1 & 4 & -1 \\ & & & & -1 & 5 \end{bmatrix} \in \mathbb{R}^{m_\ell \times m_\ell}.$$

We now define the prolongation and restriction matrices via their grid operator representations. The restriction matrix $\mathsf{R}_{\ell-1} : \mathbb{R}^{n_\ell} \to \mathbb{R}^{n_{\ell-1}}$ can be understood as an equivalent grid operator $R_{\ell-1} : \mathcal{C}_\ell^\circ \to \mathcal{C}_{\ell-1}^\circ$. We define the latter as follows: given the fine-level grid function $v_\ell \in \mathcal{C}_\ell^\circ$, the coarse-level grid function $v_{\ell-1} = R_{\ell-1}v_\ell \in \mathcal{C}_{\ell-1}^\circ$ is defined for $1 \le i,j \le m_{\ell-1}$ as

$$v_{\ell-1,i,j} := \frac{v_{\ell,2i,2j} + v_{\ell,2i-1,2j} + v_{\ell,2i,2j-1} + v_{\ell,2i-1,2j-1}}{4}. \tag{8.8.9}$$

The entries of the matrix $\mathsf{R}_{\ell-1} \in \mathbb{R}^{n_{\ell-1},n_\ell}$ can be deduced from the (forward) lexicographic correspondence.

Analogously, the prolongation matrix $\mathsf{P}_{\ell-1} : \mathbb{R}^{n_{\ell-1}} \to \mathbb{R}^{n_\ell}$ can be understood as an equivalent grid operator $P_{\ell-1} : \mathcal{C}_{\ell-1}^\circ \to \mathcal{C}_\ell^\circ$. For the coarse-grid function $v_{\ell-1} \in \mathcal{C}_{\ell-1}^\circ$, it produces $v_\ell = P_{\ell-1}v_{\ell-1} \in \mathcal{C}_\ell^\circ$ such that, for $1 \le i,j \le m_{\ell-1}$,

$$\begin{aligned} v_{\ell,2i,2j} &= v_{\ell-1,i,j}, \\ v_{\ell,2i-1,2j} &= v_{\ell-1,i,j}, \\ v_{\ell,2i,2j-1} &= v_{\ell-1,i,j}, \\ v_{\ell,2i-1,2j-1} &= v_{\ell-1,i,j}. \end{aligned} \tag{8.8.10}$$

The entries of the associated prolongation matrix $\mathsf{P}_{\ell-1} \in \mathbb{R}^{n_\ell \times n_{\ell-1}}$ can be determined via the (forward) lexicographic correspondence. See Problem 8.17 and Figure 8.11.

The following result shows that, as in the one-dimensional case, Assumption (G3) holds.

**Proposition 8.8.4** (Assumption (G3)). *Let $1 \le \ell \le L$, and let the restriction and prolongation operators be defined by* (8.8.9) *and* (8.8.10), *respectively. Then their respective matrix representations $\mathsf{R}_{\ell-1} \in \mathbb{R}^{n_{\ell-1} \times n_\ell}$ and $\mathsf{P}_{\ell-1} \in \mathbb{R}^{n_\ell \times n_{\ell-1}}$ satisfy*

$$\mathsf{R}_{\ell-1}\mathsf{P}_{\ell-1} = \mathsf{I}_{\ell-1}, \quad \mathsf{R}_{\ell-1} = \frac{1}{4}\mathsf{P}_{\ell-1}^\top, \quad 2\mathsf{A}_{\ell-1} = \mathsf{R}_{\ell-1}\mathsf{A}_\ell\mathsf{P}_{\ell-1},$$

*where $\mathsf{A}_\ell \in \mathbb{R}^{n_\ell \times n_\ell}$ is the stiffness matrix defined in* (8.8.8) *for $0 \le \ell \le L$. In other words, our multigrid procedure satisfies the imbalanced Galerkin condition, Assumption* (G3), *with $r = \frac{1}{4}$.*

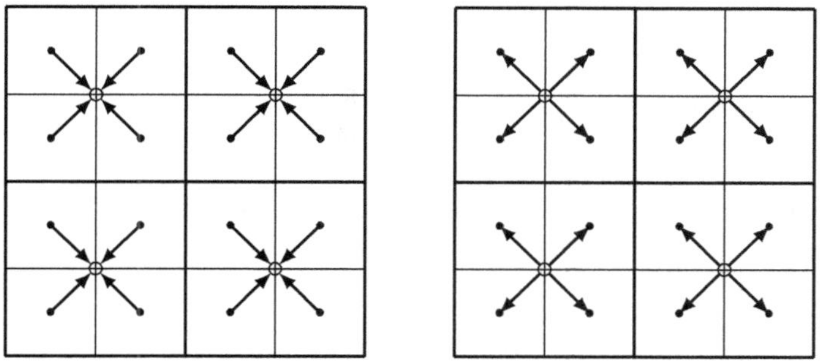

(a) Restriction (b) Prolongation

**Figure 8.11:** A graphical illustration of the (a) restriction and (b) prolongation operations. In restriction, the four values at surrounding fine-grid points (filled circles) are averaged to give the value at the coarse-grid point (unfilled circle). In prolongation, the value at the coarse-grid point is copied into the values of the fine grid function at the surrounding four fine grid cell-centered points.

*Proof.* Exercise; see Problem 8.18. It may be useful to use Figure 8.14. See also the proof of Proposition 8.8.5. □

The eigenvalues and eigenvectors of the stiffness matrix can be found by examination of the one-dimensional case. In fact, it is sufficient to note that the two-dimensional discrete Laplacian operator $-\Delta_\ell$ can be decomposed into the sum of two one-dimensional discrete Laplacian operators, one for each coordinate direction. Indeed, if $w_\ell \in \mathcal{C}_{\ell,0}$, then, for $1 \le i,j \le m_\ell$,

$$-\Delta_\ell w_{\ell,i,j} = \frac{-w_{\ell,i-1,j} + 2w_{\ell,i,j} - w_{\ell,i+1,j}}{h_\ell^2} + \frac{-w_{\ell,i,j-1} + 2w_{\ell,i,j} - w_{\ell,i,j+1}}{h_\ell^2}.$$

The one-dimensional stiffness matrix was introduced in Definition 8.1.7. Recall that its eigenvalues are given in (8.1.8), and the corresponding eigenvectors are presented in (8.1.9). On the basis of this knowledge, we can construct the eigenpairs of the stiffness matrix $\mathsf{A}_\ell$. The description will be easier in the language of grid functions. Define, for $1 \le s,t \le m_\ell$, the grid functions $w_\ell^{(s,t)} \in \mathcal{C}_\ell^\circ$ via

$$w_{\ell,i,j}^{(s,t)} = \sin(s\pi\zeta_{\ell,i})\sin(t\pi\zeta_{\ell,j}), \quad 1 \le i,j \le m_\ell.$$

It is not difficult to show that

$$-\Delta_\ell^\circ w_{\ell,i,j}^{(s,t)} = \lambda_\ell^{(s,t)} w_{\ell,i,j}^{(s,t)},$$

where

$$\lambda_\ell^{(s,t)} = \lambda_\ell^{(s)} + \lambda_\ell^{(t)} = \frac{4}{h_\ell^2}\left(\sin^2\left(\frac{s\pi h_\ell}{2}\right) + \sin^2\left(\frac{t\pi h_\ell}{2}\right)\right).$$

Clearly, the largest eigenvalue is given by

$$\lambda_\ell^{(m_\ell,m_\ell)} = \lambda_\ell^{(m_\ell)} + \lambda_\ell^{(m_\ell)} = 2\lambda_\ell^{(m_\ell)} = \frac{8}{h_\ell^2}.$$

Thus we have just shown that the spectral radius of the two-dimensional cell-centered stiffness matrix is

$$\rho(\mathsf{A}_\ell) = \frac{8}{h_\ell^2}.$$

This information allows us to construct the Λ-Richardson smoother in a precise manner.

At this point, all the ingredients are in place to construct our multigrid methods in their various flavors. To prove convergence, we need just two more ingredients, a smoothing property and an approximation property. The smoothing property is quite straightforward, provided that we use the Λ-Richardson smoother. We can follow the steps used to prove Theorem 8.6.8 and show that the $\alpha$-weak approximation property (A2) holds with $\alpha = \frac{1}{2}$. Consequently, the W-cycle multigrid method converges with a contraction factor that is independent of the number of levels $L$.

### 8.8.2 The variable-diffusivity case

To conclude this section, we describe how to proceed in the case of a problem with a variable diffusivity coefficient. In other words, we consider the two-dimensional analogue of the setting of Section 8.7.

A formally second-order cell-centered finite difference method is to find $u_\ell \in \mathcal{C}_{\ell,0}$ such that

$$-\Delta_\ell^D u_{\ell,i,j} = f_{\ell,i,j}, \quad 1 \le i,j \le m_\ell,$$

where, as before, the function $f_\ell \in \mathcal{C}_\ell^\circ$ is defined by $f_{\ell,i,j} = f(\boldsymbol{x}_{\ell,i,j})$. We define the operator $-\Delta_\ell^D : \mathcal{C}_\ell \to \mathcal{C}_\ell^\circ$ for $1 \le i,j \le m_\ell$ via

$$\begin{aligned}(-\Delta_\ell^D v_\ell)_{i,j} &:= -\Delta_\ell^D v_{\ell,i,j} \\ &:= \frac{-D_{\ell,i+1/2,j}(v_{\ell,i+1,j} - v_{\ell,i,j}) + D_{\ell,i-1/2,j}(v_{\ell,i,j} - v_{\ell,i-1,j})}{h_\ell^2} \\ &\quad + \frac{-D_{\ell,i,j+1/2}(v_{\ell,i,j+1} - v_{\ell,i,j}) + D_{\ell,i,j-1/2}(v_{\ell,i,j} - v_{\ell,i,j-1})}{h_\ell^2},\end{aligned} \tag{8.8.11}$$

for all $v_\ell \in \mathcal{C}_{\ell,0}$. See Figure 8.12. This operator is implemented in Listing 8.21.

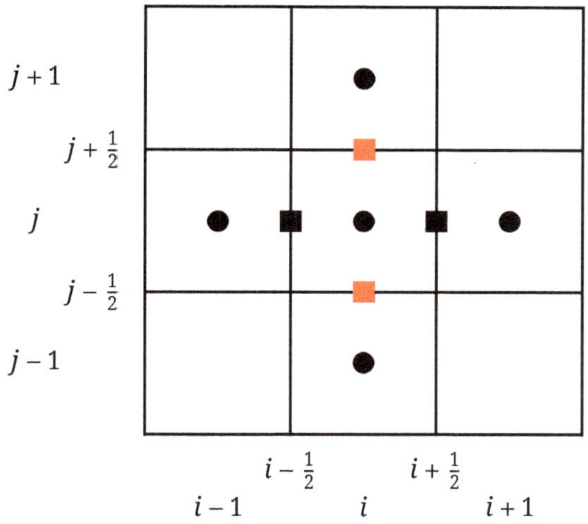


**Figure 8.12:** Grid points involved in the calculation of the discrete minus Laplacian operator, as expressed in (8.8.11), at the grid point with index $(i,j)$. The black filled squares are called east–west edge points. The red filled squares are called north–south edge points. The filled black circles are the relevant cell-centered grid points.

The discrete diffusivity values are computed recursively as follows: at the finest level $\ell = L$,

$$D_{L,i+1/2,j} := D\left(ih_L, \left(j - \frac{1}{2}\right)h_L\right), \quad 0 \le i \le m_L, \quad 1 \le j \le m_L,$$
$$D_{L,i,j+1/2} := D\left(\left(i - \frac{1}{2}\right)h_L, jh_L\right), \quad 1 \le i \le m_L, \quad 0 \le j \le m_L.$$

For $1 \le \ell \le L$, we define

$$\begin{aligned} D_{\ell-1,I+1/2,J} &:= \frac{D_{\ell,2I+1/2,2J-1} + D_{\ell,2I+1/2,2J}}{2}, \quad && 0 \le I \le m_{\ell-1}, \\ & && 1 \le J \le m_{\ell-1}, \\ D_{\ell-1,I,J+1/2} &:= \frac{D_{\ell,2I-1,2J+1/2} + D_{\ell,2I,2J+1/2}}{2}, \quad && 0 \le I \le m_{\ell-1}, \\ & && 1 \le J \le m_{\ell-1}. \end{aligned} \tag{8.8.12}$$

See Figure 8.13 and also Listing 8.20. Referring to the particular indexing in Figure 8.14, observe that

$$\begin{aligned} D_{\ell-1,I+1/2,J} &:= \frac{D_{\ell,i+3/2,j} + D_{\ell,i+3/2,j+1}}{2}, \\ D_{\ell-1,I-1/2,J} &:= \frac{D_{\ell,i-1/2,j} + D_{\ell,i-1/2,j+1}}{2}, \\ D_{\ell-1,I,J+1/2} &:= \frac{D_{\ell,i,j+3/2} + D_{\ell,i+1,j+3/2}}{2}, \\ D_{\ell-1,I,J-1/2} &:= \frac{D_{\ell,i,j-1/2} + D_{\ell,i+1,j-1/2}}{2}. \end{aligned} \tag{8.8.13}$$

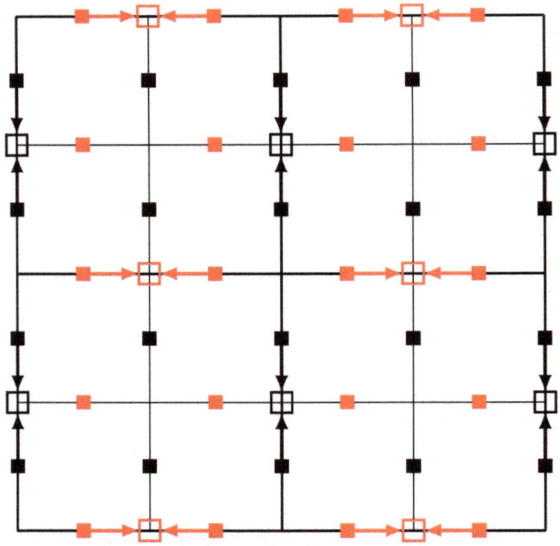

**Figure 8.13:** This diagram illustrates the computation of the diffusion coefficient at level $\ell - 1$ from that at level $\ell$ via averaging, as prescribed in (8.8.12). The diffusion coefficient $D_\ell$ is defined on level $\ell$ at the filled black and red squares. On level $\ell - 1$, the diffusion coefficient, $D_{\ell-1}$ is defined at the unfilled black and red squares.

As usual, we can use the discrete homogeneous Dirichlet boundary conditions $u_\ell = 0$ on $\partial\Omega$, defined as in (8.8.2), to eliminate degrees of freedom at the ghost cells. This allows us to define the equivalent grid operator

$$-\Delta_\ell^{D,\circ} : \mathcal{C}_\ell^\circ \to \mathcal{C}_\ell^\circ.$$

The definition of the level-$\ell$ stiffness matrix $\mathsf{A}_\ell$ follows.

As in the one-dimensional variable-coefficient case or the two-dimensional constant-coefficient case, we aim to verify Assumption (G3). This verification requires the very specific construction of the discrete diffusivity functions defined above, as we now show.

**Proposition 8.8.5** (Assumption (G3)). *Let $0 \le \ell \le L$. Assume that $\mathsf{A}_\ell \in \mathbb{R}^{n_\ell \times n_\ell}$ is defined in accordance with $-\Delta_\ell^{D,\circ}$. Then the imbalanced Galerkin condition, Assumption* (G3), *holds: for* $1 \le \ell \le L$,

$$\mathsf{R}_{\ell-1}\mathsf{A}_\ell\mathsf{P}_{\ell-1} = 2\mathsf{A}_{\ell-1},$$

*or, in terms of grid operators,*

$$-R_{\ell-1}\Delta_\ell^{D,\circ}P_{\ell-1} = -2\Delta_{\ell-1}^{D,\circ}. \tag{8.8.14}$$

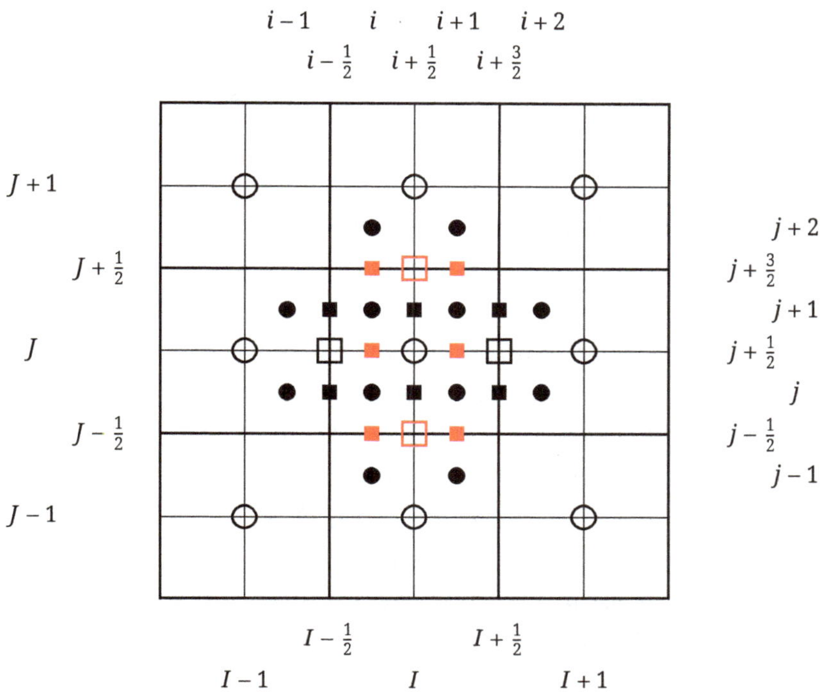


**Figure 8.14:** Fine- and coarse-grid points used in the calculations in the proof of Proposition 8.8.5. The (coarse) level-($\ell - 1$) grid points are denoted by unfilled shapes, whereas the (fine) level-$\ell$ grid points by filled shapes.

*Proof.* The proof is an exercise in juggling notation. For $v_{\ell-1} \in \mathcal{C}^{\circ}_{\ell-1}$, define

$$a_{I,J} := -4h_\ell^2 (R_{\ell-1}\Delta_\ell^{D,\circ} P_{\ell-1} v_{\ell-1})_{I,J}$$

for all $2 \le I, J \le m_{\ell-1} - 1$. We will stay away from the grid boundaries for simplicity and brevity. The reader should complete the proof by adapting the calculation appropriately at the boundaries.

Define

$$v_\ell = P_{\ell-1} v_{\ell-1} \in \mathcal{C}^{\circ}_\ell.$$

Using the definition of the prolongation operator, we observe, for example,

$$v_{\ell,i+2,j} = v_{\ell-1,I+1,J}, \quad v_{\ell,i+1,j} = v_{\ell-1,I,J},$$

et cetera, referring to Figure 8.14. Essentially, the value of $v_\ell$ at every filled black circle (fine-grid point) in Figure 8.14 is equal to the value of $v_{\ell-1}$ at the nearest unfilled black circle (coarse-grid point).

To further reduce our writing, let us introduce the difference operators

$$\delta^x v_{\ell,i+1/2,j} := v_{\ell,i+1,j} - v_{\ell,i,j}, \quad 0 \le i \le m_\ell, \quad 1 \le j \le m_\ell,$$

and

$$\delta^y v_{\ell,i,j+1/2} := v_{\ell,i,j+1} - v_{\ell,i,j}, \quad 1 \le i \le m_\ell, \quad 0 \le j \le m_\ell.$$

Then,

$$\begin{aligned}
a_{I,J} = &-D_{\ell,i+1/2,j}\delta^x v_{\ell,i+1/2,j} + D_{\ell,i-1/2,j}\delta^x v_{\ell,i-1/2,j} \\
&- D_{\ell,i,j+1/2}\delta^y v_{\ell,i,j+1/2} + D_{\ell,i,j-1/2}\delta^y v_{\ell,i,j-1/2} \\
&- D_{\ell,i+3/2,j}\delta^x v_{\ell,i+3/2,j} + D_{\ell,i+1/2,j}\delta^x v_{\ell,i+1/2,j} \\
&- D_{\ell,i+1,j+1/2}\delta^y v_{\ell,i+1,j+1/2} + D_{\ell,i+1,j-1/2}\delta^y v_{\ell,i+1,j-1/2} \\
&- D_{\ell,i+1/2,j+1}\delta^x v_{\ell,i+1/2,j+1} + D_{\ell,i-1/2,j+1}\delta^x v_{\ell,i-1/2,j+1} \\
&- D_{\ell,i,j+3/2}\delta^y v_{\ell,i,j+3/2} + D_{\ell,i,j+1/2}\delta^y v_{\ell,i,j+1/2} \\
&- D_{\ell,i+3/2,j+1}\delta^x v_{\ell,i+3/2,j+1} + D_{\ell,i+1/2,j+1}\delta^x v_{\ell,i+1/2,j+1} \\
&- D_{\ell,i+1,j+3/2}\delta^y v_{\ell,i+1,j+3/2} + D_{\ell,i+1,j+1/2}\delta^y v_{\ell,i+1,j+1/2}.
\end{aligned}$$

Canceling flux terms that are equal in magnitude but opposite in sign, we have

$$\begin{aligned}
a_{I,J} = &D_{\ell,i-1/2,j}(v_{\ell,i,j} - v_{\ell,i-1,j}) \\
&+ D_{\ell,i,j-1/2}(v_{\ell,i,j} - v_{\ell,i,j-1}) \\
&- D_{\ell,i+3/2,j}(v_{\ell,i+2,j} - v_{\ell,i+1,j}) \\
&+ D_{\ell,i+1,j-1/2}(v_{\ell,i+1,j} - v_{\ell,i+1,j-1}) \\
&+ D_{\ell,i-1/2,j+1}(v_{\ell,i,j+1} - v_{\ell,i-1,j+1}) \\
&- D_{\ell,i,j+3/2}(v_{\ell,i,j+2} - v_{\ell,i,j+1}) \\
&- D_{\ell,i+3/2,j+1}(v_{\ell,i+2,j+1} - v_{\ell,i+1,j+1}) \\
&- D_{\ell,i+1,j+3/2}(v_{\ell,i+1,j+2} - v_{\ell,i+1,j+1}).
\end{aligned}$$

Using the definition of the prolongation operator, referring to the particular 2-level indexing in Figure 8.14, and collecting like terms, we have

$$\begin{aligned}
a_{I,J} = &-(D_{\ell,i+3/2,j} + D_{\ell,i+3/2,j+1})(v_{\ell-1,I+1,J} - v_{\ell-1,I,J}) \\
&+ (D_{\ell,i-1/2,j} + D_{\ell,i-1/2,j+1})(v_{\ell-1,I,J} - v_{\ell-1,I-1,J}) \\
&- (D_{\ell,i,j+3/2} + D_{\ell,i+1,j+3/2})(v_{\ell-1,I,J+1} - v_{\ell-1,I,J}) \\
&+ (D_{\ell,i,j-1/2} + D_{\ell,i+1,j-1/2})(v_{\ell-1,I,J} - v_{\ell-1,I,J-1}).
\end{aligned}$$

Finally, defining the discrete coefficients at the coarse-grid level via (8.8.13), we find

$$\begin{aligned} a_{I,J} &= 2[-D_{\ell-1,I+1/2,J}(v_{\ell-1,I+1,J} - v_{\ell-1,I,J}) \\ &\quad + D_{\ell-1,I-1/2,J}(v_{\ell-1,I,J} - v_{\ell-1,I-1,J}) \\ &\quad - D_{\ell-1,I,J+1/2}(v_{\ell-1,I,J+1} - v_{\ell-1,I,J}) \\ &\quad + D_{\ell-1,I,J-1/2}(v_{\ell-1,I,J} - v_{\ell-1,I,J-1})] \\ &= -2h_{\ell-1}^2(\Delta_{\ell-1}^{D,\circ} v_{\ell-1})_{I,J}. \end{aligned}$$

This yields (8.8.14). □

We have just proven that Assumption (G3) holds. To establish convergence of various multigrid methods, we need to establish smoothing and approximation properties. However, the proof of a result analogous to Theorem 8.6.8 is nontrivial and beyond the scope of this text.

### 8.8.3 Smoothers in two dimensions

To conclude this section, let us describe two simple smoothing strategies that work well in the cell-centered finite difference setting, the damped quasi-Jacobi smoother, and the damped quasi-Gauss–Seidel smoother. To keep the discussion general, we will consider the variable-diffusivity case.

#### Damped quasi-Jacobi smoother

Let us first describe a damped quasi-Jacobi method. Let $u_\ell^\sigma \in \mathcal{C}_\ell^\circ$. It is an easy task to find a unique function $u_\ell^\sigma \in \mathcal{C}_{\ell,0}$ — for which we use the same symbol, in an obvious abuse of notation — that agrees with the first at every interior cell-centered grid point but now has discrete homogeneous Dirichlet boundary conditions. Define the grid function $z_\ell \in \mathcal{C}_\ell^\circ$ as follows: for $1 \le i,j \le m_\ell$,

$$\begin{aligned} f_{\ell,i,j} &= \frac{-D_{\ell,i+1/2,j}(u_{\ell,i+1,j}^\sigma - z_{\ell,i,j}) + D_{\ell,i-1/2,j}(z_{\ell,i,j} - u_{\ell,i-1,j}^\sigma)}{h_\ell^2} \\ &\quad + \frac{-D_{\ell,i,j+1/2}(u_{\ell,i,j+1}^\sigma - z_{\ell,i,j}) + D_{\ell,i,j-1/2}(z_{\ell,i,j} - u_{\ell,i,j-1}^\sigma)}{h_\ell^2}. \end{aligned}$$

We can explicitly solve for $z_{\ell,i,j}$ at each index $i$ and $j$, obtaining

$$\begin{aligned} z_{\ell,i,j} &= \frac{h_\ell^2 f_{\ell,i,j} + D_{\ell,i+1/2,j}u_{\ell,i+1,j}^\sigma + D_{\ell,i-1/2,j}u_{\ell,i-1,j}^\sigma}{D_{\ell,i+1/2,j} + D_{\ell,i-1/2,j} + D_{\ell,i,j+1/2} + D_{\ell,i,j-1/2}} \\ &\quad + \frac{D_{\ell,i,j+1/2}u_{\ell,i,j+1}^\sigma + D_{\ell,i,j-1/2}u_{\ell,i,j-1}^\sigma}{D_{\ell,i+1/2,j} + D_{\ell,i-1/2,j} + D_{\ell,i,j+1/2} + D_{\ell,i,j-1/2}}. \end{aligned}$$

Observe that it does not matter in what order we compute the entries of the grid function $z_\ell$. Once all the entries of $z_\ell$ are found, we compute the update via

$$u_\ell^{\sigma+1} := \omega z_\ell + (1-\omega)u_\ell^\sigma \in \mathcal{C}_\ell^\circ,$$

where $\omega \in (0,1]$ is a damping parameter. This smoother is implemented in Listing 8.22.

This present method is called *quasi-Jacobi*, because unlike the true Jacobi method, we do not modify the stencil near the boundary to take into account the discrete homogeneous Dirichlet boundary conditions. For the one hand, in the formula for $z_{\ell,i,j}$ above, we always divide by

$$D_{\ell,i+1/2,j} + D_{\ell,i-1/2,j} + D_{\ell,i,j+1/2} + D_{\ell,i,j-1/2},$$

regardless of the indices. In the true Jacobi method the denominator would be modified at the boundary and corner cells. More precisely, the true damped Jacobi method is expressed (in matrix–vector formulation) by

$$\boldsymbol{u}_\ell^{\sigma+1} = \boldsymbol{u}_\ell^\sigma + \omega \mathsf{D}_\ell^{-1}(\boldsymbol{f}_\ell - \mathsf{A}_\ell \boldsymbol{u}_\ell^\sigma), \tag{8.8.15}$$

where $\mathsf{A}_\ell$ is the stiffness matrix (which we have not exhibited for the nonconstant diffusivity case), and $\mathsf{D}_\ell$ is its diagonal part. For the grid points satisfying $2 \le i,j \le m_\ell - 1$, (8.8.15) will result in the same update formula for $u_{\ell,i,j}^{\sigma+1}$ as from the quasi-Jacobi method. However, the two update strategies will lead to two slightly different formulae at the boundary and corner points. We have already seen this in a much simpler setting in Section 8.2.3.

Using $\omega = 1$, which corresponds to no damping, results in a method that does not adequately smooth high-frequency errors, and the multigrid methods will fail to converge. See Section 8.2 for a related discussion. Choosing a value in the range $\omega \in [\frac{1}{2}, \frac{2}{3}]$ usually provides good multigrid convergence, as we will see.

### Forward damped quasi-Gauss–Seidel smoother

Let $u_\ell^\sigma \in \mathcal{C}_\ell^\circ$. Denote by the same symbol $u_\ell^\sigma \in \mathcal{C}_{\ell,0}$ the unique homogeneous Dirichlet cell-centered grid function that agrees with the first at every interior cell-centered grid point. Suppose that the indices $i,j \in \{1,\dots,m_\ell\}$ are traversed in *forward lexicographic order*. Then, for $1 \le i,j \le m_\ell$, define $a_{F,i,j}$ via

$$\begin{aligned} f_{\ell,i,j} &= \frac{-D_{\ell,i+1/2,j}(u_{\ell,i+1,j}^\sigma - a_{F,i,j}) + D_{\ell,i-1/2,j}(a_{F,i,j} - u_{\ell,i-1,j}^{\sigma+1})}{h_\ell^2} \\ &\quad + \frac{-D_{\ell,i,j+1/2}(u_{\ell,i,j+1}^\sigma - a_{F,i,j}) + D_{\ell,i,j-1/2}(a_{F,i,j} - u_{\ell,i,j-1}^{\sigma+1})}{h_\ell^2}. \end{aligned}$$

Solving for $a_{F,i,j}$, we have

$$\begin{aligned} a_{F,i,j} &= \frac{h_\ell^2 f_{\ell,i,j} + D_{\ell,i+1/2,j} u_{\ell,i+1,j}^{\sigma} + D_{\ell,i-1/2,j} u_{\ell,i-1,j}^{\sigma+1}}{D_{\ell,i+1/2,j} + D_{\ell,i-1/2,j} + D_{\ell,i,j+1/2} + D_{\ell,i,j-1/2}} \\ &\quad + \frac{D_{\ell,i,j+1/2} u_{\ell,i,j+1}^{\sigma} + D_{\ell,i,j-1/2} u_{\ell,i,j-1}^{\sigma+1}}{D_{\ell,i+1/2,j} + D_{\ell,i-1/2,j} + D_{\ell,i,j+1/2} + D_{\ell,i,j-1/2}}. \end{aligned}$$

Before moving to the next index in the forward lexicographic order, we update via the formula

$$u_{\ell,i,j}^{\sigma+1} := \omega a_{F,i,j} + (1-\omega) u_{\ell,i,j}^{\sigma},$$

where $\omega \in (0,1]$ is a fixed damping parameter. This smoother is implemented in Listing 8.23.

Notice the difference between the formula for $a_{F,i,j}$ and that for $z_{\ell,i,j}$ in the damped quasi-Jacobi method. Observe that in the forward Gauss–Seidel framework, as soon an updated value of $u_{\ell,i,j}^{\sigma+1}$ is obtained at $(i,j)$, it is used in the calculations of $u_{\ell,i',j'}^{\sigma+1}$, where

$$F(i,j) < F(i',j').$$

See Figure 8.15.

The name *quasi-Gauss–Seidel* is used because we do not modify the stencil at boundary and corner cells, as would be done by the actual Gauss–Seidel method.

### Backward damped quasi-Gauss–Seidel smoother

The idea for the backward smoother is analogous to the forward version, but we traverse the indices in the opposite order. Let $u_\ell^{\sigma} \in \mathcal{C}_\ell^{\circ}$. Denote by the same symbol $u_\ell^{\sigma} \in \mathcal{C}_{\ell,0}$ the unique homogeneous Dirichlet cell-centered grid function that agrees with the first at every interior cell-centered grid point. Suppose that the indices $i,j \in \{1,\dots,m_\ell\}$ are traversed in *backward lexicographic order*. Then, for $1 \le i,j \le m_\ell$, define $a_{B,i,j}$ via

$$\begin{aligned} f_{\ell,i,j} &= \frac{-D_{\ell,i+1/2,j}(u_{\ell,i+1,j}^{\sigma+1} - a_{B,i,j}) + D_{\ell,i-1/2,j}(a_{B,i,j} - u_{\ell,i-1,j}^{\sigma})}{h_\ell^2} \\ &\quad + \frac{-D_{\ell,i,j+1/2}(u_{\ell,i,j+1}^{\sigma+1} - a_{B,i,j}) + D_{\ell,i,j-1/2}(a_{B,i,j} - u_{\ell,i,j-1}^{\sigma})}{h_\ell^2}. \end{aligned}$$

Solving for $a_{B,i,j}$, we have

$$\begin{aligned} a_{B,i,j} &= \frac{h_\ell^2 f_{\ell,i,j} + D_{\ell,i+1/2,j} u_{\ell,i+1,j}^{\sigma+1} + D_{\ell,i-1/2,j} u_{\ell,i-1,j}^{\sigma}}{D_{\ell,i+1/2,j} + D_{\ell,i-1/2,j} + D_{\ell,i,j+1/2} + D_{\ell,i,j-1/2}} \\ &\quad + \frac{D_{\ell,i,j+1/2} u_{\ell,i,j+1}^{\sigma+1} + D_{\ell,i,j-1/2} u_{\ell,i,j-1}^{\sigma}}{D_{\ell,i+1/2,j} + D_{\ell,i-1/2,j} + D_{\ell,i,j+1/2} + D_{\ell,i,j-1/2}}. \end{aligned}$$

Before moving to the next index in the backward lexicographic order, we update via the formula

$$u_{\ell,i,j}^{\sigma+1} := \omega a_{B,i,j} + (1-\omega)u_{\ell,i,j}^{\sigma},$$

where $\omega \in (0,1]$ is a fixed damping parameter. This smoother is implemented in Listing 8.23, the same code that implements the forward quasi-Gauss–Seidel algorithm.

In the backward Gauss–Seidel framework, as soon as an updated value of $u_{\ell,i,j}^{\sigma+1}$ is obtained at $(i,j)$, it is used in the calculations of all future values, that is, $u_{\ell,i',j'}^{\sigma+1}$, where

$$B(i,j) < B(i',j').$$

See Figure 8.15.

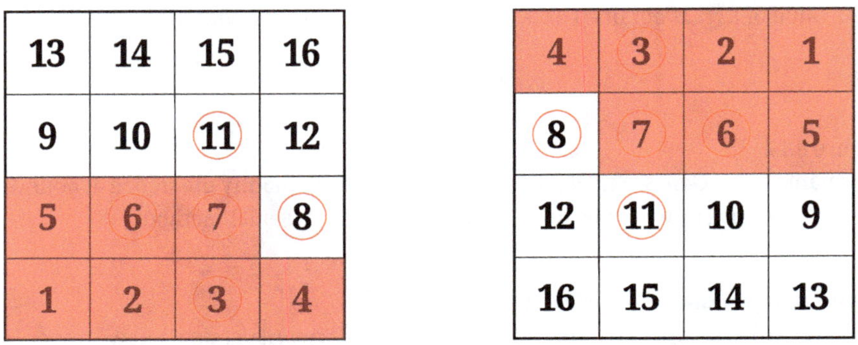

(a) Forward lexicographic (b) Backward lexicographic

**Figure 8.15:** Illustration of (a) forward Gauss–Seidel and (b) backward Gauss–Seidel updates of the grid function up to the current cell, cell 7 of 16, in the smoothing process on a $4 \times 4$ grid. The red-shaded cells are those already updated. The unfilled red circles indicate the cells involved in the update of cell 7.

## 8.9 Some computational experiments

To conclude this chapter, we now present some computational experiments that illustrate the predicted convergence results of our various multigrid algorithms for cell-centered finite difference discretizations of linear elliptic problems in one and two dimensions.

We point out that the code listed below can handle problems that are much more general than those we will experiment with. In particular, the code can handle non-constant diffusion coefficients, more general boundary conditions, etc. The interested reader is challenged to do some independent experiments.

We also point out that for simplicity, we deviate from the standard multigrid algorithms in one major way. Specifically, at the coarsest level, rather than performing an exact solution, we simply conduct smoothing. Therefore the codes may not perform optimally, but many lines of code are spared with this simplified approach. Moreover, the error contraction factors are still quite good. The reader should modify the codes to implement the true multigrid algorithms and then compare the performance.

### 8.9.1 A one-dimensional problem

To test the multigrid algorithm in the one-dimensional setting, we will consider the model problem (1.7.1) and its cell-centered finite difference discretization. As we have discussed, for a given maximum level $L$, we end up having to solve the linear system (8.1.6), that is,

$$\mathsf{A}_L \boldsymbol{u}_L = \boldsymbol{f}_L.$$

Here $\mathsf{A}_L$ is the stiffness matrix defined in (8.1.7). In our experiments, we specify the exact solution to be

$$[\boldsymbol{u}_L^{\mathrm{E}}]_i = u_{L,i}^{\mathrm{E}} = \exp(\sin(3.0\pi x_{L,i})) - 1.$$

Note that this solution satisfies the homogeneous Dirichlet boundary conditions. The right-hand side vector is manufactured by setting $\boldsymbol{f}_L := \mathsf{A}_L \boldsymbol{u}_L^{\mathrm{E}}$.

The initial approximation $\boldsymbol{u}_L^{(0)}$ is chosen via pseudorandom number generation. The MATLAB© codes implementing the multigrid algorithms are given in Listings 8.2–8.14, where further details can be found.

In Figure 8.16 the error $\|\boldsymbol{u}_L^{\mathrm{E}} - \boldsymbol{u}_L^k\|_L$ is plotted as a function of $k$ for the multigrid V-cycle algorithm ($p = 1$) using the damped quasi-Jacobi smoother with parameters $\omega = \frac{1}{2}$, $m_1 = m_2 = 3$, $n_L = 1024$, and $L = 10$. Note that we use a slight variant of the Euclidean norm on level $L$ for computing the errors. Specifically, we use a scaled Euclidean norm described in Listing 8.14, because it is more standard in the cell-centered finite difference setting. This use does not affect the rate of convergence. In this section only, we will abuse notation and use the same symbol $\|\cdot\|_L$ for the scaled norm.

The error contraction factor $\gamma_{\mathrm{comp}}$, shown on the plot in Figure 8.16, is computed using a log-linear fit of the last four error values:

$$\|\boldsymbol{u}_L^{\mathrm{E}} - \boldsymbol{u}_L^k\|_L \approx \gamma_{\mathrm{comp}} \|\boldsymbol{u}_L^{\mathrm{E}} - \boldsymbol{u}_L^{k-1}\|_L.$$

The stopping tolerance

$$\|\boldsymbol{u}_L^{\mathrm{E}} - \boldsymbol{u}_L^k\|_L \le 10^{-11}$$

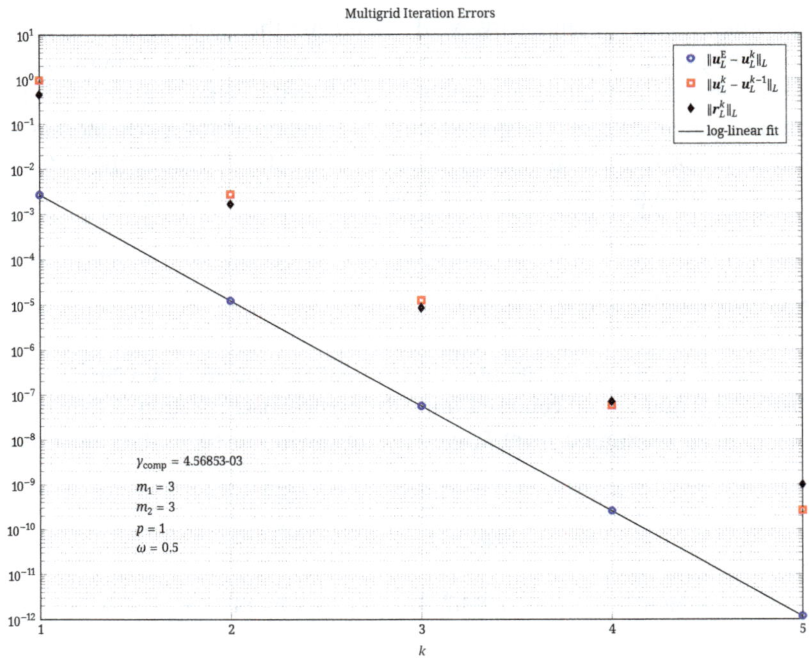


**Figure 8.16:** The error reduction for the multigrid V-cycle algorithm, using the parameters $\omega = \frac{1}{2}$, $m_1 = m_2 = 3$, and $L = 10$. Notice that in this case we have $n_L = 1024$. The factor $\gamma_{\text{comp}}$ is computed using a log-linear fit of the last four error values. Note that $\|\boldsymbol{u}_L^k - \boldsymbol{u}_L^{k-1}\|_L$ is a good indicator of the error.

is used. Although there is no theory for this method (see Remark 6.9.16), we observe a proper linear rate of convergence.

In Table 8.1, we report the results of several computational experiments for the V-cycle algorithm. Nevertheless, it is interesting to note that the convergence rate seems independent of the number of levels. Indeed, the factor $\gamma_{\text{comp}}$ does not change substantially as we vary $L$ between five (5) and eleven (11).

Finally, a parameter study was conducted to determine the relationship between the damping parameter $\omega$ and the multigrid error contraction factor $\gamma_{\text{comp}}$. For this test, the number of pre- and postsmoothing steps was fixed at $m_1 = m_2 = 3$ with $L = 9$ levels for a finest grid size of $n_L = 512$. The parameter $\omega$ was then varied over the interval $[0.3, 0.9]$ for the damped quasi-Jacobi and damped quasi-Gauss–Seidel smoothers. Figure 8.17 illustrates the resulting contraction factor for each value of $\omega$.

**Table 8.1:** Computed V-cycle multigrid error contraction factors $\gamma_{\text{comp}}$ for various parameter choices. The damped quasi-Jacobi smoother is used with a stopping tolerance of $10^{-11}$. The factor $\gamma_{\text{comp}}$ is computed using a log-linear fit of the last four error values $\|u_L^E - u_L^k\|_L$.

| $n_L$ | $L$ | $\omega$ | $m_1$ | $m_2$ | $p$ | $\gamma_{\text{comp}}$ |
|---|---|---|---|---|---|---|
| 64 | 6 | 1/2 | 3 | 3 | 1 | $4.71 \times 10^{-03}$ |
| 128 | 7 | 1/2 | 3 | 3 | 1 | $4.57 \times 10^{-03}$ |
| 256 | 8 | 1/2 | 3 | 3 | 1 | $4.57 \times 10^{-03}$ |
| 512 | 9 | 1/2 | 3 | 3 | 1 | $4.57 \times 10^{-03}$ |
| 1024 | 10 | 1/2 | 3 | 3 | 1 | $4.57 \times 10^{-03}$ |
| 2048 | 11 | 1/2 | 3 | 3 | 1 | $4.57 \times 10^{-03}$ |
| 2048 | 11 | 1/2 | 2 | 2 | 1 | $3.03 \times 10^{-02}$ |
| 2048 | 11 | 1/2 | 3 | 3 | 1 | $4.57 \times 10^{-03}$ |
| 2048 | 11 | 1/2 | 4 | 4 | 1 | $6.93 \times 10^{-04}$ |
| 512 | 9 | 0.40 | 3 | 3 | 1 | $3.35 \times 10^{-02}$ |
| 512 | 9 | 0.45 | 3 | 3 | 1 | $1.05 \times 10^{-02}$ |
| 512 | 9 | 0.50 | 3 | 3 | 1 | $4.57 \times 10^{-03}$ |
| 512 | 9 | 0.55 | 3 | 3 | 1 | $2.46 \times 10^{-03}$ |
| 512 | 9 | 0.60 | 3 | 3 | 1 | $5.97 \times 10^{-03}$ |
| 512 | 9 | 0.65 | 3 | 3 | 1 | $1.64 \times 10^{-02}$ |
| 512 | 9 | 0.70 | 3 | 3 | 1 | $3.07 \times 10^{-02}$ |
| 512 | 9 | 0.75 | 3 | 3 | 1 | $6.52 \times 10^{-02}$ |
| 512 | 9 | 0.80 | 3 | 3 | 1 | $1.29 \times 10^{-01}$ |

### 8.9.2 A two-dimensional elliptic problem with variable diffusivity

To extend our investigation to higher dimensions and assess the performance of multigrid algorithms in the case of a problem with variable coefficients, we consider the following two-dimensional elliptic boundary value problem:

$$\begin{cases} -\nabla \cdot (D\nabla u) + Cu = f & \text{in } \Omega, \\ \dfrac{\partial u}{\partial n} = 0 & \text{on } \partial\Omega. \end{cases} \tag{8.9.1}$$

Here $\frac{\partial u}{\partial n}$ denotes the normal derivative of $u$ on the boundary $\partial\Omega$, that is,

$$\frac{\partial u}{\partial n} = \boldsymbol{n} \cdot \nabla u,$$

where $\boldsymbol{n}$ denotes outward-pointing unit normal vector on $\partial\Omega$. The variable diffusion coefficient $D$ is chosen as

$$D(x,y) = 1 + \frac{1}{2}\sin(2\pi x)\cos(2\pi y),$$

and the reaction coefficient is set to $C = 2$.

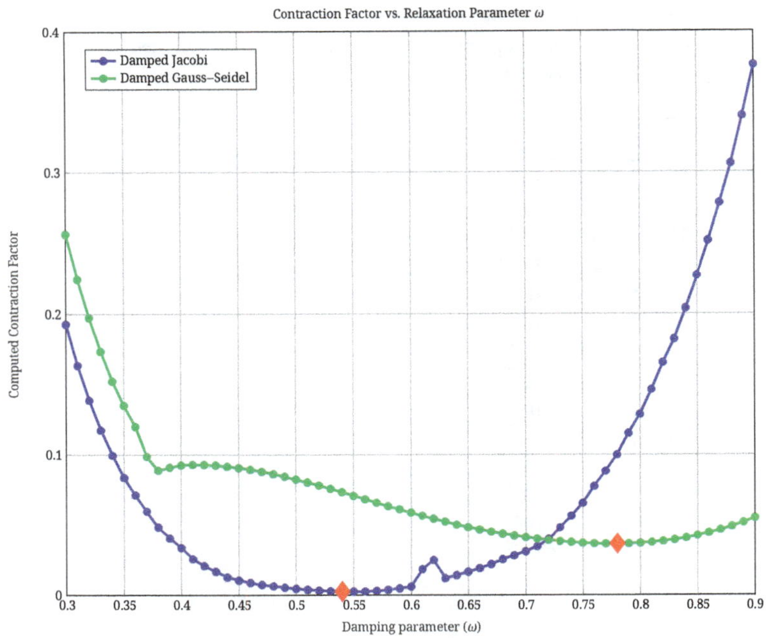


**Figure 8.17:** Convergence performance of the V-cycle multigrid solver as a function of the damping parameter $\omega$, shown for both damped quasi-Jacobi and damped quasi-Gauss–Seidel smoothers. All tests were performed with $m_1 = m_2 = 3$ smoothing steps, $L = 9$ grid levels, and a finest grid size of $n_L = 512$. The factor $\gamma_{\text{comp}}$ is computed using a log-linear fit of the last four error values. The optimal values are denoted by the red diamonds.

We discretize the problem using the cell-centered finite method defined earlier in the chapter. For a given total number of levels $L$, we solve the resulting linear system

$$\mathsf{A}_L \boldsymbol{u}_L = \boldsymbol{f}_L,$$

where $\mathsf{A}_L$ is the stiffness matrix.

In our computational experiments, we assume that the computational domain is $\Omega = (0,1)^2$ and define the exact solution to be

$$[\boldsymbol{u}_L^{\mathrm{E}}]_{i,j} = u_{L,i,j}^{\mathrm{E}} = \exp(\cos(3.0\pi x_{L,i})\cos(3.0\pi y_{L,j})),$$

where $x_{L,i} = (i - \frac{1}{2})h_L$ and $y_{L,j} = (j - \frac{1}{2})h_L$ are the cell-centered coordinates. This solution satisfies discrete homogeneous Neumann boundary conditions. The right-hand side is defined as $\boldsymbol{f}_L = \mathsf{A}_L \boldsymbol{u}_L^{\mathrm{E}}$.

The initial approximation $\boldsymbol{u}_L^0$ is generated using pseudorandom numbers. The MATLAB© codes implementing the two-dimensional multigrid algorithm are provided in the listings, and all the details of the implementation can be found there. See Listings 8.15–8.26.

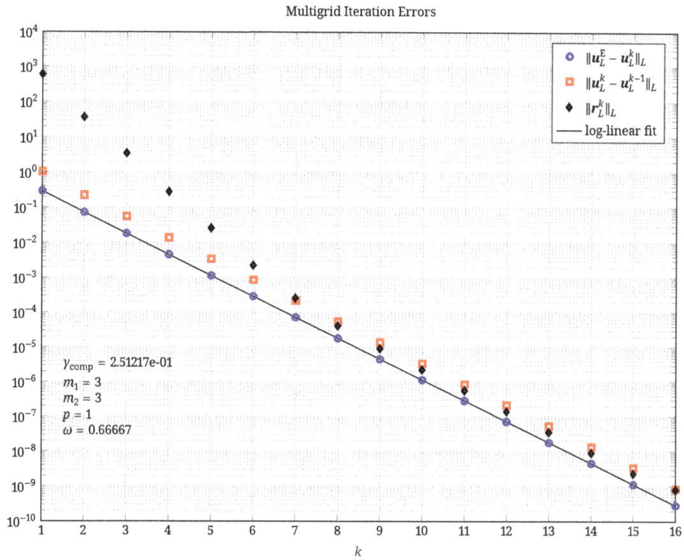


**Figure 8.18:** The error reduction for the 2D multigrid V-cycle algorithm with variable diffusion coefficient using the parameters $\omega = \frac{2}{3}$, $m_1 = m_2 = 3$, and $L = 10$. The finest grid has $n_L = 1024 \times 1024$ interior cells. The factor $\gamma_{comp}$ is computed using a log-linear fit of the last four exact error values.

Figure 8.18 displays the error $\|\boldsymbol{u}_L^E - \boldsymbol{u}_L^k\|_L$ as a function of $k$ for the multigrid V-cycle algorithm ($p = 1$) applied to the two-dimensional variable coefficient problem. The computation uses the parameters $\omega = \frac{2}{3}$ and $m_1 = m_2 = 3$. The finest grid is of size

$$m_L \times m_L = 1024 \times 1024,$$

and the number of multigrid levels is $L = 10$. The number of unknowns in the calculation is

$$n_L = m_L^2 = 1,048,576.$$

Despite the presence of variable coefficients, we observe robust linear convergence with a computed error contraction factor of $\gamma_{comp} = 2.51 \times 10^{-1}$.

The results demonstrate that the multigrid method maintains its effectiveness when applied to problems with spatially varying diffusion coefficients. The observed convergence rate is comparable to the constant coefficient case, indicating that the restriction and prolongation operators, along with the smoothing scheme, handle the coefficient variation adequately. This confirms the robustness of the cell-centered multigrid approach for realistic applications where material properties vary throughout the computational domain.

## 8.10 Problems

8.1 Prove Proposition 8.1.6.

8.2 Prove Proposition 8.1.10.

8.3 Prove Theorem 8.1.11.

8.4 Prove Theorem 8.3.1.

8.5 Prove Theorem 8.3.3.

8.6 Prove Proposition 8.4.3.

8.7 Prove Proposition 8.4.4.

8.8 Finish the proof and calculations of Proposition 8.4.5.

8.9 Prove Corollary 8.4.6.

8.10 Prove Proposition 8.5.2.

8.11 Prove Lemma 8.5.7.

8.12 Prove Proposition 8.5.6.

8.13 Prove Proposition 8.5.8.

8.14 Suppose that $f \in C([0,1])$, $\ell \in \mathbb{N}$, and $\boldsymbol{f}_\ell \in \mathbb{R}^{n_\ell}$ is the vector with components

$$f_{\ell,i} = f(x_{\ell,i}), \quad i = 1, \dots, n_\ell,$$

where $C_\ell = \{x_{\ell,i}\}_{i=1}^{n_\ell}$ is the usual set of cell-centered points for level $\ell$. Find/construct a function $\tilde{f} \in L^2(0,1)$ such that

$$Q_\ell \tilde{f} \in V_\ell \overset{\mathcal{B}_\ell}{\longleftrightarrow} \boldsymbol{f}_\ell \in \mathbb{R}^{n_\ell}.$$

Find/construct a function $\hat{f} \in C([0,1])$ such that

$$Q_\ell \hat{f} \in V_\ell \overset{\mathcal{B}_\ell}{\longleftrightarrow} \boldsymbol{f}_\ell \in \mathbb{R}^{n_\ell}.$$

8.15 Prove Theorem 8.6.8.

8.16 Show that (8.7.3) is indeed a formally second-order approximation of equation (8.7.1).

8.17 Using the forward lexicographic ordering, find $\mathsf{R}_{\ell-1} \in \mathbb{R}^{n_{\ell-1} \times n_\ell}$ and $\mathsf{P}_{\ell-1} \in \mathbb{R}^{n_\ell \times n_{\ell-1}}$, the respective matrix representations of the two-dimensional cell-centered restriction and prolongation operators.

8.18 Complete the proof of Proposition 8.8.4.

8.19 Let $\Omega = (0,1)$ and consider the following boundary value problem:

$$\begin{cases} -u'' + 5u = f & \text{in } \Omega, \\ u' = 0 & \text{on } \partial\Omega. \end{cases} \tag{8.10.1}$$

Modify the codes in Listings 8.2–8.14, where appropriate, to solve problem (8.10.1). Use the following exact solution for error comparison:

$$[\boldsymbol{u}_L^{\mathrm{E}}]_i = u_{L,i}^{\mathrm{E}} = \cos^3(2.0\pi x_{L,i}).$$

# Listings

## One-dimensional codes

```
./1DCC/EllipticNCDiff/
    elliptic1DMGCCDiff.m
    multiGridSolver.m
    MGOperator.m
    prolongation.m
    restriction.m
    restrictDiff.m
    FDOperator.m
    smoothQJacDamped.m
    smoothRichardson.m
    smoothQGSDamped.m
    getResidual.m
    applyBCs.m
    normScaledL2.m
```

**Listing 8.2:** `elliptic1DMGCCDiff.m`: Calling script for the multigrid solver for a 1D cell-centered elliptic equation with nonconstant diffusion coefficient.

```
%
%   elliptic1DMGCCDiff.m
%
% Calling script for solving the 2D elliptic equation
%
%   -(D u')' + rC u = f
%
% on an interval, where D is a possibly non-constant,
% everywhere positive diffusion coefficient, and rC is a
% non-negative constant. The boundary conditions may be
% taken to be (i) homogeneous Dirichlet,
% (ii) homogeneous Neumann, or (iii) periodic.
% See the function applyBCs.
%
% Discretization is accomplished using the cell-centered
% finite difference method. The cell-centered multigrid
% method is used as the solver.
%
% One can set rC = 0.0, provided that homogeneous
% Dirichlet boundary conditions are chosen.
```

```
%
%% Initialization and Parameter

% Clear command line window and workspace.
clear; clc;

% Tolerance of residual to stop multigrid iterations.
tol = 1.0e-11;

% Number of interior cells at finest level.
nL = 1024;

% Number of multigrid levels.
L = 10;

% Define the domain of the equation.
xLower = 0.0;
xUpper = 1.0;

hL = (xUpper-xLower)/nL;

% Parameters for the multigrid solver.
pCycle = 1;
m1     = 3;
m2     = 3;
omega  = 1/2;
kMax   = 100;
rC     = 0.0;

%% Setting up the multigrid solver

% Check that nL is divisible by 2^L.
nl = nL;
for k = L-1:-1:0
  if any(mod(nl,2) ~= 0)
    error('nL not divisible by 2^L.  Coarsening error.');
  end
  nl = nl/2;
end

MGParam.nL     = nL;
MGParam.xLower = xLower;
```

```
MGParam.xUpper = xUpper;
MGParam.L      = L;
MGParam.pCycle = pCycle;
MGParam.m1     = m1;
MGParam.m2     = m2;
MGParam.omega  = omega;
MGParam.kMax   = kMax;
MGParam.tol    = tol;
MGParam.rC     = rC;

% Allocate solution arrays (with ghost layers).
u      = zeros(nL+2,1);
uExact = zeros(nL+2,1);
f      = zeros(nL  ,1);
xCC    = zeros(nL  ,1);
DeW    = zeros(nL+1,1);

% Build an exact solution and define f accordingly, so we
% can test the true algebraic error.
for i = 1:nL
  x = (i-0.5)*hL+xLower;
  xArg = 3.0*pi*x/xUpper;
% uExact(i+1) = exp(cos(xArg));
  uExact(i+1) = exp(sin(xArg))-1.0;
  xCC(i) = x;
end

% We next build the finest-grid diffusion array DeW:
for i = 1:nL+1
  xEdge = (i-1.0)*hL+xLower;
  xArg = 2.0*pi*x/xUpper;

% Evaluate D at the east-west edge point:
% DeW(i) = 1.0+0.5*sin(xArg);
  DeW(i) = 1.0;
end

% Apply the BCs to uExact, then compute f by calling
% FDOperator.
uExact = applyBCs(uExact);
f = FDOperator(uExact,DeW,rC,hL);

```

```
%% Call the Multigrid solver.

tic;
[u,errVals,kStop] = multiGridSolver(u,f,DeW,hL,...
                                        MGParam,uExact);
toc;

%% Plotting the numerical solution and error plots.

uPlot(1:nL) = u(2:nL+1);

figure(1)
clf

plot(xCC,uPlot,'b-','LineWidth',1.5);
title('Numerical Solution u');
xlabel('x');
ylabel('u');

figure(2)
clf
semilogy(errVals(1:kStop,3),'bo','LineWidth',1.5)
hold on
semilogy(errVals(1:kStop,2),'rs','LineWidth',1.5)
hold on
semilogy(errVals(1:kStop,1),'kd','LineWidth',1.5)
hold on

rate = 1.0;
if kStop >= 4
  kv = kStop-3:kStop;
  le = log(errVals(kStop-3:kStop,3));
  p1 = polyfit(kv,le,1);
  rate = exp(p1(1));
  p1k = polyval(p1,1:kStop);
  semilogy(1:kStop,exp(p1k),'-k')
else
  semilogy(errVals(1:kStop,1),'.','LineWidth',1.5)
end

xlabel('$k$','Interpreter','latex');
title('Multigrid Iteration Errors',...
```

```
  'Interpreter','latex');
legend(...
  '$\left\|{\bf u}_L^{\rm E}-{\bf u}_L^k\right\|_L$',...
  '$\left\|{\bf u}_L^k-{\bf u}_L^{k-1} \right\|_L$',...
  '$\left\|{\bf r}_L^k\right\|_L$',...
  'log-linear fit','Interpreter','latex');
text(1.5,128*MGParam.tol,strcat(...
  '$\gamma_{\rm comp} =\hspace{.1cm}$',...
  num2str(rate,'%10.5e')),'FontSize',14,...
  'Interpreter','latex')
text(1.5,32*MGParam.tol,strcat('$m_1 =\hspace{.1cm}$',...
  num2str(MGParam.m1)),'FontSize',14,...
  'Interpreter','latex')
text(1.5,8*MGParam.tol,strcat('$m_2 =\hspace{.1cm}$',...
  num2str(MGParam.m2)),'FontSize',14,...
  'Interpreter','latex')
text(1.5,2*MGParam.tol,strcat('$p =\hspace{.1cm}$',...
  num2str(MGParam.pCycle)),'FontSize',14,...
  'Interpreter','latex')
text(1.5,0.5*MGParam.tol,...
  strcat('$\omega =\hspace{.1cm}$',...
  num2str(MGParam.omega)),'FontSize',14,...
  'Interpreter','latex')
printstr = strcat('Err_nL_',num2str(MGParam.nL(1)),...
  '_m1_',num2str(MGParam.m1),'.pdf');
exportgraphics(gca, printstr)
hold off
```

**Listing 8.3:** `multiGridSolver.m`: Multigrid solver for a 1D cell-centered elliptic equation with nonconstant diffusion coefficient.

```
function [u,errVals,kStop] =...
              multiGridSolver(u0,f,DeW,hf,MGParam,uExact)

nf = length(f);
k = 0;
rC = MGParam.rC;
errorVal = MGParam.tol+1.0;
errVals = zeros(MGParam.kMax,3);

uo = u0;

```

```
while ((k < MGParam.kMax) && (errorVal > MGParam.tol))

  % Apply one multigrid cycle.
  u = MGOperator(MGParam.L,f,DeW,uo,hf,MGParam);

  % Evaluate various norms.
  residErr = normScaledL2(getResidual(u,f,DeW,rC,hf));
  correErr = normScaledL2(uo(2:nf+1)-u(2:nf+1));
  exactErr = normScaledL2(uExact(2:nf+1)-u(2:nf+1));

  k = k+1;
  errVals(k,1) = residErr;
  errVals(k,2) = correErr;
  errVals(k,3) = exactErr;

  fprintf('Multigrid iteration %d:\n', k);
  fprintf('  residual   = %.6e\n', residErr);
  fprintf('  correction = %.6e\n', correErr);
  fprintf('  true error = %.6e\n', exactErr);

  % Stopping criterion (artificial, for demonstration).
  errorVal = exactErr;
  uo = u;
end

kStop = k;

end
```

**Listing 8.4:** `MGOperator.m`: Multigrid operator for the multigrid solver for a 1D cell-centered elliptic equation with nonconstant diffusion coefficient.

```
function u = MGOperator(level,f,DeW,u,hf,MGParam)

rC = MGParam.rC;

% Pre-smoothing
u = smoothQJacDamped(f,DeW,rC,u,hf,MGParam.m1,...
                     MGParam.omega);

% Alternative pre-smoothing options:
% u = smoothRichardson(f,DeW,rC,u,hf,MGParam.m1,...
```

```
%                      MGParam.omega);
% u = smoothQGSDamped(f,DeW,rC,u,hf,MGParam.m1,...
%                      MGParam.omega,'fwd');

if level > 0
  hc = 2*hf;
  nfPlusGhost = length(u);
  nf = nfPlusGhost-2;

  % Coarse grid interior dimension is nc.
  nc = nf/2;

  % Compute the residual and restric to the coarse grid.
  cGr = restriction(getResidual(u,f,DeW,rC,hf));

  % Restrict the diffusion coefficients to the coarse
  % grid.
  cDeW = restrictDiff(DeW);

  % Approximate on the coarse grid using recursive MG.
  cGc = zeros(nc+2,1);
  for s = 1:MGParam.pCycle
    cGc = MGOperator(level-1,cGr,cDeW,cGc,hc,MGParam);
  end

  % Prolongate the coarse grid correction and update the
  % fine grid approximation.
  u(2:nf+1) = u(2:nf+1)+prolongation(cGc(2:nc+1));

  % Post-smoothing.
  u = smoothQJacDamped(f,DeW,rC,u,hf,MGParam.m2,...
                     MGParam.omega);

  % Alternative post-smoothing options:
  % u = smoothRichardson(f,DeW,rC,u,hf,MGParam.m2,...
  %                      MGParam.omega);
  % u = smoothQGSDamped(f,DeW,rC,u,hf,MGParam.m2,...
  %                      MGParam.omega,'bwd');
end

end
```

**Listing 8.5:** `prolongation.m`: Prolongation operator for the 1D multigrid solver.

```
function presult = prolongation(u)

nc = length(u);
nf = nc*2;
presult = zeros(nf,1);

presult(2:2:nf  ) = u(1:nc(1));
presult(1:2:nf-1) = u(1:nc(1));

end
```

**Listing 8.6:** `restriction.m`: Restriction operator for the 1D multigrid solver.

```
function restrictionResult = restriction(u)

nf = length(u);
nc = nf/2;

restrictionResult = 0.5*(u(1:2:nf-1)+u(2:2:nf));

end
```

**Listing 8.7:** `restrictDiff.m`: Function to restrict the nonconstant 1D diffusion coefficient from a finer to a coarser level.

```
function cDeW = restrictDiff(DeW)

nfEw = length(DeW);
nf = nfEw-1;
nc = nf/2;
cDeW = zeros(nc+1,1);

cDeW(1:nc+1) = DeW(1:2:nf+1);

end
```

**Listing 8.8:** `FDOperator.m`: Finite difference operator for the multigrid solver for a 1D cell-centered elliptic equation with nonconstant diffusion coefficient.

```
function operatorResult = FDOperator(u,DeW,rC,hf)

nfPlusGhostLayers = length(u);
nf = nfPlusGhostLayers-2;
feW = zeros(nf+1,1);
operatorResult = zeros(nf,1);

hf2 = hf*hf;

% Vectorized code for the operator:
% - (D u')' + rC u

% Compute east-west numerical flux:
feW(1:nf+1) = DeW(1:nf+1).*(u(2:nf+2)-u(1:nf+1));

operatorResult(1:nf) = -(feW(2:nf+1)-feW(1:nf))/hf2...
                       +rC*u(2:nf+1);

end
```

**Listing 8.9:** `smoothQJacDamped.m`: Damped quasi-Jacobi smoothing operator for the multigrid solver for a 1D cell-centered elliptic equation with nonconstant diffusion coefficient.

```
function u = smoothQJacDamped(f,DeW,rC,u,hf,m,omega)

nfPlusGhostLayers = length(u);
nf = nfPlusGhostLayers-2;

z = zeros(nf,1);

hf2 = hf*hf;
omegaPrime = 1.0-omega;

u = applyBCs(u);

for sweep = 1:m
  for i = 1:nf

    z(i) = (hf2*f(i)+u(i+2)*DeW(i+1)+u(i)*DeW(i))...
```

```
                /(DeW(i+1)+DeW(i)+rC*hf2);

    end

    u(2:nf+1) = omega*z(1:nf)+omegaPrime*u(2:nf+1);
    u = applyBCs(u);

  end

end
```

**Listing 8.10:** `smoothRichardson.m`: Richardson's smoothing operator for the multigrid solver for a 1D cell-centered elliptic equation with nonconstant diffusion coefficient.

```
function u = smoothRichardson(f,DeW,rC,u,hf,m,omega)

nfPlusGhostLayers = length(u);
nf = nfPlusGhostLayers-2;
hf2 = hf*hf;
r = zeros(nf,1);

u = applyBCs(u);

for sweep = 1:m
  r(1:nf) = getResidual(u,f,DeW,rC,hf);
  u(2:nf+1) = u(2:nf+1)+omega*(hf2/2.0)*r(1:nf);
  u = applyBCs(u);
end

end
```

**Listing 8.11:** `smoothQGSDamped.m`: Damped quasi-Gauss–Seidel smoothing operator for the multigrid solver for a 1D cell-centered elliptic equation with nonconstant diffusion coefficient.

```
function u = smoothQGSDamped(f,DeW,rC,u,hf,m,omega,...
                               sweepDirection)

nfPlusGhostLayers = length(u);
nf = nfPlusGhostLayers-2;

hf2 = hf*hf;
```

```
omegaPrime = 1.0-omega;

switch sweepDirection
  case 'fwd'
    sweepIndex = [1:nf];
  otherwise
    sweepIndex = [nf:-1:1];
end

u = applyBCs(u);

for sweep = 1:m
  for i = sweepIndex

    tmp = (hf2*f(i)+u(i+2)*DeW(i+1)+u(i)*DeW(i))...
          /(DeW(i+1)+DeW(i)+rC*hf2);
    u(i+1) = omega*tmp+omegaPrime*u(i+1);

  end
  u = applyBCs(u);

end

end
```

**Listing 8.12:** `getResidual.m`: Residual calculation for a 1D cell-centered elliptic equation with nonconstant diffusion coefficient.

```
function residualResult = getResidual(u,f,DeW,rC,hf)

% Apply the boundary conditions before residual
% computation.
u = applyBCs(u);
nf = length(f);
residualResult = zeros(nf,1);

residualResult = f-FDOperator(u,DeW,rC,hf);

end
```

**Listing 8.13:** `applyBCs.m`: Function to apply various 1D boundary conditions. It is easy to swap the conditions in cell-centered finite difference solvers.

```
function u = applyBCs(u)

nfPlusGhostLayers = length(u);
nf = nfPlusGhostLayers-2;

% Homogeneous Dirichlet BC:

u(   1) = -u(   2);
u(nf+2) = -u(nf+1);

% Homogeneous Neumann BC:

% u(   1) = u(   2);
% u(nf+2) = u(nf+1);

% Periodic BC:

% u(   1) = u(nf+1);
% u(nf+2) = u(   2);

end
```

**Listing 8.14:** `normScaledL2.m`: Function to compute a scaled $L^2$ norm.

```
function normResult = normScaledL2(u)

nf = length(u);
normResult = sqrt(sum(u.*u)/nf);

end
```

## Two-dimensional codes

```
./2DCC/EllipticNCDiff/
    elliptic2DMGCCDiff.m
    multiGridSolver.m
    MGOperator.m
    prolongation.m
    restriction.m
    restrictDiff.m
    FDOperator.m
    smoothQJacDamped.m
    smoothQGSDamped.m
    getResidual.m
    applyBCs.m
    normScaledL2.m
```

**Listing 8.15:** `elliptic2DMGCCDiff.m`: Calling script for the cell-centered multigrid solver for a 2D elliptic equation with nonconstant diffusivity.

```
%
%   elliptic2DMGCCDiff.m
%
% Calling script for solving the 2D elliptic equation
%
%   -div(D grad(u)) + rC u = f
%
% on a rectangular domain, where D is a possibly
% non-constant, everywhere positive diffusion
% coefficient, and rC is a non-negative constant. The
% boundary conditions may be taken to be
% (i) homogeneous Dirichlet, (ii) homogeneous Neumann,
% or (iii) periodic. See the function applyBCs.
%
% Discretization is accomplished using the cell-centered
% finite difference method. The cell-centered multigrid
% method is used as the solver.
%
% One can set rC = 0.0, provided that homogeneous
% Dirichlet boundary conditions are chosen.
%
%% Initialization and Parameter

```

```
% Clear command line window and workspace.
clear; clc;

% Tolerance of residual to stop multigrid iterations.
tol = 1.0e-09;

% Number of interior cells in x,y at finest level.
nL = [1024,1024];

% Number of multigrid levels.
L = 10;

% Define the domain of the equation.
xLower = [0.0,0.0];
xUpper = [1.0,1.0];

hLvec = (xUpper-xLower)./nL;

% Parameters for the multigrid solver.
pCycle = 1;
m1     = 3;
m2     = 3;
omega  = 2/3;
kMax   = 100;
rC     = 2.0;

%% Setting up the multigrid solver

% Check mesh spacing.
if abs(hLvec(1)-hLvec(2)) > 1.0e-14
  error('Grid spacing mismatch in x vs y');
end
hL = hLvec(1);

% Check that nL is divisible by 2^(L-1).
nl = nL;
for k = L-1:-1:0
  if any(mod(nl,2) ~= 0)
    error('nL not divisible by 2^L.  Coarsening error.');
  end
  nl = nl/2;
end
```

```
MGParam.nL      = nL;
MGParam.xLower = xLower;
MGParam.xUpper = xUpper;
MGParam.L       = L;
MGParam.pCycle = pCycle;
MGParam.m1      = m1;
MGParam.m2      = m2;
MGParam.omega  = omega;
MGParam.kMax   = kMax;
MGParam.tol    = tol;
MGParam.rC     = rC;

% Allocate solution arrays (with ghost layers).
u      = zeros(nL(1)+2,nL(2)+2);
uExact = zeros(nL(1)+2,nL(2)+2);
f      = zeros(nL(1)  ,nL(2)  );
xCC    = zeros(nL(1)  ,nL(2)  );
yCC    = zeros(nL(1)  ,nL(2)  );
DeW    = zeros(nL(1)+1,nL(2)  );
DnS    = zeros(nL(1)  ,nL(2)+1);

% Build an exact solution and define f accordingly,
% so we can test the true algebraic error.
for i = 1:nL(1)
  x = (i-0.5)*hL+xLower(1);
  xArg = 3.0*pi*x/xUpper(1);
  for j = 1:nL(2)
    y = (j-0.5)*hL+xLower(2);
    yArg = 3.0*pi*y/xUpper(2);
    uExact(i+1,j+1) = exp(cos(xArg)*cos(yArg));
    xCC(i,j) = x;
    yCC(i,j) = y;
  end
end

% We next build the finest-grid diffusion
% arrays DeW, DnS:
for i = 1:nL(1)+1
  xEdge = (i-1.0)*hL+xLower(1);
  xArg = 2.0*pi*xEdge/xUpper(1);
  for j = 1:nL(2)
```

```
      yEdge = (j-0.5)*hL+xLower(2);
      yArg = 2.0*pi*yEdge/xUpper(2);

  % Evaluate D at the east-west edge point:
      DeW(i,j) = 1.0+0.5*sin(xArg)*cos(yArg);
    end
  end

  for i = 1:nL(1)
    xEdge = (i-0.5)*hL+xLower(1);
    xArg = 2.0*pi*xEdge/xUpper(1);
    for j = 1:nL(2)+1
      yEdge = (j-1.0)*hL+xLower(2);
      yArg = 2.0*pi*yEdge/xUpper(2);

  % Evaluate D at the north-south edge point:
      DnS(i,j) = 1.0+0.5*sin(xArg)*cos(yArg);
    end
  end

  % Apply the BCs to uExact, then compute f by calling
  % FDOperator.
  uExact = applyBCs(uExact);
  f = FDOperator(uExact,DeW,DnS,rC,hL);

  %% Call the Multigrid solver.

  tic;
  [u,errVals,kStop] = multiGridSolver(u,f,DeW,DnS,hL,...
                                          MGParam,uExact);
  toc;

  %% Plotting the numerical solution and error plots.

  uPlot(1:nL(1),1:nL(2)) = u(2:nL(1)+1,2:nL(2)+1);

  figure(1)
  clf

  surf(xCC,yCC,uPlot,'EdgeColor','none');
  title('Numerical Solution u');
  xlabel('x');
```

```
ylabel('y');
zlabel('u(x,y)');
colorbar;
view(2);
axis equal tight;

figure(2)
clf
semilogy(errVals(1:kStop,3),'bo','LineWidth',1.5)
hold on
semilogy(errVals(1:kStop,2),'rs','LineWidth',1.5)
hold on
semilogy(errVals(1:kStop,1),'kd','LineWidth',1.5)
hold on

rate = 1.0;
if kStop >= 4
  kv = kStop-3:kStop;
  le = log(errVals(kStop-3:kStop,3));
  p1 = polyfit(kv,le,1);
  rate = exp(p1(1));
  p1k = polyval(p1,1:kStop);
  semilogy(1:kStop,exp(p1k),'-k')
else
  semilogy(errVals(1:kStop,1),'.','LineWidth',1.5)
end

xlabel('$k$','Interpreter','latex');
title('Multigrid Iteration Errors',...
  'Interpreter','latex');
legend(...
  '$\left\|{\bf u}_L^{\rm E}-{\bf u}_L^k\right\|_L$',...
  '$\left\|{\bf u}_L^k-{\bf u}_L^{k-1} \right\|_L$',...
  '$\left\|{\bf r}_L^k\right\|_L$',...
  'log-linear fit','Interpreter','latex');
text(1.5,128*MGParam.tol,strcat(...
  '$\gamma_{\rm comp} =\hspace{.1cm}$',...
  num2str(rate,'%10.5e')),'FontSize',14,...
  'Interpreter','latex')
text(1.5,32*MGParam.tol,strcat('$m_1 =\hspace{.1cm}$',...
  num2str(MGParam.m1)),'FontSize',14,...
  'Interpreter','latex')
```

```
text(1.5,8*MGParam.tol,strcat('$m_2 =\hspace{.1cm}$',...
  num2str(MGParam.m2)),'FontSize',14,...
  'Interpreter','latex')
text(1.5,2*MGParam.tol,strcat('$p =\hspace{.1cm}$',...
  num2str(MGParam.pCycle)),'FontSize',14,...
  'Interpreter','latex')
text(1.5,0.5*MGParam.tol,...
  strcat('$\omega =\hspace{.1cm}$',...
  num2str(MGParam.omega)),'FontSize',14,...
  'Interpreter','latex')
printstr = strcat('Err_nL_',num2str(MGParam.nL(1)),...
  '_m1_',num2str(MGParam.m1),'.pdf');
exportgraphics(gca, printstr)
hold off
```

**Listing 8.16:** `multiGridSolver.m`: Multigrid solver for a 2D cell-centered elliptic equation with nonconstant diffusivity.

```
function [u,errVals,kStop] =...
          multiGridSolver(u0,f,DeW,DnS,hf,MGParam,uExact)

nf = size(f);
k = 0;
rC = MGParam.rC;
errorVal = MGParam.tol+1.0;
errVals = zeros(MGParam.kMax,3);

uo = u0;

while ((k < MGParam.kMax) && (errorVal > MGParam.tol))

  % Apply one multigrid cycle.
  u = MGOperator(MGParam.L,f,DeW,DnS,uo,hf,MGParam);

  % Evaluate various norms.
  residErr =...
            normScaledL2(getResidual(u,f,DeW,DnS,rC,hf));
  correErr = normScaledL2(uo(2:nf(1)+1,2:nf(2)+1)...
                         -u(2:nf(1)+1,2:nf(2)+1));
  exactErr = normScaledL2(uExact(2:nf(1)+1,2:nf(2)+1)...
                         -u(2:nf(1)+1,2:nf(2)+1));
```

```
  k = k+1;
  errVals(k,1) = residErr;
  errVals(k,2) = correErr;
  errVals(k,3) = exactErr;

  fprintf('Multigrid iteration %d:\n', k);
  fprintf('  residual   = %.6e\n', residErr);
  fprintf('  correction = %.6e\n', correErr);
  fprintf('  true error = %.6e\n', exactErr);

  % Stopping criterion (artificial, for demonstration).
  errorVal = exactErr;
  uo = u;
end

kStop = k;

end
```

**Listing 8.17:** MGOperator.m: Multigrid operator for a 2D cell-centered elliptic equation with nonconstant diffusivity.

```
function u = MGOperator(level,f,DeW,DnS,u,hf,MGParam)

rC = MGParam.rC;

% Pre-smoothing
u = smoothQJacDamped(f,DeW,DnS,rC,u,hf,MGParam.m1,...
                     MGParam.omega);

% Alternative pre-smoothing option:
% u = smoothQGSDamped(f,DeW,DnS,rC,u,hf,MGParam.m1,...
%                     MGParam.omega,'fwd');

if level > 0
  hc = 2*hf;
  nfPlusGhost = size(u);
  nf = nfPlusGhost-2;

  % Coarse grid interior dimensions are [nc(1),nc(2)].
```

```
  nc = nf/2;

  % Compute the residual and restric to the coarse grid.
  cGr = restriction(getResidual(u,f,DeW,DnS,rC,hf));

  % Restrict the diffusion coefficients to the coarse
  % grid.
  [cDeW,cDnS] = restrictDiff(DeW,DnS);

  % Approximate on the coarse grid using recursive MG.
  cGc = zeros(nc(1)+2,nc(2)+2);
  for s = 1:MGParam.pCycle
    cGc =...
        MGOperator(level-1,cGr,cDeW,cDnS,cGc,hc,MGParam);
  end

  % Prolongate the coarse grid correction and update the
  % fine grid approximation.
  u(2:nf(1)+1,2:nf(2)+1) = u(2:nf(1)+1,2:nf(2)+1)...
                 +prolongation(cGc(2:nc(1)+1,2:nc(2)+1));

  % Post-smoothing.
  u = smoothQJacDamped(f,DeW,DnS,rC,u,hf,MGParam.m2,...
                       MGParam.omega);

  % Alternative post-smoothing option:
  % u = smoothQGSDamped(f,DeW,DnS,rC,u,hf,MGParam.m2,...
  %                      MGParam.omega,'bwd');
end

end
```

**Listing 8.18:** `prolongation.m`: Prolongation operator for the 2D cell-centered multigrid solver.

```
function presult = prolongation(u)

nc = size(u);
nf = nc*2;
presult = zeros(nf);

presult(2:2:nf(1)  ,2:2:nf(2)  ) = u(1:nc(1),1:nc(2));
presult(1:2:nf(1)-1,2:2:nf(2)  ) = u(1:nc(1),1:nc(2));
```

```
presult(2:2:nf(1)   ,1:2:nf(2)-1) = u(1:nc(1),1:nc(2));
presult(1:2:nf(1)-1,1:2:nf(2)-1) = u(1:nc(1),1:nc(2));

end
```

**Listing 8.19:** `restriction.m`: Restriction operator for the 2D cell-centered multigrid solver.

```
function restrictionResult = restriction(u)

nf = size(u);
nc = nf/2;
restrictionResult...
  = 0.25*(u(1:2:nf(1)-1,1:2:nf(2)-1)...
    +     u(2:2:nf(1)   ,1:2:nf(2)-1)...
    +     u(1:2:nf(1)-1,2:2:nf(2)   )...
    +     u(2:2:nf(1)   ,2:2:nf(2)   ));

end
```

**Listing 8.20:** `restrictDiff.m`: Function to restrict the nonconstant 2D diffusion coefficient from a finer to a coarser level.

```
function [cDeW,cDnS] = restrictDiff(DeW,DnS)

[nxEW,ny] = size(DeW);
[nx,nyNS] = size(DnS);

ncx = nx/2;
ncy = ny/2;

cDeW(1:ncx+1,1:ncy) =...
        0.5*(DeW(1:2:nx+1,1:2:ny-1)+DeW(1:2:nx+1,2:2:ny));

cDnS(1:ncx,1:ncy+1) =...
        0.5*(DnS(1:2:nx-1,1:2:ny+1)+DnS(2:2:nx,1:2:ny+1));

end
```

**Listing 8.21:** `FDOperator.m`: Finite difference operator for a 2D cell-centered elliptic equation with nonconstant diffusion coefficient.

```
function operatorResult = FDOperator(u,DeW,DnS,rC,hf)

nfPlusGhostLayers = size(u);
nf = nfPlusGhostLayers-2;

nx = nf(1);
ny = nf(2);
hf2 = hf*hf;

% Vectorized Code for the operator:
%
% - div(D grad(u)) + rC u
%
% Compute east-west and north-south fluxes:
feW(1:nx+1,1:ny) = DeW(1:nx+1,1:ny).*(u(2:nx+2,2:ny+1)...
                   -u(1:nx+1,2:ny+1));
fnS(1:nx,1:ny+1) = DnS(1:nx,1:ny+1).*(u(2:nx+1,2:ny+2)...
                   -u(2:nx+1,1:ny+1));

% Compute the divergence of the fluxes and add reaction
% term:
operatorResult(1:nx,1:ny) =...
                 -(feW(2:nx+1,1:ny)-feW(1:nx,1:ny)...
                 +fnS(1:nx,2:ny+1)-fnS(1:nx,1:ny))/hf2...
                 +rC*u(2:nx+1,2:ny+1);

end
```

**Listing 8.22:** `smoothQJacDamped.m`: Damped quasi-Jacobi smoothing operator for a 2D elliptic equation with nonconstant diffusivity.

```
function u = smoothQJacDamped(f,DeW,DnS,rC,u,hf,m,omega)

nfPlusGhostLayers = size(u);
nf = nfPlusGhostLayers-2;

z = zeros(nf(1),nf(2));

```

```
hf2 = hf*hf;
omegaPrime = 1.0-omega;

u = applyBCs(u);

for sweep = 1:m
  for j = 1:nf(2)
    for i = 1:nf(1)

      z(i,j) = (hf2*f(i,j)...
               +u(i+2,j+1)*DeW(i+1,j  )...
               +u(i  ,j+1)*DeW(i  ,j  )...
               +u(i+1,j+2)*DnS(i  ,j+1)...
               +u(i+1,j  )*DnS(i  ,j  ))...
               /(DeW(i+1,j)+DeW(i,j)+DnS(i,j+1)...
               +DnS(i,j)+ rC*hf2);

    end
  end

  u(2:nf(1)+1,2:nf(2)+1) = omega*z(1:nf(1),1:nf(2))...
    + omegaPrime*u(2:nf(1)+1,2:nf(2)+1);
  u = applyBCs(u);

end

end
```

**Listing 8.23:** smoothQGSDamped.m: Damped quasi-Gauss–Seidel smoothing operator for a 2D elliptic equation with nonconstant diffusivity.

```
function u = smoothQGSDamped(f,DeW,DnS,rC,u,hf,m,...
  omega,sweepDirection)

nfPlusGhostLayers = size(u);
nf = nfPlusGhostLayers-2;

hf2 = hf*hf;
omegaPrime = 1.0-omega;

```

```
switch sweepDirection
  case 'fwd'
    xSweepIndex = [1:nf(1)];
    ySweepIndex = [1:nf(2)];
  otherwise
    xSweepIndex = [nf(1):-1:1];
    ySweepIndex = [nf(2):-1:1];
end

u = applyBCs(u);

for sweep = 1:m
  for j = ySweepIndex
    for i = xSweepIndex

      tmp = (hf2*f(i,j)...
           +u(i+2,j+1)*DeW(i+1,j  )...
           +u(i  ,j+1)*DeW(i  ,j  )...
           +u(i+1,j+2)*DnS(i  ,j+1)...
           +u(i+1,j  )*DnS(i  ,j  ))...
           /(DeW(i+1,j)+DeW(i,j)+DnS(i,j+1)+DnS(i,j)...
           +rC*hf2);
      u(i+1,j+1) = omega*tmp+omegaPrime*u(i+1,j+1);

    end
  end

  u = applyBCs(u);

end

end
```

**Listing 8.24:** `getResidual.m`: Residual calculation for a 2D cell-centered elliptic equation with nonconstant diffusivity.

```
function residualResult = getResidual(u,f,DeW,DnS,rC,hf)

% Apply the boundary conditions
u = applyBCs(u);
```

```
residualResult = f-FDOperator(u,DeW,DnS,rC,hf);

end
```

**Listing 8.25:** `applyBCs.m`: Function to apply various 2D boundary conditions. It is easy to swap the boundary conditions in cell-centered finite difference solvers.

```
function u = applyBCs(u)

nfPlusGhostLayers = size(u);
nf = nfPlusGhostLayers-2;

% Homogeneous Dirichlet BC:

% u(2:nf(1)+1,          1) = -u(2:nf(1)+1,          2);
% u(2:nf(1)+1,    nf(2)+2) = -u(2:nf(1)+1,    nf(2)+1);
% u(          1,1:nf(2)+2) = -u(          2,1:nf(2)+2);
% u(   nf(1)+2,1:nf(2)+2) = -u(   nf(1)+1,1:nf(2)+2);

% Homogeneous Neumann BC:

u(2:nf(1)+1,          1) = u(2:nf(1)+1,          2);
u(2:nf(1)+1,    nf(2)+2) = u(2:nf(1)+1,    nf(2)+1);
u(          1,1:nf(2)+2) = u(          2,1:nf(2)+2);
u(   nf(1)+2,1:nf(2)+2) = u(   nf(1)+1,1:nf(2)+2);

% Periodic BC:

% u(2:nf(1)+1,          1) = u(2:nf(1)+1,    nf(2)+1);
% u(2:nf(1)+1,    nf(2)+2) = u(2:nf(1)+1,          2);
% u(          1,1:nf(2)+2) = u(   nf(1)+1,1:nf(2)+2);
% u(   nf(1)+2,1:nf(2)+2) = u(          2,1:nf(2)+2);

end
```

**Listing 8.26:** `normScaledL2.m`: Function to compute a scaled $L^2$ norm.

```
function normResult = normScaledL2(u)

nf = size(u);
normResult = sqrt(sum(sum(u.*u))/(nf(1)*nf(2)));

end
```

# 9 Multigrid as a multiplicative process

In this chapter, we reformulate some multigrid algorithms using objects called T-matrices. We use the same conforming finite element setting as in Chapter 7. However, here we generalize the ideas of that chapter. This reformulation will make it obvious that the standard multigrid methods are multiplicative GLIS methods; see Section 1.5. Many of the ideas for this chapter come from the presentation in the book by Bramble [6].

Throughout this chapter, as the title indicates, we will deal with products of matrices or operators, and to shorten notation, we will use the product notation $\prod$. However, since the product of matrices does not commute, the order matters! We need to specify in which order a product is to be expanded. For definiteness, we make the convention that the expansion takes place from left to right.

Let $\{\mathsf{B}_j\}_{j=\alpha}^{\beta}$ be a family of matrices. If $\alpha > \beta$, then by definition we set $\prod_{j=\alpha}^{\beta} \mathsf{B}_j := \mathrm{I}$, the identity matrix. If $\alpha \le \beta$, then we set

$$\prod_{j=\alpha}^{\beta} \mathsf{B}_j := \mathsf{B}_\alpha \prod_{j=\alpha+1}^{\beta} \mathsf{B}_j.$$

For instance,

$$\prod_{j=2}^{6} \mathsf{B}_j = \mathsf{B}_2\mathsf{B}_3\mathsf{B}_4\mathsf{B}_5\mathsf{B}_6.$$

## 9.1 Multilevel matrices

**Definition 9.1.1** (Multilevel prolongation matrix). Let $\ell \in \mathbb{N}$ and $j \in \mathbb{N}_0$, with $0 \le j < \ell$. The **multilevel prolongation matrix** $\mathsf{P}_{j,\ell}$ is defined as

$$\mathsf{P}_{j,\ell} := \prod_{k=1}^{\ell-j} \mathsf{P}_{\ell-k} \in \mathbb{R}^{n_\ell \times n_j}.$$

In particular,

$$\mathsf{P}_{\ell-1,\ell} = \mathsf{P}_{\ell-1} \in \mathbb{R}^{n_\ell \times n_{\ell-1}}.$$ ■

**Remark 9.1.2** (Expansion). Notice that

$$\mathsf{P}_{j,\ell} = \mathsf{P}_{\ell-1}\mathsf{P}_{\ell-2}\cdots\mathsf{P}_j,$$

and this product is well defined, since the matrices are compatible for multiplication in any order that we choose. ◂

https://doi.org/10.1515/9783111354880-009

**Proposition 9.1.3** (Prolongation representation). *Let $\ell \in \mathbb{N}$ and $j \in \mathbb{N}_0$, with $0 \le j < \ell$, and let $V_0 \subset V_1 \subset \cdots \subset V_\ell$ be the usual nested finite element spaces. Suppose that $v_j \in V_j \overset{\mathcal{B}_j}{\longleftrightarrow} \mathbf{v}_j \in \mathbb{R}^{n_j}$. Then*

$$v_j \in V_j \subset V_\ell \overset{\mathcal{B}_\ell}{\longleftrightarrow} \mathsf{P}_{j,\ell}\mathbf{v}_j \in \mathbb{R}^{n_\ell}.$$

*Proof.* Exercise; see Problem 9.1. □

**Definition 9.1.4** (Multilevel restriction matrix). Let $\ell \in \mathbb{N}$ and $j \in \mathbb{N}_0$, with $0 \le j < \ell$. The **multilevel restriction matrix** $\mathsf{R}_{j,\ell} \in \mathbb{R}^{n_j \times n_\ell}$ is defined as

$$\mathsf{R}_{j,\ell} := \mathsf{P}_{j,\ell}^{\top}.$$ ■

Notice that, by definition,

$$\begin{aligned}\mathsf{R}_{j,\ell} &= \left(\prod_{k=1}^{\ell-j} \mathsf{P}_{\ell-k}\right)^{\top} \\ &= \prod_{k=j}^{\ell-1} \mathsf{R}_k.\end{aligned}$$

**Proposition 9.1.5** (Galerkin condition). *Let $\ell \in \mathbb{N}$ and $j \in \mathbb{N}_0$, with $0 \le j < \ell$. With the usual construction for conforming finite element spaces, we have*

$$\mathsf{A}_j = \mathsf{R}_{j,\ell}\mathsf{A}_\ell\mathsf{P}_{j,\ell} \in \mathbb{R}^{n_j \times n_j}.$$

*Proof.* This follows from the strong Galerkin condition (G0), which holds in the finite element setting:

$$\begin{aligned}\mathsf{A}_j &= \mathsf{R}_j\mathsf{A}_{j+1}\mathsf{P}_j \\ &= \mathsf{R}_j\mathsf{R}_{j+1}\mathsf{A}_{j+2}\mathsf{P}_{j+1}\mathsf{P}_j \\ &\vdots \\ &= \prod_{k=j}^{\ell-1} \mathsf{R}_k \mathsf{A}_\ell \prod_{k=1}^{\ell-j} \mathsf{P}_{\ell-k} \\ &= \mathsf{R}_{j,\ell}\mathsf{A}_\ell\mathsf{P}_{j,\ell}.\end{aligned}$$ □

**Definition 9.1.6** (Multilevel correction). Let $\ell \in \mathbb{N}$ and $j \in \mathbb{N}_0$, with $0 \le j < \ell$. We define the matrix

$$\Pi_{j,\ell} := \mathsf{A}_j^{-1}\mathsf{R}_{j,\ell}\mathsf{A}_\ell \in \mathbb{R}^{n_j \times n_\ell}.$$ ■

**Proposition 9.1.7** (Multilevel correction). *Let $\ell \in \mathbb{N}$ and $j \in \mathbb{N}_0$, with $0 \le j < \ell$. Then*

$$\Pi_{j,\ell} = \prod_{k=j}^{\ell-1} \Pi_k.$$

*Proof.* The matrix product on the right-hand side is

$$\begin{aligned}\Pi_j \cdots \Pi_{\ell-1} &= \mathsf{A}_j^{-1}\mathsf{R}_j\mathsf{A}_{j+1}\mathsf{A}_{j+1}^{-1}\mathsf{R}_{j+1}\mathsf{A}_{j+2}\cdots\mathsf{A}_{\ell-1}^{-1}\mathsf{R}_{\ell-1}\mathsf{A}_\ell \\ &= \mathsf{A}_j^{-1}\prod_{k=j}^{\ell-1}\mathsf{R}_k\mathsf{A}_\ell \\ &= \Pi_{j,\ell}.\end{aligned}$$

□

**Definition 9.1.8** (Multilevel Ritz projection matrix). Let $\ell \in \mathbb{N}$ and $j \in \mathbb{N}_0$, with $0 \le j < \ell$. The **multilevel Ritz projection matrix** is

$$\tilde{\Pi}_{j,\ell} := \mathsf{P}_{j,\ell}\Pi_{j,\ell} \in \mathbb{R}^{n_\ell \times n_\ell}.$$

In particular,

$$\tilde{\Pi}_{\ell-1,\ell} = \tilde{\Pi}_\ell \in \mathbb{R}^{n_\ell \times n_\ell}.$$

■

**Theorem 9.1.9** (Basis representations). *Let $\mathcal{A}(\cdot,\cdot)$ be defined in (7.1.2), let $\ell \in \mathbb{N}$ and $j \in \mathbb{N}_0$, with $0 \le j < \ell$, and let $u_\ell \in V_\ell \overset{\mathcal{B}_\ell}{\longleftrightarrow} \boldsymbol{u}_\ell \in \mathbb{R}^{n_\ell}$. Set*

$$u_j' = \mathcal{R}_j u_\ell \in V_j \overset{\mathcal{B}_j}{\longleftrightarrow} \boldsymbol{u}_j' \in \mathbb{R}^{n_j},$$

*where the Ritz projection $\mathcal{R}_j : H_0^1(\Omega) \to V_j$ is as introduced in Definition 7.4.8. Then*

$$\boldsymbol{u}_j' = \Pi_{j,\ell}\boldsymbol{u}_\ell \in \mathbb{R}^{n_j}.$$

*Further,*

$$\mathcal{R}_j u_\ell \in V_j \subset V_\ell \overset{\mathcal{B}_\ell}{\longleftrightarrow} \tilde{\Pi}_{j,\ell}\boldsymbol{u}_\ell \in \mathbb{R}^{n_\ell}.$$

*Proof.* Let $u_\ell \in V_\ell$. Then $\mathcal{R}_j u_\ell \in V_j$ is the unique solution to

$$\mathcal{A}(\mathcal{R}_j u_\ell, v_j) = \mathcal{A}(u_\ell, v_j), \quad \forall v_j \in V_j.$$

Now consider

$$(\boldsymbol{u}_j', \boldsymbol{v}_j)_{\mathsf{A}_j} = \mathcal{A}(\mathcal{R}_j u_\ell, v_j)$$

and

$$\mathcal{A}(u_\ell, v_j) = (\boldsymbol{u}_\ell, \mathsf{P}_{j,\ell}\boldsymbol{v}_j)_{\mathsf{A}_\ell},$$

where, as usual,

$$\boldsymbol{v}_j \in \mathbb{R}^{n_j} \overset{\mathcal{B}_j}{\longleftrightarrow} v_j \in V_j.$$

We may now gather these three identities to infer that

$$\begin{aligned}(\mathsf{A}_j\boldsymbol{u}_j', \boldsymbol{v}_j)_j &= (\boldsymbol{u}_j', \boldsymbol{v}_j)_{\mathsf{A}_j}\\ &= (\boldsymbol{u}_\ell, \mathsf{P}_{j,\ell}\boldsymbol{v}_j)_{\mathsf{A}_\ell}\\ &= (\mathsf{A}_\ell\boldsymbol{u}_\ell, \mathsf{P}_{j,\ell}\boldsymbol{v}_j)_\ell\\ &= (\mathsf{R}_{j,\ell}\mathsf{A}_\ell\boldsymbol{u}_\ell, \boldsymbol{v}_j)_j.\end{aligned}$$

Therefore, since $\boldsymbol{v}_j \in \mathbb{R}^{n_j}$ is arbitrary,

$$\mathsf{A}_j\boldsymbol{u}_j' = \mathsf{R}_{j,\ell}\mathsf{A}_\ell\boldsymbol{u}_\ell,$$

or

$$\boldsymbol{u}_j' = \mathsf{A}_j^{-1}\mathsf{R}_{j,\ell}\mathsf{A}_\ell\boldsymbol{u}_\ell = \Pi_{j,\ell}\boldsymbol{u}_\ell.$$

The second part follows from Proposition 9.1.3. □

We are now ready to define our multilevel T-matrices.

**Definition 9.1.10** (Multilevel T-matrix). Let $\ell \in \mathbb{N}$ and $j \in \mathbb{N}_0$, with $0 \le j < \ell$. Recall that

$$\mathsf{K}_j = \mathsf{I}_j - \mathsf{S}_j\mathsf{A}_j.$$

For $m \in \mathbb{N}_0$, set

$$\mathsf{T}_{j,\ell}(m) := \Pi_{j,\ell} - \mathsf{K}_j^m\Pi_{j,\ell} \in \mathbb{R}^{n_j\times n_\ell}.$$

The **multilevel T-matrix** is defined as

$$\tilde{\mathsf{T}}_{j,\ell}(m) := \mathsf{P}_{j,\ell}\mathsf{T}_{j,\ell}(m) \in \mathbb{R}^{n_\ell\times n_\ell}.$$ ■

The name and notation for T-matrices suggests the intuition behind their definition. To see this, it suffices to apply this matrix to the error at level $\ell$, which, as we recall, satisfies

$$\mathsf{A}_\ell\boldsymbol{e}_\ell = \boldsymbol{r}_\ell,$$

where $\boldsymbol{r}_\ell$ is the residual. Then

$$\begin{aligned}\Pi_{j,\ell}\boldsymbol{e}_\ell &= \mathsf{A}_j^{-1}\mathsf{R}_{j,\ell}\mathsf{A}_\ell\boldsymbol{e}_\ell \\ &= \mathsf{A}_j^{-1}\mathsf{R}_{j,\ell}\boldsymbol{r}_\ell \\ &= \mathsf{A}_j^{-1}\widehat{\boldsymbol{r}}_j \\ &= \widehat{\boldsymbol{e}}_j,\end{aligned}$$

where $\widehat{\boldsymbol{r}}_j$ is an approximation of the residual at level $j$, and, consequently, $\widehat{\boldsymbol{e}}_j$ is an approximation of the level-$\ell$ error at level $j$. In summary,

$$\mathsf{T}_{j,\ell}(m)\boldsymbol{e}_\ell = \widehat{\boldsymbol{e}}_j - \widetilde{\boldsymbol{e}}_j,$$

with $\widetilde{\boldsymbol{e}}_j$ being the "smoothed out" error, that is, the one obtained after $m$ applications of the smoother. By this logic, $\tilde{\mathsf{T}}_{j,\ell}\boldsymbol{e}_\ell$ is the representation at level $\ell$ of this difference. Finally, compare this to the action of the error transfer operator $\mathsf{T}$ of a GLIS.

**Remark 9.1.11** (Notation). In Definition 9.1.10 the parameter $m$ is the number of smoothing steps. Whenever it is understood from context, we write $\mathsf{T}_{j,\ell}$ and $\tilde{\mathsf{T}}_{j,\ell}$ instead of $\mathsf{T}_{j,\ell}(m)$ and $\tilde{\mathsf{T}}_{j,\ell}(m)$, respectively. ◂

## 9.2 Properties of the multilevel matrices

Let us investigate some properties of the multilevel matrices.

**Proposition 9.2.1** (Projection). *Let $\ell \in \mathbb{N}$ and $j \in \mathbb{N}_0$, with $0 \le j < \ell$. Then*

$$\Pi_{j,\ell}\mathsf{P}_{j,\ell} = \mathrm{I}_j, \tag{9.2.1}$$

*and*

$$\tilde{\Pi}_{j,\ell}^2 = \tilde{\Pi}_{j,\ell}.$$

*Proof.* Recall that Proposition 9.1.5 gives a sort of multilevel Galerkin condition in the sense that

$$\mathsf{A}_j = \mathsf{R}_{j,\ell}\mathsf{A}_\ell\mathsf{P}_{j,\ell}.$$

Moreover, by definition,

$$\Pi_{j,\ell} = \mathsf{A}_j^{-1}\mathsf{R}_{j,\ell}\mathsf{A}_\ell,$$

so that

$$\Pi_{j,\ell}\mathsf{P}_{j,\ell} = \mathsf{A}_j^{-1}\mathsf{R}_{j,\ell}\mathsf{A}_\ell\mathsf{P}_{j,\ell}$$

$$\begin{aligned} &= \mathsf{A}_j^{-1}\mathsf{A}_j \\ &= \mathrm{I}_j. \end{aligned}$$

In addition,

$$\begin{aligned} \tilde{\Pi}_{j,\ell}^2 &= \mathsf{P}_{j,\ell}\Pi_{j,\ell}\mathsf{P}_{j,\ell}\Pi_{j,\ell} \\ &= \mathsf{P}_{j,\ell}\Pi_{j,\ell} \\ &= \tilde{\Pi}_{j,\ell}. \end{aligned}$$

□

**Definition 9.2.2** (T′). Let $\ell \in \mathbb{N}$ and $j \in \mathbb{N}_0$, with $0 \le j < \ell$. Recall that

$$\mathsf{K}_j^* = \mathrm{I}_j - \mathsf{S}_j^\top \mathsf{A}_j.$$

For $m \in \mathbb{N}_0$, we define

$$\mathsf{T}'_{j,\ell} := \Pi_{j,\ell} - (\mathsf{K}_j^*)^m \Pi_{j,\ell}$$

and

$$\tilde{\mathsf{T}}'_{j,\ell} := \mathsf{P}_{j,\ell}\mathsf{T}'_{j,\ell}.$$

■

**Proposition 9.2.3** (Self-adjoint). *Let $\ell \in \mathbb{N}$ and $j \in \mathbb{N}_0$, with $0 \le j < \ell$. The operator $\tilde{\Pi}_{j,\ell}$ introduced in Definition* 9.1.8 *is self-adjoint in the sense that*

$$\tilde{\Pi}_{j,\ell}^* = \tilde{\Pi}_{j,\ell}.$$

*On the other hand, the operators $\tilde{\mathsf{T}}_{j,\ell}$ and $\tilde{\mathsf{T}}'_{j,\ell}$ of Definitions* 9.1.10 *and* 9.2.2, *respectively, satisfy*

$$\tilde{\mathsf{T}}_{j,\ell}^* = \tilde{\mathsf{T}}'_{j,\ell}.$$

*Proof.* Recall that, by definition, for all $\boldsymbol{u}_\ell, \boldsymbol{v}_\ell \in \mathbb{R}^{n_\ell}$,

$$(\tilde{\Pi}_{j,\ell}\boldsymbol{u}_\ell, \boldsymbol{v}_\ell)_{\mathsf{A}_\ell} = (\boldsymbol{u}_\ell, \tilde{\Pi}_{j,\ell}^*\boldsymbol{v}_\ell)_{\mathsf{A}_\ell}.$$

Then, using the definition of $\tilde{\Pi}_{j,\ell}$, we have

$$\begin{aligned} (\tilde{\Pi}_{j,\ell}\boldsymbol{u}_\ell, \boldsymbol{v}_\ell)_{\mathsf{A}_\ell} &= (\mathsf{P}_{j,\ell}\Pi_{j,\ell}\boldsymbol{u}_\ell, \mathsf{A}_\ell\boldsymbol{v}_\ell)_\ell \\ &= (\Pi_{j,\ell}\boldsymbol{u}_\ell, \mathsf{R}_{j,\ell}\mathsf{A}_\ell\boldsymbol{v}_\ell)_j \\ &= (\mathsf{A}_j^{-1}\mathsf{R}_{j,\ell}\mathsf{A}_\ell\boldsymbol{u}_\ell, \mathsf{R}_{j,\ell}\mathsf{A}_\ell\boldsymbol{v}_\ell)_j \\ &= (\mathsf{R}_{j,\ell}\mathsf{A}_\ell\boldsymbol{u}_\ell, \mathsf{A}_j^{-1}\mathsf{R}_{j,\ell}\mathsf{A}_\ell\boldsymbol{v}_\ell)_j \end{aligned}$$

$$\begin{aligned} &= (\mathsf{A}_\ell \boldsymbol{u}_\ell, \tilde{\Pi}_{j,\ell} \boldsymbol{v}_\ell)_\ell \\ &= (\boldsymbol{u}_\ell, \tilde{\Pi}_{j,\ell} \boldsymbol{v}_\ell)_{\mathsf{A}_\ell}. \end{aligned}$$

Thus

$$\tilde{\Pi}^*_{j,\ell} = \tilde{\Pi}_{j,\ell}.$$

With this identity at hand, we can use now the definitions of $\tilde{\mathsf{T}}_{j,\ell}$ and $\tilde{\Pi}_{j,\ell}$ to get

$$\begin{aligned} (\tilde{\mathsf{T}}_{j,\ell} \boldsymbol{u}_\ell, \boldsymbol{v}_\ell)_{\mathsf{A}_\ell} &= (\tilde{\Pi}_{j,\ell} \boldsymbol{u}_\ell, \boldsymbol{v}_\ell)_{\mathsf{A}_\ell} - (\mathsf{P}_{j,\ell} \mathsf{K}_j^m \Pi_{j,\ell} \boldsymbol{u}_\ell, \boldsymbol{v}_\ell)_{\mathsf{A}_\ell} \\ &= (\boldsymbol{u}_\ell, \tilde{\Pi}_{j,\ell} \boldsymbol{v}_\ell)_{\mathsf{A}_\ell} - (\mathsf{P}_{j,\ell} \mathsf{K}_j^m \Pi_{j,\ell} \boldsymbol{u}_\ell, \mathsf{A}_\ell \boldsymbol{v}_\ell)_\ell. \end{aligned}$$

Let us consider the second term on the right-hand side separately:

$$\begin{aligned} (\mathsf{P}_{j,\ell} \mathsf{K}_j^m \Pi_{j,\ell} \boldsymbol{u}_\ell, \mathsf{A}_\ell \boldsymbol{v}_\ell)_\ell &= (\mathsf{K}_j^m \Pi_{j,\ell} \boldsymbol{u}_\ell, \mathsf{R}_{j,\ell} \mathsf{A}_\ell \boldsymbol{v}_\ell)_j \\ &= (\mathsf{K}_j^m \Pi_{j,\ell} \boldsymbol{u}_\ell, \mathsf{A}_j \mathsf{A}_j^{-1} \mathsf{R}_{j,\ell} \mathsf{A}_\ell \boldsymbol{v}_\ell)_j \\ &= (\mathsf{K}_j^m \Pi_{j,\ell} \boldsymbol{u}_\ell, \mathsf{A}_j^{-1} \mathsf{R}_{j,\ell} \mathsf{A}_\ell \boldsymbol{v}_\ell)_{\mathsf{A}_j} \\ &= (\Pi_{j,\ell} \boldsymbol{u}_\ell, (\mathsf{K}_j^m)^* \Pi_{j,\ell} \boldsymbol{v}_\ell)_{\mathsf{A}_j} \\ &= (\Pi_{j,\ell} \boldsymbol{u}_\ell, (\mathsf{K}_j^*)^m \Pi_{j,\ell} \boldsymbol{v}_\ell)_{\mathsf{A}_j} \\ &= (\mathsf{A}_j^{-1} \mathsf{R}_{j,\ell} \mathsf{A}_\ell \boldsymbol{u}_\ell, (\mathsf{K}_j^*)^m \Pi_{j,\ell} \boldsymbol{v}_\ell)_{\mathsf{A}_j} \\ &= (\mathsf{R}_{j,\ell} \mathsf{A}_\ell \boldsymbol{u}_\ell, (\mathsf{K}_j^*)^m \Pi_{j,\ell} \boldsymbol{v}_\ell)_j \\ &= (\mathsf{A}_\ell \boldsymbol{u}_\ell, \mathsf{P}_{j,\ell} (\mathsf{K}_j^*)^m \Pi_{j,\ell} \boldsymbol{v}_\ell)_\ell \\ &= (\boldsymbol{u}_\ell, \mathsf{P}_{j,\ell} (\mathsf{K}_j^*)^m \Pi_{j,\ell} \boldsymbol{v}_\ell)_{\mathsf{A}_\ell}. \end{aligned}$$

Thus

$$\begin{aligned} (\tilde{\mathsf{T}}_{j,\ell} \boldsymbol{u}_\ell, \boldsymbol{v}_\ell)_{\mathsf{A}_\ell} &= (\boldsymbol{u}_\ell, \tilde{\Pi}_{j,\ell} \boldsymbol{v}_\ell)_{\mathsf{A}_\ell} - (\boldsymbol{u}_\ell, \mathsf{P}_{j,\ell} (\mathsf{K}_j^*)^m \Pi_{j,\ell} \boldsymbol{v}_\ell)_{\mathsf{A}_\ell} \\ &= (\boldsymbol{u}_\ell, \tilde{\mathsf{T}}'_{j,\ell} \boldsymbol{v}_\ell)_{\mathsf{A}_\ell}, \end{aligned}$$

meaning that

$$\tilde{\mathsf{T}}^*_{j,\ell} = \tilde{\mathsf{T}}'_{j,\ell},$$

as claimed. □

**Remark 9.2.4** (Not a projection). We note that, in general,

$$\tilde{\mathsf{T}}^2_{j,\ell} \neq \tilde{\mathsf{T}}_{j,\ell}.$$

In other words, the T-matrix $\tilde{\mathsf{T}}_{j,\ell}$ is not a projection matrix. ◂

**Proposition 9.2.5** (Relations between $\tilde{\mathsf{T}}_{j,\ell}$ and $\tilde{\Pi}_{j,\ell}$). *Let $\ell \in \mathbb{N}$ and $j \in \mathbb{N}_0$, with $0 \le j < \ell$. The operators $\tilde{\mathsf{T}}_{j,\ell}$ and $\tilde{\Pi}_{j,\ell}$ satisfy*

1. $$(\mathrm{I}_\ell - \tilde{\Pi}_{j,\ell})(\mathrm{I}_\ell - \tilde{\mathsf{T}}_{j,\ell}) = \mathrm{I}_\ell - \tilde{\Pi}_{j,\ell}; \tag{9.2.2}$$

2. $$(\mathrm{I}_\ell - \tilde{\mathsf{T}}^*_{j,\ell})(\mathrm{I}_\ell - \tilde{\Pi}_{j,\ell}) = \mathrm{I}_\ell - \tilde{\Pi}_{j,\ell}; \tag{9.2.3}$$

3. $$\mathrm{I}_\ell - \tilde{\Pi}_{j,\ell} = (\mathrm{I}_\ell - \tilde{\mathsf{T}}^*_{j,\ell})(\mathrm{I}_\ell - \tilde{\Pi}_{j,\ell})(\mathrm{I}_\ell - \tilde{\mathsf{T}}_{j,\ell}). \tag{9.2.4}$$

*Proof.* We prove each property separately.

1. For brevity, denote the left-hand side of (9.2.2) by

$$\mathsf{M}_\ell := \mathrm{I}_\ell - \tilde{\Pi}_{j,\ell} - \tilde{\mathsf{T}}_{j,\ell} + \tilde{\Pi}_{j,\ell}\tilde{\mathsf{T}}_{j,\ell}.$$

   By definition,

$$\begin{aligned}\tilde{\Pi}_{j,\ell}\tilde{\mathsf{T}}_{j,\ell} &= \mathsf{P}_{j,\ell}\Pi_{j,\ell}\mathsf{P}_{j,\ell}\mathsf{T}_{j,\ell} \\ &\overset{(9.2.1)}{=} \mathsf{P}_{j,\ell}\mathrm{I}_j\mathsf{T}_{j,\ell} \\ &= \tilde{\mathsf{T}}_{j,\ell}.\end{aligned}$$

   Consequently,

$$\mathsf{M}_\ell = \mathrm{I}_\ell - \tilde{\Pi}_{j,\ell},$$

   as claimed.

2. Similarly, denote the left-hand side of (9.2.3) by

$$\mathsf{M}'_\ell = \mathrm{I}_\ell - \tilde{\mathsf{T}}^*_{j,\ell} - \tilde{\Pi}_{j,\ell} + \tilde{\mathsf{T}}^*_{j,\ell}\tilde{\Pi}_{j,\ell},$$

   and observe that by the previous result,

$$\begin{aligned}\tilde{\mathsf{T}}^*_{j,\ell}\tilde{\Pi}_{j,\ell} &= \mathsf{P}_{j,\ell}\mathsf{T}'_{j,\ell}\mathsf{P}_{j,\ell}\Pi_{j,\ell} \\ &= \mathsf{P}_{j,\ell}(\Pi_{j,\ell} - (\mathsf{K}^*_j)^m\Pi_{j,\ell})\mathsf{P}_{j,\ell}\Pi_{j,\ell} \\ &\overset{(9.2.1)}{=} \mathsf{P}_{j,\ell}(\mathrm{I}_j - (\mathsf{K}^*_j)^m)\Pi_{j,\ell} \\ &= \tilde{\mathsf{T}}^*_{j,\ell}.\end{aligned}$$

   Therefore

$$\mathsf{M}'_\ell = \mathrm{I}_\ell - \tilde{\Pi}_{j,\ell},$$

   as stated.

3. As Proposition 9.2.1 shows, $\tilde{\Pi}_{j,\ell}$ is a projection matrix. This immediately implies that

$$(\mathrm{I}_\ell - \tilde{\Pi}_{j,\ell})^2 = \mathrm{I}_\ell - \tilde{\Pi}_{j,\ell}.$$

By the last two results,

$$\begin{aligned}
\mathrm{I}_\ell - \tilde{\Pi}_{j,\ell} &= (\mathrm{I}_\ell - \tilde{\Pi}_{j,\ell})(\mathrm{I}_\ell - \tilde{\Pi}_{j,\ell}) \\
&\overset{(9.2.2)}{=} (\mathrm{I}_\ell - \tilde{\Pi}_{j,\ell})(\mathrm{I}_\ell - \tilde{\Pi}_{j,\ell})(\mathrm{I}_\ell - \tilde{\mathsf{T}}_{j,\ell}) \\
&\overset{(9.2.3)}{=} (\mathrm{I}_\ell - \tilde{\mathsf{T}}_{j,\ell}^*)(\mathrm{I}_\ell - \tilde{\Pi}_{j,\ell})(\mathrm{I}_\ell - \tilde{\Pi}_{j,\ell})(\mathrm{I}_\ell - \tilde{\mathsf{T}}_{j,\ell}) \\
&= (\mathrm{I}_\ell - \tilde{\mathsf{T}}_{j,\ell}^*)(\mathrm{I}_\ell - \tilde{\Pi}_{j,\ell})(\mathrm{I}_\ell - \tilde{\mathsf{T}}_{j,\ell}).
\end{aligned}$$

□

**Proposition 9.2.6** (Commutation). *Let $\ell \in \mathbb{N}$ and $i, j \in \mathbb{N}_0$, with $0 \le i < j < \ell$. We have*

$$\mathsf{P}_{j,\ell}(\mathrm{I}_j - \tilde{\mathsf{T}}_{i,j}) = (\mathrm{I}_\ell - \tilde{\mathsf{T}}_{i,\ell})\mathsf{P}_{j,\ell}. \tag{9.2.5}$$

*Proof.* The proof is a direct computation:

$$\begin{aligned}
\mathsf{P}_{j,\ell}(\mathrm{I}_j - \tilde{\mathsf{T}}_{i,j}) &\overset{(9.2.1)}{=} \mathsf{P}_{j,\ell}(\mathrm{I}_j - \tilde{\mathsf{T}}_{i,j})\Pi_{j,\ell}\mathsf{P}_{j,\ell} \\
&= \{\mathsf{P}_{j,\ell}\Pi_{j,\ell} - \mathsf{P}_{j,\ell}\mathsf{P}_{i,j}\mathsf{T}_{i,j}\Pi_{j,\ell}\}\mathsf{P}_{j,\ell} \\
&= \{\tilde{\Pi}_{j,\ell} - \mathsf{P}_{i,\ell}(\Pi_{i,j} - \mathsf{K}_i^m\Pi_{i,j})\Pi_{j,\ell}\}\mathsf{P}_{j,\ell} \\
&= \{\tilde{\Pi}_{j,\ell} - \mathsf{P}_{i,\ell}\Pi_{i,\ell} + \mathsf{P}_{i,\ell}\mathsf{K}_i^m\Pi_{i,\ell}\}\mathsf{P}_{j,\ell} \\
&= \{\tilde{\Pi}_{j,\ell} - \tilde{\mathsf{T}}_{i,\ell}\}\mathsf{P}_{j,\ell} \\
&= \mathsf{P}_{j,\ell}\Pi_{j,\ell}\mathsf{P}_{j,\ell} - \tilde{\mathsf{T}}_{i,\ell}\mathsf{P}_{j,\ell} \\
&\overset{(9.2.1)}{=} (\mathrm{I}_\ell - \tilde{\mathsf{T}}_{i,\ell})\mathsf{P}_{j,\ell}.
\end{aligned}$$

□

**Proposition 9.2.7** (Commutation). *Let $\ell \in \mathbb{N}$ and $i, j \in \mathbb{N}_0$, with $0 \le i < j < \ell$. Then*

$$\mathsf{P}_{j,\ell}(\mathrm{I}_j - \tilde{\Pi}_{i,j}) = (\mathrm{I}_\ell - \tilde{\Pi}_{i,\ell})\mathsf{P}_{j,\ell}. \tag{9.2.6}$$

*Proof.* The proof is similar to that of Proposition 9.2.6 and is left as an exercise; see Problem 9.2. □

## 9.3 Multigrid error transfer matrices in multiplicative forms

Now using the definitions and properties of the multilevel matrices, we can rewrite the error transfer matrices of some common multigrid algorithms.

**Theorem 9.3.1** (V-cycle). *Let $\mathcal{A}(\cdot,\cdot)$ be defined in* (7.1.2), *let $1 \le \ell \le L$, and let $V_\ell$ be as in Definition* 7.2.1. *Consider the symmetric V-cycle algorithm, that is, $m_1 = m_2 = m > 0$ and*

$p = 1$. *The error transfer matrix can be expressed as*

$$\mathsf{E}_\ell = (\mathsf{K}_\ell^*)^m \left( \prod_{j=1}^{\ell-1} (\mathrm{I}_\ell - \tilde{\mathsf{T}}_{j,\ell}) \right)^* (\mathrm{I}_\ell - \tilde{\Pi}_{0,\ell}) \left( \prod_{j=1}^{\ell-1} (\mathrm{I}_\ell - \tilde{\mathsf{T}}_{j,\ell}) \right) (\mathsf{K}_\ell)^m. \tag{9.3.1}$$

*Proof.* For $0 \le j < \ell$, define the quantity

$$\mathsf{M}_{j,\ell} := \mathrm{I}_\ell - \tilde{\Pi}_{j,\ell} + \mathsf{P}_{j,\ell}\mathsf{E}_j\Pi_{j,\ell},$$

and observe that for $j = 0$,

$$\mathsf{M}_{0,\ell} = \mathrm{I}_\ell - \tilde{\Pi}_{0,\ell},$$

since $\mathsf{E}_0 = \mathsf{O}_0$. Now, by Theorem 6.1.10,

$$\begin{aligned} \mathsf{M}_{j,\ell} &= \mathrm{I}_\ell - \tilde{\Pi}_{j,\ell} + \mathsf{P}_{j,\ell}(\mathsf{K}_j^*)^m(\mathrm{I}_j - \tilde{\Pi}_{j-1,j} + \mathsf{P}_{j-1,j}\mathsf{E}_{j-1}\Pi_{j-1,j})\mathsf{K}_j^m\Pi_{j,\ell} \\ &= \mathrm{I}_\ell - \tilde{\Pi}_{j,\ell} + \mathsf{P}_{j,\ell}(\mathsf{K}_j^*)^m\mathsf{M}_{j-1,j}\mathsf{K}_j^m\Pi_{j,\ell}. \end{aligned} \tag{9.3.2}$$

Now observe that

$$\begin{aligned} \mathsf{P}_{j,\ell}(\mathsf{K}_j^*)^m &\overset{(9.2.1)}{=} \mathsf{P}_{j,\ell}(\mathsf{K}_j^*)^m\Pi_{j,\ell}\mathsf{P}_{j,\ell} \\ &= (\mathsf{P}_{j,\ell}\Pi_{j,\ell} - \mathsf{P}_{j,\ell}\Pi_{j,\ell} + \mathsf{P}_{j,\ell}(\mathsf{K}_j^*)^m\Pi_{j,\ell})\mathsf{P}_{j,\ell} \\ &= (\mathsf{P}_{j,\ell}\Pi_{j,\ell} - \tilde{\mathsf{T}}_{j,\ell}^*)\mathsf{P}_{j,\ell} \\ &= \mathsf{P}_{j,\ell}\Pi_{j,\ell}\mathsf{P}_{j,\ell} - \tilde{\mathsf{T}}_{j,\ell}^*\mathsf{P}_{j,\ell} \\ &\overset{(9.2.1)}{=} (\mathrm{I}_\ell - \tilde{\mathsf{T}}_{j,\ell}^*)\mathsf{P}_{j,\ell}, \end{aligned}$$

that is,

$$\mathsf{P}_{j,\ell}(\mathsf{K}_j^*)^m = (\mathrm{I}_\ell - \tilde{\mathsf{T}}_{j,\ell}^*)\mathsf{P}_{j,\ell}. \tag{9.3.3}$$

Similarly,

$$\begin{aligned} \mathsf{K}_j^m\Pi_{j,\ell} &= \Pi_{j,\ell} - \Pi_{j,\ell} + \mathsf{K}_j^m\Pi_{j,\ell} \\ &= \Pi_{j,\ell} - \mathsf{T}_{j,\ell} \\ &\overset{(9.2.1)}{=} \Pi_{j,\ell}\mathsf{P}_{j,\ell}(\Pi_{j,\ell} - \mathsf{T}_{j,\ell}) \\ &= \Pi_{j,\ell}(\mathsf{P}_{j,\ell}\Pi_{j,\ell} - \tilde{\mathsf{T}}_{j,\ell}) \\ &= \Pi_{j,\ell}\mathsf{P}_{j,\ell}\Pi_{j,\ell} - \Pi_{j,\ell}\tilde{\mathsf{T}}_{j,\ell} \\ &\overset{(9.2.1)}{=} \Pi_{j,\ell}(\mathrm{I}_\ell - \tilde{\mathsf{T}}_{j,\ell}), \end{aligned}$$

so that

$$\mathsf{K}_j^m \Pi_{j,\ell} = \Pi_{j,\ell}(\mathrm{I}_\ell - \tilde{\mathsf{T}}_{j,\ell}). \tag{9.3.4}$$

Putting (9.3.2)–(9.3.4) together, we have

$$\begin{aligned}\mathsf{M}_{j,\ell} &= \mathrm{I}_\ell - \tilde{\Pi}_{j,\ell} + (\mathrm{I}_\ell - \tilde{\mathsf{T}}_{j,\ell}^*)\mathsf{P}_{j,\ell}\mathsf{M}_{j-1,j}\Pi_{j,\ell}(\mathrm{I}_\ell - \tilde{\mathsf{T}}_{j,\ell}) \\ &\overset{(9.2.4)}{=} (\mathrm{I}_\ell - \tilde{\mathsf{T}}_{j,\ell}^*)(\mathrm{I}_\ell - \tilde{\Pi}_{j,\ell} + \mathsf{P}_{j,\ell}\mathsf{M}_{j-1,j}\Pi_{j,\ell})(\mathrm{I}_\ell - \tilde{\mathsf{T}}_{j,\ell}).\end{aligned}$$

Using the definition of $\mathsf{M}_{j-1,j}$ we realize that

$$\begin{aligned}\mathsf{P}_{j,\ell}\mathsf{M}_{j-1,j}\Pi_{j,\ell} &= \mathsf{P}_{j,\ell}(\mathrm{I}_j - \tilde{\Pi}_{j-1,j} + \mathsf{P}_{j-1,j}\mathsf{E}_{j-1}\Pi_{j-1,j})\Pi_{j,\ell} \\ &= \tilde{\Pi}_{j,\ell} - \tilde{\Pi}_{j-1,\ell} + \mathsf{P}_{j-1,\ell}\mathsf{E}_{j-1}\Pi_{j-1,\ell} \\ &= \tilde{\Pi}_{j,\ell} - \mathrm{I}_\ell + \mathsf{M}_{j-1,\ell}.\end{aligned}$$

We combine these two identities to realize that

$$\mathsf{M}_{j,\ell} = (\mathrm{I}_\ell - \tilde{\mathsf{T}}_{j,\ell}^*)\mathsf{M}_{j-1,\ell}(\mathrm{I}_\ell - \tilde{\mathsf{T}}_{j,\ell}).$$

Therefore

$$\begin{aligned}\mathsf{M}_{\ell-1,\ell} &= (\mathrm{I}_\ell - \tilde{\mathsf{T}}_{\ell-1,\ell}^*)\mathsf{M}_{\ell-2,\ell}(\mathrm{I}_\ell - \tilde{\mathsf{T}}_{\ell-1,\ell}) \\ &= (\mathrm{I}_\ell - \tilde{\mathsf{T}}_{\ell-1,\ell}^*)(\mathrm{I}_\ell - \tilde{\mathsf{T}}_{\ell-2,\ell}^*)\mathsf{M}_{\ell-3,\ell}(\mathrm{I}_\ell - \tilde{\mathsf{T}}_{\ell-2,\ell})(\mathrm{I}_\ell - \tilde{\mathsf{T}}_{\ell-1,\ell}) \\ &\;\;\vdots \\ &= \left(\prod_{j=1}^{\ell-1}(\mathrm{I}_\ell - \tilde{\mathsf{T}}_{\ell-j,\ell}^*)\right)(\mathrm{I}_\ell - \tilde{\Pi}_{0,\ell})\left(\prod_{j=1}^{\ell-1}(\mathrm{I}_\ell - \tilde{\mathsf{T}}_{j,\ell})\right) \\ &= \left(\prod_{j=1}^{\ell-1}(\mathrm{I}_\ell - \tilde{\mathsf{T}}_{j,\ell})\right)^*(\mathrm{I}_\ell - \tilde{\Pi}_{0,\ell})\left(\prod_{j=1}^{\ell-1}(\mathrm{I}_\ell - \tilde{\mathsf{T}}_{j,\ell})\right).\end{aligned}$$

Recall now that by Theorem 6.1.10 we have

$$\mathsf{E}_\ell = (\mathsf{K}_\ell^*)^m \mathsf{M}_{\ell-1,\ell}\mathsf{K}_\ell^m.$$

The result is proven. □

**Remark 9.3.2** (Multiplicative GLIS). The previous result gives a proof of the fact that the V-cycle multigrid method is a multiplicative GLIS, as we defined in Chapter 1. Indeed, the error transfer matrix can be written as a product of matrices of the form

$$\mathsf{E}_\ell = \prod_k (\mathrm{I}_\ell - \mathsf{B}_\ell^{(k)}\mathsf{A}_\ell),$$

where $\{\mathsf{B}_\ell^{(k)}\}_k$ is a family of iterator matrices. ◂

**Corollary 9.3.3** (One-sided V-cycle). *The error transfer of the one-sided V-cycle with only presmoothing, that is $p = 1$, $m_1 = m > 0$, and $m_2 = 0$, satisfies*

$$\mathsf{E}_\ell^{\mathrm{pre}} = (\mathrm{I}_\ell - \tilde{\Pi}_{0,\ell})\left(\prod_{j=1}^{\ell-1}(\mathrm{I}_\ell - \tilde{\mathsf{T}}_{j,\ell})\right)\mathsf{K}_\ell^m. \tag{9.3.5}$$

*For the V-cycle with only postsmoothing, that is, $p = 1$, $m_2 = m > 0$, and $m_1 = 0$, we have*

$$\mathsf{E}_\ell^{\mathrm{post}} = (\mathsf{K}_\ell^*)^m\left(\prod_{j=1}^{\ell-1}(\mathrm{I}_\ell - \tilde{\mathsf{T}}_{j,\ell})\right)^*(\mathrm{I}_\ell - \tilde{\Pi}_{0,\ell}).$$

*Therefore, for the symmetric V-cycle,*

$$\mathsf{E}_\ell = \mathsf{E}_\ell^{\mathrm{post}}\mathsf{E}_\ell^{\mathrm{pre}}.$$

*Furthermore, if $m_1 = m_2 = m$, then*

$$(\mathsf{E}_\ell^{\mathrm{post}})^* = \mathsf{E}_\ell^{\mathrm{pre}}.$$

*Finally,*

$$\mathsf{E}_\ell^* = \mathsf{E}_\ell = (\mathsf{E}_\ell^{\mathrm{pre}})^*\mathsf{E}_\ell^{\mathrm{pre}}$$

*is SPSD.*

*Proof.* Exercise; see Problem 9.3. □

**Theorem 9.3.4** (Convergence). *Let $m > 0$ and assume that Richardson's method is used for smoothing. Then both one-sided V-cycle methods converge.*

*Proof.* We have shown (see Theorem 6.8.2) that there is $C_0$ such that

$$\|\mathsf{E}_\ell \boldsymbol{u}_\ell\|_{\mathsf{A}_\ell} \le \frac{C_0}{m + C_0}\|\boldsymbol{u}_\ell\|_{\mathsf{A}_\ell},$$

for all $\boldsymbol{u}_\ell \in \mathbb{R}^{n_\ell}$. We wish to prove that there is $0 \le \gamma < 1$ for which

$$\|\mathsf{E}_\ell^{\mathrm{pre}}\boldsymbol{u}_\ell\|_{\mathsf{A}_\ell} \le \gamma\|\boldsymbol{u}_\ell\|_{\mathsf{A}_\ell}.$$

Observe that

$$\begin{aligned}\|\mathsf{E}_\ell^{\mathrm{pre}}\boldsymbol{u}_\ell\|_{\mathsf{A}_\ell}^2 &= (\mathsf{E}_\ell^{\mathrm{pre}}\boldsymbol{u}_\ell, \mathsf{E}_\ell^{\mathrm{pre}}\boldsymbol{u}_\ell)_{\mathsf{A}_\ell}\\ &= (\boldsymbol{u}_\ell, (\mathsf{E}_\ell^{\mathrm{pre}})^*\mathsf{E}_\ell^{\mathrm{pre}}\boldsymbol{u}_\ell)_{\mathsf{A}_\ell}\\ &= (\boldsymbol{u}_\ell, \mathsf{E}_\ell\boldsymbol{u}_\ell)_{\mathsf{A}_\ell}\end{aligned}$$

$$\begin{aligned}
&\leq \|\boldsymbol{u}_\ell\|_{\mathsf{A}_\ell} \|\mathsf{E}_\ell \boldsymbol{u}_\ell\|_{\mathsf{A}_\ell} \\
&\leq \|\boldsymbol{u}_\ell\|_{\mathsf{A}_\ell} \frac{C_0}{m + C_0} \|\boldsymbol{u}_\ell\|_{\mathsf{A}_\ell} \\
&\leq \frac{C_0}{m + C_0} \|\boldsymbol{u}_\ell\|_{\mathsf{A}_\ell}^2.
\end{aligned}$$

Thus, taking square roots, we have

$$\|\mathsf{E}_\ell^{\text{pre}} \boldsymbol{u}_\ell\|_{\mathsf{A}_\ell} \leq \sqrt{\frac{C_0}{m + C_0}} \|\boldsymbol{u}_\ell\|_{\mathsf{A}_\ell}.$$

We leave the case with only postsmoothing as an exercise for the reader; see Problem 9.4. □

**Theorem 9.3.5** (One-sided W-cycle). *Consider the W-cycle algorithm with only presmoothing, that is $p = 2$, $m_1 = m > 0$, and $m_2 = 0$. The error transfer matrix may be expressed as*

$$\mathsf{E}_\ell^{W,\text{pre}} = \mathsf{F}_\ell \mathsf{E}_\ell^{\text{pre}},$$

*where $\mathsf{E}_\ell^{\text{pre}}$ is defined above, and $\mathsf{F}_\ell \in \mathbb{R}^{n_\ell \times n_\ell}$ is a matrix with*

$$\|\mathsf{F}_\ell\|_{\mathsf{A}_\ell} \leq 1.$$

*Consequently, the one-sided W-cycle method with presmoothing converges for all $m > 0$.*

*Proof.* Exercise; see Problem 9.5. □

**Corollary 9.3.6** (Symmetric W-cycle). *Consider the symmetric W-cycle algorithm, that is, $p = 2$ and $m_1 = m_2 = m > 0$. The error transfer matrix is*

$$\mathsf{E}_\ell^W = (\mathsf{E}_\ell^{\text{pre}})^* \mathsf{D}_\ell \mathsf{E}_\ell^{\text{pre}},$$

*where $\mathsf{E}_\ell^{\text{pre}}$ was defined in (9.3.5), and*

$$\|\mathsf{D}_\ell\|_{\mathsf{A}_\ell} \leq 1, \quad \forall \ell \geq 1.$$

*Consequently, the algorithm converges if the symmetric V-cycle algorithm converges with a uniform contraction factor that may depend on m.*

*Proof.* Exercise; see Problem 9.6. □

## 9.4 Problems

9.1 Prove Proposition 9.1.3.

9.2 Prove Proposition 9.2.7.

9.3 Prove Corollary 9.3.3.

9.4 Prove the second part of Theorem 9.3.4.

9.5 Prove Theorem 9.3.5.

9.6 Prove Corollary 9.3.6.

# 10 Additive preconditioners based on subspace decompositions

In the last chapter, we discovered that the common multigrid algorithms can be classified as multiplicative linear iterative methods, in the sense of Definition 1.5.1. In this chapter, we study an alternative to these methods, which belong to the class of additive methods. These methods will be developed not as stand-alone solvers but as preconditioners or preconditioning strategies. The advantage over multiplicative methods can be seen by comparing (1.5.1) and (1.5.2). The order in which the terms that comprise (1.5.1) is computed and accumulated does not matter. Therefore additive methods easily lend themselves to parallelization.

When designing preconditioners, the idea is to use them in conjunction with the preconditioned conjugate gradient (PCG) method detailed in Chapter 2. These preconditioners $\mathsf{C}$ should be cheap to apply, and, ideally, $\mathsf{CA}$ should have a condition number that is close to one.

The preconditioners we build in this chapter are based on the theory of subspace decomposition. We thus begin this chapter with a review of that material from a matrix-based point of view. A more abstract operator interpretation can be found in the book by Brenner and Scott [10] and later in Chapter 12.

## 10.1 Subspace decompositions

**Definition 10.1.1** (Prolongation matrices). Let $n \in \mathbb{N}$, and let $\{m_j\}_{j=0}^{L} \subset \mathbb{N}$ with

$$0 < m_0 \leq m_1 \leq \cdots \leq m_j \leq \cdots \leq m_L \leq n.$$

Matrices

$$\mathsf{Q}_j \in \mathbb{R}^{n \times m_j}, \quad j \in \{0, 1, \ldots, L\},$$

are called **ordered prolongation matrices** iff $\operatorname{rank}(\mathsf{Q}_j) = m_j$, for all $j \in \{0, 1, \ldots, L\}$. ■

**Assumption 10.1.2** (SS1). Let $\{\mathsf{Q}_j\}_{j=0}^{L}$ be a set of ordered prolongation matrices. We say that the set $\{\mathsf{Q}_j\}_{j=0}^{L}$ **supports a subspace decomposition of** $\mathbb{R}^n$ iff, for every $\boldsymbol{u} \in \mathbb{R}^n$, there exist, not necessarily unique, vectors

$$\boldsymbol{w}_j \in \mathbb{R}^{m_j}, \quad 0 \leq j \leq L,$$

such that

$$\boldsymbol{u} = \sum_{j=0}^{L} \mathsf{Q}_j \boldsymbol{w}_j. \tag{SS1a}$$

•

https://doi.org/10.1515/9783111354880-010

**Example 10.1.3** (Simplest space decomposition). The simplest example of a space decomposition might also seem like the most obvious one. Consider $\mathbb{R}^n$, set $L = n - 1$, and

$$m_0 = m_1 = \cdots = m_{n-1} = 1.$$

Define $\mathsf{Q}_j \in \mathbb{R}^{n\times 1}$ as

$$\mathsf{Q}_j = \hat{\boldsymbol{e}}_{j+1} = \begin{bmatrix} 0 \\ \vdots \\ 0 \\ 1 \\ 0 \\ \vdots \\ 0 \end{bmatrix},$$

where the nonzero entry is located at the $(j + 1)$th position. For $\boldsymbol{u} \in \mathbb{R}^n$, define

$$\boldsymbol{w}_j = [u_{j+1}] \in \mathbb{R}^1, \quad 0 \le j \le L = n - 1,$$

where $[\boldsymbol{u}]_i = u_i$. It is easy to see that (SS1) holds. ▲

**Example 10.1.4.** Suppose that

$$\mathbb{R}^n = V_0 + V_1 + \cdots + V_L,$$

where each $V_i \subseteq \mathbb{R}^n$ is a nontrivial vector subspace. This is an example of an explicit subspace decomposition of $\mathbb{R}^n$. With the present example, we will show that this notion of subspace decomposition is equivalent to our definition based upon a set of ordered prolongation matrices. Let

$$m_i := \dim(V_i), \quad 0 \le i \le L.$$

Then

$$1 \le m_i \le n, \quad 0 \le i \le L,$$

and

$$\sum_{i=0}^{L} m_i \ge n.$$

Without loss of generality, we can assume that the spaces are ordered so that

$$0 < m_0 \le m_1 \le \cdots \le m_j \le \cdots \le m_L \le n.$$

Next, we want to construct a set of ordered prolongation matrices $\{\mathsf{Q}_j\}_{j=0}^L$ subordinate to this explicit subspace decomposition. Suppose that

$$\mathcal{B}_i = \{\boldsymbol{x}_{i,1}, \ldots, \boldsymbol{x}_{i,m_i}\} \subset \mathbb{R}^n, \quad 0 \le i \le L,$$

is a basis for $V_i$. Let $\boldsymbol{u} \in \mathbb{R}^n$. By assumption,

$$\boldsymbol{u} = \sum_{i=0}^{L} \boldsymbol{v}_i, \quad \boldsymbol{v}_i \in V_i, \quad 0 \le i \le L,$$

though this decomposition is not necessarily unique. For each $i$, $0 \le i \le L$, upon fixing the basis $\mathcal{B}_i$, there are unique scalars

$$w_{i,1}, \ldots, w_{i,m_i} \in \mathbb{R}$$

such that

$$\boldsymbol{v}_i = \sum_{j=1}^{m_i} w_{i,j} \boldsymbol{x}_{i,j}.$$

Define, for all $0 \le i \le L$,

$$\mathsf{Q}_i := \begin{bmatrix} | & | & & | \\ \boldsymbol{x}_{i,1} & \boldsymbol{x}_{i,2} & \cdots & \boldsymbol{x}_{i,m_i} \\ | & | & & | \end{bmatrix} \in \mathbb{R}^{n \times m_i},$$

$$\boldsymbol{w}_i := \begin{bmatrix} w_{i,1} \\ w_{i,2} \\ \vdots \\ w_{i,m_i} \end{bmatrix} \in \mathbb{R}^{m_i}.$$

Then it follows that

$$\boldsymbol{u} = \sum_{i=0}^{L} \mathsf{Q}_i \boldsymbol{w}_i.$$

Note that our construction of $\mathsf{Q}_i$ is independent of $\boldsymbol{u}$. On the other hand, the vectors $\boldsymbol{w}_i$ do depend upon $\boldsymbol{u}$ via the explicit subspace decomposition. More such examples will be discussed in Chapter 12. ▲

**Example 10.1.5** (Hierarchical basis decomposition). In this example, we describe the so-called hierarchical basis decomposition. Let $n = 2^3 - 1 = 7$. Set $L = 3 - 1 = 2$ and

$$m_0 = 2^0 = 1, \quad m_1 = 2^1 = 2, \quad m_2 = 2^2 = 4.$$

Note that in this case,

$$\begin{aligned}\sum_{j=0}^{L} m_j &= m_0 + m_1 + m_2 \\ &= 1 + 2 + 4 \\ &= 7 \\ &= n,\end{aligned}$$

which is a typical feature of hierarchical basis decompositions. Let $\{\hat{\boldsymbol{e}}_j\}_{j=1}^7 \subset \mathbb{R}^7$ be the canonical basis of $\mathbb{R}^7$. Define

$$\mathsf{Q}_0 = [\hat{\boldsymbol{e}}_4] \in \mathbb{R}^{7\times 1}, \quad \mathsf{Q}_1 = [\hat{\boldsymbol{e}}_2\ \hat{\boldsymbol{e}}_6] \in \mathbb{R}^{7\times 2}, \quad \mathsf{Q}_2 = [\hat{\boldsymbol{e}}_1\ \hat{\boldsymbol{e}}_3\ \hat{\boldsymbol{e}}_5\ \hat{\boldsymbol{e}}_7] \in \mathbb{R}^{7\times 4}.$$

Let now $\boldsymbol{u} \in \mathbb{R}^7$. Set

$$\boldsymbol{w}_0 = [u_4] \in \mathbb{R}^1, \quad \boldsymbol{w}_1 = \begin{bmatrix} u_2 \\ u_6 \end{bmatrix} \in \mathbb{R}^2, \quad \boldsymbol{w}_2 = \begin{bmatrix} u_1 \\ u_3 \\ u_5 \\ u_7 \end{bmatrix} \in \mathbb{R}^4.$$

Then, clearly,

$$\begin{aligned}\boldsymbol{u} &= \sum_{j=0}^{2} \mathsf{Q}_j \boldsymbol{w}_j \\ &= \mathsf{Q}_0 \boldsymbol{w}_0 + \mathsf{Q}_1 \boldsymbol{w}_1 + \mathsf{Q}_2 \boldsymbol{w}_2.\end{aligned}$$

▲

**Example 10.1.6** (Hierarchical basis). Here is another hierarchical basis decomposition. The reader is challenged to determine the general pattern; see Problem 10.1. Suppose that $n = 3 \cdot 2^3 - 1 = 23$. Set $L = 3$ and

$$m_0 = 2, \quad m_1 = 3, \quad m_2 = 6, \quad m_3 = 12.$$

Again,

$$\begin{aligned}\sum_{j=0}^{L} m_j &= m_0 + m_1 + m_2 + m_3 \\ &= 2 + 3 + 6 + 12 \\ &= 23 \\ &= n.\end{aligned}$$

We let $\{\hat{\boldsymbol{e}}_j\}_{j=1}^{23} \subset \mathbb{R}^{23}$ be the canonical basis of $\mathbb{R}^{23}$. Define

$$\begin{aligned}
\mathsf{Q}_0 &= [\hat{\boldsymbol{e}}_8\ \hat{\boldsymbol{e}}_{16}] \in \mathbb{R}^{23\times 2},\\
\mathsf{Q}_1 &= [\hat{\boldsymbol{e}}_4\ \hat{\boldsymbol{e}}_{12}\ \hat{\boldsymbol{e}}_{20}] \in \mathbb{R}^{23\times 3},\\
\mathsf{Q}_2 &= [\hat{\boldsymbol{e}}_2\ \hat{\boldsymbol{e}}_6\ \hat{\boldsymbol{e}}_{10}\ \hat{\boldsymbol{e}}_{14}\ \hat{\boldsymbol{e}}_{18}\ \hat{\boldsymbol{e}}_{22}] \in \mathbb{R}^{23\times 6},\\
\mathsf{Q}_3 &= [\hat{\boldsymbol{e}}_1\ \hat{\boldsymbol{e}}_3\ \hat{\boldsymbol{e}}_5\ \cdots\ \hat{\boldsymbol{e}}_{19}\ \hat{\boldsymbol{e}}_{21}\ \hat{\boldsymbol{e}}_{23}] \in \mathbb{R}^{23\times 12}.
\end{aligned}$$

For arbitrary $\boldsymbol{u} \in \mathbb{R}^{23}$, set

$$\boldsymbol{w}_0 = \begin{bmatrix} u_8 \\ u_{16} \end{bmatrix} \in \mathbb{R}^2, \qquad \boldsymbol{w}_1 = \begin{bmatrix} u_4 \\ u_{12} \\ u_{20} \end{bmatrix} \in \mathbb{R}^3,$$

$$\boldsymbol{w}_2 = \begin{bmatrix} u_2 \\ u_6 \\ u_{10} \\ u_{14} \\ u_{18} \\ u_{22} \end{bmatrix} \in \mathbb{R}^6, \qquad \boldsymbol{w}_3 = \begin{bmatrix} u_1 \\ u_3 \\ u_5 \\ \vdots \\ u_{19} \\ u_{21} \\ u_{23} \end{bmatrix} \in \mathbb{R}^{12}.$$

Then

$$\begin{aligned}
\boldsymbol{u} &= \sum_{j=0}^{3} \mathsf{Q}_j \boldsymbol{w}_j\\
&= \mathsf{Q}_0\boldsymbol{w}_0 + \mathsf{Q}_1\boldsymbol{w}_1 + \mathsf{Q}_2\boldsymbol{w}_2 + \mathsf{Q}_3\boldsymbol{w}_3.
\end{aligned}$$

▲

**Definition 10.1.7** (Additive subspace preconditioner). Let $\{\mathsf{Q}_j\}_{j=0}^{L}$ be a set of ordered prolongation matrices for which Assumption (SS1) holds. A matrix $\mathsf{C} \in \mathbb{R}^{n\times n}$ is called an **additive subspace preconditioner** with respect to $\{\mathsf{Q}_j\}_{j=0}^{L}$ if there are SPD matrices $\mathsf{C}_\ell \in \mathbb{R}^{m_\ell \times m_\ell}$, $0 \le \ell \le L$, such that

$$\mathsf{C} = \sum_{\ell=0}^{L} \mathsf{Q}_\ell \mathsf{C}_\ell^{-1} \mathsf{Q}_\ell^\top. \tag{10.1.1}$$

■

**Remark 10.1.8** (Transposes). The matrix $\mathsf{Q}_\ell^\top \in \mathbb{R}^{m_\ell \times n}$ is the adjoint of $\mathsf{Q}_\ell$ with respect to the Euclidean inner product. From this moment on, $(\cdot,\cdot)$ represents the Euclidean inner product on $\mathbb{R}^n$, and $(\cdot,\cdot)_\ell$ is the Euclidean inner product on $\mathbb{R}^{m_\ell}$, $0 \le \ell \le L$. Thus, for instance,

$$(\mathsf{Q}_\ell \boldsymbol{u}_\ell, \boldsymbol{v}) = (\boldsymbol{u}_\ell, \mathsf{Q}_\ell^\top \boldsymbol{v})_\ell, \quad \forall\, \boldsymbol{u}_\ell \in \mathbb{R}^{m_\ell}, \forall\, \boldsymbol{v} \in \mathbb{R}^n.$$

Similarly, $\| \cdot \|$ is the Euclidean norm on $\mathbb{R}^n$, whereas $\| \cdot \|_\ell$ is the Euclidean norm on $\mathbb{R}^{m_\ell}$, $0 \le \ell \le L$. ◂

**Lemma 10.1.9** (C is SPD). *Let $\{\mathsf{Q}_j\}_{j=0}^L$ be a set of ordered prolongation matrices for which Assumption* (SS1) *holds, and let $\mathsf{C} \in \mathbb{R}^{n\times n}$ be an additive subspace preconditioner with respect to this family. Then* C *is SPD with respect to $(\cdot,\cdot)$, and, consequently, if $\mathsf{A} \in \mathbb{R}^{n\times n}$ is also SPD with respect to $(\cdot,\cdot)$, then* CA *is SPD with respect to $(\cdot,\cdot)_{\mathsf{C}^{-1}}$ and $(\cdot,\cdot)_{\mathsf{A}}$.*

*Proof.* The matrix C is clearly symmetric, since each $\mathsf{C}_\ell^{-1}$ is, by definition, symmetric. To show positive definiteness, for $\boldsymbol{u} \in \mathbb{R}^n$, we compute

$$\begin{aligned}
(\boldsymbol{u}, \mathsf{C}\boldsymbol{u}) &\overset{(10.1.1)}{=} \left(\boldsymbol{u}, \sum_{\ell=0}^{L} \mathsf{Q}_\ell \mathsf{C}_\ell^{-1} \mathsf{Q}_\ell^\top \boldsymbol{u}\right) \\
&= \sum_{\ell=0}^{L} (\boldsymbol{u}, \mathsf{Q}_\ell \mathsf{C}_\ell^{-1} \mathsf{Q}_\ell^\top \boldsymbol{u}) \\
&= \sum_{\ell=0}^{L} (\mathsf{Q}_\ell^\top \boldsymbol{u}, \mathsf{C}_\ell^{-1} \mathsf{Q}_\ell^\top \boldsymbol{u})_\ell \\
&\ge 0,
\end{aligned}$$

since, for $0 \le \ell \le L$, $\mathsf{C}_\ell^{-1}$ is SPD with respect to $(\cdot,\cdot)_\ell$. Suppose now that equality holds. Since, again, $\mathsf{C}_\ell^{-1}$ is SPD, we must have

$$\mathsf{Q}_\ell^\top \boldsymbol{u} = \boldsymbol{0}, \quad 0 \le \ell \le L. \tag{10.1.2}$$

In this case, since (SS1) holds, we have

$$\begin{aligned}
\|\boldsymbol{u}\|^2 &= (\boldsymbol{u}, \boldsymbol{u}) \\
&\overset{(SS1)}{=} \left(\boldsymbol{u}, \sum_{\ell=0}^{L} \mathsf{Q}_\ell \boldsymbol{w}_\ell\right) \\
&= \sum_{\ell=0}^{L} (\boldsymbol{u}, \mathsf{Q}_\ell \boldsymbol{w}_\ell) \\
&= \sum_{\ell=0}^{L} (\mathsf{Q}_\ell^\top \boldsymbol{u}, \boldsymbol{w}_\ell)_\ell \\
&\overset{(10.1.2)}{=} \sum_{\ell=0}^{L} (\boldsymbol{0}, \boldsymbol{w}_\ell)_\ell \\
&= 0.
\end{aligned}$$

This implies that $\boldsymbol{u} = \boldsymbol{0}$, which shows that C is SPD with respect to $(\cdot,\cdot)$. The statements about CA follow from Proposition 1.1.12. ☐

**Definition 10.1.10** (Q-Class). Let $\{Q_j\}_{j=0}^{L}$ be a set of ordered prolongation matrices for which Assumption (SS1) holds. For $\boldsymbol{u} \in \mathbb{R}^n$, the set

$$Q[\boldsymbol{u}] := \left\{ (\boldsymbol{w}_j)_{j=0}^{L} \;\middle|\; \boldsymbol{w}_j \in \mathbb{R}^{m_j},\ 0 \le j \le L,\ \boldsymbol{u} = \sum_{j=0}^{L} Q_j \boldsymbol{w}_j \right\}$$

is called the **Q-class of** $\boldsymbol{u}$. We write

$$(\boldsymbol{w}_j)_{j=0}^{L} \in Q[\boldsymbol{u}],$$

or simply

$$(\boldsymbol{w}_j) \in Q[\boldsymbol{u}],$$

and we remark that $(\boldsymbol{w}_j)_{j=0}^{L}$ is an ordered tuple. ■

**Theorem 10.1.11** (Norm representation). *Let $\{Q_j\}_{j=0}^{L}$ be a set of ordered prolongation matrices for which Assumption* (SS1) *holds, and let $C \in \mathbb{R}^{n\times n}$ be an additive subspace preconditioner with respect to this family. Then*

$$(\boldsymbol{u}, \boldsymbol{u})_{C^{-1}} = \min_{(\boldsymbol{w}_\ell)\in Q[\boldsymbol{u}]} \sum_{\ell=0}^{L} (C_\ell \boldsymbol{w}_\ell, \boldsymbol{w}_\ell)_\ell, \quad \forall\, \boldsymbol{u} \in \mathbb{R}^n. \tag{10.1.3}$$

*Proof.* Since, for $0 \le \ell \le L$, each $C_\ell \in \mathbb{R}^{m_\ell\times m_\ell}$ is SPD with respect to the Euclidean product, so is its inverse $C_\ell^{-1}$, and, consequently, $(\cdot,\cdot)_{C_\ell^{-1}}$ is an inner product on $\mathbb{R}^{m_\ell\times m_\ell}$. Therefore, for all $\boldsymbol{u}_\ell, \boldsymbol{v}_\ell \in \mathbb{R}^{m_\ell}$,

$$\begin{aligned}(C_\ell^{-1}\boldsymbol{u}_\ell, \boldsymbol{v}_\ell)_\ell &= (\boldsymbol{u}_\ell, \boldsymbol{v}_\ell)_{C_\ell^{-1}} \\ &\le \|\boldsymbol{u}_\ell\|_{C_\ell^{-1}} \|\boldsymbol{v}_\ell\|_{C_\ell^{-1}} \\ &= \sqrt{(\boldsymbol{u}_\ell, \boldsymbol{u}_\ell)_{C_\ell^{-1}}}\sqrt{(\boldsymbol{v}_\ell, \boldsymbol{v}_\ell)_{C_\ell^{-1}}}.\end{aligned}$$

Let now $\boldsymbol{u} \in \mathbb{R}^n$. By definition,

$$\boldsymbol{u} \overset{\text{(SS1)}}{=} \sum_{\ell=0}^{L} Q_\ell \boldsymbol{w}_\ell, \quad \forall\, (\boldsymbol{w}_\ell) \in Q[\boldsymbol{u}].$$

Then we have

$$\begin{aligned}(\boldsymbol{u}, \boldsymbol{u})_{C^{-1}} &= \sum_{\ell=0}^{L} (\boldsymbol{u}, Q_\ell \boldsymbol{w}_\ell)_{C^{-1}} \\ &= \sum_{\ell=0}^{L} (Q_\ell^\top C^{-1}\boldsymbol{u}, \boldsymbol{w}_\ell)_\ell\end{aligned}$$

$$
\begin{aligned}
&= \sum_{\ell=0}^{L} (\mathsf{Q}_\ell^\top \mathsf{C}^{-1}\boldsymbol{u}, \mathsf{C}_\ell \boldsymbol{w}_\ell)_{\mathsf{C}_\ell^{-1}} \\
&\le \sum_{\ell=0}^{L} \|\mathsf{Q}_\ell^\top \mathsf{C}^{-1}\boldsymbol{u}\|_{\mathsf{C}_\ell^{-1}} \|\mathsf{C}_\ell \boldsymbol{w}_\ell\|_{\mathsf{C}_\ell^{-1}} \\
&\le \left( \sum_{\ell=0}^{L} \|\mathsf{Q}_\ell^\top \mathsf{C}^{-1}\boldsymbol{u}\|_{\mathsf{C}_\ell^{-1}}^2 \right)^{1/2} \left( \sum_{\ell=0}^{L} \|\mathsf{C}_\ell \boldsymbol{w}_\ell\|_{\mathsf{C}_\ell^{-1}}^2 \right)^{1/2},
\end{aligned}
$$

where, in the last step, we used the Cauchy–Schwarz inequality. Using the definition of the $\mathsf{C}_\ell^{-1}$-energy norm, we continue our estimation as follows:

$$
\begin{aligned}
(\boldsymbol{u}, \boldsymbol{u})_{\mathsf{C}^{-1}} &\le \left( \sum_{\ell=0}^{L} (\mathsf{Q}_\ell^\top \mathsf{C}^{-1}\boldsymbol{u}, \mathsf{Q}_\ell^\top \mathsf{C}^{-1}\boldsymbol{u})_{\mathsf{C}_\ell^{-1}} \right)^{1/2} \left( \sum_{\ell=0}^{L} (\mathsf{C}_\ell \boldsymbol{w}_\ell, \mathsf{C}_\ell \boldsymbol{w}_\ell)_{\mathsf{C}_\ell^{-1}} \right)^{1/2} \\
&= \left( \sum_{\ell=0}^{L} (\mathsf{Q}_\ell^\top \mathsf{C}^{-1}\boldsymbol{u}, \mathsf{C}_\ell^{-1}\mathsf{Q}_\ell^\top \mathsf{C}^{-1}\boldsymbol{u})_{\ell} \right)^{1/2} \left( \sum_{\ell=0}^{L} (\mathsf{C}_\ell \boldsymbol{w}_\ell, \boldsymbol{w}_\ell)_{\ell} \right)^{1/2} \\
&\overset{(10.1.1)}{=} (\mathsf{C}^{-1}\boldsymbol{u}, \mathsf{C}\mathsf{C}^{-1}\boldsymbol{u})^{1/2} \left( \sum_{\ell=0}^{L} (\mathsf{C}_\ell \boldsymbol{w}_\ell, \boldsymbol{w}_\ell)_{\ell} \right)^{1/2} \\
&= \|\boldsymbol{u}\|_{\mathsf{C}^{-1}} \left( \sum_{\ell=0}^{L} (\mathsf{C}_\ell \boldsymbol{w}_\ell, \boldsymbol{w}_\ell)_{\ell} \right)^{1/2}.
\end{aligned}
$$

In conclusion, for every $(\boldsymbol{w}_\ell) \in \mathsf{Q}[\boldsymbol{u}]$,

$$
(\boldsymbol{u}, \boldsymbol{u})_{\mathsf{C}^{-1}} \le \sum_{\ell=0}^{L} (\mathsf{C}_\ell \boldsymbol{w}_\ell, \boldsymbol{w}_\ell)_{\ell}.
$$

On the other hand, choosing

$$
\boldsymbol{w}_\ell = \mathsf{C}_\ell^{-1}\mathsf{Q}_\ell^\top \mathsf{C}^{-1}\boldsymbol{u} \in \mathbb{R}^{m_\ell}, \quad 0 \le \ell \le L,
$$

we have $(\boldsymbol{w}_\ell) \in \mathsf{Q}[\boldsymbol{u}]$; see Problem 10.3. Then

$$
\begin{aligned}
\sum_{\ell=0}^{L} (\mathsf{C}_\ell \boldsymbol{w}_\ell, \boldsymbol{w}_\ell)_{\ell} &= \sum_{\ell=0}^{L} (\mathsf{C}_\ell \mathsf{C}_\ell^{-1}\mathsf{Q}_\ell^\top \mathsf{C}^{-1}\boldsymbol{u}, \mathsf{C}_\ell^{-1}\mathsf{Q}_\ell^\top \mathsf{C}^{-1}\boldsymbol{u})_{\ell} \\
&= \sum_{\ell=0}^{L} (\mathsf{C}^{-1}\boldsymbol{u}, \mathsf{Q}_\ell \mathsf{C}_\ell^{-1}\mathsf{Q}_\ell^\top \mathsf{C}^{-1}\boldsymbol{u}) \\
&\overset{(10.1.1)}{=} (\mathsf{C}^{-1}\boldsymbol{u}, \mathsf{C}\mathsf{C}^{-1}\boldsymbol{u}) \\
&= (\boldsymbol{u}, \boldsymbol{u})_{\mathsf{C}^{-1}}.
\end{aligned}
$$

□

Next, let us write two seemingly complicated formulas for the largest and smallest eigenvalues of CA. However, in light of the last result, the results are actually rather trivial.

**Theorem 10.1.12** (Eigenvalues of CA). *Let* $\{\mathsf{Q}_j\}_{j=0}^{L}$ *be a set of ordered prolongation matrices for which Assumption* (SS1) *holds, and let* $\mathsf{C} \in \mathbb{R}^{n\times n}$ *be as in* (10.1.1). *Assume that* $\mathsf{A} \in \mathbb{R}^{n\times n}$ *is SPD with respect to the Euclidean inner product. Then the eigenvalues of* CA *are positive. Moreover,*

$$\lambda_{\max}(\mathsf{CA}) = \max_{\boldsymbol{u}\in\mathbb{R}^n_\star} \frac{(\mathsf{A}\boldsymbol{u},\boldsymbol{u})}{\min\limits_{(\boldsymbol{w}_\ell)\in\mathsf{Q}[\boldsymbol{u}]} \sum_{\ell=0}^{L} (\mathsf{C}_\ell \boldsymbol{w}_\ell, \boldsymbol{w}_\ell)_\ell}, \tag{10.1.4}$$

$$\lambda_{\min}(\mathsf{CA}) = \min_{\boldsymbol{u}\in\mathbb{R}^n_\star} \frac{(\mathsf{A}\boldsymbol{u},\boldsymbol{u})}{\min\limits_{(\boldsymbol{w}_\ell)\in\mathsf{Q}[\boldsymbol{u}]} \sum_{\ell=0}^{L} (\mathsf{C}_\ell \boldsymbol{w}_\ell, \boldsymbol{w}_\ell)_\ell}. \tag{10.1.5}$$

*Proof.* Recall that the matrix CA is SPD with respect to $(\cdot,\cdot)_{\mathsf{C}^{-1}}$. Thus the eigenvalues of CA are real and positive, and the corresponding eigenvectors form a basis of $\mathbb{R}^n$ and may be chosen to be orthonormal with respect to $(\cdot,\cdot)_{\mathsf{C}^{-1}}$. Moreover, by properties of the Rayleigh quotient we have

$$\begin{aligned}\lambda_{\max}(\mathsf{CA}) &\overset{(7.3.7)}{=} \max_{\boldsymbol{u}\in\mathbb{R}^n_\star} \frac{(\mathsf{CA}\boldsymbol{u},\boldsymbol{u})_{\mathsf{C}^{-1}}}{(\boldsymbol{u},\boldsymbol{u})_{\mathsf{C}^{-1}}} \\ &\overset{(10.1.3)}{=} \max_{\boldsymbol{u}\in\mathbb{R}^n_\star} \frac{(\mathsf{A}\boldsymbol{u},\boldsymbol{u})}{\min\limits_{(\boldsymbol{w}_\ell)\in\mathsf{Q}[\boldsymbol{u}]} \sum_{\ell=0}^{L} (\mathsf{C}_\ell \boldsymbol{w}_\ell, \boldsymbol{w}_\ell)_\ell}.\end{aligned}$$

The formula for $\lambda_{\min}(\mathsf{CA})$ is established similarly. □

## 10.2 Hierarchical bases

In this section, we examine a particular subspace decomposition suitable for the finite element setting, based on a hierarchical decomposition of the finite element bases. Recall that nested families of finite element spaces were introduced in Definition 7.2.1 and that the Lagrange nodal basis is described in Definition 7.2.2.

**Definition 10.2.1** (Hierarchical finite element space). In the setting and notation of Definitions 7.2.1 and 7.2.2, we define the **level-$\ell$ hierarchical finite element space** as $W_0 := V_0$ for $l = 0$ and

$$W_\ell := \left\{ v \in V_\ell \mid v(\boldsymbol{N}_{\ell-1,j}) = 0,\ \forall\, 1 \le j \le n_{\ell-1} \right\} \quad \text{for } 1 \le \ell \le L. \tag{10.2.1}$$

Its dimension will be denoted by $m_\ell := \dim(W_\ell)$. ■

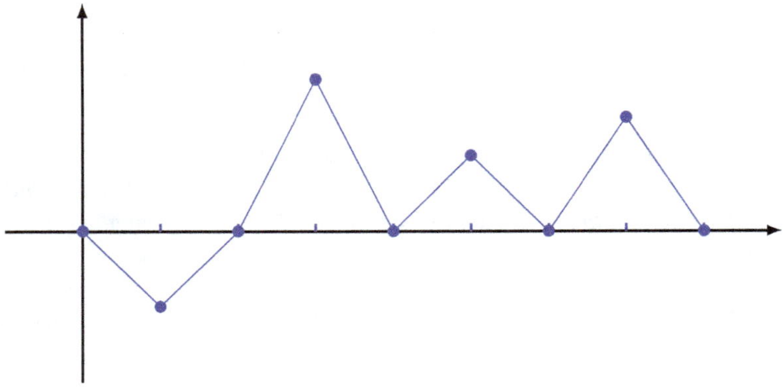

**Figure 10.1:** A sample function $v \in W_\ell$. Observe that $v(\boldsymbol{N}_{\ell-1,j}) = 0$, for all $1 \le j \le n_{\ell-1}$.

A generic $v \in W_\ell$ is depicted in Figure 10.1.

**Lemma 10.2.2** (Direct sum). *For the hierarchical spaces $W_\ell \subset V_\ell$ of Definition* 10.2.1, *we have*

$$V_\ell = V_{\ell-1} \oplus W_\ell, \quad 1 \le \ell \le L, \tag{10.2.2}$$

*and therefore*

$$V_L = W_0 \oplus W_1 \oplus \cdots \oplus W_L. \tag{10.2.3}$$

*Proof.* Recall that the Lagrange nodal interpolation operator $\mathcal{I}_\ell$ was introduced in Definition 7.4.2. Let now $v_\ell \in V_\ell$. We write

$$v_\ell = \mathcal{I}_{\ell-1}(v_\ell) + \big(v_\ell - \mathcal{I}_{\ell-1}(v_\ell)\big).$$

Clearly,

$$\mathcal{I}_{\ell-1}(v_\ell) \in V_{\ell-1}, \quad v_\ell - \mathcal{I}_{\ell-1}(v_\ell) \in W_\ell.$$

We now show that this decomposition is unique. To see this, we assume that $v_{\ell-1} \in V_{\ell-1}$ and $w_\ell \in W_\ell$ are such that

$$v_{\ell-1} + w_\ell = 0.$$

Let now $\boldsymbol{N}_{\ell-1,j}$ be a node of $\mathcal{T}_{\ell-1}$. We must have

$$v_{\ell-1}(\boldsymbol{N}_{\ell-1,j}) = -w_\ell(\boldsymbol{N}_{\ell-1,j}) = 0.$$

This implies that $v_{\ell-1} = 0$, and this in turn implies $w_\ell = 0$. This proves (10.2.2). Identity (10.2.3) follows from (10.2.2). □

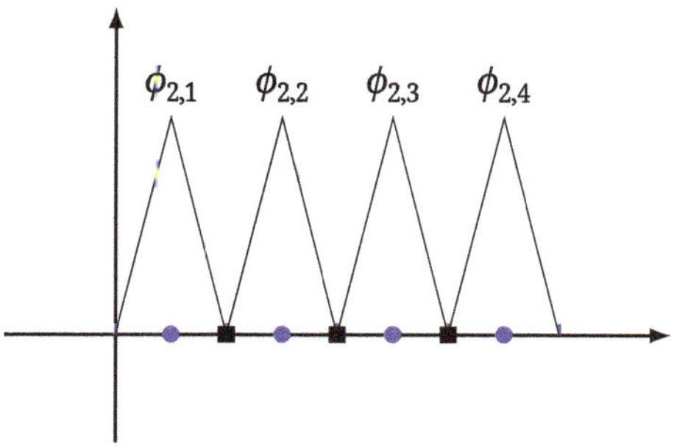


**Figure 10.2:** The nodal basis functions for $W_2$ in one dimension: $\Omega = (0,1)$, $\ell = 2$, $n_1 = 3$, $m_2 = 4$, $n_2 = 7$.

**Definition 10.2.3** (Basis of $W_\ell$). Let $1 \le \ell \le L$, and let

$$\{\boldsymbol{N}^W_{\ell,j}\}_{j=1}^{m_\ell} := \{\boldsymbol{N}_{\ell,j}\}_{j=1}^{n_\ell} \setminus \{\boldsymbol{N}_{\ell-1,j}\}_{j=1}^{n_{\ell-1}}.$$

Define $\mathcal{B}^W_\ell := \{\phi_{\ell,i}\}_{i=1}^{m_\ell} \subset W_\ell$ by

$$\phi_{\ell,i}(\boldsymbol{N}^W_{\ell,j}) = \delta_{i,j}, \quad 1 \le i,j \le m_\ell.$$

For $\ell = 0$, we set $\mathcal{B}^W_0 := \mathcal{B}_0$. ■

**Lemma 10.2.4** ($\mathcal{B}^W_\ell$ is a basis). *For each $0 \le \ell \le L$, the set $\mathcal{B}^W_\ell$ is a basis for $W_\ell$.*

*Proof.* Exercise; see Problem 10.4. □

The elements of $\mathcal{B}^W_\ell$ are illustrated in Figure 10.2.

**Definition 10.2.5** (Hierarchical basis). Assume that, for $0 \le \ell \le L$, the spaces $W_\ell \subset V_\ell$ are as in Definition 10.2.1. For every $0 \le \ell \le L$, the **hierarchical basis** of $V_\ell$ is

$$\mathcal{H}_\ell := \bigcup_{j=0}^{\ell} \mathcal{B}^W_j. \qquad ■$$

**Lemma 10.2.6** ($\mathcal{H}_\ell$ is a basis). *For $0 \le \ell \le L$, the set $\mathcal{H}_\ell$ is a basis of $V_\ell$.*

*Proof.* The result will be proved if we show that $\mathcal{H}_\ell$ is linearly independent and

$$\operatorname{span} \mathcal{H}_\ell = V_\ell.$$

Let $v_\ell \in V_\ell$. For $0 \le j \le \ell$, there are unique $w_j \in W_j$ such that

$$v_\ell = w_0 + w_1 + \cdots + w_\ell.$$

In fact, we can write down this decomposition explicitly. Observe that

$$v_\ell = \mathcal{I}_0 v_\ell + (\mathcal{I}_1 v_\ell - \mathcal{I}_0 v_\ell) + (\mathcal{I}_2 v_\ell - \mathcal{I}_1 v_\ell) + \cdots + (v_\ell - \mathcal{I}_{\ell-1} v_\ell)$$

and that $\mathcal{I}_\ell v_\ell = v_\ell$. Setting

$$w_j := \begin{cases} \mathcal{I}_0 v_\ell, & j = 0, \\ \mathcal{I}_j v_\ell - \mathcal{I}_{j-1} v_\ell, & 1 \leq j \leq \ell, \end{cases}$$

gives the result. Since $w_j \in W_j$ for $0 \leq j \leq \ell$, there are unique coefficients $\{c_{\ell,k}\}_{\ell=0,k=1}^{L,m_\ell} \in \mathbb{R}$ such that

$$w_j = \sum_{k=1}^{m_j} c_{j,k} \phi_{j,k}.$$

Hence

$$v_\ell = \sum_{j=0}^{\ell} w_j = \sum_{j=0}^{\ell} \sum_{k=1}^{m_j} c_{j,k} \phi_{j,k}.$$

Consequently, $V_\ell \subset \operatorname{span} \mathcal{H}_\ell$.

On the other hand, since

$$\#(\mathcal{H}_\ell) = \#(\mathcal{B}_\ell),$$

it follows that $\mathcal{H}_\ell$ is linearly independent, since $\mathcal{B}_\ell$ is a basis. □

The reader is referred to Figure 10.3 for a comparison between the nodal and hierarchical bases, $\mathcal{B}_\ell$ and $\mathcal{H}_\ell$, respectively.

## 10.3 Hierarchical basis preconditioner

We now connect the spaces $W_j$ to $V_\ell$, where $0 \leq j \leq \ell$. By doing this, we will have the tools to build a preconditioner based on the hierarchical basis. Be careful, as the number of indices in this section can become overwhelming.

**Proposition 10.3.1** (Hierarchical prolongation). *Let $0 \leq j \leq \ell$. By $\mathcal{B}_j^W = \{\phi_{j,i}\}_{i=1}^{m_j}$ and $\mathcal{B}_\ell = \{\psi_{\ell,i}\}_{i=1}^{n_\ell}$ denote the usual bases for $W_j$ and $V_\ell$, respectively. There are unique numbers*

$$q_{j,k,i}^{\ell} \in \mathbb{R}, \quad 1 \leq k \leq n_\ell, \quad 1 \leq i \leq m_j,$$

*such that*

$$\phi_{j,i} = \sum_{k=1}^{n_\ell} q_{j,k,i}^{\ell} \psi_{\ell,k}. \tag{10.3.1}$$

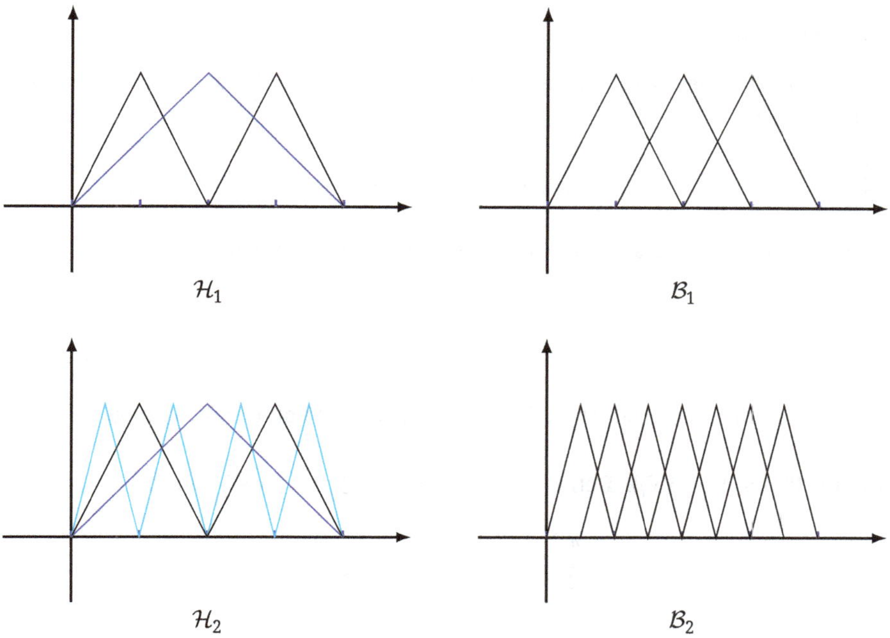


**Figure 10.3:** Hierarchical versus standard Lagrange nodal bases.

*Proof.* Exercise; see Problem 10.5. □

**Definition 10.3.2** (Hierarchical prolongation matrix). For $0 \le j \le \ell$, define the matrix $\mathsf{Q}_j^\ell \in \mathbb{R}^{n_\ell \times m_j}$ by

$$[\mathsf{Q}_j^\ell]_{k,i} := q_{j,k,i}^\ell, \quad 1 \le k \le n_\ell, \quad 1 \le i \le m_j,$$

where the numbers $q_{j,k,i}^\ell$ are defined in Proposition 10.3.1. We call $\mathsf{Q}_j^\ell$ a **hierarchical prolongation matrix.** ■

**Lemma 10.3.3** (Coordinate vectors). *Let $\mathsf{Q}_j^\ell \in \mathbb{R}^{n_\ell \times m_j}$ be a hierarchical prolongation matrix. Then*

$$\operatorname{rank}(\mathsf{Q}_j^\ell) = m_j.$$

*Let now $w_j \in W_j$, and assume that $\boldsymbol{w}_j \in \mathbb{R}^{m_j}$ is its coordinate vector with respect to the basis $\mathcal{B}_j^W$. Then its coordinate vector with respect to the basis $\mathcal{B}_\ell$ is*

$$\mathsf{Q}_j^\ell \boldsymbol{w}_j \in \mathbb{R}^{n_\ell}.$$

*In other words,*

$$\boldsymbol{w}_j \in \mathbb{R}^{m_j} \overset{\mathcal{B}_j^W}{\longleftrightarrow} w_j \in W_j \subset V_\ell \overset{\mathcal{B}_\ell}{\longleftrightarrow} \mathsf{Q}_j^\ell \boldsymbol{w}_j \in \mathbb{R}^{n_\ell}.$$

*Proof.* Exercise; see Problem 10.6. □

**Remark 10.3.4** (Not nested). Note that the family of spaces $\{W_j\}_{j=0}^L$ is hierarchical, but it is not nested, i. e.,

$$W_0 \not\subset W_1 \not\subset W_2 \not\subset \cdots.$$

Furthermore, it makes no sense to compose the prolongation matrices as we did in the past:

$$\mathsf{Q}_j^\ell \neq \mathsf{Q}_k^\ell \mathsf{Q}_j^k$$

for $j < k < \ell$. In fact, the product on the right-hand side is not usually defined. ◀

**Definition 10.3.5** (Matrix $\mathsf{C}_j$). Define the bilinear form $\mathcal{Y}_j : W_j \times W_j \to \mathbb{R}$ by

$$\mathcal{Y}_j(w_j, v_j) := \sum_{r=1}^{m_j} w_j(\boldsymbol{N}_{j,r}^W) v_j(\boldsymbol{N}_{j,r}^W), \quad \forall\, w_j, v_j \in W_j.$$

Let $\mathcal{B}_j^W = \{\phi_{j,i}\}_{i=1}^{m_j}$ be a basis of $W_j$. Define $\mathsf{C}_j \in \mathbb{R}^{m_j \times m_j}$ by

$$\begin{aligned}[\mathsf{C}_j]_{i,k} &:= \mathcal{Y}_j(\phi_{j,k}, \phi_{j,i}) \\ &= \sum_{r=1}^{m_j} \phi_{j,i}(\boldsymbol{N}_{j,r}^W)\phi_{j,k}(\boldsymbol{N}_{j,r}^W) \\ &= \sum_{r=1}^{m_j} \delta_{ir}\delta_{kr} \\ &= \delta_{ik}.\end{aligned} \tag{10.3.2}$$

■

**Definition 10.3.6** (Hierarchical basis preconditioner). Let $\mathcal{B}_L^V = \{\psi_{L,i}\}_{i=1}^{n_L}$ be the Lagrange nodal basis for the finite element space $V_L$. Recall that the bilinear form $\mathcal{A} : H_0^1(\Omega) \times H_0^1(\Omega) \to \mathbb{R}$ was defined in (7.1.2) and that the stiffness matrix $\mathsf{A}_L \in \mathbb{R}^{n_L \times n_L}$ satisfies

$$[\mathsf{A}_L]_{i,j} = \mathcal{A}(\psi_{L,j}, \psi_{L,i}), \quad 1 \le i, j \le n_L.$$

The **hierarchical basis preconditioner for** $\mathsf{A}_L$ is

$$\begin{aligned}\mathsf{C}_{\mathrm{H}} &= \sum_{\ell=0}^{L} \mathsf{Q}_\ell^L \mathsf{C}_\ell^{-1} (\mathsf{Q}_\ell^L)^\top \\ &= \sum_{\ell=0}^{L} \mathsf{Q}_\ell^L (\mathsf{Q}_\ell^L)^\top,\end{aligned} \tag{10.3.3}$$

where $\mathsf{C}_\ell$ is as in (10.3.2), and $\mathsf{Q}_\ell^L \in \mathbb{R}^{n_L \times m_\ell}$ is the hierarchical prolongation matrix from Definition 10.3.2. ∎

**Remark 10.3.7** (Notation). The hierarchical basis preconditioner was developed and analyzed by *H. Yserentant*; see [2, 39, 40]. As a homage to its origins, we use the symbol $\mathcal{Y}$ to denote the bilinear forms that define this preconditioner. ◀

**Lemma 10.3.8** (Hierarchical space decomposition). *Assumption* (SS1) *holds for the hierarchical basis decomposition. In particular, given*

$$\boldsymbol{u}_L \in \mathbb{R}^{n_L} \overset{\mathcal{B}_L}{\longleftrightarrow} u_L \in V_L,$$

*there are unique*

$$\boldsymbol{w}_\ell \in \mathbb{R}^{m_\ell} \overset{\mathcal{B}_\ell^W}{\longleftrightarrow} w_\ell \in W_\ell, \quad 0 \le \ell \le L,$$

*such that*

$$\boldsymbol{u}_L = \sum_{\ell=0}^{L} \mathsf{Q}_\ell^L \boldsymbol{w}_\ell \in \mathbb{R}^{n_L} \overset{\mathcal{B}_L}{\longleftrightarrow} u_L = \sum_{\ell=0}^{L} w_\ell \in V_L.$$

*Furthermore, the hierarchical basis preconditioner* $\mathsf{C}_\mathrm{H}$*, defined in* (10.3.3)*, is SPD.*

*Proof.* This follows from Lemmas 10.1.9, 10.2.2, and 10.2.6. The details are left as an exercise; see Problem 10.7. □

Our next goal is to show, using Theorem 10.1.12, bounds on the smallest and largest eigenvalues of $\mathsf{C}_\mathrm{H}\mathsf{A}_L$. We will only be able to do this in two dimensions, i. e., $d = 2$. For the three-dimensional case, see, for instance, [28].

In what follows, if $D \subset \mathbb{R}^d$ is a bounded domain and $v \in L^1(D)$, then we denote its average by

$$\langle v \rangle_D := \frac{1}{|D|} \int_D v(\boldsymbol{x})\, \mathrm{d}\boldsymbol{x}. \tag{10.3.4}$$

To continue, we need a series of technical results. For more details and proofs, the reader is referred, for instance, to [10].

**Theorem 10.3.9** (Mean-zero Poincaré). *Let* $d \in \mathbb{N}$*, and let* $\Omega$ *be a bounded polytopal domain in* $\mathbb{R}^d$*. There is a constant* $C > 0$*, depending only on* $\Omega$*, such that, for every* $u \in H^1(\Omega)$*,*

$$\|u - \langle u \rangle_\Omega\|_{L^2(\Omega)} \le C|u|_{H^1(\Omega)}.$$

We saw a particular version of the following result in Chapter 7, specifically, in Lemma 7.3.4. To state it, we recall that if $D \subset \mathbb{R}^d$ is a bounded domain, then for $1 \le q \le \infty$ and $m \in \mathbb{N}_0$, we define

$$W^{m,q}(D) := \{v \in L^q(D) \mid \partial^{\boldsymbol{\alpha}} v \in L^q(D),\ \forall \boldsymbol{\alpha} \in \mathbb{N}_0^d\ |\boldsymbol{\alpha}| \le m\},$$

where

$$\partial^{\boldsymbol{\alpha}} v = \frac{\partial^{|\boldsymbol{\alpha}|} v}{\partial x_1^{\alpha_1} \cdots \partial x_d^{\alpha_d}}$$

and

$$|\boldsymbol{\alpha}| = \sum_{i=1}^{d} \alpha_i.$$

**Theorem 10.3.10** (Inverse inequality). *Suppose that $\Omega$ is a bounded polytopal domain in $\mathbb{R}^d$, $\{\mathcal{T}_\ell\}_{\ell=0}^L$ is a nested family of triangulations of $\Omega$, and $\{S_\ell\}_{\ell=0}^L$ are the associated finite element spaces of piecewise linear functions. Let $1 \le p, q \le \infty$ and $0 \le m \le j$. Then there is a constant $C > 0$, depending only on the shape regularity of $\{\mathcal{T}_\ell\}_{\ell=0}^L$, such that, for all $v \in S_\ell$ and $K \in \mathcal{T}_\ell$,*

$$\|v\|_{W^{j,p}(K)} \le C h_\ell^{m-j+d/p-d/q} \|v\|_{W^{m,q}(K)}. \tag{10.3.5}$$

*Proof.* See [10, Section 5.3]. □

The following result is known as a *trace estimate* and allows us to control the values of a function on the boundary by its function values and derivatives in the interior of the domain.

**Theorem 10.3.11** (Trace estimate). *Let $\Omega$ be a bounded polytopal domain in $\mathbb{R}^d$, and let $1 \le p \le \infty$. There is a constant $C > 0$, depending only on $\Omega$, such that, for all $v \in W^{1,p}(\Omega)$,*

$$\|v\|_{L^p(\partial\Omega)} \le C \|v\|_{L^p(\Omega)}^{1-1/p} \|v\|_{W^{1,p}(\Omega)}^{1/p}. \tag{10.3.6}$$

*Proof.* We refer the reader to [10, Section 1.6] for a proof in a simplified setting. For the full details, we refer the reader, for instance, to [22, Theorem 1.5.1.10]. □

**Theorem 10.3.12** (Interpolation error). *Suppose that $\Omega$ is a bounded polytopal domain in $\mathbb{R}^d$, $\{\mathcal{T}_\ell\}_{\ell=0}^L$ is a nested family of triangulations of $\Omega$, and $\{S_\ell\}_{\ell=0}^L$ are the associated finite element spaces of piecewise linear functions. There is a constant $C > 0$ such that, for all $\ell \ge 1$ and $v_\ell \in S_\ell$, we have*

$$\left\|v_\ell - \mathcal{I}_{\ell-1}^{+} v_\ell\right\|_{L^2(\Omega)} \le C h_\ell |v_\ell|_{H^1(\Omega)}, \tag{10.3.7}$$

*Proof.* Note that estimate (7.4.5) from Theorem 7.4.4 does not cover this case, since $v_\ell$ does not belong to $H^2(\Omega)$. However, the stated result still holds since we interpolate a very specific class of functions. The proof of the one-dimensional case is left as an exercise; see Problem 10.8. □

In two space dimensions the embedding $H^1(\Omega) \hookrightarrow L^p(\Omega)$ holds for all $1 \le p < \infty$ but is not valid for $p = \infty$. This means that if we consider finite element functions, then the norm equivalence constant must depend on the mesh size and blow up when the mesh size goes to zero. This, weak, dependence is quantified in the next result.

**Theorem 10.3.13** (Almost embedding). *Suppose that $\Omega$ is a bounded polygonal domain in $\mathbb{R}^2$, $\{\mathcal{T}_\ell\}_{\ell=0}^L$ is a nested family of triangulations of $\Omega$, and $\{V_\ell\}_{\ell=0}^L$ and $\{S_\ell\}_{\ell=0}^L$ are the associated finite element spaces of piecewise linear functions. There is a constant $C > 0$ such that, for all $\ell \ge 0$ and $v_\ell \in V_\ell$, we have*

$$\|v_\ell\|_{L^\infty(\Omega)} \le C\sqrt{1 + |\log(h_\ell)|}|v_\ell|_{H^1(\Omega)},$$

*for some constant $C > 0$, which is independent of $\ell$ but depends upon the shape of $\Omega$. Further, there is another constant $C > 0$, depending on the shape regularity of $\{\mathcal{T}_\ell\}_{\ell=0}^L$, such that, for all $v_\ell \in S_\ell$ and $K \in \mathcal{T}_\ell$,*

$$\|v_\ell - \langle v_\ell \rangle_K\|_{L^\infty(K)} \le C\sqrt{1 + |\log(h_\ell)|}|v_\ell|_{H^1(K)},$$

*where we recall that the average was defined in* (10.3.4).

*Proof.* See [10, Section 4.9]. □

The following result can be proven from those stated above.

**Lemma 10.3.14** (Multilevel almost embedding). *Suppose that $\Omega$ is a bounded polygonal domain in $\mathbb{R}^2$, $\{\mathcal{T}_\ell\}_{\ell=0}^L$ is a nested family of triangulations of $\Omega$, and $\{S_\ell\}_{\ell=0}^L$ are the associated finite element spaces of piecewise linear functions. There is a constant $C > 0$ such that, for all $\ell > 0$, $0 \le j < \ell$, $v_\ell \in S_\ell$, and $K_j \in \mathcal{T}_j$,*

$$\|v_\ell - \langle v_\ell \rangle_{K_j}\|_{L^\infty(K_j)} \le C\sqrt{1 + \left|\log\left(\frac{h_j}{h_\ell}\right)\right|}|v_\ell|_{H^1(K_j)}, \tag{10.3.8}$$

*where we recall that $\langle v_\ell \rangle_{K_j}$ was defined in* (10.3.4).

*Proof.* Exercise; see Problem 10.9. □

**Lemma 10.3.15** (Interpolation error estimate). *Let $\Omega \subset \mathbb{R}^2$ be a bounded polygonal domain. Recall that, for $0 \le \ell \le L$, $\mathcal{I}_\ell$ is the Lagrange nodal interpolation operator. Set $\mathcal{I}_{-1} \equiv 0$. Then there is a constant $C > 0$, independent of $L$ and $\ell$, such that, for all $u_L \in V_L$,*

$$\|\mathcal{I}_\ell u_L - \mathcal{I}_{\ell-1} u_L\|_{L^2(\Omega)} \le C h_\ell (1 + \sqrt{L - \ell}) |u_L|_{H^1(\Omega)}. \tag{10.3.9}$$

*Proof.* For $u_L \in V_L$, define the piecewise constant function $\bar{u}_L^\ell$ as

$$\bar{u}_L^\ell|_K \coloneqq \langle u_L \rangle_K, \quad \forall K \in \mathcal{T}_\ell.$$

Then

$$\begin{aligned}
\|\mathcal{I}_\ell u_L - \mathcal{I}_{\ell-1} u_L\|_{L^2(\Omega)}^2 &= \|\mathcal{I}_\ell u_L - \mathcal{I}_{\ell-1}[\mathcal{I}_\ell[u_L]]\|_{L^2(\Omega)}^2 \\
&\overset{(10.3.7)}{\leq} Ch_\ell^2 |\mathcal{I}_\ell[u_L]|_{H^1(\Omega)}^2 \\
&= Ch_\ell^2 \sum_{K \in \mathcal{T}_\ell} |\mathcal{I}_\ell[u_L]|_{H^1(K)}^2 \\
&= Ch_\ell^2 \sum_{K \in \mathcal{T}_\ell} |\mathcal{I}_\ell u_L - \bar{u}_L^\ell|_{H^1(K)}^2 \\
&\overset{(10.3.5)}{\leq} Ch_\ell^2 \sum_{K \in \mathcal{T}_\ell} \|\mathcal{I}_\ell u_L - \bar{u}_L^\ell\|_{L^\infty(K)}^2 \\
&\leq Ch_\ell^2 \sum_{K \in \mathcal{T}_\ell} \|u_L - \bar{u}_L^\ell\|_{L^\infty(K)}^2 \\
&\overset{(10.3.8)}{\leq} Ch_\ell^2 \sum_{K \in \mathcal{T}_\ell} \left(1 + \left|\log\left(\frac{h_\ell}{h_L}\right)\right|\right) |u_L|_{H^1(K)}^2 \\
&= Ch_\ell^2 \left(1 + \left|\log\left(\frac{h_\ell}{h_L}\right)\right|\right) |u_L|_{H^1(\Omega)}^2 .
\end{aligned}$$

Now notice that

$$h_\ell = h_0 2^{-\ell} \quad 1 \leq \ell \leq L.$$

So

$$\log(h_\ell / h_L) = \log(2^{L-\ell}) = (L - \ell)\log(2).$$

The result follows. □

**Lemma 10.3.16** (Smallest eigenvalue bound). *Let $d = 2$. There is a constant $C > 0$, independent of $L$, such that*

$$\lambda_{\min}(\mathsf{C}_{\mathrm{H}} \mathsf{A}_L) \geq \frac{C}{1 + |\log(h_L)|^2}. \tag{10.3.10}$$

*Proof.* Recall that, by definition, for all $w_{\ell,1}, w_{\ell,2} \in W_\ell$,

$$\mathcal{Y}_\ell(w_{\ell,1}, w_{\ell,2}) = \sum_{i=1}^{m_\ell} w_{\ell,1}(\boldsymbol{N}_{\ell,i}^W) w_{\ell,2}(\boldsymbol{N}_{\ell,i}^W).$$

Let

$$\boldsymbol{w}_{\ell,\alpha} \in \mathbb{R}^{m_\ell} \overset{\mathcal{B}_\ell^W}{\longleftrightarrow} w_{\ell,\alpha} \in W_\ell, \quad \alpha = 1, 2.$$

Then

$$
\begin{aligned}
(\mathsf{C}_\ell \boldsymbol{w}_{\ell,1}, \boldsymbol{w}_{\ell,2})_\ell &= \sum_{i=1}^{m_\ell} [\boldsymbol{w}_{\ell,1}]_i [\boldsymbol{w}_{\ell,2}]_i \\
&= \sum_{i=1}^{m_\ell} w_{\ell,1}(\boldsymbol{N}_{\ell,i}^W) w_{\ell,2}(\boldsymbol{N}_{\ell,i}^W) \\
&= \mathcal{Y}_\ell(w_{\ell,1}, w_{\ell,2}).
\end{aligned}
$$

This last object is like a mass-lumped inner product. All that is missing is a factor of $h_\ell^2$. As in Lemma 7.3.3, there are constants $C_1 > 0$ and $C_2 > 0$ such that, for $0 \le \ell \le L$ and $w_\ell \in W_\ell$,

$$
C_1 h_\ell^2 \mathcal{Y}_\ell(w_\ell, w_\ell) \le \|w_\ell\|_{L^2(\Omega)}^2 \le C_2 h_\ell^2 \mathcal{Y}_\ell(w_\ell, w_\ell). \tag{10.3.11}
$$

Therefore, for all $w_\ell \in W_\ell \xleftrightarrow{\mathcal{B}_\ell^W} \boldsymbol{w}_\ell \in \mathbb{R}^{m_\ell}$,

$$
\begin{aligned}
(\mathsf{C}_\ell \boldsymbol{w}_\ell, \boldsymbol{w}_\ell)_\ell &= h_\ell^{-2} h_\ell^2 \mathcal{Y}_\ell(w_\ell, w_\ell) \\
&\overset{(10.3.11)}{\le} C_1^{-1} h_\ell^{-2} \|w_\ell\|_{L^2(\Omega)}^2 \\
&= C_1^{-1} h_\ell^{-2} \|w_\ell - \mathcal{I}_{\ell-1} w_\ell\|_{L^2(\Omega)}^2 \\
&\overset{(7.4.4)}{\le} C_1^{-1} C |w_\ell|_{H^1(\Omega)}^2 \\
&\overset{(10.3.5)}{\le} C h_\ell^{-2} \|w_\ell\|_{L^2(\Omega)}^2 \\
&\overset{(10.3.11)}{\le} C (\mathsf{C}_\ell \boldsymbol{w}_\ell, \boldsymbol{w}_\ell)_\ell.
\end{aligned} \tag{10.3.12}
$$

Since the constants in (10.3.12) are independent of $\ell$, we can add (10.3.12) for $0 \le \ell \le L$ to conclude that, for any collection $(w_\ell)$ with $w_\ell \in W_\ell$, we have the equivalence

$$
C_3 \sum_{\ell=0}^{L} |w_\ell|_{H^1(\Omega)}^2 \le \sum_{\ell=0}^{L} (\mathsf{C}_\ell \boldsymbol{w}_\ell, \boldsymbol{w}_\ell)_\ell \le C_4 \sum_{\ell=0}^{L} |w_\ell|_{H^1(\Omega)}^2. \tag{10.3.13}
$$

Let now $u_L \in V_L \xleftrightarrow{\mathcal{B}_\ell} \boldsymbol{u}_L \in \mathbb{R}^{n_L}$. Recall that, for $0 \le \ell \le L$, there are unique $w_\ell \in W_\ell$ such that

$$
u_L = \sum_{\ell=0}^{L} w_\ell.
$$

Namely,

$$
w_\ell = \mathcal{I}_\ell u_L - \mathcal{I}_{\ell-1} u_L, \quad 1 \le \ell \le L,
$$

and

$$w_0 = \mathcal{I}_0 u_L.$$

We make the usual identification $w_\ell \in W_\ell \overset{\mathcal{B}_\ell^W}{\longleftrightarrow} \boldsymbol{w}_\ell \in \mathbb{R}^{m_\ell}$ and observe that

$$(\boldsymbol{w}_\ell)_{\ell=0}^L \in \mathsf{Q}[\boldsymbol{u}_L]$$

with respect to the hierarchical prolongation matrices from Definition 10.3.2. Then from (10.3.12) we have

$$\begin{aligned}
\sum_{\ell=0}^{L} (\mathsf{C}_\ell \boldsymbol{w}_\ell, \boldsymbol{w}_\ell)_\ell &\le C \sum_{\ell=0}^{L} h_\ell^{-2} \|w_\ell\|_{L^2(\Omega)}^2 \\
&\overset{(10.3.9)}{\le} C \sum_{\ell=0}^{L} (1 + \sqrt{L-\ell})^2 |u_L|_{H^1(\Omega)}^2 \\
&\le C \sum_{\ell=0}^{L} (1 + L - \ell) |u_L|_{H^1(\Omega)}^2 \\
&\le C(1 + L + L^2) |u_L|_{H^1(\Omega)}^2 \\
&\le C L^2 |u_L|_{H^1(\Omega)}^2,
\end{aligned}$$

where the last estimate follows from $L \ge 1$. Observe now that from

$$\begin{aligned}
|\log(h_L)|^2 &= |\log(h_0 2^{-L})|^2 \\
&= |\log(h_0) - L\log(2)|^2 \\
&= \log^2(h_0) - 2\log(h_0) L \log(2) + L^2 \log^2(2)
\end{aligned}$$

it follows that there is $C > 0$ such that

$$L^2 \le C(1 + |\log(h_L)|^2).$$

This, combined with

$$|u_L|_{H^1(\Omega)}^2 = \mathcal{A}(u_L, u_L) = (\mathsf{A}_L \boldsymbol{u}_L, \boldsymbol{u}_L)_L,$$

allows us to conclude that

$$\sum_{\ell=0}^{L} (\mathsf{C}_\ell \boldsymbol{w}_\ell, \boldsymbol{w}_\ell) \le C(1 + |\log(h_L)|^2)(\mathsf{A}_L \boldsymbol{u}_L, \boldsymbol{u}_L)_L.$$

This last inequality, combined with Theorem 10.1.12, yields

$$\lambda_{\min}(\mathsf{C}_{\mathrm{H}} \mathsf{A}_L) \ge C(1 + |\log(h_L)|^2)^{-1},$$

where the constant $C > 0$ is independent of $L$. □

Our next objective is to obtain an upper bound for the largest eigenvalue. First, we need a string of technical results. We start with a new trace-type estimate.

**Lemma 10.3.17** (Trace-type estimate). *Let $\Omega$ be a polytopal domain in $\mathbb{R}^d$, let $\{\mathcal{T}_\ell\}_{\ell=0}^L$ be a nested family of triangulations of $\Omega$, and let $\{S_\ell\}_{\ell=0}^L$ be the ensuing family of finite element spaces. There is a constant $C > 0$, depending only on the shape regularity of the triangulations, such that, for all $K \in \mathcal{T}_\ell$ and $v_\ell, w_\ell \in S_\ell$, we have*

$$\int_{\partial K} \frac{\partial v_\ell}{\partial n} w_\ell \, \mathrm{d}\boldsymbol{s} \le C h_\ell^{-d/2} |v_\ell|_{H^1(K)} \|w_\ell\|_{L^1(\partial K)}. \tag{10.3.14}$$

*Proof.* We first apply Young's inequality and then a trace inequality, followed by an inverse inequality, to obtain the result

$$\begin{aligned} \int_{\partial K} \frac{\partial v_\ell}{\partial n} w_\ell \, \mathrm{d}\boldsymbol{s} &\le \left\| \frac{\partial v_\ell}{\partial n} \right\|_{L^\infty(\partial K)} \|w_\ell\|_{L^1(\partial K)} \\ &\overset{(10.3.6)}{\le} C |v_\ell|_{W^{1,\infty}(K)} \|w_\ell\|_{L^1(\partial K)} \\ &\overset{(10.3.5)}{\le} C h_\ell^{-d/2} |v_\ell|_{H^1(K)} \|w_\ell\|_{L^1(\partial K)}. \end{aligned}$$

□

**Lemma 10.3.18** (Strengthened Cauchy–Schwarz inequality I). *Let $\Omega \subset \mathbb{R}^d$ be a bounded polytopal domain in $\mathbb{R}^d$. There is a constant $C > 0$ such that, for $0 \le \ell \le k \le L$, $v_\ell \in V_\ell$, and $v_k \in V_k$, we have*

$$\int_{\Omega} \nabla v_\ell \cdot \nabla v_k \, \mathrm{d}\boldsymbol{x} \le 2^{(\ell-k)/2} C |v_\ell|_{H^1(\Omega)} \big( h_k^{-1} \|v_k\|_{L^2(\Omega)} \big). \tag{10.3.15}$$

*Proof.* For any $K \in \mathcal{T}_\ell$, since $\Delta v_\ell|_K \equiv 0$,

$$\begin{aligned} \mathrm{I} := \int_K \nabla v_\ell \cdot \nabla v_k \, \mathrm{d}\boldsymbol{x} &= \int_{\partial K} \frac{\partial v_\ell}{\partial n} v_k \, \mathrm{d}\boldsymbol{s} \\ &\overset{(10.3.14)}{\le} C h_\ell^{-d/2} |v_\ell|_{H^1(K)} \int_{\partial K} |v_k| \, \mathrm{d}\boldsymbol{s}. \end{aligned}$$

We next observe that since $K \in \mathcal{T}_\ell$, its boundary is the union of boundaries of some $K' \in \mathcal{T}_k$. Over each one of these components, the function $v_k$ is linear, and so we can approximate the integral by a quadrature, i. e.,

$$\int_{\partial K} |v_k| \, \mathrm{d}\boldsymbol{s} \le C h_k^{d-1} \sum_{\boldsymbol{N}_k \in \partial K} |v_k(\boldsymbol{N}_k)|.$$

Then we may continue our estimation to get

$$\begin{aligned} \mathrm{I} &\le Ch_\ell^{-d/2}|v_\ell|_{H^1(K)}\left(h_k^{d-1}\sum_{N_k\in\partial K}|v_k(N_k)|\right) \\ &\le C\left(\frac{h_k}{h_\ell}\right)^{d/2}|v_\ell|_{H^1(K)}h_k^{-1}\left(\frac{h_\ell}{h_k}\right)^{\frac{d-1}{2}}\left(h_k^d\sum_{N_k\in\partial K}|v_k(N_k)|^2\right)^{1/2} \\ &\overset{(7.3.2)}{\le} C\left(\frac{h_k}{h_\ell}\right)^{1/2}|v_\ell|_{H^1(K)}h_k^{-1}\|v_k\|_{L^2(K)}, \end{aligned}$$

where we also used that, due to quasi-uniformity,

$$\#\left(\partial K\bigcap\{N_{k,j}\}_{j=1}^{n_k}\right)\le\left(\frac{h_\ell}{h_k}\right)^{d-1}.$$

Thus

$$\begin{aligned} \mathrm{I} &= \sum_{K\in\mathcal{T}_\ell}\int_K\nabla v_\ell\cdot\nabla v_k\,\mathrm{d}\boldsymbol{x} \\ &\le C2^{(\ell-k)/2}\sum_{K\in\mathcal{T}_\ell}|v_\ell|_{H^1(K)}h_k^{-1}\|v_k\|_{L^2(K)} \\ &\le C2^{(\ell-k)/2}|v_\ell|_{H^1(\Omega)}h_k^{-1}\|v_k\|_{L^2(\Omega)}. \end{aligned}$$

□

**Lemma 10.3.19** (Strengthened Cauchy–Schwarz inequality II). *In the setting of Lemma 10.3.18, there is a constant $C > 0$ such that, for all $0 \le \ell \le k \le L$, $w_\ell \in W_\ell$, and $w_k \in W_k$,*

$$\int_\Omega\nabla w_\ell\cdot\nabla w_k\,\mathrm{d}\boldsymbol{x}\le 2^{(\ell-k)/2}C|w_\ell|_{H^1(\Omega)}|w_k|_{H^1(\Omega)}. \tag{10.3.16}$$

*Proof.* Observe that, for every $w_k \in W_k$,

$$w_k = w_k - \mathcal{I}_{k-1}w_k.$$

Then we use the interpolation error estimate of (10.3.7) to conclude that

$$\|w_k\|_{L^2(\Omega)} \le Ch_k|w_k|_{H^1(\Omega)}.$$

Now, since $w_\ell \in V_\ell$ and $w_k \in V_k$, we use (10.3.15) to conclude

$$\begin{aligned} \int_\Omega\nabla w_\ell\cdot\nabla w_k\,\mathrm{d}\boldsymbol{x} &\le C2^{(\ell-k)/2}|w_\ell|_{H^1(\Omega)}h_k^{-1}\|w_k\|_{L^2(\Omega)} \\ &\le 2^{(\ell-k)/2}C|w_\ell|_{H^1(\Omega)}|w_k|_{H^1(\Omega)}. \end{aligned}$$

□

The next result is a particular case of the so-called Young inequality for convolutions.

**Lemma 10.3.20** (Discrete convolution estimate). *For two sequences of nonnegative numbers* $\{a_j\}_{j\in\mathbb{Z}}$ *and* $\{b_j\}_{j\in\mathbb{Z}} \subset \mathbb{R}$ *such that*

$$s_1 := \sum_{j=-\infty}^{\infty} a_j < \infty$$

*and*

$$s_2 := \sum_{j=-\infty}^{\infty} b_j^2 < \infty,$$

*we have the following estimate:*

$$\sum_{j=-\infty}^{\infty} \left( \sum_{k=-\infty}^{\infty} a_{j-k} b_k \right)^2 \le s_1^2 s_2. \tag{10.3.17}$$

*Proof.* Exercise; see Problem 10.10. □

**Lemma 10.3.21** (Largest eigenvalue bound). *Let $d \in \mathbb{N}$. There is a constant $C > 0$, independent of $L$, such that*

$$\lambda_{\max}(\mathsf{C}_{\mathrm{H}}\mathsf{A}_L) \le C.$$

*Proof.* Let $v_L \in V_L$, and let

$$v_L \in V_L \overset{\mathcal{B}_L}{\longleftrightarrow} \boldsymbol{v}_L \in \mathbb{R}^{n_L}.$$

There exist unique $w_\ell \in W_\ell \overset{\mathcal{B}_\ell^W}{\longleftrightarrow} \boldsymbol{w}_\ell \in \mathbb{R}^{m_\ell}$, $\ell = 0, \dots, L$, such that

$$v_L = \sum_{\ell=0}^{L} w_\ell \overset{\mathcal{B}_L}{\longleftrightarrow} \boldsymbol{v}_L = \sum_{\ell=0}^{L} \mathsf{Q}_\ell^L \boldsymbol{w}_\ell.$$

Then

$$\begin{aligned}
(\boldsymbol{v}_L, \boldsymbol{v}_L)_{\mathsf{A}_L} &= \mathcal{A}\left( \sum_{\ell=0}^{L} w_\ell, \sum_{k=0}^{L} w_k \right) \\
&= \sum_{\ell=0}^{L} \sum_{k=0}^{L} \int_\Omega \nabla w_\ell \cdot \nabla w_k \, \mathrm{d}\boldsymbol{x} \\
&\overset{(10.3.16)}{\le} C \sum_{\ell,k=0}^{L} 2^{-|\ell-k|/2} |w_\ell|_{H^1(\Omega)} |w_k|_{H^1(\Omega)}
\end{aligned}$$

$$
\begin{aligned}
&\leq C \sum_{\ell=0}^{L} \left( \sum_{k=0}^{L} 2^{-|\ell-k|/2} |w_k|_{H^1(\Omega)} \right) |w_\ell|_{H^1(\Omega)} \qquad (10.3.18)\\
&\leq C \left( \sum_{\ell=0}^{L} \left( \sum_{k=0}^{L} 2^{-|\ell-k|/2} |w_k|_{H^1(\Omega)} \right)^2 \right)^{1/2} \left( \sum_{\ell=0}^{L} |w_\ell|^2_{H^1(\Omega)} \right)^{1/2}\\
&\overset{(10.3.17)}{\leq} C \left( \sum_{\ell=0}^{L} |w_\ell|^2_{H^1(\Omega)} \right)^{1/2} \left( \sum_{\ell=0}^{L} |w_\ell|^2_{H^1(\Omega)} \right)^{1/2}\\
&= C \sum_{\ell=0}^{L} |w_\ell|^2_{H^1(\Omega)}\\
&\overset{(10.3.13)}{\leq} \frac{C}{C_3} \sum_{\ell=0}^{L} (\boldsymbol{w}_\ell, \boldsymbol{w}_\ell)_{\mathsf{C}_\ell}.
\end{aligned}
$$

Recall now that since hierarchical decompositions are unique,

$$
\begin{aligned}
\lambda_{\max}(\mathsf{C}_{\mathrm{H}}\mathsf{A}_L) &\overset{(10.1.4)}{=} \max_{\boldsymbol{v}_L \in \mathbb{R}^{n_L}_\star} \frac{(\boldsymbol{v}_L, \boldsymbol{v}_L)_{\mathsf{A}_L}}{\sum_{\ell=0}^{L} (\boldsymbol{w}_\ell, \boldsymbol{w}_\ell)_{\mathsf{C}_\ell}}\\
&\overset{(10.3.18)}{\leq} \max_{\boldsymbol{v}_L \in \mathbb{R}^{n_L}_\star} \frac{\frac{C}{C_3} \sum_{\ell=0}^{L} (\boldsymbol{w}_\ell, \boldsymbol{w}_\ell)_{\mathsf{C}_\ell}}{\sum_{\ell=0}^{L} (\boldsymbol{w}_\ell, \boldsymbol{w}_\ell)_{\mathsf{C}_\ell}}\\
&= \frac{C}{C_3}\\
&= C,
\end{aligned}
$$

as claimed. □

We finally provide a condition number estimate for the hierarchical basis preconditioner.

**Theorem 10.3.22** (Condition number estimate). *Let $d = 2$. There is a constant $C > 0$, independent of $L$, such that*

$$
\kappa(\mathsf{C}_{\mathrm{H}}\mathsf{A}_L) = \frac{\lambda_{\max}}{\lambda_{\min}} \leq C(1 + |\log(h_L)|^2).
$$

*Proof.* The result follows from Lemmas 10.3.16 and 10.3.19. □

**Remark 10.3.23** (Condition number estimate). The previous result has shown that, for $d = 2$,

$$
\kappa(\mathsf{C}_{\mathrm{H}}\mathsf{A}_L) \leq C(1 + |\log(h_L)|^2).
$$

This estimate is quite useful, since the logarithmic dependence on $h_L$ is rather weak. For example, suppose

$$h_L = \frac{1}{2^L}.$$

Then

$$|\log(h_L)|^2 = L^2 |\log(1/2)|^2.$$

The analysis that we provided, however, will only work in the two-dimensional case. In three and higher space dimensions, we lose the nice logarithmic dependence in the lower bound, and the condition number estimate worsens; see, for instance, [28]. ◂

## 10.4 The BPX preconditioner

In this section, we investigate the BPX preconditioner introduced by Bramble, Pasciak, and Xu [7]. This preconditioner has better performance than that based on hierarchical bases, in the sense that the dependence on $h_L$ can be removed.

**Definition 10.4.1** (BPX preconditioner). For $0 \le \ell \le L$, let $\mathcal{B}_\ell = \{\psi_{\ell,j}\}_{j=1}^{n_\ell}$ be the Lagrange nodal basis for the space $V_\ell$. Define the bilinear form $\mathcal{J}_\ell : V_\ell \times V_\ell \to \mathbb{R}$ by

$$\mathcal{J}_\ell(w_\ell, v_\ell) = h_\ell^{d-2} \sum_{i=1}^{n_\ell} w_\ell(\boldsymbol{N}_{\ell,i}) v_\ell(\boldsymbol{N}_{\ell,i}), \quad \forall\, w_\ell, v_\ell \in V_\ell.$$

The matrix $\mathsf{C}_\ell \in \mathbb{R}^{n_\ell \times n_\ell}$ is defined as

$$\begin{aligned}[\mathsf{C}_\ell]_{j,k} &= \mathcal{J}_\ell(\psi_{\ell,j}, \psi_{\ell,k}) \\ &= h_\ell^{d-2} \delta_{j,k}\end{aligned}$$

for $1 \le j, k \le n_\ell$. The **BPX preconditioner** is

$$\begin{aligned}\mathsf{C}_{\text{BPX}} &:= \sum_{\ell=0}^{L} \mathsf{P}_{\ell,L} \mathsf{C}_\ell^{-1} \mathsf{P}_{\ell,L}^\top \\ &= \sum_{\ell=0}^{L} h_\ell^{2-d} \mathsf{P}_{\ell,L} \mathsf{P}_{\ell,L}^\top,\end{aligned}$$

where $\mathsf{P}_{\ell,L} \in \mathbb{R}^{n_L \times n_\ell}$ is the standard multilevel prolongation matrix. ■

**Remark 10.4.2** (Notation). As mentioned above, the BPX preconditioner is named after *James Bramble*, *Joseph Pasciak*, and *Jinchao Xu*, who developed it in [7]. As a homage to its origins, we denote by $\mathcal{J}$ the bilinear form that defines it. ◂

**Remark 10.4.3** (Subspace decomposition). In the BPX framework, we effectively take

$$\mathsf{Q}_\ell^L = \mathsf{P}_{\ell,L}. \qquad ◂$$

**Lemma 10.4.4** (Subspace decomposition). *Assumption* (SS1) *holds for the BPX framework, that is, for every $u_L \in V_L$, there exist $v_\ell \in V_\ell$, $0 \le \ell \le L$, such that*

$$u_L = \sum_{\ell=0}^{L} v_\ell$$

*or, equivalently,*

$$\boldsymbol{u}_L = \sum_{\ell=0}^{L} \mathsf{P}_{\ell,L} \boldsymbol{v}_\ell$$

*with*

$$v_\ell \in V_\ell \stackrel{\mathcal{B}_\ell}{\longleftrightarrow} \boldsymbol{v}_\ell \in \mathbb{R}^{n_\ell}$$

*and*

$$u_L \in V_L \stackrel{\mathcal{B}_L}{\longleftrightarrow} \boldsymbol{u}_L \in \mathbb{R}^{n_L}.$$

*This decomposition is not unique.*

*Proof.* The existence of the decomposition follows from the nestedness of the spaces

$$V_0 \subset V_1 \subset V_2 \subset \cdots \subset V_{L-1} \subset V_L.$$

The details are left for an exercise; see Problem 10.11. □

**Lemma 10.4.5** (Strengthened Cauchy–Schwarz III). *There is a constant $C > 0$ such that, for $0 \le j \le \ell$, $v_j \in V_j$, and $v_\ell \in V_\ell$, we have*

$$\int_\Omega \nabla v_j \cdot \nabla v_\ell \, \mathrm{d}\boldsymbol{x} \le C 2^{-(\ell-j)/2} \frac{\|v_j\|_{L^2(\Omega)}}{h_j} \frac{\|v_\ell\|_{L^2(\Omega)}}{h_\ell}. \tag{10.4.1}$$

*Proof.* This follows from (10.3.15) and the inverse inequality

$$|v_j|_{H^1(\Omega)} \le C h_j^{-1} \|v_j\|_{L^2(\Omega)} \quad \forall\, v_j \in V_j.$$ □

We now provide an upper bound on the largest eigenvalue.

**Lemma 10.4.6** (Largest eigenvalue bound). *There is $C > 0$, independent of $L$, such that*

$$\lambda_{\max}(\mathsf{C}_{\mathrm{BPX}} \mathsf{A}_L) \le C.$$

*Proof.* Let $u_L \in V_L$. There exist $v_\ell \in V_\ell$, $0 \le \ell \le L$, such that

$$u_L = \sum_{\ell=0}^{L} v_\ell,$$

or

$$\boldsymbol{u}_L = \sum_{\ell=0}^{L} \mathsf{P}_{\ell,L} \boldsymbol{v}_\ell,$$

where

$$v_\ell \in V_\ell \overset{\mathcal{B}_\ell}{\longleftrightarrow} \boldsymbol{v}_\ell \in \mathbb{R}^{n_\ell}.$$

Although the decomposition is not unique, we still write

$$(\boldsymbol{v}_\ell) \in \mathsf{Q}[\boldsymbol{u}_L].$$

Then

$$\begin{aligned}
(\boldsymbol{u}_L, \boldsymbol{u}_L)_{\mathsf{A}_L} &= \mathcal{A}\left(\sum_{j=0}^{L} v_j, \sum_{\ell=0}^{L} v_\ell\right) \\
&= \sum_{\ell=0}^{L}\sum_{j=0}^{L} \mathcal{A}(v_j, v_\ell) \\
&\overset{(10.4.1)}{\leq} C \sum_{\ell=0}^{L}\sum_{j=0}^{L} 2^{-|j-\ell|/2} h_j^{-1} \|v_j\|_{L^2(\Omega)} h_\ell^{-1} \|v_\ell\|_{L^2(\Omega)} \\
&\leq C\left[\sum_{\ell=0}^{L}\left(\sum_{j=0}^{L} 2^{-\frac{|\ell-j|}{2}} \frac{\|v_j\|_{L^2(\Omega)}}{h_j}\right)^2\right]^{1/2} \left[\sum_{\ell=0}^{L}\left(\frac{\|v_\ell\|_{L^2(\Omega)}}{h_\ell}\right)^2\right]^{1/2} \\
&\overset{(10.3.17)}{\leq} C \sum_{j=0}^{L} h_j^{-2} \|v_j\|_{L^2(\Omega)}^2 \\
&\overset{(7.3.2)}{\leq} C \sum_{j=0}^{L} h_j^{d-2} (\boldsymbol{v}_j, \boldsymbol{v}_j)_j \\
&= C \sum_{j=0}^{L} (\mathsf{C}_j \boldsymbol{v}_j, \boldsymbol{v}_j)_j.
\end{aligned}$$

Now recall that

$$\lambda_{\max}(\mathsf{C}_{\mathrm{BPX}} \mathsf{A}_L) \overset{(10.1.4)}{=} \max_{\boldsymbol{u}_L \in \mathbb{R}_\star^{n_L}} \frac{(\boldsymbol{u}_L, \boldsymbol{u}_L)_{\mathsf{A}_L}}{\min\limits_{(\boldsymbol{w}_\ell) \in \mathsf{Q}[\boldsymbol{u}_L]} \sum_{\ell=0}^{L} (\boldsymbol{w}_\ell, \boldsymbol{w}_\ell)_{\mathsf{C}_\ell}}$$

and that by Theorem 10.1.11 the minimum in the denominator is attained for some $(\boldsymbol{w}_\ell) \in \mathsf{Q}[\boldsymbol{u}_L]$. We may then set $\boldsymbol{v}_\ell = \boldsymbol{w}_\ell$ to estimate

$$
\begin{aligned}
\lambda_{\max}(\mathsf{C}_{\mathrm{BPX}}\mathsf{A}_L) &\overset{(10.1.4)}{=} \max_{\boldsymbol{u}_L\in\mathbb{R}_\star^{n_L}} \frac{(\boldsymbol{u}_L,\boldsymbol{u}_L)_{\mathsf{A}_L}}{\min\limits_{(\boldsymbol{w}_\ell)\in\mathrm{Q}[\boldsymbol{u}_L]} \sum_{\ell=0}^{L}(\boldsymbol{w}_\ell,\boldsymbol{w}_\ell)_{\mathsf{C}_\ell}} \\
&= \max_{\boldsymbol{u}_L\in\mathbb{R}_\star^{n_L}} \frac{(\boldsymbol{u}_L,\boldsymbol{u}_L)_{\mathsf{A}_L}}{\sum_{\ell=0}^{L}(\boldsymbol{w}_\ell,\boldsymbol{w}_\ell)_{\mathsf{C}_\ell}} \\
&\le \max_{\boldsymbol{u}_L\in\mathbb{R}_\star^{n_L}} \frac{C\sum_{\ell=0}^{L}(\mathsf{C}_\ell\boldsymbol{v}_\ell,\boldsymbol{v}_\ell)_\ell}{\sum_{\ell=0}^{L}(\boldsymbol{w}_\ell,\boldsymbol{w}_\ell)_{\mathsf{C}_\ell}} \\
&\le C.
\end{aligned}
$$

□

Before we establish a lower bound for the smallest eigenvalue, we need another technical result.

**Lemma 10.4.7** (Ritz projection). *Let $u \in H_0^1(\Omega)$. Then, for all $1 \le \ell \le L$,*

$$\mathcal{R}_{\ell-1}u = \mathcal{R}_{\ell-1}(\mathcal{R}_\ell u), \tag{10.4.2}$$

*where for $0 \le \ell \le L$, $\mathcal{R}_\ell : H_0^1(\Omega) \to V_\ell$ is the Ritz projection. In other words, $\mathcal{R}_{\ell-1} = \mathcal{R}_{\ell-1}\mathcal{R}_\ell$.*

*Proof.* By definition,

$$\mathcal{A}(\mathcal{R}_{\ell-1}(\mathcal{R}_\ell u), w_{\ell-1}) = \mathcal{A}(\mathcal{R}_\ell u, w_{\ell-1}), \quad \forall\, w_{\ell-1} \in V_{\ell-1}.$$

Since $V_{\ell-1} \subset V_\ell$, we also have

$$\mathcal{A}(\mathcal{R}_\ell u, w_{\ell-1}) = \mathcal{A}(u, w_{\ell-1}), \quad \forall\, w_{\ell-1} \in V_{\ell-1}.$$

Also observe that

$$\mathcal{A}(\mathcal{R}_{\ell-1} u, w_{\ell-1}) = \mathcal{A}(u, w_{\ell-1}), \quad \forall\, w_{\ell-1} \in V_{\ell-1}.$$

Hence

$$\mathcal{A}(\mathcal{R}_{\ell-1}(\mathcal{R}_\ell u), w_{\ell-1}) = \mathcal{A}(\mathcal{R}_{\ell-1} u, w_{\ell-1}), \quad \forall\, w_{\ell-1} \in V_{\ell-1},$$

and we conclude that $\mathcal{R}_{\ell-1} = \mathcal{R}_{\ell-1}\mathcal{R}_\ell$ since

$$\mathcal{R}_{\ell-1}(\mathcal{R}_\ell u),\ \mathcal{R}_{\ell-1} u \in V_{\ell-1}.$$

□

**Lemma 10.4.8** (Smallest eigenvalue bound). *Assume that the standard regularity condition of Definition 7.1.2 holds. Then there is a constant $C > 0$, independent of $L$, such that*

$$\lambda_{\min}(\mathsf{C}_{\mathrm{BPX}}\mathsf{A}_L) \ge C.$$

*Proof.* Let $u_L \in V_L$ and set, for $0 \le \ell \le L$,

$$v_\ell := \mathcal{R}_\ell u_L - \mathcal{R}_{\ell-1} u_L,$$

where we defined $\mathcal{R}_{-1} \equiv 0$. Since

$$\mathcal{R}_L u_L = u_L,$$

it follows that

$$u_L = \sum_{\ell=0}^{L} v_\ell \in V_L \overset{\mathcal{B}_\ell}{\longleftrightarrow} \boldsymbol{u}_L = \sum_{\ell=0}^{L} \mathsf{P}_{\ell,L} \boldsymbol{v}_\ell \in \mathbb{R}^{n_L},$$

where

$$v_\ell \in V_\ell \overset{\mathcal{B}_\ell}{\longleftrightarrow} \boldsymbol{v}_\ell \in \mathbb{R}^{n_\ell}.$$

Observe now that

$$\mathcal{A}(v_j, v_\ell) = 0, \quad 0 \le j \ne \ell \le L. \tag{10.4.3}$$

To see this, recall that, in general,

$$\mathcal{A}(\mathcal{R}_\ell u_L, w_\ell) = \mathcal{A}(u_L, w_\ell), \quad \forall\, w_\ell \in V_\ell.$$

Suppose $j < \ell$ for definiteness. Then, since $V_j \subset V_\ell$,

$$\mathcal{A}(\mathcal{R}_\ell u_L, w_j) = \mathcal{A}(u_L, w_j), \quad \forall\, w_j \in V_j.$$

In particular, since

$$v_j := \mathcal{R}_j u_L - \mathcal{R}_{j-1} u_L \in V_j \subset V_\ell,$$

it follows that

$$\mathcal{A}(\mathcal{R}_\ell u_L, v_j) = \mathcal{A}(u_L, v_j).$$

Likewise,

$$\mathcal{A}(\mathcal{R}_{\ell-1} u_L, v_j) = \mathcal{A}(u_L, v_j).$$

Subtracting, we have

$$\mathcal{A}(\mathcal{R}_\ell u_L - \mathcal{R}_{\ell-1} u_L, v_j) = 0.$$

Since we have assumed that the standard regularity condition holds, we may use Theorem 7.4.7 to estimate, for $1 \le \ell \le L$,

$$\begin{aligned} h_\ell^{-2}\|v_\ell\|_{L^2(\Omega)}^2 &= h_\ell^{-2}\|\mathcal{R}_\ell u_L - \mathcal{R}_{\ell-1}u_L\|_{L^2(\Omega)}^2 \\ &\overset{(10.4.2)}{=} h_\ell^{-2}\|\mathcal{R}_\ell u_L - \mathcal{R}_{\ell-1}(\mathcal{R}_\ell u_L)\|_{L^2(\Omega)}^2 \\ &\overset{(7.4.6)}{\le} Ch_\ell^{-2}h_{\ell-1}^2|\mathcal{R}_\ell u_L - \mathcal{R}_{\ell-1}(\mathcal{R}_\ell u_L)|_{H^1(\Omega)}^2 \\ &= C|\mathcal{R}_\ell u_L - \mathcal{R}_{\ell-1}(\mathcal{R}_\ell u_L)|_{H^1(\Omega)}^2 \\ &= C|v_\ell|_{H^1(\Omega)}^2. \end{aligned} \tag{10.4.4}$$

Estimate (10.4.4) holds trivially for $\ell = 0$. Therefore

$$\begin{aligned} \sum_{\ell=0}^{L}(\mathsf{C}_\ell \boldsymbol{v}_\ell, \boldsymbol{v}_\ell)_\ell &= \sum_{\ell=0}^{L} h_\ell^{d-2}\|\boldsymbol{v}_\ell\|_\ell^2 \\ &\overset{(7.3.2)}{\le} C\sum_{\ell=0}^{L} h_\ell^{-2}\|v_\ell\|_{L^2(\Omega)}^2 \\ &\overset{(10.4.4)}{\le} C\sum_{\ell=0}^{L} |v_\ell|_{H^1(\Omega)}^2 \\ &\overset{(10.4.3)}{=} C|u_L|_{H^1(\Omega)}^2. \end{aligned} \tag{10.4.5}$$

Finally,

$$\begin{aligned} \lambda_{\min}(\mathsf{C}_{\mathrm{BPX}}\mathsf{A}_L) &= \min_{\boldsymbol{u}_L \in \mathbb{R}_\star^{n_L}} \frac{(\boldsymbol{u}_L, \boldsymbol{u}_L)_{\mathsf{A}_L}}{\min_{(\boldsymbol{w}_\ell)\in \mathsf{Q}[\boldsymbol{u}_L]} \sum_{\ell=0}^{L}(\boldsymbol{w}_\ell, \boldsymbol{w}_\ell)_{\mathsf{C}_\ell}} \\ &\overset{(10.4.5)}{\ge} \min_{\boldsymbol{u}_L \in \mathbb{R}_\star^{n_L}} \frac{(\mathsf{A}_L \boldsymbol{u}_L, \boldsymbol{u}_L)_L}{C|u_L|_{H^1(\Omega)}^2} \\ &= C. \end{aligned}$$

□

**Theorem 10.4.9** (Condition number estimate). *There is a constant $C > 0$, independent of L, such that*

$$\kappa(\mathsf{C}_{\mathrm{BPX}}\mathsf{A}_L) = \frac{\lambda_{\max}(\mathsf{C}_{\mathrm{BPX}}\mathsf{A}_L)}{\lambda_{\min}(\mathsf{C}_{\mathrm{BPX}}\mathsf{A}_L)} \le C.$$

*Proof.* The result follows from Lemmas 10.4.6 and 10.4.8. □

## 10.5 Some details on implementation

In this section, which is heavily inspired by [8, Section III.15], we discuss some details on the implementation of the BPX preconditioner $\mathsf{C}_{\mathrm{BPX}}$ of Definition 10.4.1. First of all, it clearly makes no sense to construct either the matrices $\mathsf{P}_{\ell,L}$ or their transposes. This is especially the case if this is going to be used as a preconditioner in the PCG scheme detailed in Chapter 2; see (2.5.1) and (2.5.5) of Definition 2.5.1. In fact, all we need in this case is to compute, for $\boldsymbol{r}_L \in \mathbb{R}^{n_L}$,

$$\boldsymbol{z}_L \coloneqq \mathsf{C}_{\mathrm{BPX}}\boldsymbol{r}_L = \sum_{\ell=0}^{L} \mathsf{P}_{\ell,L}(h_\ell^{2-d}\mathsf{P}_{\ell,L}^\top \boldsymbol{r}_L).$$

We can separate this construction into four steps:

- Multilevel decomposition: For each $0 \le \ell \le L$, compute

$$\boldsymbol{r}_\ell = \mathsf{P}_{\ell,L}^\top \boldsymbol{r}_L \in \mathbb{R}^{n_\ell}.$$

- Scaling: For each $0 \le \ell \le L$, define

$$\boldsymbol{p}_\ell \coloneqq h_\ell^{2-d}\boldsymbol{r}_\ell \in \mathbb{R}^{n_\ell}. \tag{10.5.1}$$

- Prolongation: For each $0 \le \ell \le L$, compute

$$\boldsymbol{z}_{\ell,L} = \mathsf{P}_{\ell,L}\boldsymbol{p}_\ell \in \mathbb{R}^{n_L}.$$

- Accumulation: Compute $\sum_{\ell=0}^{L} \boldsymbol{z}_{\ell,L}$.

The scaling step (10.5.1) poses no difficulty, so we focus our attention on the implementation of the other steps.

It is perhaps easiest to understand the prolongation and accumulation steps together. The trick is to reorder the terms in the sum. Indeed, following Definition 9.1.1 we may write

$$\begin{aligned}
\boldsymbol{z}_L &= \sum_{\ell=0}^{L}\prod_{k=1}^{L-\ell} \mathsf{P}_{L-k}\boldsymbol{p}_\ell \\
&= \boldsymbol{p}_L + \sum_{\ell=0}^{L-1}\prod_{k=1}^{L-\ell} \mathsf{P}_{L-k}\boldsymbol{p}_\ell \\
&= \boldsymbol{p}_L + \mathsf{P}_{L-1}\sum_{\ell=0}^{L-1}\prod_{k=2}^{L-\ell} \mathsf{P}_{L-k}\boldsymbol{p}_\ell \\
&= \boldsymbol{p}_L + \mathsf{P}_{L-1}\sum_{\ell=0}^{L-1}\prod_{k=1}^{(L-1)-\ell} \mathsf{P}_{(L-1)-k}\boldsymbol{p}_\ell
\end{aligned}$$

$$
\begin{aligned}
&= \boldsymbol{p}_L + \mathsf{P}_{L-1}\left( \boldsymbol{p}_{L-1} + \sum_{\ell=0}^{L-2} \prod_{k=1}^{(L-1)-\ell} \mathsf{P}_{(L-1)-k} \boldsymbol{p}_\ell \right) \\
&= \boldsymbol{p}_L + \mathsf{P}_{L-1}\left( \boldsymbol{p}_{L-1} + \mathsf{P}_{L-2} \sum_{\ell=0}^{L-2} \prod_{k=2}^{(L-1)-\ell} \mathsf{P}_{(L-1)-k} \boldsymbol{p}_\ell \right) \\
&= \boldsymbol{p}_L + \mathsf{P}_{L-1}\left( \boldsymbol{p}_{L-1} + \mathsf{P}_{L-2} \sum_{\ell=0}^{L-2} \prod_{k=1}^{(L-2)-\ell} \mathsf{P}_{(L-2)-k} \boldsymbol{p}_\ell \right) \\
&= \cdots .
\end{aligned}
$$

This greatly resembles the well-known Horner method for the evaluation of polynomials and the construction of a Lagrange interpolating polynomial in Newton form; see [31, Example 9.6]. The vectors $\boldsymbol{p}_\ell$ may be thought of as the coefficients of the polynomial, whereas instead of multiplying by (the value of) the free variable, we apply the operators $\mathsf{P}_{L-\ell}$. This can be easily described algorithmically. Let $\{\boldsymbol{p}_\ell \in \mathbb{R}^{n_\ell}\}_{\ell=0}^{L}$ and set $\boldsymbol{z}_0 := \boldsymbol{p}_0$. For $\ell = 1$ to $L$,

$$
\boldsymbol{z}_\ell := \boldsymbol{p}_\ell + \mathsf{P}_\ell \boldsymbol{z}_{\ell-1} \in \mathbb{R}^{n_\ell}. \tag{10.5.2}
$$

The previous considerations show that to accomplish the prolongation and accumulation steps together, we merely need to understand the action of a prolongation matrix. To do so, we invoke Lemma 7.2.10: the columns of the matrix $\mathsf{P}_{\ell-1}$ are nothing but the coefficient vectors of the standard Lagrange basis of $V_{\ell-1}$ in the standard Lagrange basis of $V_\ell$. Then this can be easily related to point values. Let

$$
\boldsymbol{v}_\ell = \mathsf{P}_{\ell-1}\boldsymbol{v}_{\ell-1}, \quad v_\ell \in V_\ell \overset{\mathcal{B}_\ell}{\longleftrightarrow} \boldsymbol{v}_\ell \in \mathbb{R}^{n_\ell}, \quad v_{\ell-1} \in V_{\ell-1} \overset{\mathcal{B}_{\ell-1}}{\longleftrightarrow} \boldsymbol{v}_{\ell-1} \in \mathbb{R}^{n_{\ell-1}}.
$$

If there is a vertex $\boldsymbol{N}_{\ell,i}$ that belongs to both $\mathcal{T}_\ell$ and $\mathcal{T}_{\ell-1}$, then $v_\ell(\boldsymbol{N}_{\ell,i}) = v_{\ell-1}(\boldsymbol{N}_{\ell,i})$. On the other hand, if $\boldsymbol{N}_{\ell,i}$ is a newly created vertex, then it must be the midpoint of an edge in $\mathcal{T}_{\ell-1}$. In other words, we must have

$$
\boldsymbol{N}_{\ell,i} = \frac{1}{2}(\boldsymbol{N}_{\ell-1,\alpha} + \boldsymbol{N}_{\ell-1,\beta}),
$$

for some $\boldsymbol{N}_{\ell-1,\alpha}$ and $\boldsymbol{N}_{\ell-1,\beta}$ that are vertices of $\mathcal{T}_{\ell-1}$. Since the corresponding basis functions are piecewise linear, we must necessarily have that

$$
\psi_{\ell-1,\alpha}(\boldsymbol{N}_{\ell,i}) = \psi_{\ell-1,\beta}(\boldsymbol{N}_{\ell,i}) = \frac{1}{2}.
$$

Consequently,

$$
[\boldsymbol{v}_\ell]_i = \frac{1}{2}([\boldsymbol{v}_{\ell-1}]_\alpha + [\boldsymbol{v}_{\ell-1}]_\beta).
$$

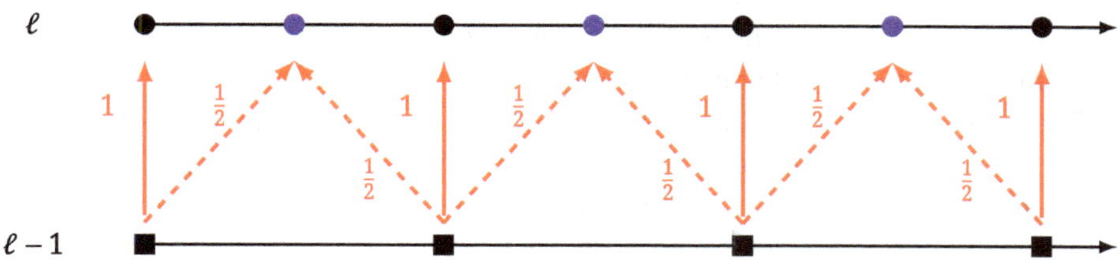


**Figure 10.4:** Flow of information for the prolongation between levels. Arrows depict the flow of information. The weights of the arrows indicate the necessary scaling.

Using the adjacency graphs of $\mathcal{T}_\ell$ and $\mathcal{T}_{\ell-1}$, it is possible to quickly find the indices $\alpha$ and $\beta$ corresponding to $i$. The flow of information can be easily represented graphically; see Figures 3.3 and 10.4 for the depiction in one dimension.

Then it remains to understand the implementation of the multilevel decomposition $\{\boldsymbol{r}_\ell\}_{\ell=0}^{L}$. Its definition, together with Definition 9.1.4, yields

$$
\begin{aligned}
\boldsymbol{r}_{L-1} &= \mathsf{R}_{L-1,L}\boldsymbol{r}_L = \mathsf{R}_{L-1}\boldsymbol{r}_L,\\
\boldsymbol{r}_{L-2} &= \mathsf{R}_{L-2,L}\boldsymbol{r}_L = \prod_{k=L-2}^{L-1}\mathsf{R}_k\boldsymbol{r}_L = \mathsf{R}_{L-2}\mathsf{R}_{L-1}\boldsymbol{r}_L = \mathsf{R}_{L-2}\boldsymbol{r}_{L-1},\\
\boldsymbol{r}_{L-2} &= \mathsf{R}_{L-3,L}\boldsymbol{r}_L = \prod_{k=L-3}^{L-1}\mathsf{R}_k\boldsymbol{r}_L = \mathsf{R}_{L-3}\prod_{k=L-2}^{L-1}\mathsf{R}_k\boldsymbol{r}_L = \mathsf{R}_{L-3}\boldsymbol{r}_{L-2},\\
&\;\;\vdots\\
\boldsymbol{r}_{\ell-1} &= \mathsf{R}_{\ell-1}\boldsymbol{r}_\ell,\\
&\;\;\vdots\\
\boldsymbol{r}_0 &= \mathsf{R}_0\boldsymbol{r}_1.
\end{aligned}
$$

Clearly, the algorithm to compute the multilevel decomposition is the following. Let $\boldsymbol{r}_L \in \mathbb{R}^{n_L}$. For $\ell = L-1$ down to 0,

$$\boldsymbol{r}_\ell = \mathsf{R}_\ell\boldsymbol{r}_{\ell+1} \in \mathbb{R}^{n_\ell}. \tag{10.5.3}$$

Finally, since, for every $0 \le \ell \le L-1$, it follows that $\mathsf{R}_\ell = \mathsf{P}_\ell^\top$, the action of a restriction matrix $\mathsf{R}_\ell$ is obtained by simply reversing the direction of the arrows in Figure 10.4; see also Figure 3.2.

## 10.6 A computational experiment

We conclude this chapter with a computational experiment illustrating the performance of the BPX preconditioner when used within the preconditioned conjugate gradient

(PCG) method. We consider the one-dimensional Poisson problem (1.7.1) on $\Omega = (0,1)$ with homogeneous Dirichlet boundary conditions.

The problem is discretized using a standard finite difference scheme. We choose $L$, and for $\ell = 0, \ldots, L$, we construct a finite difference scheme like (1.7.3) with $n = n_\ell$. Then we obtain, at every level, a linear system of the form

$$\mathsf{A}_\ell \boldsymbol{u}_\ell = \boldsymbol{f}_\ell,$$

where the stiffness matrix $\mathsf{A}_\ell \in \mathbb{R}^{n_\ell \times n_\ell}$ is given by

$$[\mathsf{A}_\ell]_{i,i} = \frac{2}{h_\ell}, \quad [\mathsf{A}_\ell]_{i,j} = -\frac{1}{h_\ell}, \quad i,j = 1, \ldots, n_\ell, \ |i-j| \le 1,$$

and the right-hand side $\boldsymbol{f}_\ell \in \mathbb{R}^{n_\ell}$ is manufactured in terms of the exact solution.

We set $L = 9$, so that $n_L = 1023$. We specify the exact solution

$$[\boldsymbol{u}_L^{\mathrm{E}}]_i = u_{L,i}^{\mathrm{E}} = \exp(\sin(3\pi x_{L,i})) - 1,$$

which satisfies the homogeneous boundary conditions. The initial approximation $\boldsymbol{u}_L^0$ for the PCG iteration is chosen using pseudorandom numbers.

The linear system is solved using the PCG algorithm; see Listing 2.4. The preconditioner is the BPX preconditioner $\mathsf{C}_{\mathrm{BPX}}$, described in Definition 10.4.1 and implemented as detailed in Section 10.5; see Listings 10.2, 4.7, and 4.8. The PCG iteration is stopped when

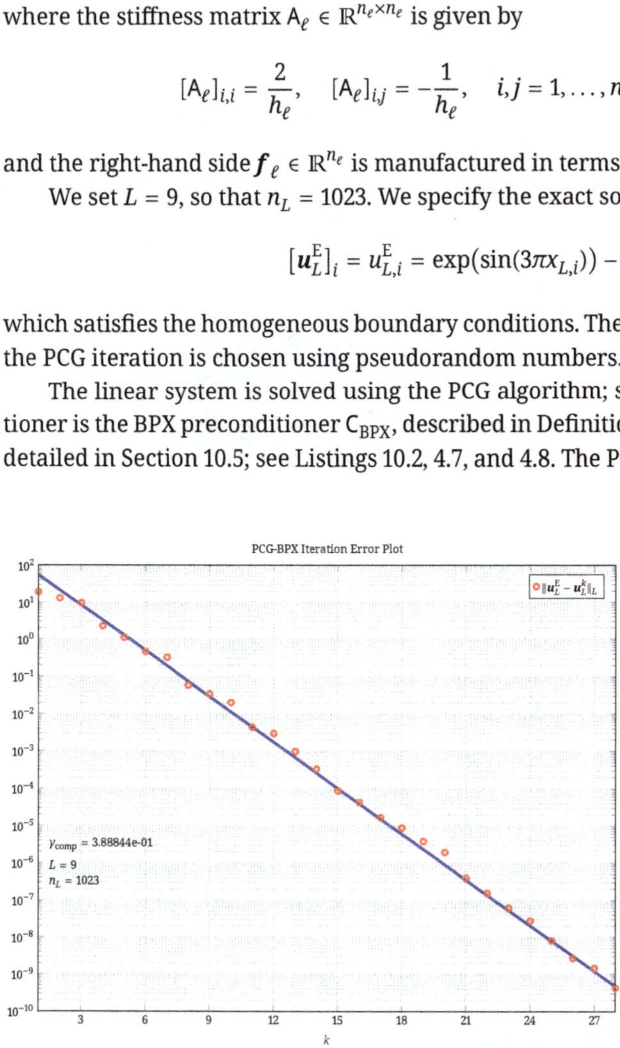


**Figure 10.5:** The true error $\|\boldsymbol{u}_L^{\mathrm{E}} - \boldsymbol{u}_L^k\|_L$ versus iteration number $k$ for the BPX-preconditioned PCG method. The parameters used are $n_L = 1023$ and $L = 9$ levels for the preconditioner. The computed convergence factor, based on a log-linear fit of the last four error values, is $\gamma_{\mathrm{comp}} \approx 0.389$.

the Euclidean norm of the true error $\|\boldsymbol{u}_L^{\mathrm{E}} - \boldsymbol{u}_L^k\|$ falls below the tolerance `tol` $= 10^{-9}$, or after a maximum of 100 iterations. The main script orchestrating this experiment is given in Listing 10.1.

The convergence history of the true error is shown in Figure 10.5. We observe that the BPX-preconditioned PCG method exhibits linear convergence, as indicated by the straight line on the semilogarithmic plot. The computed convergence factor is approximately $\gamma_{\mathrm{comp}} \approx 0.389$. This rapid convergence is consistent with the theory presented, which shows that the condition number $\kappa(\mathsf{C}_{\mathrm{BPX}}\mathsf{A}_L)$ is bounded independently of the mesh size $h_L$ (and thus $n_L$) and the number of levels $L$, leading to an iteration count for PCG that is largely independent of the problem size.

## 10.7 Problems

10.1 Determine a general structure for the hierarchical basis decomposition of $\mathbb{R}^n$, where $n \in \mathbb{N}$ is odd, using Examples 10.1.5 and 10.1.6 as guides.

10.2 Exhibit a set of ordered prolongation matrices $\{\mathsf{Q}_j\}_{j=0}^{L}$ that satisfies Assumption (SS1) but, for every $\boldsymbol{u} \in \mathbb{R}^n$, the Q-class $\mathsf{Q}[\boldsymbol{u}]$ always has more than one element. In other words, furnish an example for which, given $\{\mathsf{Q}_j\}_{j=0}^{L}$ and $\boldsymbol{u} \in \mathbb{R}^n$, the decomposition

$$\boldsymbol{u} = \sum_{j=0}^{L} \mathsf{Q}_j \boldsymbol{w}_j$$

is never unique.

10.3 Complete the proof of Theorem 10.1.11.

10.4 Prove Lemma 10.2.4.

10.5 Prove Proposition 10.3.1.

10.6 Prove Lemma 10.3.3.

10.7 Prove Lemma 10.3.8.

10.8 Prove Theorem 10.3.12 in the case that $d = 1$ and the multilevel grid is nested and uniform.

10.9 Prove Lemma 10.3.14.

10.10 Prove Lemma 10.3.20.

10.11 Prove Lemma 10.4.4.

# Listings

```
./1DFEMBPX/
    poisson1DBPX.m
    BPXPreconditioner.m
    PCG.m (see Listing 2.4)
    prolongation.m (see Listing 4.7)
    restriction.m (see Listing 4.8)
```

**Listing 10.1:** `poisson1DBPX.m`: MATLAB© driver script for solving the 1D finite element Poisson equation using PCG preconditioned by the BPX method.

```
%
% poisson1DBPX.m
%
% Calling script for solving the 1D Poisson equation
%
%    -u'' = f
%
% on an interval with homogeneous Dirichlet boundary
% conditions using Preconditioned Conjugate Gradient
% (PCG) method with BPX (Bramble-Pasciak-Xu)
% preconditioner.
%
%% Initialization and Parameters

% Clear command line window and workspace.
clear; clc;

% Tolerance for the true error ||xExact - x||.
tol = 1.0e-09;

% Number of interior points at finest level.
nL = 1023;

% Number of multigrid levels.
L = 9;

% Define the domain of the equation.
xLower = 0.0;
xUpper = 1.0;

```

```
% Grid spacing at finest level.
hL = 1/(nL+1);

% Check that nL+1 is divisible by 2^(L+1) for coarsening.
if mod(nL+1,2^(L+1)) ~= 0
  error('Coarsening error: L is too large.');
end

% Initialize our approximate solution (e.g. random).
rng(1234);
u0 = rand(nL,1)-0.5;
px = xUpper-xLower;

% Build exact solution and define f accordingly.
x = linspace(xLower+hL,xUpper-hL,nL)';
AL = gallery('tridiag',nL,-1/hL,2/hL,-1/hL);
uExact = exp(sin(3*pi*x/px))-1;
f = AL*uExact;

% MGParam structure.
MGParam.nL    = nL;
MGParam.L     = L;
MGParam.kMax = 100;

% Preconditioner function handle.
precond = @(r) BPXPreconditioner(r,MGParam);

%% Calling the PCG solver

% Solve using PCG with BPX preconditioner.
tic;
[u,iter,errVecFull] = PCG(AL,u0,f,uExact,MGParam.kMax,...
                          tol, precond);
toc;

% Trim the error vector to the actual number of
% iterations.
if iter == 0
  errVec = errVecFull(1);
  iter = 1;
else
  errVec = errVecFull(1:iter);
```

```
end

% We can do a quick final L2 error check.
finalErr = errVec(end);
fprintf('Final error in L2 sense = %g\n',finalErr);
fprintf('Code terminated after %d iterations\n',iter);

%% Plotting

figure(1);
clf;
plot(x,u,'k-','LineWidth',1.5);
xlabel('$x$','Interpreter','latex');
ylabel('$u_L(x)$','Interpreter','latex');
title('Numerical Solution with BPX Preconditioned PCG');
axis([xLower,xUpper,-1.0,2.0]);

figure(2);
clf;
% Plot the true error history up to the stopping
% iteration.
semilogy(1:iter,errVec,'ro','LineWidth',1.5);
xlabel('$k$','Interpreter','latex');
% Corrected title for true error plot.
title('PCG-BPX Iteration Error Plot',...
  'Interpreter','latex');
hold on;

% Calculate convergence rate from last few iterations.
rate = NaN;
rateStr = 'NaN';

% Check if enough iterations were performed for rate
% calculation.
if iter >= 4

  % Indices for the last 4 iterations performed.
  kv = (iter-3):iter;

  % Corresponding error values.
  last_errors = errVec(kv);

```

```
  % Ensure the errors used for log are strictly positive.
  if all(last_errors > 0)
    le = log(last_errors);
    p1 = polyfit(kv',le,1);
    rate = exp(p1(1));
    rateStr = num2str(rate,'%10.5e');

    % Plot the fitted line based on calculated rate.
    p1k = polyval(p1,1:iter);
    semilogy(1:iter,exp(p1k),'b','LineWidth',2)
  end
end

legend(['$\left\|{\bf u}_{\rm exact} - ',...
        '{\bf u}^k\right\|_2$'],...
        'Interpreter','latex');

textYbase = min(errVec(errVec>0));
if isempty(textYbase) || textYbase < tol
  textYbase = tol;
end
textYrate = max(textYbase*100,tol*256);
textYL = textYrate/8;
textYnL = textYrate/32;

% Display the calculated rate.
text(1.5,textYrate,...
  strcat('$\gamma_{\rm comp} =\hspace{.1cm}$',...
  rateStr),'FontSize',14,'Interpreter', 'latex');
text(1.5, textYL, strcat('$L =\hspace{.1cm}$',...
  num2str(L)),'FontSize',14,'Interpreter','latex')
text(1.5, textYnL, strcat('$n_L =\hspace{.1cm}$',...
  num2str(nL)),'FontSize',14,'Interpreter','latex')
printstr = strcat('TrueErr_nL_',...
  num2str(nL),'_L_',num2str(L),'.pdf');
grid on
grid minor
exportgraphics(gca,printstr)
hold off
```

**Listing 10.2:** `BPXPreconditioner.m`: MATLAB© function that applies the BPX preconditioner $C_{BPX}$ to a given vector, following the algorithm described in Section 10.5.

```
function z = BPXPreconditioner(r,MGParam)

% Getting number of levels.
L = MGParam.L;

% Phase 1: Restriction.
% Since the residuals at different levels are of
% different lengths, we need to use cell instead of
% array.
residuals = cell(L+1,1);
residuals{L+1} = r;
for l = L:-1:1
  residuals{l} = restriction(residuals{l+1});
end

% Phase 2: Scaling.
scaled = cell(L+1,1);
for l = 1:(L+1)
  nl = length(residuals{l});
  hl = 1/(nl+1);
  scaled{l} = hl*residuals{l};
end

% Phase 3: Accumulation using Horner-like scheme.
zTemp = scaled{1};
for l = 2:(L+1)
  zTemp = scaled{l}+prolongation(zTemp);
end
z = zTemp;

end
```

# 11 Convex optimization

The purpose of this chapter is to present some background information regarding basic (linear) functional analysis, convex functions, and their minimization. We significantly deviate from our "mostly matrix-based" approach in this chapter. We will set up the language and tools for the next chapter, which deals with sophisticated methods for convex optimization problems based on subspace decompositions.

## 11.1 Analysis on vector spaces

Let us begin by recalling some basic facts about linear functional analysis. For further developments, we refer the reader, for instance, to [11, 15]. We recall that a Banach space is a pair $(\mathcal{V}, \|\cdot\|_{\mathcal{V}})$, where $\mathcal{V}$ is a vector space, $\|\cdot\|_{\mathcal{V}}$ is a *norm* on $\mathcal{V}$, and this space is *complete* under this norm. In addition, if this norm is induced by an *inner product* $(\cdot,\cdot)_{\mathcal{V}}$, then we call the space a Hilbert space.

**Example 11.1.1** (Finite-dimensional spaces). Let $n \in \mathbb{N}$ and consider $\mathbb{R}^n$. For $p \in [1,\infty]$ the $\ell^p$-norms, introduced in Definition 1.1.3, are such that $(\mathbb{R}^n, \|\cdot\|_{\ell^p})$ is a Banach space. In addition, for $p = 2$, we clearly see that this is the Euclidean norm, and so this is a Hilbert space. ▲

It is well known that in finite dimensions, all norms are equivalent. By the previous example any finite-dimensional vector space is complete under any norm. The following example shows that in infinite dimensions the choice of norm matters.

**Example 11.1.2** (Continuous functions). Let $d \in \mathbb{N}$, and let $\Omega \subset \mathbb{R}^d$ be a bounded domain. The space of functions $C(\overline{\Omega})$ can be endowed with the norm

$$\|f\|_{L^2(\Omega)} := \left( \int_{\Omega} |f(\boldsymbol{x})|^2 \, \mathrm{d}\boldsymbol{x} \right)^{1/2}. \tag{11.1.1}$$

However, normed this way, the space is not complete. On the other hand, the norm

$$\|f\|_{L^\infty(\Omega)} := \sup_{\boldsymbol{x} \in \overline{\Omega}} |f(\boldsymbol{x})|$$

makes this space complete. ▲

**Example 11.1.3.** Let $d \in \mathbb{N}$, and let $\Omega \subset \mathbb{R}^d$ be a bounded domain. The space of Lebesgue-measurable real-valued square-integrable functions on $\Omega$ is commonly denoted $L^2(\Omega)$, or $L^2(\Omega; \mathbb{R})$ whenever the range is emphasized. $L^2(\Omega)$ is a vector space and, when equipped with the norm (11.1.1), is a Hilbert space. ▲

https://doi.org/10.1515/9783111354880-011

As it is customary, in what follows, if it is clear from the context, then we will refer to a Banach or Hilbert space by simply mentioning the vector space but not necessarily the norm or inner product.

We now consider the natural mappings between normed vector spaces, that is, the continuous linear mappings.

**Definition 11.1.4** (Continuous linear mappings). Let $\mathcal{X}$ and $\mathcal{Y}$ be normed spaces. The collection of all mappings with domain $\mathcal{X}$ and range $\mathcal{Y}$ that are both linear and continuous is denoted by

$$\mathcal{L}(\mathcal{X}, \mathcal{Y}).$$

A **functional** on $\mathcal{X}$ is a mapping $\mathcal{X} \to \mathbb{R}$. The **dual** space to $\mathcal{X}$ is the space

$$\mathcal{X}' := \mathcal{L}(\mathcal{X}, \mathbb{R}).$$ ∎

The following result details some properties of the continuous linear mappings.

**Proposition 11.1.5** (Properties of $\mathcal{L}(\mathcal{X}, \mathcal{Y})$). *Let $\mathcal{X}$ and $\mathcal{Y}$ be normed spaces, and let $T : \mathcal{X} \to \mathcal{Y}$ be linear. The following statements are equivalent:*
1. $T \in \mathcal{L}(\mathcal{X}, \mathcal{Y})$.
2. *$T$ is continuous at zero, that is, for every $\{v_n\}_{n=1}^{\infty} \subset \mathcal{X}$ such that $v_n \to 0$ in the norm of $\mathcal{X}$, we have*

$$Tv_n \to 0$$

*in the norm of $\mathcal{Y}$.*
3. *The mapping $T$ is bounded, that is, it maps bounded sets in $\mathcal{X}$ to bounded sets in $\mathcal{Y}$, that is, there is a constant $M > 0$ such that, for every $v \in \mathcal{X}$,*

$$\|Tv\|_{\mathcal{Y}} \leq M\|v\|_{\mathcal{X}}. \tag{11.1.2}$$

*Proof.* Exercise. See Problem 11.1. □

Let us now introduce the so-called induced norm.

**Definition 11.1.6** (Induced norm). Let $\mathcal{X}$ and $\mathcal{Y}$ be normed spaces. For $T \in \mathcal{L}(\mathcal{X}, \mathcal{Y})$, the quantity

$$\|T\|_{\mathcal{L}(\mathcal{X},\mathcal{Y})} := \sup_{v \in \mathcal{X}\setminus\{0\}} \frac{\|Tv\|_{\mathcal{Y}}}{\|v\|_{\mathcal{X}}} \tag{11.1.3}$$

is called the **induced** or **operator norm** on $\mathcal{L}(\mathcal{X}, \mathcal{Y})$. ∎

Note that for the particular case that $T \in \mathcal{X}'$, the operator norm is simply

$$\|T\|_{\mathcal{X}'} = \sup_{v \in \mathcal{X} \setminus \{0\}} \frac{|Tv|}{\|v\|_{\mathcal{X}}}. \tag{11.1.4}$$

**Proposition 11.1.7** ($\mathcal{L}(\mathcal{X}, \mathcal{Y})$ is Banach). *Let $\mathcal{X}$ and $\mathcal{Y}$ be normed spaces. The set $\mathcal{L}(\mathcal{X}, \mathcal{Y})$ is a vector space, and the induced norm is indeed a norm on this space. In addition, if $\mathcal{Y}$ is Banach, then so is $\mathcal{L}(\mathcal{X}, \mathcal{Y})$ under the induced norm. Consequently, the dual space $\mathcal{X}'$ to any normed space $\mathcal{X}$ is always complete.*

*Proof.* Exercise; see Problem 11.2. □

**Example 11.1.8** (Finite-dimensional spaces). Let $m, n \in \mathbb{N}$. The space of matrices $\mathbb{R}^{m \times n}$ can be identified with the space $\mathcal{L}(\mathbb{R}^n, \mathbb{R}^m)$ under the action of matrix vector multiplication. If we endow $\mathbb{R}^m$ and $\mathbb{R}^n$ with the Euclidean norm, then $\mathbb{R}^{m \times n}$ is normed with the induced norm, which is defined as in (11.1.3).

The dual space to $\mathbb{R}^n$ can be identified with $\mathbb{R}^n$ itself. If $F$ is a continuous linear functional on $\mathbb{R}^n$, then it can be uniquely written as

$$F(\boldsymbol{u}) = \boldsymbol{v}^\top \boldsymbol{u},$$

for some $\boldsymbol{v} \in \mathbb{R}^n$. ▲

The following examples show that, again, in infinite dimensions, things become more delicate.

**Example 11.1.9** (Not bounded). Consider the following normed linear spaces:

$$(\mathcal{X}, \|\cdot\|_{\mathcal{X}}) = (C^1([0,1]), \|\cdot\|_{L^\infty(0,1)})$$

and

$$(\mathcal{Y}, \|\cdot\|_{\mathcal{Y}}) = (C([0,1]), \|\cdot\|_{L^\infty(0,1)}).$$

The $\mathcal{X}$ space with this norm is not a Banach space, but the $\mathcal{Y}$ space is. Define $T : \mathcal{X} \to \mathcal{Y}$ as

$$(Tf)(x) = f'(x). \tag{11.1.5}$$

Clearly, this is a linear mapping. However, it is not bounded. To see this, consider, for $N \in \mathbb{N}$, the family of functions

$$f_N(x) = \sin(2\pi N x).$$

Clearly, for all $N \in \mathbb{N}$,

$$\|f_N\|_{L^\infty(0,1)} = 1,$$

whereas

$$f_N'(x) = 2\pi N \sin(2\pi N x),$$

which implies that

$$\|f_N'\|_{L^\infty(0,1)} = 2\pi N.$$

Since $N$ can be arbitrarily large, the mapping $T$ is not bounded.

If, on the other hand, we consider the same mapping but with a different domain, then this mapping may be bounded. For instance, for every $u \in C^1([0,1])$, define

$$\|u\|_{C^1([0,1])} := \|u\|_{L^\infty(0,1)} + \|u'\|_{L^\infty(0,1)},$$

and consider

$$(\mathcal{X}, \|\cdot\|_{\mathcal{X}}) = (C^1([0,1]), \|\cdot\|_{C^1([0,1])}).$$

Incidentally, this version of the space $\mathcal{X}$ is a Banach space. Furthermore, $T : \mathcal{X} \to \mathcal{Y}$, defined as in (11.1.5), is now a bounded linear operator. ▲

**Example 11.1.10** (Linear functional). Fix $g \in L^2(0,1)$. The mapping

$$L^2(0,1) \ni f \mapsto T_g(f) := (f,g)_{L^2(0,1)} := \int_0^1 f(x)g(x)\,\mathrm{d}x$$

defines a bounded linear functional on $L^2(0,1)$. ▲

Examples 11.1.8 and 11.1.10 suggest a similarity between the action of a bounded linear functional and an inner product. This similarity is the motivation for the following notation.

**Definition 11.1.11** (Duality pairing). Let $\mathcal{X}$ be a Banach space, and let $\mathcal{X}'$ be its dual. The **duality pairing** between $\mathcal{X}$ and $\mathcal{X}'$ is the mapping

$$\mathcal{X} \times \mathcal{X}' \ni (u, F) \mapsto \langle F, u\rangle_{\mathcal{X}',\mathcal{X}} \in \mathbb{R},$$

where

$$\langle F, u\rangle_{\mathcal{X}',\mathcal{X}} := Fu.$$

If the space $\mathcal{X}$ is understood from the context, then we will suppress the subindices $\mathcal{X}', \mathcal{X}$ from this notation. ■

For Hilbert spaces, this duality pairing turns out to essentially coincide with the inner product.

**Theorem 11.1.12** (Riesz representation). *Let $\mathcal{V}$ be a Hilbert space. There is a linear isometry between $\mathcal{V}$ and its dual $\mathcal{V}'$. In other words, for every $F \in \mathcal{V}'$, there is a unique $\mathfrak{R}_{\mathcal{V}} F \in \mathcal{V}$ such that*

$$\langle F, u \rangle_{\mathcal{V}',\mathcal{V}} = (u, \mathfrak{R}_{\mathcal{V}} F)_{\mathcal{V}}, \quad \forall u \in \mathcal{V}.$$

*Moreover,*

$$\|\mathfrak{R}_{\mathcal{V}} F\|_{\mathcal{V}} = \|F\|_{\mathcal{V}'}, \quad \forall F \in \mathcal{V}'.$$

*Proof.* See Problem 11.3 for the particular case $\mathcal{V} = \mathbb{R}^n$. □

**Definition 11.1.13** (Riesz map). The mapping $\mathfrak{R}_{\mathcal{V}} \in \mathcal{L}(\mathcal{V}', \mathcal{V})$ of Theorem 11.1.12 is called the **canonical Riesz map.** If it is clear from the context then the subindex $\mathcal{V}$ may be suppressed. ■

The final notion we introduce related to the dual space is that of weak convergence.

**Definition 11.1.14** (Weak convergence). Let $\mathcal{X}$ be a Banach space, and let $\mathcal{X}'$ be its dual. We say that the sequence $\{v_n\}_{n=1}^{\infty} \subset \mathcal{X}$ **converges weakly** to $v \in \mathcal{X}$ if

$$\lim_{n \to \infty} \langle F, v_n - v \rangle = 0, \quad \forall F \in \mathcal{X}'.$$

We will denote this convergence by $v_n \rightharpoonup v$. A set $K \subseteq \mathcal{X}$ is called **weakly closed** if it contains all its weak limit points. A set $O \subseteq \mathcal{X}$ is called **weakly open** if $\mathcal{X} \setminus O$ is weakly closed. ■

**Example 11.1.15** (Weak convergence). Consider $\{v_n\}_{n=1}^{\infty} \subset L^2(-\pi, \pi)$ defined by

$$v_n(x) = \sin(nx).$$

Let us show that this sequence is not convergent in norm. If this were the case, then this sequence would be Cauchy. However, if $m, n \in \mathbb{N}$ with $m \neq n$, then

$$\|v_m - v_n\|_{L^2(-\pi,\pi)}^2 = 2\pi,$$

as the reader may easily verify; see Problem 11.4.

On the other hand, we will now show that $v_n \rightharpoonup 0$. To obtain this, we invoke the Riesz representation Theorem 11.1.12 and recall that any bounded linear functional has the form

$$v \mapsto \int_{-\pi}^{\pi} f(x) v(x) \, \mathrm{d}x,$$

for some $f \in L^2(-\pi,\pi)$. We will additionally use the fact that the step functions are *dense* in $L^2(-\pi,\pi)$. Fix $f \in L^2(-\pi,\pi)$ and $\varepsilon > 0$. There are $m \in \mathbb{N}$, $\{c_i\}_{i=0}^m \subset [-\pi,\pi]$, and $\{A_i\}_{i=1}^m \subset \mathbb{R}$ with

$$-\pi \le c_0 < c_2 < \cdots < c_m \le \pi$$

such that for

$$f_\varepsilon(x) = \sum_{i=1}^m A_i \chi_{(c_{i-1},c_i]}(x),$$

we have

$$\|f - f_\varepsilon\|_{L^2(-\pi,\pi)} < \frac{\varepsilon}{2\sqrt{2\pi}}.$$

Define

$$I_n = \left| \int_{-\pi}^{\pi} f(x) v_n(x)\, \mathrm{d}x \right|$$

and consider

$$\begin{aligned} I_n &\le \|f - f_\varepsilon\|_{L^2(-\pi,\pi)} \|v_n\|_{L^2(-\pi,\pi)} + \left| \int_{-\pi}^{\pi} f_\varepsilon(x) v_n(x)\, \mathrm{d}x \right| \\ &\le \frac{\varepsilon}{2} + \left| \sum_{i=1}^m A_i \int_{c_{i-1}}^{c_i} v_n(x) \mathrm{d}x \right| \\ &\le \frac{\varepsilon}{2} + m \max_{i=1}^m |A_i| \max_{i=1}^m \left| \int_{c_{i-1}}^{c_i} v_n(x) \mathrm{d}x \right|. \end{aligned} \tag{11.1.6}$$

As a side computation, for $c_L, c_R \in [-\pi,\pi]$ with $c_L < c_R$, we estimate

$$\begin{aligned} \left| \int_{c_L}^{c_R} \sin(nx)\, \mathrm{d}x \right| &= \frac{1}{n} |\cos(nc_L) - \cos(nc_R)| \\ &\le \frac{2}{n}. \end{aligned}$$

This shows that if

$$n \ge n_0 := \left\lceil \frac{4m \max_{i=1}^m |A_i|}{\varepsilon} \right\rceil,$$

then $I_n < \varepsilon$, which shows the claimed weak convergence. ▲

**Definition 11.1.16** (Closure). Let $K$ be a subset of the Hilbert space $\mathcal{V}$. Define

$$L_s = \{u \in \mathcal{V} \mid \exists \{u_n\}_{n=1}^{\infty} \subset K : u_n \to u\}$$

and

$$L_w = \{u \in \mathcal{V} \mid \exists \{u_n\}_{n=1}^{\infty} \subset K : u_n \rightharpoonup u\}.$$

The set

$$\mathrm{clo}_s(K) = K \cup L_s$$

is called the **closure of** $K$, and the set

$$\mathrm{clo}_w(K) = K \cup L_w$$

is called the **weak closure of** $K$. ∎

These last two sets are the smallest closed and weakly closed sets, respectively, that contain $K$, as we might expect.

Next, let us remind the reader about subsequences, as these play an important role in the concept of compactness and the coming new concept of weak compactness.

**Definition 11.1.17** (Subsequence). Let $\{n_k\}_{k=1}^{\infty} \subseteq \mathbb{N}$ be a strictly increasing sequence of numbers, and let $\{u_n\}_{n=1}^{\infty} \subset \mathcal{V}$. The sequence $\{u_{n_k}\}_{k=1}^{\infty}$ is called a **subsequence** of $\{u_n\}_{n=1}^{\infty}$, and we write $\{u_{n_k}\}_{k=1}^{\infty} \subseteq \{u_n\}_{n=1}^{\infty}$. ∎

**Definition 11.1.18** (Sequential compactness). We say that a set $K \subset \mathcal{V}$ is **sequentially compact** if every sequence $\{u_n\}_{n=1}^{\infty}$ has a subsequence $\{u_{n_k}\}_{k=1}^{\infty} \subseteq \{u_n\}_{n=1}^{\infty}$ that converges to a point $u \in K$. A set $K \subset \mathcal{V}$ is called **weakly sequentially compact** if every sequence $\{u_n\}_{n=1}^{\infty}$ has a subsequence $\{u_{n_k}\}_{k=1}^{\infty} \subseteq \{u_n\}_{n=1}^{\infty}$ that converges weakly to a point $u \in K$. ∎

Now we can state the so-called weak compactness property, a fundamental attribute of Hilbert spaces. The proof of this result is beyond the scope of the text.

**Theorem 11.1.19** (Weak compactness). *Let $\mathcal{V}$ be a Hilbert space. Then every bounded sequence $\{u_n\}_{n=1}^{\infty} \subset \mathcal{V}$ has a weakly convergent subsequence, that is, there are an element $u \in \mathcal{V}$ and a strictly increasing sequence of indices $\{n_k\}_{k=1}^{\infty} \subset \mathbb{N}$ such that $u_{n_k} \rightharpoonup u$ as $k \to \infty$.*

Although much more could be said about Banach and Hilbert spaces, linear mappings between them, dual spaces, and their representations, we now turn our attention to nonlinear mappings and how we measure their smoothness.

**Definition 11.1.20** (Lower semicontinuity). Let $\mathcal{V}$ be a Hilbert space, and let $E : \mathcal{V} \to \mathbb{R}$. We say that $E$ is **lower semicontinuous (LSC)** at $v \in \mathcal{V}$ if whenever $v_n \to v$ in $\mathcal{V}$, we have

$$E(v) \leq \liminf_{n\to\infty} E(v_n).$$

Similarly, we say that $E$ is **weakly lower semicontinuous (wLSC)** at $v \in \mathcal{V}$ if whenever $v_n \rightharpoonup v$ in $\mathcal{V}$, we have

$$E(v) \leq \liminf_{n\to\infty} E(v_n).$$

If $K \subset \mathcal{V}$ and $E$ is (weakly) lower semicontinuous for every $v \in K$, then we say that $E$ is (weakly) lower semicontinuous on $K$. ■

An example of an LSC function is given in Figure 11.1. It turns out that LSC has an interesting geometric interpretation. To state it, we introduce the so-called epigraph and sublevel sets of a functional.

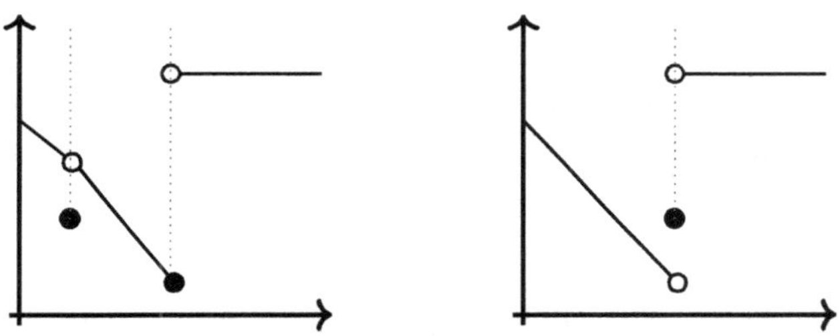

**Figure 11.1:** The figure on the left depicts the graph of a lower semicontinuous function, whereas the right is the graph of a function that is not lower semicontinuous. As Proposition 11.1.23 shows, the epigraph of the function on the left is closed. This can be seen by the dotted blue lines.

**Definition 11.1.21** (Epigraph). Let $\mathcal{V}$ be a Hilbert space, and let $E : \mathcal{V} \to \mathbb{R}$. The **epigraph** of $E$ is the set

$$\operatorname{epi}(E) := \{(v, \alpha) \in \mathcal{V} \times \mathbb{R} \mid E(v) \leq \alpha\}.$$ ■

**Definition 11.1.22** (Sublevel set). Let $\mathcal{V}$ be a Hilbert space, let $E : \mathcal{V} \to \mathbb{R}$, and let $\alpha \in \mathbb{R}$. The **sublevel set** at height $\alpha$ of $E$ is

$$\operatorname{level}_\alpha(E) := \{v \in \mathcal{V} \mid E(v) \leq \alpha\}.$$ ■

**Proposition 11.1.23** (wLSC). *Let $\mathcal{V}$ be a Hilbert space, and let $E : \mathcal{V} \to \mathbb{R}$. The functional $E$ is (weakly) LSC iff its epigraph is (weakly) closed iff, for every $\alpha \in \mathbb{R}$, the set* $\operatorname{level}_\alpha E$ *is (weakly) closed.*

*Proof.* Exercise; see Problem 11.5. □

We now turn our attention to differentiable mappings.

**Definition 11.1.24** (Fréchet differentiability). Let $\mathcal{X}$ and $\mathcal{Y}$ be Banach spaces, and let $F : \mathcal{X} \to \mathcal{Y}$. We say that $F$ is **Fréchet differentiable** at $v \in \mathcal{X}$ if there is a mapping $T \in \mathcal{L}(\mathcal{X}, \mathcal{Y})$ such that

$$\lim_{\|h\|_{\mathcal{X}} \to 0} \frac{\|F(v+h) - F(v) - Th\|_{\mathcal{Y}}}{\|h\|_{\mathcal{X}}} = 0.$$

The necessarily unique $T$ is called the **Fréchet derivative** of $F$ at $v$ and is denoted by $\mathrm{D}F(v) \in \mathcal{L}(\mathcal{X}, \mathcal{Y})$. If $K \subseteq \mathcal{X}$ and $F$ is Fréchet differentiable at every $v \in K$, then we say that $F$ is Fréchet differentiable on $K$. ∎

**Remark 11.1.25** (Derivative of a functional). Let $\mathcal{V}$ be a Hilbert space, let $E : \mathcal{V} \to \mathbb{R}$ be a functional, and let $v \in \mathcal{V}$. The Fréchet derivative of $E$ at $v$, if it exists, is an element of $\mathcal{V}'$, and so we write

$$\langle \mathrm{D}E(v), w \rangle := \mathrm{D}E(v)w, \quad \forall\, w \in \mathcal{V}.$$ ◀

**Definition 11.1.26.** Let $\mathcal{X}$ and $\mathcal{Y}$ be Banach spaces, and let $F : \mathcal{X} \to \mathcal{Y}$. Suppose that $F$ is Fréchet differentiable at every point of the set $K \subseteq \mathcal{X}$. The derivative defines a mapping

$$\mathcal{X} \supseteq K \ni v \mapsto \mathrm{D}F(v) \in \mathcal{L}(\mathcal{X}, \mathcal{Y}),$$

which is naturally denoted $\mathrm{D}F : K \to \mathcal{L}(\mathcal{X}, \mathcal{Y})$. If this mapping is bounded and continuous, then we say that $F$ **continuously Fréchet differentiable** on $K$. ∎

**Definition 11.1.27** (Convex sets). Let $\mathcal{X}$ be a vector space, and let $K \subseteq \mathcal{X}$. We say that $K$ is **convex** iff, for all $v, w \in K$ and $t \in [0,1]$, we have

$$tv + (1-t)w \in K.$$ ∎

If $F$ is continuously Fréchet differentiable on a convex set $K$, then we have a nice analogue of Taylor's theorem with integral remainder.

**Proposition 11.1.28** (Taylor). *Let $\mathcal{X}$ and $\mathcal{Y}$ be Banach spaces, and let $F : \mathcal{X} \to \mathcal{Y}$ be continuously Fréchet differentiable on a convex subset $K \subseteq \mathcal{X}$. For all $v, w \in K$,*

$$F(w) - F(v) = \int_0^1 \mathrm{D}F(tw + (1-t)v)(w - v)\,\mathrm{d}t.$$

In a similar way that we can study the continuity of the derivative mapping, we can study its differentiability. Although this can be developed in full generality, we restrict our attention to the case of functionals, which will be sufficient for our purposes.

**Definition 11.1.29** (Second derivative). Let $\mathcal{V}$ be a Hilbert space, and let $E : \mathcal{V} \to \mathbb{R}$ be a continuously Fréchet differentiable functional on the subset $K \subseteq \mathcal{V}$. If the mapping

$$\mathrm{D}E : K \to \mathcal{V}', \quad \mathcal{V} \supseteq K \ni v \mapsto \mathrm{D}E(v) \in \mathcal{V}'$$

is Fréchet differentiable for all $v \in K \subseteq \mathcal{V}$, then its derivative

$$\mathrm{D}^2E(v) \in \mathcal{L}(\mathcal{V}, \mathcal{L}(\mathcal{V}, \mathbb{R})) = \mathcal{L}(\mathcal{V}, \mathcal{V}')$$

is called the **second Fréchet derivative** of $E$ at $v \in K$, and we say that $E$ is **twice Fréchet differentiable** on $K$. ∎

As before, we can investigate when a functional is twice Fréchet differentiable on a set or when it is twice continuously Fréchet differentiable at a point or on a set.

Next, we observe that since $\mathrm{D}^2E(v) \in \mathcal{L}(\mathcal{V}, \mathcal{V}')$, for $w_1, w_2 \in \mathcal{V}$, we have

$$\mathrm{D}^2E(v)w_1 \in \mathcal{V}', \quad \mathrm{D}^2E(v)(w_1, w_2) := \langle \mathrm{D}^2E(v)w_1, w_2 \rangle.$$

In other words, $\mathrm{D}^2E(v)$ defines a bilinear form on $\mathcal{V}$. The following result shows that this bilinear form is symmetric.

**Proposition 11.1.30** (Symmetry). *Let $\mathcal{V}$ be a Hilbert space, and let $E : \mathcal{V} \to \mathbb{R}$ be twice Fréchet differentiable at $v \in \mathcal{V}$. Then*

$$\mathrm{D}^2E(v)(w_1, w_2) = \mathrm{D}^2E(v)(w_2, w_1), \quad \forall\, w_1, w_2 \in \mathcal{V}.$$

We now characterize the *growth* of a functional.

**Definition 11.1.31** (Coercivity). Let $\mathcal{V}$ be a Hilbert space, and let $E : \mathcal{V} \to \mathbb{R}$. We say that $E$ is **coercive** if whenever $\|v_n\|_{\mathcal{V}} \to \infty$, we have

$$E(v_n) \to \infty.$$ ∎

Finally, we present a second-order version of Taylor's theorem.

**Proposition 11.1.32** (Second-order Taylor). *Let $\mathcal{V}$ be a Hilbert space, and let $E : \mathcal{V} \to \mathbb{R}$ be twice continuously Fréchet differentiable on a convex subset $K \subseteq \mathcal{V}$. Then, for all $v, w \in K$, we have*

$$E(w) - E(v) = \langle \mathrm{D}E(v), w - v \rangle + \int_0^1 (1-t)\mathrm{D}^2E(v + t(w-v))(w-v, w-v)\,\mathrm{d}t.$$

*Moreover, for every $h \in \mathcal{V}$,*

$$\langle \mathrm{D}E(w) - \mathrm{D}E(v), h \rangle = \int_0^1 \mathrm{D}^2E(v + t(w-v))(w-v, h)\,\mathrm{d}t.$$

As an application, we give a sufficient condition for the coercivity of a functional.

**Corollary 11.1.33** (Coercivity). *Let $\mathcal{V}$ be a Hilbert space, and let $E : \mathcal{V} \to \mathbb{R}$ be twice continuously differentiable on $\mathcal{V}$. If there is a constant $\mu > 0$ such that*

$$\mathrm{D}^2E(v)(w,w) \geq c\|w\|_{\mathcal{V}}^2, \quad \forall\, v, w \in \mathcal{V},$$

*then E is coercive.*

*Proof.* Let $w$ be arbitrary. By Proposition 11.1.32 we have

$$\begin{aligned} E(w) &= E(0) + \langle \mathrm{D}E(0), w\rangle + \int_0^1 (1-t)\mathrm{D}^2E(tw)(w,w)\mathrm{d}t \\ &\geq E(0) - \|\mathrm{D}E(0)\|_{\mathcal{V}'}\|w\|_{\mathcal{V}} + \frac{\mu}{2}\|w\|_{\mathcal{V}}^2, \end{aligned}$$

where, to obtain the lower bound, we used the assumption on the second derivative. This clearly shows that $E(w) \to \infty$ as $\|w\|_{\mathcal{V}} \to \infty$, and thus the functional $E$ is coercive. □

**Example 11.1.34.** Suppose that $n \in \mathbb{N}$ and $\mathsf{A} \in \mathbb{R}^{n\times n}$ with $\mathsf{A}^\top = \mathsf{A}$. For every $\boldsymbol{x} \in \mathbb{R}^n$, define the functional

$$E(\boldsymbol{x}) := \frac{1}{2}\boldsymbol{x}^\top \mathsf{A}\boldsymbol{x} + \frac{1}{4}\sum_{i=1}^n (x_i)^4.$$

This functional is twice continuously Fréchet differentiable. Recall that the gradient of $E$ and the Hessian matrix of $E$ a point $\boldsymbol{z} \in \mathbb{R}^n$ are defined, respectively, as

$$\nabla E(\boldsymbol{z}) = \mathsf{A}\boldsymbol{z} + \begin{bmatrix} z_1^3 \\ z_2^3 \\ \vdots \\ z_n^3 \end{bmatrix}$$

and

$$\mathsf{H}_E(\boldsymbol{z}) = \mathsf{A} + 3\,\mathrm{diag}[z_1^2, z_2^2, \ldots, z_n^2].$$

We can prove that

$$\langle \mathrm{D}E(\boldsymbol{z}), \boldsymbol{x}\rangle = \nabla E(\boldsymbol{z})^\top \boldsymbol{x}, \quad \forall\, \boldsymbol{x} \in \mathbb{R}^n$$

and

$$\langle \mathrm{D}^2E(\boldsymbol{z})\boldsymbol{x}, \boldsymbol{y}\rangle = \boldsymbol{y}^\top \mathsf{H}_E(\boldsymbol{z})\boldsymbol{x}, \quad \forall\, \boldsymbol{x}, \boldsymbol{y} \in \mathbb{R}^n.$$

The matrix $\mathsf{H}_E(\boldsymbol{z})$ is symmetric. Furthermore, if $\mathsf{A}$ is positive definite, then $\mathsf{H}_E(\boldsymbol{z})$ is also positive definite at every point $\boldsymbol{z} \in \mathbb{R}^n$. In this case, $E$ is strictly convex and coercive. ▲

## 11.2 Convex sets and functions

We now turn our attention more fully to the fundamental notion of convexity. For further details, we refer to [16, 18]. First, let us define the sum of two sets.

**Definition 11.2.1** (Minkowski sum). Let $\mathcal{X}$ be a vector space, and let $A, B \subseteq \mathcal{X}$. The **(Minkowski) sum** of these sets is

$$A + B := \{a + b \in \mathcal{X} \mid a \in A, b \in B\}.$$

Similarly, for $\lambda \in \mathbb{R}$, we define

$$\lambda A := \{\lambda a \in \mathcal{X} \mid a \in A\}.$$

■

**Proposition 11.2.2** (Properties of convex sets). *Let $\mathcal{X}$ be a vector space.*

1 *Let $\mathcal{A}$ be an index set, and let $K_\alpha \subset \mathcal{X}$ be convex for every $\alpha \in \mathcal{A}$. Then*

$$\bigcap_{\alpha \in \mathcal{A}} K_\alpha$$

*is convex.*

2 *Let $\{K_i\}_{i=1}^{\infty}$ be an increasing family of convex subsets of $\mathcal{X}$, that is, $K_1 \subset K_2 \subset \cdots \subset \mathcal{X}$. Then*

$$\bigcup_{i=1}^{\infty} K_i$$

*is convex.*

3 *Let $N \in \mathbb{N}$, let $\{K_i\}_{i=1}^{N}$ be a (finite) family of convex subsets of $\mathcal{X}$, and let $\{\lambda_i\}_{i=1}^{N} \subset \mathbb{R}$. Then*

$$\sum_{i=1}^{N} \lambda_i K_i$$

*is convex.*

4 *Let $\mathcal{Y}$ be another vector space, and let $T : \mathcal{X} \to \mathcal{Y}$ be a linear, not necessarily bounded, map. The set $K \subset \mathcal{X}$ is convex iff $T(K) \subset \mathcal{Y}$ is convex.*

*Proof.* We prove each statement separately.

1. Denote by $W$ the intersection of the family of sets. Let $v, w \in W$ and $t \in [0,1]$. By definition of intersection, for every $\alpha \in \mathcal{A}$, we have $v, w \in K_\alpha$. Since $K_\alpha$ is convex, $tv + (1-t)w \in K_\alpha$. Thus $tv + (1-t)w \in W$.
2. Denote by $W$ the union of the increasing family of sets. Let $v, w \in W$ and $t \in [0,1]$. Then there are $N_v, N_w \in \mathbb{N}$ for which $v \in K_{N_v}$ and $w \in K_{N_w}$, respectively. Set $N =$

$\max\{N_v, N_w\}$. Since the family of sets is increasing, we have that

$$K_{N_v} \cup K_{N_w} \subset K_N.$$

Thus $v, w \in K_N$. Since $K_N$ is convex, $tv+(1-t)w \in K_N$, which implies $tv+(1-t)w \in W$.

3. Exercise; see Problem 11.6.
4. Exercise; see Problem 11.7. □

Before we state the next result, we observe that if $\mathcal{V}$ is a Hilbert space and $v_n \to v$, then $v_n \rightharpoonup v$. Indeed, if $F \in \mathcal{V}'$, then

$$\left|\langle F, v_n - v\rangle\right| \leq \|F\|_{\mathcal{V}'}\|v_n - v\|_{\mathcal{V}} \to 0.$$

This shows that if a set is weakly closed, then it is closed. The converse, however, is in general false; see Example 11.1.15. The following result shows that the converse is also true if the set is convex.

**Lemma 11.2.3** (Mazur). *Let $\mathcal{V}$ be a Hilbert space, and let $K \subseteq \mathcal{V}$. If $K$ is convex and closed, then it is weakly closed.*

We now turn our attention to convex functions. We restrict ourselves to the case we are interested here, functionals on Hilbert spaces.

**Definition 11.2.4** (Convex functional). Let $\mathcal{V}$ be a Hilbert space, and let $E : \mathcal{V} \to \mathbb{R}$. We say that $E$ is **convex** if for all $v, w \in \mathcal{V}$ and $t \in [0,1]$, we have

$$E\big(tv + (1-t)w\big) \leq tE(v) + (1-t)E(w).$$

We say that $E$ is **strictly convex** if the inequality is strict for $v \neq w$ and $t \in (0,1)$. ■

**Example 11.2.5** (Convex functionals). The following are examples of convex functionals:
- If $\mathcal{V}$ is a Hilbert space, then the norm is convex.
- Let $n \in \mathbb{N}$, and let $\mathsf{A} \in \mathbb{R}^{n\times n}$ be SPSD. The function

$$\Phi(\boldsymbol{x}) = \frac{1}{2}\boldsymbol{x}^\top \mathsf{A}\boldsymbol{x}$$

  is convex. If $\mathsf{A}$ is SPD, then $\Phi$ is strictly convex.
- Let $n \in \mathbb{N}$ and $K \subset \mathbb{R}^n$. Its support function is

$$\mathbb{1}_K^\star(\boldsymbol{w}) = \sup_{\boldsymbol{v}\in K} \boldsymbol{w}^\top \boldsymbol{v}.$$

  This function is convex. ▲

Let us now present some properties of convex functionals.

**Proposition 11.2.6** (Properties of convex functionals). *Let $\mathcal{V}$ be a Hilbert space.*

1 *The functional $E : \mathcal{V} \to \mathbb{R}$ is convex iff* $\operatorname{epi}(E)$ *is convex.*
2 *If the functional $E : \mathcal{V} \to \mathbb{R}$ is convex, then for every $\alpha \in \mathbb{R}$, the set* $\operatorname{level}_\alpha(E)$ *is convex.*
3 *Let $\alpha \in \mathcal{A}$ be an index set, and let $E_\alpha : \mathcal{V} \to \mathbb{R}$, $\alpha \in \mathcal{A}$, be functionals. Define*

$$E(v) := \sup_{\alpha \in \mathcal{A}} \{E_\alpha(v)\}, \quad \forall v \in \mathcal{V}.$$

*If every $E_\alpha$ is convex, then $E$ is convex.*
4 *If a functional $E : \mathcal{V} \to \mathbb{R}$ is convex and LSC, then it is wLSC.*
5 *Let $E : \mathcal{V} \to \mathbb{R}$ be a continuously Fréchet differentiable functional. It is convex iff*

$$E(v) \geq E(w) + \langle \mathrm{D}E(w), v - w \rangle, \quad \forall v, w \in \mathcal{V}.$$

*In addition, $E$ is strictly convex iff the inequality is strict for $v \neq w$.*
6 *If the functional $E : \mathcal{V} \to \mathbb{R}$ is convex and Fréchet differentiable, then its Fréchet derivative is monotone, that is,*

$$\langle \mathrm{D}E(v) - \mathrm{D}E(w), v - w \rangle \geq 0, \quad \forall v, w \in \mathcal{V}.$$

*If in addition, $E$ is strictly convex, then its derivative is strictly monotone, that is, the inequality is strict for $v \neq w$.*
7 *Let $E : \mathcal{V} \to \mathbb{R}$ be a twice continuously Fréchet differentiable functional. If the second Fréchet derivative is positive semidefinite, that is,*

$$\mathrm{D}^2 E(v)(w, w) \geq 0, \quad \forall v, w \in \mathcal{V},$$

*then $E$ is convex. In addition, if the second derivative is positive definite, that is, the inequality above is strict for $w \neq 0$, then $E$ is strictly convex.*

Note that the converse of Proposition 11.2.6.2 is false; see Problem 11.8.

We now introduce a notion that makes strict convexity more quantitative.

**Definition 11.2.7** (Strong convexity). Let $\mathcal{V}$ be a Hilbert space, and let $E : \mathcal{V} \to \mathbb{R}$ be Fréchet differentiable. We say that $E$ is **strongly convex** with constant $\mu > 0$ if

$$\mu \|w - v\|_{\mathcal{V}}^2 \leq \langle \mathrm{D}E(w) - \mathrm{D}E(v), w - v \rangle, \tag{11.2.1}$$

for all $v, w \in \mathcal{V}$. ■

**Remark 11.2.8** (Terminology). Some references, like [14], use the term *elliptic* for strong convexity. ◂

**Theorem 11.2.9** (Strong convexity). *Let $\mathcal{V}$ be a Hilbert space, and let $E$ be a strongly convex functional with constant $\mu$. Then*

$$E(w) - E(v) \geq \langle \mathrm{D}E(v), w - v\rangle + \frac{\mu}{2}\|w - v\|_{\mathcal{V}}^2, \qquad \forall\, v, w \in \mathcal{V}. \tag{11.2.2}$$

*Consequently, $E$ is strictly convex and coercive.*

*Proof.* Using the Taylor theorem with integral remainder (Proposition 11.1.28), we have

$$\begin{aligned}
E(w) - E(v) &= \int_0^1 \langle \mathrm{D}E(v + t(w - v)), w - v\rangle \,\mathrm{d}t \\
&= \langle \mathrm{D}E(v), w - v\rangle \\
&\quad + \int_0^1 \frac{1}{t}\langle \mathrm{D}E(v + t(w - v)) - \mathrm{D}E(v), t(w - v)\rangle \,\mathrm{d}t \\
&\geq \langle \mathrm{D}E(v), w - v\rangle + \int_0^1 \frac{1}{t}\mu\|t(w - v)\|_{\mathcal{V}}^2 \,\mathrm{d}t \\
&= \langle \mathrm{D}E(v), w - v\rangle + \frac{\mu}{2}\|w - v\|_{\mathcal{V}}^2,
\end{aligned}$$

which proves (11.2.2).

Next, we see that for $v \neq w$, (11.2.2) implies

$$E(w) > E(v) + \langle \mathrm{D}E(v), w - v\rangle,$$

which by Proposition 11.2.6 shows that $E$ is strictly convex.

Finally, setting $v = 0$ in (11.2.2) and using Young's inequality, we get

$$\begin{aligned}
E(w) &\geq E(0) + \langle \mathrm{D}E(0), w\rangle + \frac{\mu}{2}\|w\|_{\mathcal{V}}^2 \\
&\geq E(0) - \|\mathrm{D}E(0)\|_{\mathcal{V}'}\|w\|_{\mathcal{V}} + \frac{\mu}{2}\|w\|_{\mathcal{V}}^2 \\
&\geq E(0) - \frac{1}{\mu}\|\mathrm{D}E(0)\|_{\mathcal{V}'}^2 - \frac{\mu}{4}\|w\|_{\mathcal{V}}^2 + \frac{\mu}{2}\|w\|_{\mathcal{V}}^2 \\
&\geq C_1 + C_2\|w\|_{\mathcal{V}}^2
\end{aligned}$$

with

$$C_1 = E(0) - \frac{1}{\mu}\|\mathrm{D}E(0)\|_{\mathcal{V}'}^2, \quad C_2 = \frac{\mu}{4}.$$

Since $C_2 > 0$, this clearly implies the coercivity of $E$. □

The last notion we introduce is the Lipschitz smoothness.

**Definition 11.2.10** (Lipschitz smoothness). Let $\mathcal{V}$ be a Hilbert space, and let $E : \mathcal{V} \to \mathbb{R}$ be Fréchet differentiable. We say that $E$ is **locally Lipschitz smooth** on $\mathcal{V}$ iff, for every bounded set $B \subset \mathcal{V}$, there is a constant $L = L_B$ such that

$$\|\mathrm{D}E(v) - \mathrm{D}E(w)\|_{\mathcal{V}'} \le L\|v - w\|_{\mathcal{V}}, \quad \forall\, v, w \in B.$$ ∎

We now present some useful properties that combine strong convexity and local Lipschitz smoothness.

**Lemma 11.2.11** (Two-sided bounds). *Let $\mathcal{V}$ be a Hilbert space, and let $E : \mathcal{V} \to \mathbb{R}$. If $E$ is strongly convex and locally Lipschitz smooth, then, for all $v, w \in \mathcal{V}$, we have*

$$\mu\|w - v\|_{\mathcal{V}}^2 \le \langle \mathrm{D}E(w) - \mathrm{D}E(v), w - v\rangle.$$

*Let now $\alpha \in \mathbb{R}$. Then, for all $\alpha \in \mathbb{R}$ and $v, w \in \mathrm{level}_\alpha(E)$,*

$$\mu\|w - v\|_{\mathcal{V}}^2 \le \langle \mathrm{D}E(w) - \mathrm{D}E(v), w - v\rangle \le L\|w - v\|_{\mathcal{V}}^2,$$

*where $L = L(E, \alpha)$ is the local Lipschitz constant of $\mathrm{D}E$ on the set $\mathrm{level}_\alpha(E)$.*

*Proof.* The lower bound is just strong convexity. To get the upper bound, observe that the (local) Lipschitz smoothness of $E$ implies that, for all $w, v \in \mathrm{level}_\alpha(E)$ and $z \in \mathcal{V}$,

$$\begin{aligned} |\langle \mathrm{D}E(w) - \mathrm{D}E(v), z\rangle| &\le \|\mathrm{D}E(w) - \mathrm{D}E(v)\|_{\mathcal{V}'}\|z\|_{\mathcal{V}} \\ &\le L\|w - v\|_{\mathcal{V}}\|z\|_{\mathcal{V}}. \end{aligned}$$

Setting $z = w - v$ gives the desired estimate. □

## 11.3 Minimization of convex functionals

We now combine all the previously introduced notions and study optimization problems. In other words, if $\mathcal{V}$ is a Hilbert space and $E : \mathcal{V} \to \mathbb{R}$, then we look for $u \in \mathcal{V}$ such that

$$u \in \operatorname*{argmin}_{v \in \mathcal{V}} E(v).$$

If this problem has a solution, then we call such $u \in \mathcal{V}$ a *minimizer* of $E$. Our goal here is to provide sufficient conditions for the existence and uniqueness of a minimizer, as well as its characterization. Again, we only present a basic introduction, and we refer, for instance, to [26, 18, 14] for further insight.

The following two results are particular cases of what is commonly referred to as the *direct method of calculus of variations*. They guarantee the existence of minimizers.

**Theorem 11.3.1** (Existence I). *Let $\mathcal{V}$ be a Hilbert space, let $K \subset \mathcal{V}$ be bounded and weakly closed, and let $E : \mathcal{V} \to \mathbb{R}$ be wLSC. Then the set*

$$\operatorname*{argmin}_{v \in K} E(v)$$

*is not empty.*

*Proof.* Set $\alpha := \inf_{v\in K} E(v)$ and assume that $\alpha > -\infty$. We will show that $\alpha > -\infty$ is the only possibility. There exists a sequence $\{u_n\}_{n=1}^{\infty} \subset K$ such that

$$E(u_n) \to \alpha.$$

The sequence $\{u_n\}_{n=1}^{\infty} \subset K$ is called an infinimizing sequence. For completeness, let us prove its existence. Since $\alpha$ is the greatest lower bound, we know that for each $n \in \mathbb{N}$, there is an element $y_n \in E(K)$, where $E(K) \subseteq \mathbb{R}$ denotes the image of $K$ under $E$, such that

$$\alpha < y_n < \alpha + \frac{1}{n}.$$

Otherwise, $\alpha$ would not be the greatest lower bound. Now take $u_n \in K$ such that $y_n = E(u_n)$. This proves the existence of an infinimizing sequence.

Having shown the existence of an infinimizing sequence, we observe that since $K$ is bounded, $\{u_n\}_{n=1}^{\infty}$ is bounded. Therefore by Theorem 11.1.19 there are a subsequence $\{u_{n_k}\}_{k=1}^{\infty} \subseteq \{u_n\}_{n=1}^{\infty}$ and a point $u \in \mathcal{V}$ such that $u_{n_k} \rightharpoonup u$ as $k \to \infty$. Since $K$ is weakly closed, $u \in K$. Since $E$ is wLSC,

$$E(u) \le \liminf_{k\to\infty} E(u_{n_k}).$$

However, since $E(u_n) \to \alpha$,

$$\liminf_{k\to\infty} E(u_{n_k}) = \lim_{k\to\infty} E(u_{n_k}) = \alpha,$$

which implies that

$$E(u) \le \alpha.$$

There are two possibilities, $E(u) < \alpha$ and $E(u) = \alpha$. If $E(u) < \alpha$, then we arrive at a contradiction to the fact that $\alpha := \inf_{v\in K} E(v)$ and $u \in K$. Thus $E(u) = \alpha$.

Finally, if $\alpha = -\infty$, then we will arrive at a contradiction. By a slightly different argument from that above we can establish the existence of an infinimizing sequence in this case. The rest of the argument is the same as above, but $E(u) = -\infty$ would clearly be problematical, since $E$ is real valued. □

**Theorem 11.3.2** (Existence II). *Let $\mathcal{V}$ be a Hilbert space, and let $E : \mathcal{V} \to \mathbb{R}$ be wLSC and coercive in the sense of Definition* 11.1.31. *Then the set*

$$\operatorname*{argmin}_{v \in \mathcal{V}} E(v)$$

*is not empty.*

*Proof.* Define

$$\mathcal{B}_0 := \operatorname{level}_{E(0)}(E).$$

By Proposition 11.1.23 this set is weakly closed. Moreover, due to the coercivity of $E$, this set is bounded. Thus we may invoke Theorem 11.3.1 to guarantee the existence of an element of

$$\operatorname*{argmin}_{v \in \mathcal{B}_0} E(v).$$

On the other hand, if $w \notin \mathcal{B}_0$, then we have

$$E(w) \geq E(0) \geq \inf_{v \in \mathcal{B}_0} E(v).$$

Thus

$$\inf_{v \in \mathcal{B}_0} E(v) \leq \inf_{v \in \mathcal{V}} E(v),$$

and the minimizer over $\mathcal{B}_0$ minimizes $E$ globally, that is, over $\mathcal{V}$. □

We now turn our attention to the question of uniqueness and characterization of minimizers.

**Corollary 11.3.3** (Existence and uniqueness). *Let $\mathcal{V}$ be a Hilbert space, and let $E : \mathcal{V} \to \mathbb{R}$ be convex, LSC, and coercive. Then the set*

$$\operatorname*{argmin}_{v \in \mathcal{V}} E(v)$$

*is not empty. In addition, if $E$ is strictly convex, then it is a singleton, that is, there is a unique $u \in \mathcal{V}$ such that*

$$E(u) = \min_{v \in \mathcal{V}} E(v).$$

*Proof.* Since $E$ is convex and LSC, its epigraph is convex and closed; see Propositions 11.1.23 and 11.2.6. This implies, by Mazur's Lemma 11.2.3, that its epigraph is weakly closed. We, again, invoke Proposition 11.1.23 to conclude that $E$ is wLSC. Since $E$ is coercive, Theorem 11.3.2 shows the existence of a minimizer.

We now turn our attention to uniqueness. Assuming that $u_1, u_2 \in \mathcal{V}$ with $u_1 \neq u_2$ are minimizers, by strict convexity we realize that

$$E\left(\frac{u_1 + u_2}{2}\right) < \frac{1}{2}(E(u_1) + E(u_2)) \\ = \min_{v \in \mathcal{V}} E(v),$$

thus contradicting that $u_1$ and $u_2$ are minimizers. This shows the uniqueness. □

Now that we can guarantee the existence and uniqueness of a minimizer, we aim to characterize it.

**Theorem 11.3.4** (Euler–Lagrange equation). *Let $\mathcal{V}$ be a Hilbert space, and let $E : \mathcal{V} \to \mathbb{R}$ be strongly convex and Fréchet differentiable. Then the functional $E$ has a unique minimizer $u \in \mathcal{V}$. Moreover, this minimizer is uniquely characterized by the so-called* Euler–Lagrange equation

$$\langle \mathrm{D}E(u), v\rangle = 0, \quad \forall\, v \in \mathcal{V}. \tag{11.3.1}$$

*Proof.* Obviously, the Fréchet differentiability of $E$ implies that $E$ is continuous, which, along with convexity, implies that it is wLSC. Next, we invoke Theorem 11.2.9 to see that $E$ is coercive. This guarantees the existence and uniqueness of a minimizer.

We now prove that the Euler–Lagrange equation characterizes the minimizer. First, if $u \in \mathcal{V}$ solves the Euler–Lagrange equation then, by strong convexity,

$$E(v) \geq E(u) + \langle \mathrm{D}E(u), v - u\rangle + \frac{\mu}{2}\|v - u\|_{\mathcal{V}}^2 \\ > E(u)$$

for $v \neq u$. This shows that $u$ is the unique minimizer of $E$. On the other hand, if $u \in \mathcal{V}$ is the (unique) minimizer of $E$, then for $t \in \mathbb{R}_\star$ and $v \in \mathcal{V} \setminus \{0\}$,

$$E(u + tv) - E(u) > 0.$$

Since $E$ is Fréchet differentiable, we have

$$0 \leq \lim_{t \downarrow 0} \frac{1}{t}(E(u + tv) - E(u)) = \langle \mathrm{D}E(u), v\rangle$$

and

$$0 \geq \lim_{t \uparrow 0} \frac{1}{t}(E(u + tv) - E(u)) = \langle \mathrm{D}E(u), v\rangle,$$

so that $\langle \mathrm{D}E(u), v\rangle = 0$, for all $v \in \mathcal{V}$. This shows that the Euler–Lagrange equation is satisfied. □

Having guaranteed the existence and uniqueness of a minimizer, as well as its characterization via the Euler–Lagrange equations, we provide some additional facts about this minimizer.

The first result provides a relation between the energy and norm centered at the minimizer. The following estimates can be easily proved using Taylor's theorem with integral remainder, as above; see, e. g., [26] for further details.

**Lemma 11.3.5** (Quadratic trap). *Let $\mathcal{V}$ be a Hilbert space, and let $E : \mathcal{V} \to \mathbb{R}$ be strongly convex and locally Lipschitz smooth. Then, for all $v, w \in \mathcal{V}$, we have*

$$\frac{\mu}{2}\|w-v\|_{\mathcal{V}}^2 + \langle \mathrm{D}E(v), w-v\rangle \le E(w) - E(v).$$

*In addition, for every $\alpha \in \mathbb{R}$, we have that if $v, w \in \operatorname{level}_\alpha(E)$, then*

$$\begin{aligned} \frac{\mu}{2}\|w-v\|_{\mathcal{V}}^2 + \langle \mathrm{D}E(v), w-v\rangle &\le E(w) - E(v) \\ &\le \langle \mathrm{D}E(v), w-v\rangle + \frac{L}{2}\|w-v\|_{\mathcal{V}}^2, \end{aligned} \tag{11.3.2}$$

*where $L = L(\alpha)$ is the local Lipschitz constant of $\mathrm{D}E$. Finally, if $u \in \mathcal{V}$ is the minimizer of $E$, then, for all $\alpha \in \mathbb{R}$ and $w \in \operatorname{level}_\alpha(E)$,*

$$\frac{\mu}{2}\|w-u\|_{\mathcal{V}}^2 \le E(w) - E(u) \le \frac{L}{2}\|w-u\|_{\mathcal{V}}^2. \tag{11.3.3}$$

*Again, the lower bound holds for all $w \in \mathcal{V}$.*

*Proof.* The lower bound for (11.3.2) is already established in Proposition 11.2.9. For the upper bound, we again apply Taylor's theorem with integral remainder and use the fact that $\operatorname{level}_\alpha(E)$ is convex. The details are left to the reader; see Problem 11.9. □

**Example 11.3.6.** The function $E : \mathbb{R} \to \mathbb{R}$ defined by

$$E(v) = \frac{1}{4}v^4, \quad \forall\, v \in \mathbb{R},$$

is strictly convex and coercive but not strongly convex. It is possible to find quadratic functions that bound $E$ from above and below on any bounded interval of $\mathbb{R}$ but not as tightly as suggested in (11.3.3).

On the other hand, suppose that $\lambda \in \mathbb{R}$ and the function $E : \mathbb{R} \to \mathbb{R}$ is defined by

$$E(v) = \frac{1}{4}v^4 + \frac{\lambda}{2}v^2, \quad \forall\, v \in \mathbb{R}.$$

Then, for all $\lambda > 0$, $E$ is both strictly convex and strongly convex. In fact, we can prove that $\mu = \lambda$ can be taken as the "best" strong convexity constant. It is not difficult to show that the global minimizer of $E$ is $u = 0$ and, for all $w \in (-r, r)$,

$$\frac{\lambda}{2}(w-0)^2 \le E(w) - E(0) \le \frac{\frac{r^2}{2}+\lambda}{2}(w-0)^2. \qquad \blacktriangle$$

We can further characterize the minimizer as follows.

**Lemma 11.3.7** (Norm of derivative). *In the setting of Lemma 11.3.5, we have*

$$0 \le E(v) - E(u) \le \frac{1}{2\mu}\|\mathrm{D}E(v)\|_{\mathcal{V}'}^2, \quad \forall v \in \mathcal{V}. \tag{11.3.4}$$

*Proof.* Fix the point $v \in \mathcal{V}$. Now for any $w \in \mathcal{V}$, using the lower bound of (11.3.2), we have

$$E(w) \ge E(v) + \langle \mathrm{D}E(v), w - v\rangle + \frac{\mu}{2}\|w - v\|_{\mathcal{V}}^2.$$

Define

$$g(w) := E(v) + \langle \mathrm{D}E(v), w - v\rangle + \frac{\mu}{2}\|w - v\|_{\mathcal{V}}^2,$$

and observe that the minimizer of $g$ on $\mathcal{V}$ is

$$w_\star := v - \frac{1}{\mu}\mathfrak{R}\mathrm{D}E(v).$$

Therefore

$$\begin{aligned} E(w) &\ge g(w) \\ &\ge g(w_\star) \\ &= E(v) - \frac{1}{2\mu}\|\mathfrak{R}\mathrm{D}E(v)\|_{\mathcal{V}}^2 \\ &= E(v) - \frac{1}{2\mu}\|\mathrm{D}E(v)\|_{\mathcal{V}'}^2. \end{aligned}$$

Then we obtain (11.3.4) by letting $w = u$ in the above inequality. □

We will often use the following simple variant of Lemma 11.3.5.

**Lemma 11.3.8** (Convexity of sections). *Let $\mathcal{V}$ be a Hilbert space, and let $E : \mathcal{V} \to \mathbb{R}$ be strongly convex and locally Lipschitz smooth. Fix $\alpha \in \mathbb{R}$ and set $\mathcal{B} = \mathrm{level}_\alpha(E)$. Let $\xi \in \mathcal{B}$ be arbitrary, and let $\mathcal{W} \subseteq \mathcal{V}$ be a subspace. Define the section*

$$J_\xi(w) := E(\xi + w), \quad \forall w \in \mathcal{W}.$$

*Then $J_\xi : \mathcal{W} \to \mathbb{R}$ is differentiable, strongly convex, and locally Lipschitz smooth. Furthermore, there exists a unique element $\eta \in \mathcal{W}$ such that $\xi + \eta \in \mathcal{B}$, $\eta$ is the unique global minimizer of $J_\xi$, and*

$$\langle \mathrm{D}E(\xi+\eta), w\rangle = \langle \mathrm{D}J_\xi(\eta), w\rangle = 0, \quad \forall\, w \in \mathcal{W}.$$

*As a consequence, for all* $w \in \mathcal{W}$*, with* $w + \xi \in \mathcal{B}$*,*

$$\frac{\mu}{2}\|w-\eta\|_{\mathcal{V}}^2 \le J(w) - J(\eta) = E(\xi + w) - E(\xi+\eta) \le \frac{L}{2}\|w-\eta\|_{\mathcal{V}}^2.$$

*The lower bound holds for all* $w \in \mathcal{W}$ *without any restrictions.*

*Proof.* Exercise; see Problem 11.10. □

**Remark 11.3.9** (Condition number). For a strongly convex and a locally Lipschitz smooth functional $E : \mathcal{V} \to \mathbb{R}$, the ratio $\frac{L}{\mu}$ is sometimes referred to as the nonlinear condition number or the condition number of the derivative, D$E$; see [26, page 63]. The rate of convergence of iterative methods for solving

$$u = \operatorname*{argmin}_{v \in \mathcal{V}} E(v)$$

often depends on this condition number, as we will see. ◂

## 11.4 The gradient descent method

As a final consideration, we present two methods to approximate minimizers. Let $\mathcal{V}$ be a Hilbert space, and let $E : \mathcal{V} \to \mathbb{R}$ be a strongly convex and locally Lipschitz smooth functional. We seek its (unique) minimizer $u \in \mathcal{V}$.

Since we assume that $E$ is strongly convex, (11.3.1) holds. Thus, provided that $E$ is twice Fréchet differentiable, we can propose a version of **Newton's method** to solve (11.3.1) and thus to find the minimizer. Given $u^0 \in \mathcal{V}$, this method finds $\{u^k\}_{k=0}^\infty \subset \mathcal{V}$ as follows. For $k \ge 0$, find $\eta^k \in \mathcal{V}$ that solves the linear problem

$$\mathrm{D}^2E(u^k)\eta^k = -\mathrm{D}E(u^k).$$

Notice that this equation is understood in $\mathcal{V}'$, so we can equivalently write

$$\mathrm{D}^2E(u^k)(\eta^k, v) = -\langle \mathrm{D}E(u^k), v\rangle, \quad \forall\, v \in \mathcal{V}.$$

If a solution exists, then

$$u^{k+1} := u^k + \eta^k.$$

Many questions, like the existence and uniqueness of $\eta^k$, immediately arise. Even if this is the case, the convergence of such approach must be justified. We refer the reader to [31, Section 16.8] for a complete treatment of Newton's method.

Note, however, that in practice the functionals we wish to minimize are often not twice Fréchet differentiable, and for those that are, it may be difficult to compute their second derivative. Thus we abandon this approach. Instead, we will focus on *first-order methods*, that is, such that only use first derivatives.

**Definition 11.4.1** (Gradient descent). Let $\mathcal{V}$ be a Hilbert space, and let $E : \mathcal{V} \to \mathbb{R}$ be strongly convex and locally Lipschitz smooth. The **gradient descent** method is a scheme to approximate $u \in \mathcal{V}$, the unique minimizer of $E$ as follows. Fix a step size $\tau > 0$. Starting from arbitrary $u^0 \in \mathcal{V}$, we compute $\{u^k\}_{k=0}^{\infty} \subset \mathcal{V}$ recursively via

$$u^{k+1} = u^k - \tau \mathfrak{R} \mathrm{D}E(u^k) \tag{11.4.1}$$

for $k \geq 0$. ∎

This method can be understood as a generalization of the preconditioned steepest descent (PSD) method of Definition 2.2.1. Let us show that if the step size $\tau$ is sufficiently small, then the method converges at least linearly.

**Theorem 11.4.2** (Convergence). *Let $\mathcal{V}$ be a Hilbert space, let $E : \mathcal{V} \to \mathbb{R}$ be strongly convex and locally Lipschitz smooth, and let $u \in \mathcal{V}$ be the minimizer of $E$. Let $u^0 \in \mathcal{V}$ be arbitrary. Define*

$$\mathcal{B} = \{v \in \mathcal{V} \mid \|v - u\|_{\mathcal{V}} \leq \|u^0 - u\|_{\mathcal{V}}\},$$

*and let $L = L_{\mathcal{B}}$ be the local Lipschitz smoothness constant of $E$ on $\mathcal{B}$. Assume that the step size satisfies*

$$\tau \in \left(0, \frac{2\mu}{L^2}\right).$$

*Then the sequence $\{u^k\}_{k=0}^{\infty}$ generated by the gradient descent method of Definition* 11.4.1 *satisfies*

$$\{u^k\}_{k=0}^{\infty} \subset \mathcal{B}.$$

*Moreover,*

$$\|u - u^{k+1}\|_{\mathcal{V}} \leq \rho \|u - u^k\|_{\mathcal{V}}, \quad \rho := \sqrt{1 - 2\mu\tau + \tau^2 L^2} \in (0, 1).$$

*Consequently, the sequence $\{u^k\}_{k=0}^{\infty}$ converges at least linearly to $u$.*

*Proof.* To shorten the presentation, we introduce the error $e^k := u - u^k$. Using the Euler equation (11.3.1), we may rewrite (11.4.1) as

$$e^{k+1} = e^k - \tau(\mathfrak{R}\mathrm{D}E(u) - \mathfrak{R}\mathrm{D}E(u^k)).$$

Using the definition of the Riesz map $\mathfrak{R}$, it follows that

$$\begin{aligned}\|e^{k+1}\|_{\mathcal{V}}^2 &= \|e^k - \tau(\mathfrak{R}DE(u) - \mathfrak{R}DE(u^k))\|_{\mathcal{V}}^2 \\ &= \|e^k\|_{\mathcal{V}}^2 - 2\tau\langle DE(u) - DE(u^k), e^k\rangle + \tau^2\|DE(u) - DE(u^k)\|_{\mathcal{V}'}^2.\end{aligned}$$

By strong convexity

$$\begin{aligned}2\tau\langle DE(u) - DE(u^k), e^k\rangle &= 2\tau\langle DE(u) - DE(u^k), u - u^k\rangle \\ &\geq 2\mu\tau\|e^k\|_{\mathcal{V}}^2,\end{aligned}$$

so that

$$\|e^{k+1}\|_{\mathcal{V}}^2 \leq (1 - 2\mu\tau)\|e^k\|_{\mathcal{V}}^2 + \tau^2\|DE(u) - DE(u^k)\|_{\mathcal{V}'}^2. \tag{11.4.2}$$

Observe now that $u, u^0 \in \mathcal{B}$. We use this to inductively show that the whole sequence belongs to $\mathcal{B}$. Assume that $u^k \in \mathcal{B}$. By the local Lipschitz smoothness assumption

$$\tau^2\|DE(u) - DE(u^k)\|_{\mathcal{V}'}^2 \leq \tau^2 L^2\|e^k\|_{\mathcal{V}}^2.$$

Using this fact in (11.4.2), we conclude

$$\|e^{k+1}\|_{\mathcal{V}}^2 \leq (1 - 2\mu\tau + \tau^2 L^2)\|e^k\|_{\mathcal{V}}^2.$$

Define

$$\phi(\tau) := 1 - 2\mu\tau + \tau^2 L^2$$

and observe that, for $\tau \in (0, \frac{2\mu}{L^2})$,

$$0 < \phi(\tau) < 1.$$

Thus, for a step size in the stated range,

$$\begin{aligned}\|e^{k+1}\|_{\mathcal{V}}^2 &\leq \rho^2\|e^k\|_{\mathcal{V}}^2 \\ &\leq \|e^k\|_{\mathcal{V}}^2 \\ &\leq \|u^0 - u\|_{\mathcal{V}}^2,\end{aligned}$$

and we get that $u^{k+1} \in \mathcal{B}$.

The previous inequality,

$$\|e^{k+1}\|_{\mathcal{V}}^2 \leq \rho^2\|e^k\|_{\mathcal{V}}^2,$$

also shows the claimed convergence with linear rate. □

**Corollary 11.4.3** (Optimal rate). *In the setting of Theorem* 11.4.2 *the optimal rate of convergence is obtained for*

$$\tau_{\mathrm{opt}} = \frac{\mu}{L^2}.$$

*In fact,*

$$\|u - u^{k+1}\|_{\mathcal{V}} \le \rho_{\mathrm{opt}} \|u - u^k\|_{\mathcal{V}}, \quad \rho_{\mathrm{opt}} := \sqrt{1 - \frac{\mu^2}{L^2}}.$$

*Proof.* It suffices to optimize the function $\phi$ defined in the proof of Theorem 11.4.2. The details are left as an exercise; see Problem 11.11. □

**Remark 11.4.4** (Ill-conditioning). We now see a possible problem for this method. The convergence rate deteriorates as the nonlinear condition number $\frac{L}{\mu}$ becomes large. More efficient methods that do not experience this deterioration will be presented in the next chapter. ◂

## 11.5 Problems

11.1 Prove Proposition 11.1.5.

11.2 Prove Proposition 11.1.7.

11.3 Prove the Riesz representation Theorem 11.1.12 in the particular case $\mathcal{V} = \mathbb{R}^n$.

11.4 Complete the details of Example 11.1.15.

11.5 Prove Proposition 11.1.23.

11.6 Prove Proposition 11.2.2.3.

11.7 Prove Proposition 11.2.2.4.

11.8 A functional $E : \mathcal{V} \to \mathbb{R}$ is called *quasiconvex* iff, for every $a \in \mathbb{R}$, the sublevel set $\mathrm{level}_a(E)$ is convex. Give an example of a function $E : \mathbb{R} \to \mathbb{R}$ that is quasiconvex but not convex, thus showing that the converse to Proposition 11.2.6.2 is false.

11.9 Prove Lemma 11.3.5.

11.10 Prove Lemma 11.3.8.

11.11 Prove Corollary 11.4.3.

# 12 Axiomatic methods of subspace corrections

In this chapter, which is based on the paper [13], we build an axiomatic framework for building multiplicative multigrid methods for approximating the solution of nonlinear boundary value problems that arise as the Euler–Lagrange equations of strongly convex functionals. The method relies on subspace decompositions, as we described in Chapter 10. However, as opposed to the hierarchical basis and BPX preconditioners, these methods are designed to be stand-alone solvers.

## 12.1 Description of the problem and basic axioms

Let $\mathcal{V}$ be a Hilbert space with inner product $(\cdot,\cdot)_{\mathcal{V}}$ and norm $\|\cdot\|_{\mathcal{V}}$. Given a functional $E : \mathcal{V} \to \mathbb{R}$, which we will call an *energy* or *objective*, we consider the following minimization problem: Find $u \in \mathcal{V}$ such that

$$u = \operatorname*{argmin}_{v\in\mathcal{V}} E(v). \tag{12.1.1}$$

Often this problem is defined through the corresponding Euler–Lagrange problem: Find $u \in \mathcal{V}$ such that

$$\langle \mathrm{D}E(u), w\rangle = 0, \quad \forall w \in \mathcal{V}. \tag{12.1.2}$$

The methods we will describe hinge on several assumptions on the energy $E : \mathcal{V} \to \mathbb{R}$.

**Assumption 12.1.1** (E0). The energy $E$ is Fréchet differentiable on $\mathcal{V}$ and strongly convex with constant $\mu > 0$. •

**Assumption 12.1.2** (E1). The energy $E$ satisfies (E0) and is locally Lipschitz smooth on $\mathcal{V}$. •

In what follows, for a bounded set $\mathcal{B} \subset \mathcal{V}$, we will denote the (local) Lipschitz smoothness constant of $E$ by $L_{\mathcal{B}}$.

As Theorem 11.3.4 shows, Assumption (E0) is sufficient to guarantee that problem (12.1.1) has a unique solution, which, in addition, can be uniquely characterized as the unique solution to the Euler–Lagrange equations (12.1.2).

As we saw in Section 11.4, we can use the gradient descent method to approximate the solution to (12.1.1). The convergence, however, will be too slow for practical purposes. For this reason, here we devise other methods based on subspace decompositions.

**Definition 12.1.3** (Subspace decomposition). Let $\mathcal{V}$ be a Hilbert space, and let $N \in \mathbb{N}$, $N \geq 2$. A collection $\{\mathcal{V}_i\}_{i=1}^N$ is called a **subspace decomposition of** $\mathcal{V}$ if
(i) $\mathcal{V}_i$, $i = 1, \dots, N$, are a norm-closed subspaces of $\mathcal{V}$, and
(ii) $\mathcal{V} = \mathcal{V}_1 + \mathcal{V}_2 + \cdots + \mathcal{V}_N$. ■

https://doi.org/10.1515/9783111354880-012

At this point, it may be useful to reexamine Example 10.1.4, where we give an explicit subspace decomposition of the very particular Hilbert space $\mathbb{R}^n$.

**Assumption 12.1.4** (NS1). Let $\mathcal{V}$ be a Hilbert space, and let $\{\mathcal{V}_i\}_{i=1}^N$ be a subspace decomposition of $\mathcal{V}$. We say that the subspace decomposition is **stable** if there is a constant $C_A > 0$ such that for every $v \in \mathcal{V}$, there are $v_i \in \mathcal{V}_i$, $i = 1, \ldots, N$, such that

$$v = \sum_{i=1}^{N} v_i \tag{NS1a}$$

and

$$\sum_{i=1}^{N} \|v_i\|_{\mathcal{V}}^2 \le C_A^2 \|v\|_{\mathcal{V}}^2. \tag{NS1b}$$

•

**Assumption 12.1.5** (NS2). Let $\{\mathcal{V}_i\}_{i=1}^N$ be a subspace decomposition of $\mathcal{V}$ and suppose that $E : \mathcal{V} \to \mathbb{R}$ is Fréchet differentiable. We say that the energy functional $E$ satisfies the **strengthened Cauchy–Schwarz inequality** with respect to the subspace decomposition if, given $u_0 \in \mathcal{V}$, there is a constant $C_S = C_S(u_0) > 0$ such that if

$$u_i, v_i, w_i \in \mathcal{V}_i, \quad 1 \le i \le N, \tag{NS2a}$$

with

$$v_{i-1}, v_{i-1} + u_i \in \operatorname{level}_{E(u_0)}(E), \quad 2 \le i \le N, \tag{NS2b}$$

then

$$\left| \sum_{i=1}^{N} \sum_{j=i+1}^{N} \langle \mathrm{D}E(v_{j-1} + u_j) - \mathrm{D}E(v_{j-1}), w_i \rangle \right|^2 \le C_S^2 \sum_{i=1}^{N} \|u_i\|_{\mathcal{V}}^2 \sum_{i=1}^{N} \|w_i\|_{\mathcal{V}}^2. \tag{NS2c}$$

•

**Remark 12.1.6** (Quadratic energies). Notice that if $E$ is not quadratic, then in general we cannot assume that the strengthened Cauchy–Schwarz inequality holds without restriction to some bounded convex subset, such as the sublevel set $\operatorname{level}_{E(u_0)}(E)$ used in (NS2b). ◀

**Remark 12.1.7** (Terminology). Note that the constant $C_S > 0$ in (NS2) is related to the local Lipschitz constant $L$. In particular, suppose that $E$ satisfies (E1) and $L$ is the local Lipschitz constant relative to $\operatorname{level}_{E(u_0)}(E)$. Since we assume (NS2b), it follows that

$$\left| \langle \mathrm{D}E(v_{j-1} + u_j) - \mathrm{D}E(v_{j-1}), w_i \rangle \right| \le L \|u_j\|_{\mathcal{V}} \|w_i\|_{\mathcal{V}}.$$

Setting S equal to the left-hand side (NS2c), we have

$$
\begin{aligned}
S &\le L^2 \Big( \sum_{j=1}^{N} \|u_j\|_{\mathcal{V}} \Big)^2 \Big( \sum_{i=1}^{N} \|w_i\|_{\mathcal{V}} \Big)^2 \\
&\le L^2 N^2 \Big( \sum_{j=1}^{N} \|u_j\|_{\mathcal{V}}^2 \Big) \Big( \sum_{i=1}^{N} \|w_i\|_{\mathcal{V}}^2 \Big),
\end{aligned}
$$

which yields the naïve estimate $C_S = LN$. In the case where the dimension of the ambient space $\mathcal{V}$ is finite, we typically have

$$C_1 \dim(\mathcal{V}) \le N \le C_2 \dim(\mathcal{V})$$

for some constants $0 < C_1 < C_2$. Thus we could take $C_S = C_2 L \dim(\mathcal{V})$. Obviously, if $\dim(\mathcal{V})$ is large, then $C_S$ could be undesirably large, if estimated in this crude way.

On the other hand, if $E$ is twice continuously differentiable, then we may write

$$\langle \mathrm{D}E(v_{j-1} + u_j) - \mathrm{D}E(v_{j-1}), w_i \rangle = M_j(u_j, w_i),$$

where the bilinear form is

$$M_j = \int_0^1 \mathrm{D}^2 E(v_{j-1} + t u_j)\, \mathrm{d}t \in \mathcal{L}(\mathcal{V}, \mathcal{V}').$$

If this bilinear form is uniformly spectrally equivalent to the $\mathcal{V}$-inner product, then we can obtain a better constant $C_S$, in particular a dimension-independent one. This is the reason we call (NS2) the strengthened Cauchy–Schwarz inequality. ◂

**Remark 12.1.8** (Minimal assumptions). Because of the connection mentioned in the last remark, we will not always need to assume (E1). Often, (E0) is sufficient, or (E0) along with the more subtle Assumption (NS2) will be required. ◂

## 12.2 Successive subspace optimization methods

We have reached the point where it is meaningful to describe the first method that we will employ. It is called the successive subspace optimization method, and it was originally proposed in [34].

**Definition 12.2.1** (SSO). Suppose that Assumption (E0) holds for $E$, and let $u \in \mathcal{V}$ be the unique solution to (12.1.1). The **successive subspace optimization (SSO)** method is an algorithm to approximate $u$ that computes a sequence $\{u^k\}_{k=0}^{\infty} \subset \mathcal{V}$ recursively, $u^{k+1} := \mathrm{SSO}(u^k)$, as follows. Let $u^0 \in \mathcal{V}$ be arbitrary. Then, for $k \ge 0$,

- Set $v_0^k = u^k$.
- For $i = 1$ to $N$:
  - Define an energy section along $\mathcal{V}_i$:
$$J_i^k(w) := E(v_{i-1}^k + w), \quad \forall\, w \in \mathcal{V}_i.$$
  - Compute the subspace correction:
$$e_i^k = \operatorname*{argmin}_{w \in \mathcal{V}_i} J_i^k(w). \tag{12.2.1}$$
  - Apply the subspace correction:
$$v_i^k = v_{i-1}^k + e_i^k. \tag{12.2.2}$$
- $u^{k+1} = v_N^k$. ■

**Remark 12.2.2** (Fundamental orthogonality). Note that due to the strong convexity inherited by the energy section $J_i^k$, the correction $e_i^k$ computed using (12.2.1) is uniquely defined. In fact, the correction satisfies

$$\langle \mathrm{D}E(v_i^k), w\rangle = \langle \mathrm{D}E(v_{i-1}^k + e_i^k), w\rangle = \langle \mathrm{D}J_i^k(e_i^k), w\rangle = 0, \quad \forall\, w \in \mathcal{V}_i.$$

The orthogonality relation satisfied by the corrected approximation $v_i^k$, specifically,

$$\langle \mathrm{D}E(v_i^k), w\rangle = 0, \quad \forall\, w \in \mathcal{V}_i, \tag{12.2.3}$$

is sometimes referred to as the *fundamental orthogonality (FO)* of the solver. In fact, $v_i^k$ can be understood as the Ritz approximation of $u$ over $\mathcal{V}_i$; see Definition 2.1.4. This orthogonality property was then already obtained in Proposition 2.1.3. ◂

**Remark 12.2.3** (Relation to Gauss–Seidel). We point out that when $\mathcal{V}_i$ is one-dimensional, the computation of the subspace correction is identical to a nonlinear Gauss–Seidel method. In fact, the SSO method can be considered as a generalization of the nonlinear Gauss–Seidel methodology. ◂

In what follows, we will show a linear reduction of the energy difference for one iteration of the SSO algorithm. In other words, there is $\rho \in (0,1)$ such that if $u$ solves (12.1.1) and $u^{k+1} = \mathrm{SSO}(u^k)$, then

$$E(u^{k+1}) - E(u) \le \rho(E(u^k) - E(u)). \tag{12.2.4}$$

Ideally, $\rho$ is independent of the size of the problem, that is, the dimension of $\mathcal{V}$. This is the reason why we merely assume that $\mathcal{V}$ is a Hilbert space, even if we have in mind discretized, that is, finite-dimensional, problems. We mention that the SSO method and its convergence properties were developed in [34, 35] in the more general case where $\mathcal{V}$ is a Banach space. Here we present a simplified version for Hilbert spaces.

In what follows, we will utilize the following simple result.

**Lemma 12.2.4** (Monotone convergence). *Let $\{d_k\}_{k=0}^{\infty} \subset \mathbb{R}_+ \cup \{0\}$. Assume that there is a constant $C > 0$ such that for every $k \geq 0$,*

$$d_{k+1} \leq C(d_k - d_{k+1}).$$

*Then, for $k \geq 0$,*

$$d_{k+1} \leq \rho d_k, \quad \rho = \frac{C}{1+C}. \tag{12.2.5}$$

*In particular, the sequence $\{d_k\}_{k=0}^{\infty}$ converges monotonically and at least linearly to zero.*

*Proof.* Since $C > 0$, inequality (12.2.5) follows from a simple algebraic manipulation. In addition, since $\rho \in (0,1)$, (12.2.5) implies that $\{d_k\}$ is decreasing. Thus the monotone convergence theorem implies that there is $\gamma \geq 0$ such that $d_k \to \gamma$ as $k \to \infty$. Suppose now, for contradiction, that $\gamma > 0$. Monotone convergence shows that there is $K \in \mathbb{N}$ such that for $k \geq K$,

$$\gamma \leq d_k < d_K \leq \gamma + \frac{(1-\rho)\gamma}{2\rho}.$$

Therefore, for $k \geq K$,

$$\begin{aligned} d_{k+1} &\leq \rho d_k \\ &\leq \rho\gamma + \frac{1}{2}(1-\rho)\gamma \\ &= \frac{1}{2}(\rho+1)\gamma \\ &< \gamma. \end{aligned}$$

This is a contradiction. The only possibility is $\gamma = 0$. □

To analyze the SSO method of Definition 12.2.1, we will assume the following.

- (Lower bound on corrections) There is $C_L > 0$ such that for all $k \geq 0$,

$$E(u^k) - E(u^{k+1}) \geq C_L \sum_{i=1}^{N} \|e_i^k\|_{\mathcal{V}}^2. \tag{12.2.6}$$

- (Upper bound on corrections) There is $C_U > 0$ such that for $k \geq 0$,

$$E(u^{k+1}) - E(u) \leq C_U \sum_{i=1}^{N} \|e_i^k\|_{\mathcal{V}}^2. \tag{12.2.7}$$

Under these assumptions, as a consequence of Lemma 12.2.4, we have that (12.2.4) holds.

**Corollary 12.2.5** (Convergence). *Suppose that Assumption* (E0) *holds for* $E : \mathcal{V} \to \mathbb{R}$, *and let* $u \in \mathcal{V}$ *be the unique solution to* (12.1.1). *Suppose that* $\{u^k\}_{k=0}^{\infty} \subset \mathcal{V}$ *is obtained by the SSO method of Definition* 12.2.1. *Assume that the lower bound* (12.2.6) *and upper bound* (12.2.7) *hold. Then the sequence of energies* $\{E(u^k)\}_{k=0}^{\infty}$ *satisfies* (12.2.4) *with*

$$\rho = \frac{C_U}{C_L + C_U}.$$

*Consequently, as* $k \to \infty$, *the sequence* $\{E(u^k)\}_{k=0}^{\infty}$ *converges monotonically and at least linearly to* $E(u)$. *Furthermore,* $\{u^k\}_{k=0}^{\infty}$ *converges at least linearly to* $u$.

*Proof.* We will use Lemma 12.2.4 with

$$d_k := E(u^k) - E(u).$$

The quantity $d_k$ is the difference between the current value of the energy and its minimum; see Figure 12.1. Thus the difference

$$d_k - d_{k+1} = E(u^k) - E(u^{k+1})$$

is the energy decrease associated with the $(k+1)$th iteration. We have

$$\begin{aligned} d_{k+1} &\overset{(12.2.7)}{\leq} C_U \sum_{i=1}^{N} \|e_i^k\|_{\mathcal{V}}^2 \\ &\overset{(12.2.6)}{\leq} \frac{C_U}{C_L}(d_k - d_{k+1}), \end{aligned}$$

so that by Lemma 12.2.4 we have (12.2.5) with

$$\rho = \frac{\frac{C_U}{C_L}}{1 + \frac{C_U}{C_L}} = \frac{C_U}{C_L + C_U}.$$

Thus the linear convergence of $\{E(u^k)\}_{k=0}^{\infty}$ to $E(u)$ at the rate $\rho$ is guaranteed by Lemma 12.2.4.

Finally, using (11.3.3) with $w = u^k$, we have

$$\frac{\mu}{2}\|u^k - u\|_{\mathcal{V}}^2 \leq E(u^k) - E(u),$$

which guarantees the linear convergence of $u^k$ to $u$. □

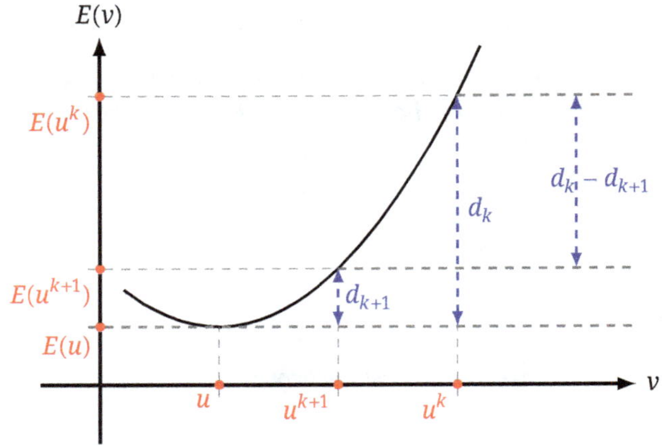


**Figure 12.1:** The sequence $\{d_k\}_{k=0}^{\infty}$ used in the proof of Corollary 12.2.5.

Our next steps are clear. We must verify the lower and upper bounds on corrections (12.2.6) and (12.2.7), respectively. Verifying the lower bound is relatively easy since $E$ is convex. Solving the convex optimization problem in each subspace will surely decrease the energy, and this decrease can be quantified in terms of the norms of the corrections. We make essential use of the fundamental orthogonality property.

**Theorem 12.2.6** (Lower bound on corrections). *Assume* (E0) *holds for the energy* $E : \mathcal{V} \to \mathbb{R}$. *Let* $\{u^k\}_{k=0}^{\infty} \subset \mathcal{V}$ *be obtained by the SSO method of Definition* 12.2.1. *Then condition* (12.2.6) *holds with* $C_L \coloneqq \frac{\mu}{2}$.

*Proof.* Fix $k \in \mathbb{N}_0$. By Lemma 11.3.8 we have that for every $i = 1, \dots, N$, the energy section $J_i^k$ is strictly convex over $\mathcal{V}_i$ and is Fréchet differentiable, as it inherits the structure of $E$. It follows that

$$\langle \mathrm{D}J_i^k(e_i^k), w\rangle = 0, \quad \forall\, w \in \mathcal{V}_i.$$

Using the definition of $J_i^k$, for all $w \in \mathcal{V}_i$, we have

$$\langle \mathrm{D}J_i^k(e_i^k), w\rangle = \langle \mathrm{D}E(v_{i-1}^k + e_i^k), w\rangle = \langle \mathrm{D}E(v_i^k), w\rangle.$$

Therefore the fundamental orthogonality $\mathrm{D}E(v_i^k) = 0$ in $\mathcal{V}_i'$, holds. As $e_i^k = v_i^k - v_{i-1}^k \in \mathcal{V}_i$, by Lemma 11.3.8 we have

$$E(v_{i-1}^k) - E(v_i^k) = J_i^k(0) - J_i^k(e_i^k) \geq \frac{\mu}{2}\|e_i^k\|_{\mathcal{V}}^2. \tag{12.2.8}$$

Summing (12.2.8) over $i = 1, \dots, N$, we see that the left-hand side telescopes, and we obtain

$$E(u^k) - E(u^{k+1}) = \sum_{i=1}^{N} (E(v_{i-1}) - E(v_i)) \geq \frac{\mu}{2} \sum_{i=1}^{N} \|e_i^k\|_{\mathcal{V}}^2. \qquad \square$$

**Remark 12.2.7** (Convexity of energy sections). In view of (12.2.8), we can relax the convexity assumption of the energy $E$ and assume only the convexity of the energy sections $J_i^k$ in each subspace $\mathcal{V}_i$. In other words, we may have a nonconvex energy $E$ that, when restricted to subspaces $\mathcal{V}_i$, is convex and the lower bound (12.2.6) still holds. A simple example of this is given in (12.2.9) below. ◀

**Example 12.2.8.** Consider the energy functional $E : \mathbb{R}^2 \to \mathbb{R}$ defined in polar coordinates as

$$E(r, \theta) = r^2 \sin(2\theta). \tag{12.2.9}$$

Setting $R = (0, \infty) \times [0, 2\pi)$, we see that any subspace of $\mathbb{R}^2$ can be described for some $c \in [0, \pi)$ as

$$L = \{(r, \theta) \in R \mid \theta = c\} \cup \{(0, 0)\} \cup \{(r, \theta) \in R \mid \theta = c + \pi\}.$$

Since $\sin(2\theta + 2\pi) = \sin(2\theta)$, we have that $J(r, \theta) = ar^2$ on $L$ for some $a \in [0, 1]$; see Figure 12.2. Thus the energy section will be convex for $\theta \in [0, \frac{\pi}{2}]$. In particular, this energy is convex when restricted to each coordinate axis. We comment that some references, like [17], refer to a function that is convex along the coordinate axes as *separately convex*. ▲

We now focus our attention on the upper bound on corrections, (12.2.7). The proof of this bound is more complicated, as it relies on the assumptions on the subspace decomposition. This is the content of the following result.

**Theorem 12.2.9** (Upper bound on corrections). *Suppose that* (E0) *holds for the energy* $E : \mathcal{V} \to \mathbb{R}$. *Given* $u^0 \in \mathcal{V}$, *let* $\{u^k\}_{k=0}^{\infty} \subset \mathcal{V}$ *be obtained by the SSO method of Definition* 12.2.1. *If the subspace decomposition satisfies* (NS1) *and if the energy satisfies* (NS2) *with respect to the same subspace decomposition, then we have* (12.2.7) *with* $C_U := \frac{C_S^2 C_A^2}{2\mu}$, *where* $C_S = C_S(u^0)$.

*Proof.* Fix $k \geq 0$. Due to Lemma 11.3.7, with the choice $v = u^{k+1}$, we have

$$E(u^{k+1}) - E(u) \leq \frac{1}{2\mu} \|\mathrm{D}E(u^{k+1})\|_{\mathcal{V}'}^2.$$

Observe next that in (12.2.1), we solve the minimization problem on each subspace exactly. Thus the energy decreases. This guarantees that if

$$\mathcal{B} = \mathrm{level}_{E(u^0)}(E),$$

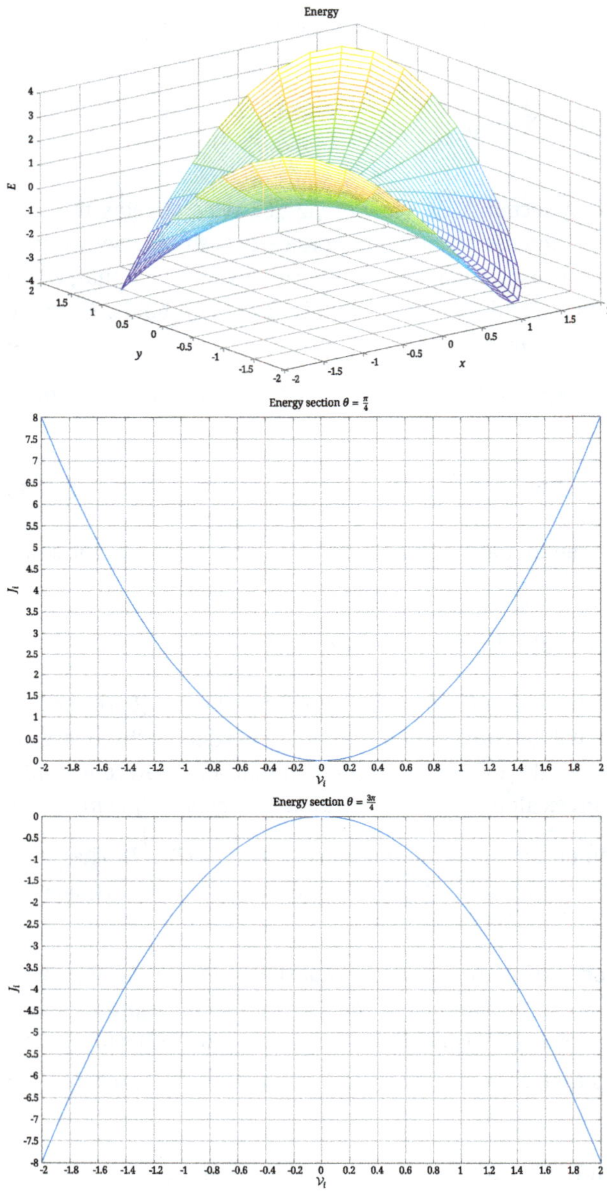


**Figure 12.2:** (Top) The energy $E$ defined in (12.2.9). (Middle) Energy section along $\theta = \frac{\pi}{4}$. (Bottom) Energy section along $\theta = \frac{3\pi}{4}$. As we see, this energy is not convex, but some of its sections are convex.

then for all $k \geq 0$ and $j = 1, \dots, N$, we have $v_j^k \in \mathcal{B}$. Moreover, $\mathrm{D}E(v_j^k) = 0$ in $\mathcal{V}_j'$. Let now $w \in \mathcal{V}$. By invoking (NS1) we may choose a stable decomposition $w = \sum_{i=1}^N w_i$. Then

$$\langle \mathrm{D}E(u^{k+1}), w \rangle = \sum_{i=1}^{N} \langle \mathrm{D}E(u^{k+1}), w_i \rangle$$

$$
\begin{aligned}
&\overset{(12.2.3)}{=} \sum_{i=1}^{N} \langle \mathrm{D}E(u^{k+1}) - \mathrm{D}E(v_i^k), w_i \rangle \\
&= \sum_{i=1}^{N} \sum_{j=i+1}^{N} \langle \mathrm{D}E(v_j^k) - \mathrm{D}E(v_{j-1}^k), w_i \rangle \\
&\overset{(12.2.2)}{=} \sum_{i=1}^{N} \sum_{j=i+1}^{N} \langle \mathrm{D}E(v_{j-1}^k + e_j^k) - \mathrm{D}E(v_{j-1}^k), w_i \rangle \\
&\overset{(\mathrm{NS2})}{=} C_S \left( \sum_{i=1}^{N} \|e_i^k\|_{\mathcal{V}}^2 \right)^{1/2} \left( \sum_{i=1}^{N} \|w_i\|_{\mathcal{V}}^2 \right)^{1/2} \\
&\overset{(\mathrm{NS1})}{\leq} C_S C_A \left( \sum_{i=1}^{N} \|e_i^k\|_{\mathcal{V}}^2 \right)^{1/2} \|w\|_{\mathcal{V}}.
\end{aligned}
$$

We may then gather the two previous estimates and use the definition of the norm of $\mathcal{V}'$ to obtain

$$
\begin{aligned}
E(u^{k+1}) - E(u) &\leq \frac{1}{2\mu} \|\mathrm{D}E(u^{k+1})\|_{\mathcal{V}'}^2 \\
&= \frac{1}{2\mu} \left( \sup_{w \in \mathcal{V} \setminus \{0\}} \frac{\langle \mathrm{D}E(u^{k+1}), w \rangle}{\|w\|_{\mathcal{V}}} \right)^2 \\
&\leq \frac{C_S^2 C_A^2}{2\mu} \sum_{i=1}^{N} \|e_i^k\|_{\mathcal{V}}^2,
\end{aligned}
$$

which finishes the proof. □

Based on the lower bound given in Theorem 12.2.6 and upper bound given in Theorem 12.2.9, we can conclude the convergence of SSO. Comparing with the results in [35], for this convergence result, we are able to use slightly weaker assumptions, and the constant $C_U$ seems to be slightly better; see [13] for the original details and arguments.

**Corollary 12.2.10** (Convergence of SSO). *Assume that the energy $E$ satisfies* (E0). *Given $u^0 \in \mathcal{V}$, let $\{u^k\}_{k=0}^{\infty} \subset \mathcal{V}$ be obtained by the SSO method of Definition* 12.2.1. *If the space decomposition satisfies* (NS1) *and* (NS2), *then the SSO method converges linearly. In other words, set*

$$
\rho = \frac{C_S^2 C_A^2}{C_S^2 C_A^2 + \mu^2}.
$$

*Then, for every $k \geq 0$, we have*

$$
E(u^{k+1}) - E(u) \leq \rho(E(u^k) - E(u)).
$$

*Proof.* This is a restatement of Corollary 12.2.5. Indeed, since (E0) holds, Theorem 12.2.6 guarantees that (12.2.6) holds with

$$C_L = \frac{\mu}{2}.$$

In addition, since we assume that (NS1) and (NS2) hold, we may invoke Theorem 12.2.9 to state that (12.2.7) is valid with

$$C_U = \frac{C_S^2 C_A^2}{2\mu}.$$

Thus we have linear convergence with contraction factor

$$\rho = \frac{C_U}{C_L + C_U} = \frac{\frac{C_S^2 C_A^2}{2\mu}}{\frac{\mu}{2} + \frac{C_S^2 C_A^2}{2\mu}} = \frac{C_S^2 C_A^2}{C_S^2 C_A^2 + \mu^2},$$

as claimed. □

**Remark 12.2.11** (Assumption (E1)). Notice that at no point in the analysis of the SSO method, assumption (E1) was explicitly used. This may lead us to erroneous belief that this is not needed. However, as we have pointed out in Remark 12.1.7, this assumption is highly related to (NS2), and the value of the constant $C_S$ may depend on the Lipschitz smoothness constant with respect to $\text{level}_{E(u^0)}(E)$. Consequently, the contraction factor $\rho$ of Corollary 12.2.10 depends on this constant as well. ◂

## 12.3 Full approximation storage method

In this section, we take a very brief detour to describe a standard two-level method for solving nonlinear elliptic equations known as the full approximation storage (FAS) method [9, 12, 37, 23]. To put things into the context of the current chapter, suppose that $\mathcal{V} = V_h$, where the latter is a finite element space, as described in Chapter 7. We wish to solve the following finite element problem: Find $u_h \in V_h$ such that

$$\langle \mathrm{D}E(u_h), w_h \rangle = 0, \quad \forall\, w_h \in V_h,$$

where $E$ is the appropriate energy for the underlying partial differential equation. We will give an example of this in the next chapter.

First, let us assume that the derivative of the energy can be represented as a linear operator of the form

$$\mathrm{D}E(u_h) = \mathcal{L}_h u_h - f_h.$$

In other words, the Euler–Lagrange equation that we need to solve can be expressed in operator form as follows: Find $u_h \in V_h$ such that

$$\mathcal{L}_h u_h = f_h.$$

As usual, $V_h$ is constructed from a conforming finite element grid $\mathcal{T}_h$. Let us suppose that $V_H \subset V_h$ is a finite element subspace of $V_h$ based on the grid $\mathcal{T}_H$. Here we tacitly assume that $\mathcal{T}_h$ is obtained from $\mathcal{T}_H$ through refinement. We refer to $\mathcal{T}_H$ as the coarse grid (relative to $\mathcal{T}_h$) and $V_H$ as the coarse space (relative to $V_h$). In the classical two-grid algorithm, we wish to use corrections from the coarse space $V_H \subset V_h$.

The two-grid algorithm is the iterative method

$$u_h^{k+1} := \mathrm{TG}(u_h^k),$$

which can be summarized in pseudocode as follows:

1. $v_0 := u_h^k$.
2. Smooth (e. g., using linear Gauss–Seidel) on the fine grid: $v_1 := Sv_0$.
3. Compute the coarse-grid residual
$$r_H := R_h(f_h - \mathcal{L}_h(v_1)) \in V_H,$$
where $R_h$ is the restriction operator $R_h : V_h \to V_H$.
4. Solve the coarse-grid correction problem: Find $s_H \in V_H$ such that
$$\mathcal{L}_H s_H = r_H,$$
where $\mathcal{L}_H$ is a version of the linear operator realized on the coarse finite element space $V_H$. Typically, $\mathcal{L}_H$ is determined via some Galerkin condition.
5. Apply the coarse-grid correction
$$v_2 := v_1 + P_H s_H,$$
where $P_H$ is the prolongation operator $P_H : V_H \to V_h$, which, as usual, is related to the restriction operator through transposition.
6. Smooth (e. g., using linear Gauss–Seidel) on the fine grid: $v_3 := Sv_2$.
7. $u_h^{k+1} := v_3$.

This design of the linear two-grid algorithm is based on several facts. Paramount among these is the fact that the error equation is equivalent to the original equation. For example, if $w_h \in V_h$ is an approximation to the true solution $u_h$, then

$$u_h = w_h + e_h,$$

where $e_h \in V_h$ is the error function satisfying

$$\mathcal{L}_h e_h = r_h := f - \mathcal{L}_h w_h.$$

Here $r_h \in V_h$ is the residual on the fine grid. We can think of $P_H s_H$ as an approximation of the error function.

Now suppose that the derivative of the energy can be represented as a nonlinear operator of the form

$$\mathrm{D}E(u_h) = \mathcal{N}_h(u_h) - f_h.$$

The Euler–Lagrange equation to be solved can be expressed in operator form as follows: Find $u_h \in V_h$ such that

$$\mathcal{N}_h(u_h) = f_h.$$

Since the equation is nonlinear, we can no longer use the linear two-grid process. This is due, essentially, to the fact that the error-residual equation is no longer equivalent to the original equation.

The full approximation storage (FAS) method, originating in [9], is the iterative method

$$u_h^{k+1} := \mathrm{FAS}(u_h^k)$$

and is summarized by the following pseudocode:

1. $v_0 := u_h^k$.
2. Smooth (e. g., using nonlinear Gauss–Seidel) on the fine grid: $v_1 := S(v_0)$.
3. Compute the coarse-grid *τ-perturbation*:

   $$\tau_H := \mathcal{N}_H(Q_h v_1) + R_h(f_h - \mathcal{N}_h(v_1)) \in V_H,$$

   where $\mathcal{N}_H$ is a version of the nonlinear operator defined on the coarse space $V_H$, $R_h$ is the restriction operator mentioned earlier, and $Q_h$ is a projection operator, $Q_h : V_h \to V_H$. Unlike $R_h$, $Q_h$ is not typically related to the prolongation operator directly.
4. Solve the coarse-grid correction problem: Find $\eta_H \in V_H$ such that

   $$\mathcal{N}_H(\eta_H) = \tau_H$$
   $$s_H := \eta_H - Q_h v_1 \in V_H.$$

5. Apply the coarse-grid correction:

   $$v_2 := v_1 + P_H s_H.$$

6. Smooth (e. g., using nonlinear Gauss–Seidel) on the fine grid: $v_3 := S(v_2)$.
7. $u_h^{k+1} := v_3$.

We observe that the FAS method is equivalent to the linear two-grid method precisely when $\mathcal{N}_h$ and $\mathcal{N}_H$ are linear.

The FAS method is analyzed in [23] and elsewhere. However, the analysis is not very general, and the convergence is not global. The initial guess $u_h^0$ needs to be very close to $u_h$ for the result to be proven. In the next section, we want to generalize this method so that the resulting method always converges, regardless of how $u_h^0$ is chosen, as long as the energy $E : \mathcal{V} \to \mathbb{R}$ and the subspace decomposition satisfy certain properties.

## 12.4 The fast subspace descent (FASD) method

Recall that at every step of the SSO method, we need to exactly solve, for $i = 1, \dots, N$, the minimization problem (12.2.1). This at least requires the evaluation of the energy $E$ and its derivative $\mathrm{D}E$ in the space $\mathcal{V}_i$. Although the size of problem (12.2.1) is reduced to $\dim \mathcal{V}_i$, such evaluations are still in the original space of size $\dim \mathcal{V}$, which may be expensive.

To better explain the issue, we first need some notation. Let $I_i : \mathcal{V}_i \hookrightarrow \mathcal{V}$ be the natural inclusion, and let $R_i := I_i^\top : \mathcal{V}' \to \mathcal{V}_i'$ be the natural restriction of functionals, where the superscript $\top$ indicates the adjoint in $\mathcal{V}$. Thus to solve (12.2.1) for $k \geq 0$ and $i = 1, \dots, N$, we must be able to evaluate, for $v, w \in \mathcal{V}_i$,

$$\begin{aligned} \langle \mathrm{D}E(I_i v + v_{i-1}^k), I_i w \rangle_{\mathcal{V}',\mathcal{V}} &= \langle I_i^\top \mathrm{D}E(I_i v + v_{i-1}^k), w \rangle_{\mathcal{V}_i',\mathcal{V}_i} \\ &= \langle R_i \mathrm{D}E(I_i v + v_{i-1}^k), w \rangle_{\mathcal{V}_i',\mathcal{V}_i}. \end{aligned}$$

So far, since this rarely leads to confusion, we have dropped the mappings $R_i$ and $I_i$, as their actions can be assumed implicitly. In summary, to solve (12.2.1), we must be able to evaluate the derivative mapping

$$\mathcal{V}_i \ni v \mapsto R_i \mathrm{D}E(I_i v + v_{i-1}^k) \in \mathcal{V}_i'$$

and possibly second derivatives if a Newton-type method is used. As we see, even though $v \in \mathcal{V}_i$, a space of possibly small dimension, to evaluate them, we must inject (prolongate) $v$ into $\mathcal{V}$, which is assumed to be of large dimension.

To circumvent the difficulty in evaluating derivatives, in this section, we introduce the so-called fast subspace descent method, originally proposed in [13]. The idea is that instead of using the original energy $E$ in (12.2.1), we use a locally defined energy $E_i : \mathcal{V}_i \to \mathbb{R}$, $i = 1, \dots, N$. In addition to prolongation and restriction operators, we will also need projection operators $Q_i : \mathcal{V} \to \mathcal{V}_i$, $i = 1, \dots, N$.

**Definition 12.4.1** (FASD). Let $E$ satisfy (E0), and let $u \in \mathcal{V}$ be the unique solution to (12.1.1). The **fast subspace descent (FASD)** method is an algorithm to approximate $u$ that computes a sequence $\{u^k\}_{k=0}^\infty \subset \mathcal{V}$ recursively,

$$u^{k+1} := \mathrm{FASD}(u^k),$$

and is defined as follows. Assume that for $i = 1, \dots, N$, we have Fréchet differentiable $E_i : \mathcal{V}_i \to \mathbb{R}$ and projections $Q_i : \mathcal{V} \to \mathcal{V}_i$. Let $u^0 \in \mathcal{V}$ be arbitrary. Then, for $k \geq 0$,

- $v_0^k := u^k$.
- For $i = 1$ to $N$:
  - Compute the so-called subspace $\tau$**-perturbation**: let $\xi_i^k := Q_i v_{i-1}^k$ and

$$\tau_i^k := \mathrm{D}E_i(\xi_i^k) - R_i \mathrm{D}E(v_{i-1}^k) \in \mathcal{V}_i'.$$

- Solve the subspace residual problem: Find $\eta_i^k \in \mathcal{V}_i$ such that

$$\langle \mathrm{D}E_i(\eta_i^k), w\rangle_{\mathcal{V}_i',\mathcal{V}} = \langle \tau_i^k, w\rangle_{\mathcal{V}_i',\mathcal{V}_i}, \quad \forall\, w \in \mathcal{V}_i. \tag{12.4.1}$$

- Compute the search direction:

$$s_i^k := \eta_i^k - \xi_i^k \in \mathcal{V}_i.$$

- Orthogonalize the subspace correction via line search: Find

$$\begin{aligned} \alpha_i^k &:= \operatorname*{argmin}_{\alpha\in\mathbb{R}} E(v_{i-1}^k + \alpha s_i^k) \\ &= \operatorname*{argzero}_{\alpha\in\mathbb{R}} \langle \mathrm{D}E(v_{i-1}^k + \alpha s_i^k), s_i^k\rangle \end{aligned}$$

and define

$$\varepsilon_i^k := \alpha_i^k s_i^k. \tag{12.4.2}$$

- Apply the subspace correction:

$$v_i^k := v_{i-1}^k + \varepsilon_i^k.$$

- $u^{k+1} := v_N^k$. ■

**Remark 12.4.2** (FAS vs. FASD). The reader should notice similarities with the FAS method of the last section and also some differences. The biggest difference is that in the FAS method, once the coarse-grid correction is found, it is applied. In FASD, there is an additional orthogonalization step. ◂

Notice that in the orthogonalization step of the FASD method, we perform a line search to find the optimal step size $\alpha_i^k$. This still requires the evaluation of the original energy $E$, which is defined in the ambient space, that is, $E(v_{i-1}^k + \alpha s_i^k)$. The cost, however, is reduced when compared to the evaluation of $E(v_{i-1}^k + w)$ for arbitrary $w \in \mathcal{V}_i$. In the next section, we will analyze a simplified version of this scheme, where the step size is found approximately. This is closer to the FAS scheme.

**Remark 12.4.3** (Reduction to SSO). Note that Definition 12.4.1 reduces to Definition 12.2.1 when the local energies are defined as

$$E_i(\eta) := E(v_{i-1}^k - Q_i v_{i-1}^k + \eta), \quad \forall\, \eta \in \mathcal{V}_i,$$

provided that $Q_i$ is a projection. As a consequence of this choice, $\tau_i^k \equiv 0$, and

$$\begin{aligned} \langle \mathrm{D}E(v_{i-1}^k + s_i^k), w\rangle &= \langle \mathrm{D}E(v_{i-1}^k - Q_i v_{i-1}^k + \eta_i^k), w\rangle \\ &= \langle \mathrm{D}E_i(\eta_i^k), w\rangle \\ &= 0, \end{aligned}$$

for all $w \in \mathcal{V}_i$.

With this choice in FASD, the orthogonalization step is redundant because upon taking $w = s_i^k$,

$$\langle DE(v_{i-1}^k + s_i^k), s_i^k \rangle = 0.$$

In other words, the orthogonality is valid with $\alpha_i^k = 1$. ◂

Let us now analyze the FASD scheme of Definition 12.4.1. We will follow the same path as we did for the SSO method. In other words, all that is needed is to verify the lower and upper bounds on corrections (12.2.6) and (12.2.7), respectively. Verifying the lower bound is relatively easy due to the line search and the convexity of $E$.

**Theorem 12.4.4** (Lower bound on corrections). *Assume that the energy $E$ satisfies* (E1) *and that the sequence $\{u^k\}_{k=0}^{\infty} \subset \mathcal{V}$ is obtained via the FASD scheme of Definition* 12.4.1. *Then*

$$E(u^k) - E(u^{k+1}) \geq \frac{\mu}{2} \sum_{i=1}^{N} \|\varepsilon_i^k\|_{\mathcal{V}}^2, \quad \forall\, k \geq 0.$$

*Proof.* We apply a similar technique as in the proof of Theorem 12.2.6. Due to the line search, we still have a rather useful orthogonality property. Namely,

$$\langle DE(v_i^k), w \rangle = 0, \quad \forall\, w \in \operatorname{span}\{s_i^k\}.$$

We may then apply Lemma 11.3.8 with the subspace $\mathcal{W} := \operatorname{span}\{s_i^k\}$ and use that

$$v_i^k - v_{i-1}^k = \varepsilon_i^k = \alpha_i^k s_i^k \in \operatorname{span}\{s_i^k\}$$

to conclude that

$$\begin{aligned} E(v_{i-1}^k) - E(v_i^k) &\geq \frac{\mu}{2} \|v_{i-1}^k - v_i^k\|_{\mathcal{V}}^2 \\ &= \frac{\mu}{2} \|\varepsilon_i^k\|_{\mathcal{V}}^2. \end{aligned}$$

Adding this inequality for $i = 1, \ldots, N$, we then obtain

$$\begin{aligned} E(u^k) - E(u^{k+1}) &= \sum_{i=1}^{N} (E(v_{i-1}^k) - E(v_i^k)) \\ &\geq \frac{\mu}{2} \sum_{i=1}^{N} \|\varepsilon_i^k\|_{\mathcal{V}}^2. \end{aligned}$$ □

The derivation of the upper bound of corrections is more complicated and requires some assumptions on the local energies $E_i : \mathcal{V}_i \to \mathbb{R}$. To properly state them, from now on, we set

$$\mathcal{B} := \text{level}_{E(u^0)}(E). \tag{12.4.3}$$

Notice that Theorem 12.4.4 implies that the energy is always decreasing between iterates. Thus $\{u^k\}_{k=0}^{\infty} \subset \mathcal{B}$. However, the search region and, e. g., the point $\xi_i^k + s_i^k$, may not be contained in $\mathcal{B}$. To be able to use Lipschitz smoothness, we introduce the larger domain

$$\mathcal{B}^+ := \{v \in \mathcal{V} \mid \text{dist}(v, \mathcal{B}) \leq \sqrt{\chi}\}, \tag{12.4.4}$$

where

$$\chi := \frac{2L^2}{\mu \min_{i=1}^N \mu_i^2}(E(u^0) - E(u)).$$

With this at hand, we assume that the local energies $E_i$, $i = 1, \dots, N$, satisfy the following assumptions.

**Assumption 12.4.5** (E2). Let $\mathcal{V}$ be a Hilbert space, let $\{\mathcal{V}_i\}_{i=1}^N$ a subspace decomposition, and let $E_i : \mathcal{V}_i \to \mathbb{R}$, $i = 1, \dots, N$. We assume that the local energies $\{E_i\}_{i=1}^N$ satisfy the following:

(E2a) The local energy $E_i$ is Fréchet differentiable on $\mathcal{V}_i$.

(E2b) (Strong convexity) There exists a constant $\mu_i$ such that

$$\langle \mathrm{D}E_i(w) - \mathrm{D}E_i(v), w - v \rangle \geq \mu_i \|w - v\|_{\mathcal{V}}^2, \quad \forall\, v, w \in \mathcal{V}_i.$$

(E2c) (Lipschitz continuity of the derivative) The derivatives of the energies $E_i$ are locally Lipschitz continuous. Let $\mathcal{B}$ and $\mathcal{B}^+$ be defined in (12.4.3) and (12.4.4), respectively. Set $\mathcal{B}_i = Q_i\mathcal{B}^+$. We denote the local constant of $E_i$ over $\mathcal{B}_i$ as $L_i$. •

The following result will be useful in the analysis that follows.

**Proposition 12.4.6** (Convexity). *The sets $\mathcal{B}^+$ and $\mathcal{B}_i$ are convex.*

*Proof.* Observe that

$$\mathcal{B}^+ = \mathcal{B} + \{v \in \mathcal{V} \mid \|v\|_{\mathcal{V}} \leq \sqrt{\chi}\}.$$

The second set is, of course, the ball of radius $\sqrt{\chi}$ centered at $0 \in \mathcal{V}$. Since both sets are convex, so is their sum; see Proposition 11.2.2.3.

Next, recall that $\mathcal{B}_i$ is the image of a convex set $\mathcal{B}^+$ through a linear map $Q_i$; see Proposition 11.2.2.4. □

Later, during the proof of convergence, we will show that for all $u^0 \in \mathcal{V}$ and $k \geq 0$, $\xi_i^k + s_i^k \in \mathcal{B}_i$. Thus (E2c) will become relevant. It is interesting to note that at this stage, there is no relation between the energy $E$ and local energies $E_i$ besides satisfying similar properties.

The following result follows directly from Lemmas 11.2.11 and 11.3.5.

**Lemma 12.4.7** (Local quadratic trap). *Assume that the local energies $E_i$, $i = 1, \dots, N$, satisfy* (E2b) *and* (E2c). *Then for all $v, w \in \mathcal{B}_i$,*

$$\mu_i \|w - v\|_{\mathcal{V}}^2 \le \langle \mathrm{D}E_i(w) - \mathrm{D}E_i(v), w - v \rangle \le L_i \|w - v\|_{\mathcal{V}}^2,$$

*and*

$$\frac{\mu_i}{2}\|w - v\|_{\mathcal{V}}^2 + \langle \mathrm{D}E_i(v), w - v \rangle \le E_i(w) - E_i(v) \le \langle \mathrm{D}E_i(v), w - v \rangle + \frac{L_i}{2}\|w - v\|_{\mathcal{V}}^2.$$

*Moreover, if $u_i \in \mathcal{V}_i$ is the minimizer of $E_i$ and $u_i \in \mathcal{B}_i$, then, for all $w \in \mathcal{B}_i$,*

$$\frac{\mu_i}{2}\|w - u_i\|_{\mathcal{V}}^2 \le E_i(w) - E_i(u_i) \le \frac{L_i}{2}\|w - u_i\|_{\mathcal{V}}^2.$$

*The lower bounds above hold for all $w \in \mathcal{V}_i$.*

*Proof.* Exercise; see Problem 12.1. □

The next point to be noticed is that in (12.4.1), we do not minimize $E_i$, which would entail solving $\mathrm{D}E_i(s_i^k + Q_i v_{i-1}^k) = 0$. Instead, a so-called $\tau$-perturbation is added to the right-hand side. The following considerations explain why this is needed.

**Lemma 12.4.8** (Descent direction). *Assume that the local energies satisfy* (E2). *Assume that for $k \ge 0$, $s_i^k$ is computed with the FASD method of Definition* 12.4.1. *Then $s_i^k$ is a descent direction in the sense that*

$$\langle -R_i \mathrm{D}E(v_{i-1}^k), s_i^k \rangle_{\mathcal{V}_i', \mathcal{V}_i} \ge \mu_i \|s_i^k\|_{\mathcal{V}}^2$$

*or, equivalently,*

$$\langle \mathrm{D}E(v_{i-1}^k), I_i s_i^k \rangle \le -\mu_i \|s_i^k\|_{\mathcal{V}}^2.$$

*Proof.* By the definition of the $\tau$-perturbation, the local problem (12.4.1) can be rewritten as follows. Find $\eta_i^k \in \mathcal{V}_i$ such that

$$\langle \mathrm{D}E_i(\eta_i^k) - \mathrm{D}E_i(\xi_i^k), w \rangle_{\mathcal{V}_i', \mathcal{V}_i} = -\langle R_i \mathrm{D}E(v_{i-1}^k), w \rangle_{\mathcal{V}_i', \mathcal{V}_i}, \quad \forall\, w \in \mathcal{V}_i \tag{12.4.5}$$

(recall that $\xi_i^k = Q_i v_{i-1}^k \in \mathcal{V}_i$ and $\eta_i^k = \xi_i^k + s_i^k$). Choose $w = s_i^k$ and use the strong convexity of $E_i$ to obtain

$$\langle -R_i \mathrm{D}E(v_{i-1}^k), s_i^k \rangle = \langle \mathrm{D}E_i(\eta_i^k) - \mathrm{D}E_i(\xi_i^k), s_i^k \rangle \ge \mu_i \|s_i^k\|_{\mathcal{V}}^2.$$

□

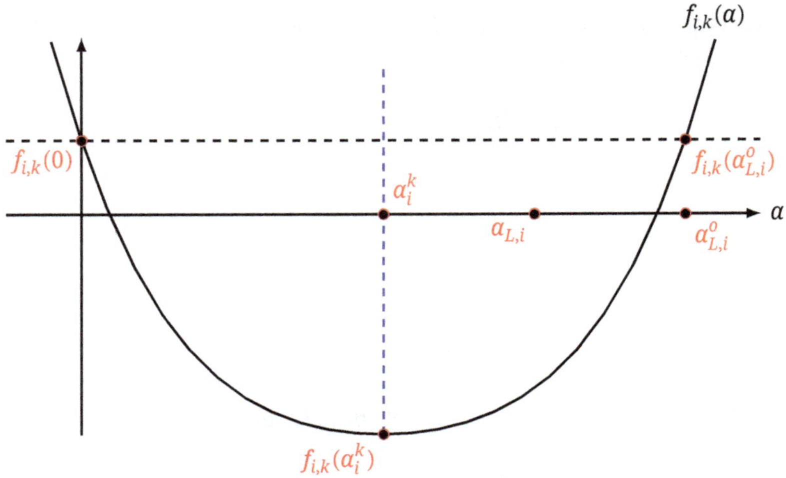


**Figure 12.3:** The function $f_{i,k}$ defined in (12.4.6); it is a one-dimensional energy section. It is straightforward to prove that its minimizer $\alpha_i^k$ is positive.

To better understand the choice of the step size, we introduce the scalar function $f_{i,k}$, see Figure 12.3, which is a one-dimensional section of the energy $E$

$$f_{i,k}(\alpha) := E(v_{i-1}^k + \alpha s_i^k). \tag{12.4.6}$$

**Proposition 12.4.9** (Energy section I). *Assume that $E$ satisfies* (E1) *and the local energies satisfy* (E2). *The one-dimensional energy section $f_{i,k}$, defined in* (12.4.6), *satisfies*

$$f_{i,k}'(0) = \langle \mathrm{D}E(v_{i-1}^k), s_i^k \rangle \le -\mu_i \|s_i^k\|_{\mathcal{V}}^2.$$

*Furthermore, $\alpha_i^k > 0$, and $f_{i,k}(\alpha) < f_{i,k}(0)$ for all $\alpha \in (0, \alpha_i^k]$.*

*Proof.* Lemma 12.4.8 implies $f_{i,k}'(0) < 0$. As $f_{i,k}'$ is continuous, we conclude that the minimizing point is positive, $\alpha_i^k > 0$, and $f_{i,k}(\alpha) < f_{i,k}(0) = E(v_{i-1}^k)$ for all $\alpha \in (0, \alpha_i^k]$. □

**Lemma 12.4.10** (Energy section II). *Assume that $E$ satisfies* (E1) *and the local energies satisfy* (E2). *Then $f_{i,k}$, defined in* (12.4.6), *is differentiable and strongly convex in the following sense: for all $\alpha, \beta \in \mathbb{R}$,*

$$(f_{i,k}'(\alpha) - f_{i,k}'(\beta))(\alpha - \beta) \ge (\alpha - \beta)^2 \mu \|s_i^k\|_{\mathcal{V}}^2.$$

*Furthermore, $f_{i,k}'$ is locally Lipschitz in the following sense: for all $0 \le \alpha, \beta \le \alpha_{L,i}$,*

$$|f_{i,k}'(\alpha) - f_{i,k}'(\beta)| \le L \|s_i^k\|_{\mathcal{V}}^2 |\alpha - \beta|,$$

*where $\alpha_{L,i} := (1 + \sqrt{\mu/L})\alpha_i^k$.*

*Proof.* The proof is based on the identity

$$f'_{i,k}(\alpha) - f'_{i,k}(\beta) = \langle DE(v^k_{i-1} + \alpha s^k_i) - DE(v^k_{i-1} + \beta s^k_i), s^k_i \rangle.$$

Then by Assumption (E0)

$$\begin{aligned}(f'_{i,k}(\alpha) - f'_{i,k}(\beta))(\alpha - \beta) &= \langle DE(v^k_{i-1} + \alpha s^k_i) - DE(v^k_{i-1} + \beta s^k_i), \alpha s^k_i - \beta s^k_i \rangle \\ &\geq \mu \|(\alpha - \beta) s^k_i\|^2_{\mathcal{V}}.\end{aligned}$$

We now turn our attention to the local Lipschitz continuity of $f'_{i,k}$. To use the Lipschitz continuity of $DE$, we need to ensure that the points of evaluation are inside the set $\mathcal{B}$, which imposes an upper bound on $\alpha$ and $\beta$. As $f'_{i,k}(0) < 0$ and $f'_{i,k}(\alpha^k_i) = 0$, by continuity there exists $\alpha^o_{L,i} > \alpha^k_i$ such that $f_{i,k}(0) = f_{i,k}(\alpha^o_{L,i})$ and $f_{i,k}(\alpha) < f_{i,k}(0)$ for all $\alpha \in (0, \alpha^o_{L,i})$. This implies $v^k_{i-1} + \alpha s^k_i \in \mathcal{B}$ for all $\alpha \in (0, \alpha^o_{L,i})$, since the energy $E$ is decreased for such values of $\alpha$.

We now estimate $\alpha^o_{L,i}$. As $f'_{i,k}(\alpha^k_i) = 0$ and $f'_{i,k}$ is Lipschitz in $(0, \alpha^o_{L,i})$, from Lemma 11.3.8 we have the bound

$$\begin{aligned} 0 &< f_{i,k}(\alpha^o_{L,i}) - f_{i,k}(\alpha^k_i) \\ &= E(v^k_{i-1} + \alpha^o_{L,i} s^k_i) - E(v^k_{i-1} + \alpha^k_i s^k_i) \\ &\leq \frac{L}{2} \|\alpha^o_{L,i} s^k_i - \alpha^k_i s^k_i\|^2_{\mathcal{V}} \\ &= (\alpha^o_{L,i} - \alpha^k_i)^2 \frac{L}{2} \|s^k_i\|^2_{\mathcal{V}}. \end{aligned}$$

On the other hand, again by Lemma 11.3.8,

$$\begin{aligned} f_{i,k}(\alpha^o_{L,i}) - f_{i,k}(\alpha^k_i) &= f_{i,k}(0) - f_{i,k}(\alpha^k_i) \\ &\geq \frac{\mu(\alpha^k_i)^2}{2} \|s^k_i\|^2_{\mathcal{V}}. \end{aligned}$$

Then the desired bound

$$\begin{aligned} \alpha^o_{L,i} &\geq \alpha_{L,i} \\ &= \left(1 + \sqrt{\frac{\mu}{L}}\right) \alpha^k_i \\ &> \alpha^k_i \\ &> 0 \end{aligned}$$

follows. To conclude, since $DE$ is Lipschitz, $f'_{i,k}$ is Lipschitz with constant $L\|s^k_i\|^2_{\mathcal{V}}$ on the interval $[0, \alpha^o_{L,i}]$: for all $\alpha, \beta \in (0, \alpha^o_{L,i})$, $\alpha \neq \beta$,

$$|f'_{i,k}(\alpha) - f'_{i,k}(\beta)| = |\langle DE(v^k_{i-1} + \alpha s^k_i) - DE(v^k_{i-1} + \beta s^k_i), s^k_i \rangle|$$

$$
\begin{aligned}
&= \frac{1}{|\alpha - \beta|} |\langle DE(v_{i-1}^k + \alpha s_i^k) - DE(v_{i-1}^k + \beta s_i^k), (\alpha - \beta) s_i^k \rangle| \\
&\le \frac{1}{|\alpha - \beta|} L \|(\alpha - \beta) s_i^k\|_{\mathcal{V}}^2 \\
&= L \|s_i^k\|_{\mathcal{V}}^2 |\alpha - \beta|.
\end{aligned}
$$

Obviously, $f'_{i,k}$ is also Lipschitz with the same constant on the smaller interval $[0, \alpha_{L,i}] \subseteq [0, \alpha_{L,i}^o]$. The proof is complete. □

**Remark 12.4.11** (Right endpoint). The importance of using the right-hand endpoint

$$
\alpha_{L,i} := \left(1 + \sqrt{\frac{\mu}{L}}\right) \alpha_i^k,
$$

rather than $\alpha_{L,i}^o$, is that $\alpha_{L,i}$ is computable, provided that $\mu$, $L$, and $\alpha_i^k$ are known. ◂

To use the local Lipschitz continuity of $DE_i$, we must have $\xi_i^k + s_i^k \in \mathcal{B}_i$. We will show this by providing a lower bound on the optimal step size.

**Lemma 12.4.12** (Lower bound on $\alpha_i^k$). *Assume that $E$ satisfies* (E1) *and the local energies satisfy* (E2). *In this setting, we have the lower bound*

$$
\frac{\mu_i}{L} \le \alpha_i^k. \tag{12.4.7}
$$

*Consequently,*

$$
\alpha_{L,i}^o > \frac{\mu_i}{L}.
$$

*Proof.* Recall that $\varepsilon_i^k = \alpha_i^k s_i^k \in \operatorname{span}\{s_i^k\}$, and due to the line search, we still have an orthogonality property that can be utilized,

$$
\langle DE(v_i^k), w \rangle = 0, \quad \forall w \in \operatorname{span}\{s_i^k\}.
$$

Thus, as an element of the dual of $\operatorname{span}\{s_i^k\}$, we have $DE(v_{i-1}^k + \varepsilon_i^k) = 0$. Using the $\tau$-perturbation in Definition 12.4.1, we conclude that

$$
-DE(v_{i-1}^k) = DE_i(\xi_i^k + s_i^k) - DE_i(\xi_i^k) \quad \text{in } \mathcal{V}_i'.
$$

Then we obtain the lower bound by the strong convexity of $E_i$ and Lipschitz continuity of $DE$:

$$
\begin{aligned}
\alpha_i^k L \|s_i^k\|_{\mathcal{V}}^2 &= \frac{1}{\alpha_i^k} L \|\varepsilon_i^k\|_{\mathcal{V}}^2 \\
&\ge \frac{1}{\alpha_i^k} \langle DE(v_{i-1}^k + \varepsilon_i^k) - DE(v_{i-1}^k), \varepsilon_i^k \rangle
\end{aligned}
$$

$$\begin{aligned}&= \langle \mathrm{D}E(v_{i-1}^k + \varepsilon_i^k) - \mathrm{D}E(v_{i-1}^k), s_i^k \rangle \\ &= -\langle \mathrm{D}E(v_{i-1}^k), s_i^k \rangle \\ &= \langle \mathrm{D}E_i(\xi_i^k + s_i^k) - \mathrm{D}E_i(\xi_i^k), s_i^k \rangle \\ &\geq \mu_i \|s_i^k\|_{\mathcal{V}}^2.\end{aligned}$$

Note that $v_{i-1}^k + \varepsilon_i^k \in \mathcal{B}$ by Lemma 12.4.10. This justifies the use of the Lipschitz continuity of $\mathrm{D}E$. Finally,

$$\begin{aligned}\alpha_{L,i}^o &\geq \alpha_{L,i} \\ &= \left(1 + \sqrt{\frac{\mu}{L}}\right)\alpha_i^k \\ &> \alpha_i^k \\ &\geq \frac{\mu_i}{L}.\end{aligned}$$

□

Next, we show that the norm of $s_i^k$ is bounded, and thus $\xi_i^k + s_i^k \in \mathcal{B}_i$.

**Lemma 12.4.13** (Bound on $s_i^k$). *In the setting of Lemma 12.4.12, we have that for all $k \geq 0$ and $i = 1, \dots, N$, the point $\xi_i^k + s_i^k$ is in the set $\mathcal{B}_i$.*

*Proof.* Notice that since $\xi_i^k + s_i^k = Q_i(v_{i-1}^k + s_i^k)$, it suffices to show that $v_{i-1}^k + s_i^k \in \mathcal{B}^+$. To start, we know that $v_{i-1}^k \in \mathcal{B}$; so by the definition of $\mathcal{B}^+$ given in (12.4.4), it suffices to prove that $\|s_i^k\|_{\mathcal{V}}^2 \leq \chi$. By Theorem 12.4.4 and Lemma 12.4.12 we have

$$\begin{aligned}\frac{\mu_i^2}{L^2}\|s_i^k\|_{\mathcal{V}}^2 &\leq (\alpha_i^k)^2\|s_i^k\|_{\mathcal{V}}^2 \\ &= \|\varepsilon_i^k\|_{\mathcal{V}}^2 \\ &\leq \frac{2}{\mu}(E(v_{i-1}^k) - E(v_i^k)) \\ &\leq \frac{2}{\mu}(E(u_0) - E(u)).\end{aligned}$$

Recalling that

$$\chi = \frac{2L^2}{\mu \min_{i=1}^N \mu_i^2}(E(u_0) - E(u)),$$

the previous computation shows that

$$\|s_i^k\|_{\mathcal{V}}^2 \leq \chi.$$

Therefore

$$\operatorname{dist}(\mathcal{B}, v_{i-1}^k + s_i^k) \le \|s_i^k\|_{\mathcal{V}},$$

and the result is proven. □

**Remark 12.4.14** (Improved bound). Note that $\chi$ can be large. However, it may be possible to improve this bound. Since for all $\alpha \in [0, \alpha_{L,i}]$, we have $v_{i-1}^k + \alpha s_i^k \in \mathcal{B}$, we get that $v_{i-1}^k + s_i^k \in \mathcal{B}$, provided that $1 \le \alpha_{L,i}$. On the other hand, if $1 > \alpha_{L,i}$, then we need to estimate the size of $(1 - \alpha_{L,i})\|s_i^k\|_{\mathcal{V}}$, since

$$v_{i-1}^k + s_i^k = v_{i-1}^k + \alpha_{L,i} s_i^k + (1 - \alpha_{L,i}) s_i^k.$$

Thus, if $1 > \alpha_{L,i}$, then

$$\begin{aligned}\operatorname{dist}(\mathcal{B}, v_{i-1}^k + s_i^k) &\le (1 - \alpha_{L,i})\|s_i^k\|_{\mathcal{V}} \\ &\le (1 - \alpha_{L,i})\sqrt{\chi}.\end{aligned}$$

The right-hand side of the last estimate may be much smaller than $\sqrt{\chi}$. ◂

Next, we present a refined upper bound on the step size $\alpha_i^k$.

**Lemma 12.4.15** (Improved upper bound). *Assume that $E$ satisfies* (E1) *and the local energies satisfy* (E2). *Then we have the upper bound*

$$\alpha_i^k \le \frac{L_i}{\mu}.$$

*Proof.* The upper bound can be proved using the strong convexity of $E$ and the Lipschitz continuity of $\mathrm{D}E_i$, knowing that $\xi_i^k, \xi_i^k + s_i^k \in \mathcal{B}_i$. Since, as an element of the dual of $\operatorname{span}\{s_i^k\}$, we have $\mathrm{D}E(v_{i-1}^k + \varepsilon_i^k) = 0$, then

$$\begin{aligned}\mu\|\varepsilon_i^k\|_{\mathcal{V}}^2 &\le \langle \mathrm{D}E(v_{i-1}^k + \varepsilon_i^k) - \mathrm{D}E(v_{i-1}^k), \varepsilon_i^k\rangle \\ &= \langle \mathrm{D}E_i(\xi_i^k + s_i^k) - \mathrm{D}E_i(\xi_i^k), \varepsilon_i^k\rangle \\ &\le \frac{L_i}{\alpha_i^k}\|\varepsilon_i^k\|_{\mathcal{V}}^2.\end{aligned}$$

□

With our estimates on the step size $\alpha_i^k$, we are now ready to establish an upper bound for the iterates in our FASD method of Definition 12.4.1.

**Theorem 12.4.16** (Upper bound on corrections). *Assume that the energy $E$ satisfies* (E1) *and the local energies $E_i$ satisfy* (E2). *Assume, in addition, that the space decomposition satisfies* (NS1) *and $E$ satisfies* (NS2) *with respect to the decomposition. Then the sequence $\{u^k\}_{k=0}^\infty$ produced by the FASD method of Definition* 12.4.1 *satisfies the upper bound*

$$E(u^{k+1}) - E(u) \le C_U \sum_{i=1}^N \|\varepsilon_i^k\|_{\mathcal{V}}^2,$$

*where*

$$C_U := \frac{C_A^2}{2\mu}\left[C_S + L\left(1 + \max_{i=1}^{N}\left\{\frac{L_i}{\mu_i}\right\}\right)\right]^2.$$

*Proof.* Let $w \in \mathcal{V}$ be arbitrary and choose a stable decomposition $w = \sum_{i=1}^N w_i$. Then

$$\begin{aligned}
\langle \mathrm{D}E(u^{k+1}), w\rangle &= \sum_{i=1}^{N}\langle \mathrm{D}E(u^{k+1}), w_i\rangle \\
&= \sum_{i=1}^{N}(\langle \mathrm{D}E(u^{k+1}) - \mathrm{D}E(v_i^k), w_i\rangle + \langle \mathrm{D}E(v_i^k), w_i\rangle) \\
&= \mathrm{I}_1 + \mathrm{I}_2,
\end{aligned}$$

where

$$\begin{aligned}
\mathrm{I}_1 &:= \sum_{i=1}^{N}\langle \mathrm{D}E(u^{k+1}) - \mathrm{D}E(v_i^k), w_i\rangle, \\
\mathrm{I}_2 &:= \sum_{i=1}^{N}\langle \mathrm{D}E(v_i^k), w_i\rangle.
\end{aligned}$$

Using the stability of the decomposition (NS1) and the strengthened Cauchy–Schwarz inequality (NS2), the term $\mathrm{I}_1$ can be estimated in exactly the same way as in Theorem 12.2.9. Therefore

$$\mathrm{I}_1 \le C_S C_A \left(\sum_{i=1}^{N}\|\varepsilon_i^k\|_{\mathcal{V}}^2\right)^{1/2}\|w\|_{\mathcal{V}}.$$

For $\mathrm{I}_2$, we insert $\tau_i^k - \mathrm{D}E_i(\xi_i^k + s_i^k)$, which is zero in $\mathcal{V}_i'$, and estimate as follows:

$$\begin{aligned}
\mathrm{I}_2 &= \sum_{i=1}^{N}\langle \mathrm{D}E(v_i^k) - \mathrm{D}E(v_{i-1}^k) - \mathrm{D}E_i(\xi_i^k + s_i^k) + \mathrm{D}E_i(\xi_i^k), w_i\rangle \\
&\overset{(\mathrm{E1})+(\mathrm{E2c})}{\le} \sum_{i=1}^{N}(L\|\varepsilon_i^k\|_{\mathcal{V}} + L_i\|s_i^k\|_{\mathcal{V}})\|w_i\|_{\mathcal{V}} \\
&\overset{(12.4.2)}{=} \sum_{i=1}^{N}\left(L\|\varepsilon_i^k\|_{\mathcal{V}} + \frac{L_i}{\alpha_i^k}\|\varepsilon_i^k\|_{\mathcal{V}}\right)\|w_i\|_{\mathcal{V}} \\
&\overset{(12.4.7)}{\le} L\sum_{i=1}^{N}\left(1 + \frac{L_i}{\mu_i}\right)\|\varepsilon_i^k\|_{\mathcal{V}}\|w_i\|_{\mathcal{V}} \\
&\le L\left(1 + \max_{i=1}^{N}\frac{L_i}{\mu_i}\right)\left(\sum_{i=1}^{N}\|\varepsilon_i^k\|_{\mathcal{V}}^2\right)^{1/2}\left(\sum_{i=1}^{N}\|w_i\|_{\mathcal{V}}^2\right)^{1/2}
\end{aligned}$$

$$\overset{\text{(NS1)}}{\le} LC_A\left(1+\max_{i=1}^{N}\frac{L_i}{\mu_i}\right)\left(\sum_{i=1}^{N}\|\varepsilon_i^k\|_{\mathcal{V}}^2\right)^{1/2}\|w\|_{\mathcal{V}}.$$

Putting the estimates for $\mathrm{I}_1$ and $\mathrm{I}_2$ together, for all $w\in\mathcal{V}$, we have

$$\langle \mathrm{D}E(u^{k+1}),w\rangle \le C_A\left[C_S+L\left(1+\max_{i=1}^{N}\frac{L_i}{\mu_i}\right)\right]\left(\sum_{i=1}^{N}\|\varepsilon_i^k\|_{\mathcal{V}}^2\right)^{1/2}\|w\|_{\mathcal{V}},$$

which implies that

$$\|\mathrm{D}E(u^{k+1})\|_{\mathcal{V}'}^2 \le C_A^2\left[C_S+L\left(1+\max_{i=1}^{N}\frac{L_i}{\mu_i}\right)\right]^2\sum_{i=1}^{N}\|\varepsilon_i^k\|_{\mathcal{V}}^2.$$

Using (11.3.4) of Lemma 11.3.7 with $v=u^{k+1}$, the result follows. □

Using Theorems 12.4.4 and 12.4.16 and Corollary 12.2.5, we obtain the following linear convergence result.

**Corollary 12.4.17** (Convergence). *Assume that the energy $E$ satisfies* (E1) *and the local energies $E_i$ satisfy* (E2). *Assume, in addition, that the space decomposition satisfies* (NS1) *and $E$ satisfies* (NS2) *with respect to the decomposition. Let $\{u^k\}_{k=0}^{\infty}\subset\mathcal{V}$ be the sequence generated by the FASD method of Definition* 12.4.1. *Then we have*

$$E(u^{k+1})-E(u)\le\rho(E(u^k)-E(u))$$

*with*

$$\rho := \frac{C_A^2[C_S+L(1+\max_i\{L_i/\mu_i\})]^2}{C_A^2[C_S+L(1+\max_i\{L_i/\mu_i\})]^2+\mu^2}.$$

*Proof.* Exercise; see Problem 12.2. □

**Remark 12.4.18** (Local energies). Our theory suggests we can simply choose, for $w\in\mathcal{V}_i$,

$$\begin{aligned}E_i(w) &:= \frac{1}{2}\|w-\xi_i^k\|_{\mathcal{V}}^2\\ &= \frac{1}{2}\|w-Q_iv_{i-1}\|_{\mathcal{V}}^2.\end{aligned}\tag{12.4.8}$$

In this case, (E2b) and (E2c) hold with $L_i=\mu_i=1$. Moreover, the local problem becomes that of the preconditioned steepest descent method (compare with Definition 2.2.1),

$$(\eta_i^k-\xi_i^k,w)_{\mathcal{V}_i} = -\langle R_i\mathrm{D}E(v_{i-1}^k),w\rangle_{\mathcal{V}_i',\mathcal{V}_i},\quad \forall w\in\mathcal{V}_i.$$

In this case, (12.4.5) has the closed-form solution

$$s_i^k := \eta_i^k-\xi_i^k = -\mathfrak{R}_iR_i\mathrm{D}E(v_{i-1}^k),$$

where $\mathfrak{R}_i$ is the Riesz map $\mathfrak{R}_i : \mathcal{V}_i' \to \mathcal{V}_i$, whose algebraic realization entails the inversion of an SPD matrix of size dim $\mathcal{V}_i$. In fact, we can even use, for any $g_i \in \mathcal{V}_i$,

$$E_i(w) = \frac{1}{2}\|w - g_i\|_{\mathcal{V}}^2, \quad \forall w \in \mathcal{V}_i,$$

and by linearity the same result holds, that is, $s_i^k = -\mathfrak{R}_i R_i DE(v_{i-1}^k)$.

Solving a linear local problem can dramatically reduce the computational cost of the FASD; see Section 13.4 for a practical discussion of this point. In this setting, FASD is closely related to the coordinate descent methods analyzed in [25]. See also, for example, [20]. Another advantage of using (12.4.8) is that the set $\mathcal{B}_i$ becomes irrelevant, as the derivative of the quadratic energy in (12.4.8) is globally Lipschitz. ◀

**Corollary 12.4.19** (Convergence for quadratic local energies). *In addition to the hypotheses of Theorem* 12.4.16, *assume that the local energies $E_i$, $i = 1, \dots, N$, are quadratic and given by* (12.4.8). *Then*

$$E(u^{k+1}) - E(u) \le \frac{C_A^2(C_S + 2L)^2}{2\mu} \sum_{i=1}^{N} \|\varepsilon_i^k\|_{\mathcal{V}}^2.$$

*As a consequence, we have*

$$E(u^{k+1}) - E(u) \le \rho_{\text{quad}}(E(u^k) - E(u))$$

*with*

$$\rho_{\text{quad}} := \frac{C_A^2(C_S + 2L)^2}{C_A^2(C_S + 2L)^2 + \mu^2}.$$

*Proof.* Exercise; see Problem 12.3. □

**Remark 12.4.20** (Contraction for quadratic local energies). We note that the linear contraction factor $\rho_{\text{quad}}$ for the FASD method with quadratic $E_i$ chosen as in (12.4.8) is larger than that for the SSO method, thus resulting in slower convergence. This is to be expected and follows from the fact that

$$C_U^{\text{SSO}} = \frac{C_A^2 C_S^2}{2\mu} < \frac{C_A^2(C_S + 2L)^2}{2\mu} = C_U^{\text{FASD}}.$$ ◀

## 12.5 FASD with approximate line search

In this section, we consider the FASD algorithm with approximate line search.

**Definition 12.5.1** (FASDALS). Let the energy $E$ satisfy (E1), and let $u \in \mathcal{V}$ be the unique solution to (12.1.1). The **fast subspace descent** method with **approximate line search**

**(FASDALS)** is an algorithm to approximate $u$ that computes a sequence $\{u^k\}_{k=0}^{\infty} \subset \mathcal{V}$ recursively, $u^{k+1} = \text{FASDALS}(u^k)$, as follows. Assume that we have Fréchet differentiable $E_i : \mathcal{V}_i \to \mathbb{R}$ and projections $Q_i : \mathcal{V} \to \mathcal{V}_i$, $i = 1, \ldots, N$. Let $u^0 \in \mathcal{V}$ be arbitrary. Then, for $k \geq 0$.

- $v_0^k := u^k$.
- For $i = 1$ to $N$:
  - Compute the subspace *τ-perturbation*: let $\xi_i^k := Q_i v_{i-1}^k$ and
    $$\tau_i^k := \mathrm{D}E_i(\xi_i^k) - R_i \mathrm{D}E(v_{i-1}^k) \in \mathcal{V}_i'.$$
  - Solve the subspace residual problem: Find $\eta_i^k \in \mathcal{V}_i$ such that
    $$\langle \mathrm{D}E_i(\eta_i^k), w\rangle = \langle \tau_i^k, w\rangle, \quad \forall\, w \in \mathcal{V}_i. \tag{12.5.1}$$
  - Compute the search direction and the quadratic step size:
    $$s_i^k := \eta_i^k - \xi_i^k \in \mathcal{V}_i,$$
    and
    $$\alpha_i^{q,k} := -\frac{\langle R_i \mathrm{D}E(v_{i-1}^k), s_i^k\rangle}{L\|s_i^k\|_{\mathcal{V}}^2}. \tag{12.5.2}$$
  - Apply the subspace correction:
    $$v_i^k := v_{i-1}^k + \alpha_i^{q,k} s_i^k.$$
- $u^{k+1} := v_N^k$. ∎

The key difference between this method and the FASD of Definition 12.4.1 is that a predetermined step size $\alpha_i^{q,k}$ is employed rather than computing $\alpha_i^k$, an optimal one obtained via line search. The main computational advantage of this approach is that there is no need to repeatedly evaluate the energy $E$ nor its derivatives in each subspace. We only need to compute $R_i DE(v_{i-1}^k)$ once for the local problems and use this in the computation of the $\tau$-perturbation and step size.

Recall the scalar function $f_{i,k}$ defined in (12.4.6) and observe that it satisfies $f_{i,k}(0) = E(v_{i-1}^k)$ and $f_{i,k}'(0) = \langle \mathrm{D}E(v_{i-1}^k), s_i^k\rangle < 0$. Using these values, we define the quadratic function

$$q_{i,k}(\alpha) := f_{i,k}(0) + f_{i,k}'(0)\alpha + \frac{L\|s_i^k\|_{\mathcal{V}}^2}{2}\alpha^2. \tag{12.5.3}$$

Instead of the optimal step size

$$\alpha_i^k = \operatorname*{argmin}_{\alpha \in \mathbb{R}} f_{i,k}(\alpha),$$

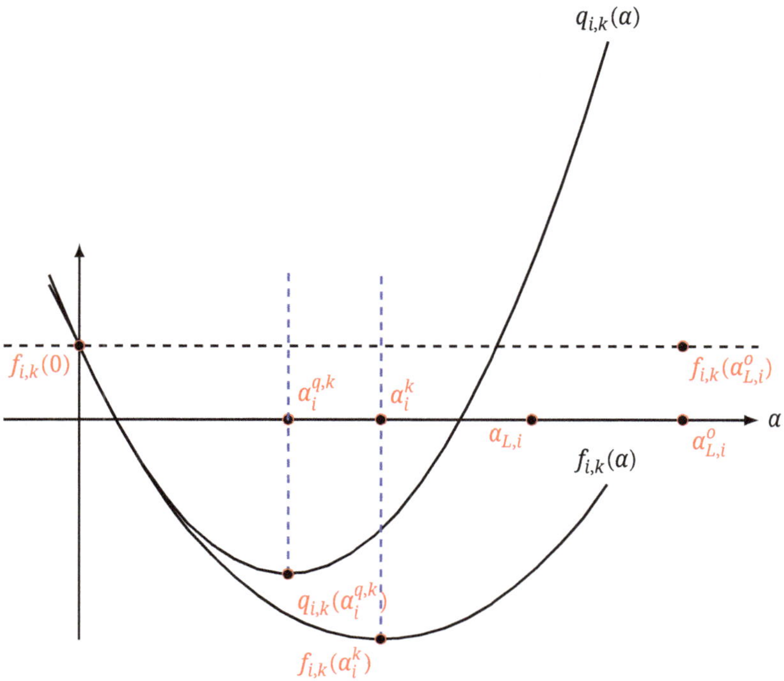


**Figure 12.4:** The function $f_{i,k}$ defined in (12.4.6) and its quadratic approximation $q_{i,k}$ defined in (12.5.3). The quadratic minimizer $\alpha_i^{q,k}$ is always to the left of $\alpha_i^k$ by construction.

the step size choice for this new algorithm is

$$\begin{aligned}\alpha_i^{q,k} &= \operatorname*{argmin}_{\alpha\in\mathbb{R}} q_{i,k}(\alpha)\\ &= -\frac{f'_{i,k}(0)}{L\|s_i^k\|_{\mathcal{V}}^2}\\ &= -\frac{\langle R_i DE(v_{i-1}^k), s_i^k\rangle}{L\|s_i^k\|_{\mathcal{V}}^2}.\end{aligned}$$

This choice satisfies the following estimates. See Figure 12.4 for an illustration.

**Lemma 12.5.2** (Lower bound). *Assume that the energy $E$ satisfies* (E1) *and the local energies $E_i$ satisfy* (E2). *Then for all $k \geq 0$ and $i = 1, \dots, N$, we have*

$$\frac{\mu_i}{L} \leq \alpha_i^{q,k} \leq \alpha_i^k.$$

*Proof.* The lower bound is obtained by the definition of $\alpha_i^{q,k}$ and Lemma 12.4.8. To prove the upper bound, we notice that due to the line search,

$$
\begin{aligned}
0 &= f'_{i,k}(\alpha_i^k) \\
&= \langle \mathrm{D}E(v_{i-1}^k + \alpha_i^k s_i^k), s_i^k \rangle,
\end{aligned}
$$

and thus

$$
\begin{aligned}
\alpha_i^{q,k} L \|s_i^k\|_{\mathcal{V}}^2 &= -\langle R_i \mathrm{D}E(v_{i-1}^k), s_i^k \rangle \\
&= \langle \mathrm{D}E(v_{i-1}^k + \alpha_i^k s_i^k) - \mathrm{D}E(v_{i-1}^k), s_i^k \rangle \\
&\le \alpha_i^k L \|s_i^k\|_{\mathcal{V}}^2.
\end{aligned}
$$

□

**Lemma 12.5.3** (Lower bound). *Suppose that the energy $E$ satisfies* (E1) *and the local energies $E_i$ satisfy* (E2). *Then for all $k \ge 0$ and $i = 1, \ldots, N$, the local energies $f_{i,k}$ defined in* (12.4.6) *satisfy*

$$
f_{i,k}(\alpha) \le q_{i,k}(\alpha), \quad \forall \alpha \in [0, \alpha_{L,i}].
$$

*Proof.* By Lemma 12.4.10, for $\alpha \in [0, \alpha_{L,i}]$, the function $f'_{i,k}$ is Lipschitz continuous with constant $L\|s_i^k\|_{\mathcal{V}}^2$. We now recall a classical result; see, for instance, [31, Theorem B.56]: since $f'_{i,k}$ is Lipschitz continuous with constant $L\|s_i^k\|_{\mathcal{V}}^2$ on $\alpha \in [0, \alpha_{L,i}]$, then

$$
|f_{i,k}(\alpha) - f_{i,k}(\beta) - (\alpha - \beta) f'_{i,k}(\beta)| \le \frac{L\|s_i^k\|_{\mathcal{V}}^2}{2} |\alpha - \beta|^2, \quad \forall \alpha, \beta \in [0, \alpha_{L,i}].
$$

Setting $\beta = 0$ above, we get, for all $\alpha \in [0, \alpha_{L,i}]$,

$$
\begin{aligned}
f_{i,k}(\alpha) - q_{i,k}(\alpha) &= f_{i,k}(\alpha) - f_{i,k}(0) - \alpha f'_{i,k}(0) - \frac{L\|s_i^k\|_{\mathcal{V}}^2}{2} \alpha^2 \\
&\le |f_{i,k}(\alpha) - f_{i,k}(0) - \alpha f'_{i,k}(0)| - \frac{L\|s_i^k\|_{\mathcal{V}}^2}{2} \alpha^2 \\
&\le \frac{L\|s_i^k\|_{\mathcal{V}}^2}{2} \alpha^2 - \frac{L\|s_i^k\|_{\mathcal{V}}^2}{2} \alpha^2 \\
&= 0,
\end{aligned}
$$

which implies the result. □

Now since the optimal linear search procedure is not employed, the orthogonality conditions with respect to the corrections do not hold, and establishing the lower bound must be done by different means.

**Theorem 12.5.4** (Lower bound on corrections). *Assume that the energy $E$ satisfies* (E1) *and the local energies $E_i$ satisfy* (E2). *Let the sequence $\{u^k\}_{k=0}^{\infty} \subset \mathcal{V}$ be obtained by the FASDALS scheme of Definition* 12.5.1. *Then for every $k \ge 0$, we have*

$$
E(u^k) - E(u^{k+1}) \ge C_L \sum_{i=1}^{N} \|\alpha_i^{q,k} s_i^k\|_{\mathcal{V}}^2
$$

*with*

$$C_L := \frac{L}{2}.$$

*Proof.* It suffices to prove

$$E(v_{i-1}^k) - E(v_i^k) = f_{i,k}(0) - f_{i,k}(\alpha_i^{q,k}) \geq \frac{L}{2}\|\alpha_i^{q,k} s_i^k\|^2.$$

By Lemma 12.5.3, $f_{i,k}(\alpha) \leq q_{i,k}(\alpha)$ for all $\alpha \in [0, \alpha_{L,i}]$. By the definition of $\alpha_i^{q,k}$ and $\alpha_i^{q,k} \leq \alpha_i^k$ (see Lemma 12.5.2) we get

$$\begin{aligned} f_{i,k}(\alpha_i^{q,k}) &\leq q_{i,k}(\alpha_i^{q,k}) \\ &= \min_{\alpha\in\mathbb{R}} q_{i,k}(\alpha) \\ &= f_{i,k}(0) - \frac{1}{2L\|s_i^k\|_{\mathcal{V}}^2} |f_{i,k}'(0)|^2 \\ &= f_{i,k}(0) - \frac{L}{2}\|\alpha_i^{q,k} s_i^k\|_{\mathcal{V}}^2. \end{aligned}$$

In the last step, we have used the definition of $\alpha_i^{q,k}$, and this completes the proof. □

The upper bound on corrections is the content of the next result.

**Theorem 12.5.5** (Upper bound on corrections). *Assume that the energy $E$ satisfies* (E1) *and the local energies $E_i$ satisfy* (E2). *Assume, in addition, that the space decomposition satisfies* (NS1) *and $E$ satisfies* (NS2) *with respect to the decomposition. Then the sequence $\{u^k\}_{k=0}^{\infty} \subset \mathcal{V}$, obtained by the FASDALS method of Definition* 12.5.1, *satisfies*

$$E(u^{k+1}) - E(u) \leq C_U \sum_{i=1}^{N} \|\alpha_i^{q,k} s_i^k\|_{\mathcal{V}}^2,$$

*where*

$$C_U := \frac{C_A^2}{2\mu}\left[C_S + L\left(1 + \max_{i=1}^{N}\left\{\frac{L_i}{\mu_i}\right\}\right)\right]^2.$$

*Proof.* Since $\alpha_i^{q,k}$ has the same lower bound as $\alpha_i^k$, we can derive an upper bound on corrections in exactly the same way as in the proof of Theorem 12.4.16. We only need to replace $\varepsilon_i^k = \alpha_i^k s_i^k$ by $\alpha_i^{q,k} s_i^k$. We leave the details to the reader; see Problem 12.4. □

These last two results imply the linear convergence of FASDALS.

**Corollary 12.5.6** (Convergence). *With the setting and assumptions of Theorems* 12.5.4 *and* 12.5.5, *we have*

$$E(u^{k+1}) - E(u) \leq \rho(E(u^k) - E(u))$$

*with*

$$\rho := \frac{C_A^2[C_S + L(1 + \max_{i=1}^N\{\frac{L_i}{\mu_i}\})]^2}{C_A^2[C_S + L(1 + \max_{i=1}^N\{\frac{L_i}{\mu_i}\})]^2 + L\mu}.$$

*Proof.* Exercise; see Problem 12.5. □

**Remark 12.5.7** (Role of $L$). We conclude by briefly mentioning that the Lipschitz constant $L$ is used in the choice of the step size $\alpha_i^k$. In the setting of FASDALS, this can be replaced by a local Lipschitz constant for the scalar function $f_{i,k}$ on $(0, \alpha_i^k)$. ◄

## 12.6 Problems

12.1 Prove Lemma 12.4.7.

12.2 Prove Corollary 12.4.17.

12.3 Prove Corollary 12.4.19.

12.4 Prove Theorem 12.5.5.

12.5 Prove Corollary 12.5.6.

# 13 FASD applied to nonlinear finite element problems

In this last chapter, we apply the axiomatic theory of Chapter 12 to a model nonlinear problem with polynomial nonlinearity and provide numerical examples to illustrate the efficiency of FASD and variants. We work exclusively in the finite element setting and refer to Chapter 7 for definitions and notation.

## 13.1 A model nonlinear problem

Let $\Omega \subset \mathbb{R}^d$ with $d = 2, 3$ be a bounded domain with a sufficiently regular boundary. Let $2 \le p < \infty$ when $d = 2$ and $2 \le p \le 6$ when $d = 3$. We consider the following problem: given $f \in L^2(\Omega)$, find $u \in H_0^1(\Omega)$ such that, for every $v \in H_0^1(\Omega)$,

$$\int_\Omega |u|^{p-2} u v \, \mathrm{d}\boldsymbol{x} + D_o \int_\Omega \nabla v^\top \nabla u \, \mathrm{d}\boldsymbol{x} = \int_\Omega f v \, \mathrm{d}\boldsymbol{x}, \tag{13.1.1}$$

where $D_o > 0$ is a parameter. As the following result shows, the unique solution of (13.1.1) is the unique minimizer of a certain strictly convex energy.

**Theorem 13.1.1** (Energy minimization). *Let $\Omega \subset \mathbb{R}^d$, $d = 2, 3$, be a bounded domain with a sufficiently regular boundary. Let $2 \le p < \infty$ when $d = 2$ and $2 \le p \le 6$ when $d = 3$. For $v \in H_0^1(\Omega)$, define the energy*

$$E(v) := \frac{1}{p} \|v\|^p_{L^p(\Omega)} + \frac{D_o}{2} \|\nabla v\|^2_{L^2(\Omega)} - (f, v)_{L^2(\Omega)}. \tag{13.1.2}$$

*This functional is twice Fréchet differentiable, strictly convex, and coercive. Therefore it has a unique global minimizer. Furthermore, $u \in H_0^1(\Omega)$ is the unique minimizer of* (13.1.2) *iff it is the solution of* (13.1.1).

*Proof.* We compute the first Fréchet derivative of $E$ at $v \in H_0^1(\Omega)$ as follows. Let $\xi \in H_0^1(\Omega)$. Then

$$\frac{\mathrm{d}}{\mathrm{d}t} E(v + t\xi)\Big|_{t=0} = \int_\Omega |v|^{p-2} v \xi \, \mathrm{d}\boldsymbol{x} + D_o \int_\Omega \nabla \xi^\top \nabla v \, \mathrm{d}\boldsymbol{x} - \int_\Omega f \xi \, \mathrm{d}\boldsymbol{x},$$

so that, as the reader may verify (see Exercise 13.1),

$$\langle \mathrm{D}E(v), \xi \rangle = \int_\Omega |v|^{p-2} v \xi \, \mathrm{d}\boldsymbol{x} + D_o \int_\Omega \nabla \xi^\top \nabla v \, \mathrm{d}\boldsymbol{x} - \int_\Omega f \xi \, \mathrm{d}\boldsymbol{x}. \tag{13.1.3}$$

https://doi.org/10.1515/9783111354880-013

Thus problem (13.1.1) is equivalent to finding $u \in H_0^1(\Omega)$ such that, for all $\xi \in H_0^1(\Omega)$, we have $\langle \mathrm{D}E(u), \xi \rangle = 0$. The first derivative of $E$ is continuous, as we can see from the following estimate:

$$\begin{aligned} \left|\langle \mathrm{D}E(v), \xi \rangle\right| &\le \|v\|_{L^p(\Omega)}^{p-1} \|\xi\|_{L^p(\Omega)} + D_o \|\nabla v\|_{L^2(\Omega)} \|\nabla \xi\|_{L^2(\Omega)} \\ &\quad + \|f\|_{L^2(\Omega)} \|\xi\|_{L^2(\Omega)} \quad \text{for } v, \xi \in H_0^1(\Omega). \end{aligned}$$

The assumed range for $p$ guarantees the embedding $H_0^1(\Omega) \hookrightarrow L^p(\Omega)$. Thus there is a constant $C_{p,\Omega} > 0$ such that

$$\|v\|_{L^p(\Omega)} \le C_{p,\Omega} \|\nabla v\|_{L^2(\Omega)}, \quad \forall\, v \in H_0^1(\Omega).$$

The Poincaré inequality of Theorem 7.1.1 and this last estimate then give

$$\left|\langle \mathrm{D}E(v), \xi \rangle\right| \le C \|\nabla \xi\|_{L^2(\Omega)}, \quad \forall\, \xi \in H_0^1(\Omega),$$

where the constant $C > 0$ depends on $v \in H_0^1(\Omega)$.

Next, we show that for $p \ge 2$, the second Fréchet derivative exists and is a continuous bilinear operator. Indeed, for $v \in H_0^1(\Omega)$, the second Fréchet derivative acts on the pair $(\xi, \eta) \in H_0^1(\Omega) \times H_0^1(\Omega)$ via (see, again, Exercise 13.1)

$$\langle \mathrm{D}^2 E(v)\xi, \eta \rangle = (p-1) \int_\Omega |v|^{p-2} \xi \eta \,\mathrm{d}\boldsymbol{x} + D_o \int_\Omega \nabla \eta^\top \nabla \xi \,\mathrm{d}\boldsymbol{x}. \tag{13.1.4}$$

The continuity of the second derivative can be proved as above.

We now observe that this operator is positive definite. In other words, for all $v \in H_0^1(\Omega)$ and $\eta \in H_0^1(\Omega)$,

$$\begin{aligned} \langle \mathrm{D}^2 E(v)\eta, \eta \rangle &= (p-1) \int_\Omega |v|^{p-2} \eta^2 \,\mathrm{d}\boldsymbol{x} + D_o \int_\Omega \|\nabla \eta\|^2 \,\mathrm{d}\boldsymbol{x} \\ &\ge D_o \|\nabla \eta\|_{L^2(\Omega)}^2. \end{aligned}$$

Therefore, due to Proposition 11.2.6.7, the energy $E$ is strictly convex. Finally, the coercivity of $E$ follows from Corollary 11.1.33.

Gathering all the observations provided above it immediately follows that there is a unique global minimizer of the energy (13.1.2),

$$u := \operatorname*{argmin}_{v \in H_0^1(\Omega)} E(v).$$

Consequently, there is a unique solution to the Euler equation (13.1.1). □

We now verify that $E$ satisfies the assumptions required for the theory of Chapter 12 to apply.

**Proposition 13.1.2** (Assumption (E1)). *Let the space $H_0^1(\Omega)$ be equipped with the norm $\|\nabla \cdot \|_{L^2(\Omega)}$. The energy functional*

$$E : H_0^1(\Omega) \to \mathbb{R},$$

*defined in* (13.1.2), *satisfies assumption (*E1*).*

*Proof.* The Fréchet differentiability and strict convexity were verified in Theorem 13.1.1.
Let $u_0 = 0$, so that $E(u_0) = 0$, and set

$$\mathcal{B} = \{v \in H_0^1(\Omega) \mid E(v) \le E(u_0)\}.$$

Recall that since $E$ is convex, so is $\mathcal{B}$. Moreover, for $v \in \mathcal{B}$, we have

$$\begin{aligned}\frac{1}{p}\|v\|_{L^p(\Omega)}^p + \frac{D_o}{2}\|\nabla v\|_{L^2(\Omega)}^2 &\le (f, v)_{L^2(\Omega)} \\ &\le \|f\|_{L^2(\Omega)}\|v\|_{L^2(\Omega)} \\ &\le C_0(D_o, C_P)\|f\|_{L^2(\Omega)}^2 + \frac{D_o}{4}\|\nabla v\|_{L^2(\Omega)}^2,\end{aligned}$$

where $C_P > 0$ is the constant in the Poincaré inequality of Theorem 7.1.1. Thus, for all $v \in \mathcal{B}$,

$$\|v\|_{L^p(\Omega)} + \|\nabla v\|_{L^2(\Omega)} \le C_1 = C_1(u_0, D_o, p, f). \tag{13.1.5}$$

An application of the mean value theorem to Proposition 11.1.32 shows that there exists $t_0 \in [0, 1]$ such that, upon defining $z = t_0 v + (1 - t_0)w$, for every $\xi \in H_0^1(\Omega)$, we have

$$\langle \mathrm{D}E(w) - \mathrm{D}E(v), \xi\rangle = \langle \mathrm{D}^2E(z)(w - v), \xi\rangle.$$

Now, since $\mathcal{B}$ is convex and $w, v \in \mathcal{B}$, we have $z \in \mathcal{B}$. This and (13.1.5) imply that $\|z\|_{L^p(\Omega)} \le C_1$. Hölder's inequality, the fact that $p \ge 2$, and (13.1.4) allow us to estimate

$$\begin{aligned}|\langle \mathrm{D}^2E(z)\xi, w - v\rangle| &\le (p-1)\|z\|_{L^p(\Omega)}^{p-2}\|\xi\|_{L^p(\Omega)}\|w - v\|_{L^p(\Omega)} \\ &\quad + D_o\|\nabla\xi\|_{L^2(\Omega)}\|\nabla(w - v)\|_{L^2(\Omega)} \\ &\le [(p-1)C_{p,\Omega}^2 C_1^{p-2} + D_o]\|\nabla\xi\|_{L^2(\Omega)}\|\nabla(w - v)\|_{L^2(\Omega)}.\end{aligned}$$

Define

$$L := (p-1)C_{p,\Omega}^2 C_1^{p-2} + D_o.$$

This shows that for $v, w \in \mathcal{B}$,

$$|\langle \mathrm{D}E(w) - \mathrm{D}E(v), \xi\rangle| = |\langle \mathrm{D}^2E(z)(w - v), \xi\rangle|$$

$$\leq L\|\nabla\xi\|_{L^2(\Omega)}\|\nabla(w-v)\|_{L^2(\Omega)},$$

that is, (E1) holds with $L_{\mathcal{B}} = L$.

We finally show strong convexity. Let $w, v \in H_0^1(\Omega)$. Using, once again, the mean value theorem in Proposition 11.1.32 shows that there is $z \in H_0^1(\Omega)$ for which

$$\begin{aligned}\langle \mathrm{D}E(w) - \mathrm{D}E(v), w - v\rangle &= \langle \mathrm{D}^2E(z)(w-v), w-v\rangle \\ &= (p-1)(|z|^{p-2}(w-v), w-v)_{L^2(\Omega)} \\ &\quad + D_o(\nabla(w-v), \nabla(w-v))_{L^2(\Omega)} \\ &\geq D_o\|\nabla(w-v)\|^2_{L^2(\Omega)}.\end{aligned}$$

In other words, strong convexity holds with constant $\mu = D_o$. □

## 13.2 Finite element discretization and space decomposition

Assume now that $\Omega \subset \mathbb{R}^d$ is polytopal and that $\mathcal{T}_H$ is a conforming simplicial triangulation of $\Omega$. Let $\mathcal{T}_h$ be the triangulation obtained by one uniform refinement of $\mathcal{T}_H$. We refer the reader to Sections 7.2 and 7.4 for details and notation. In particular, we recall that the finite element space $S_h$ consists of continuous functions that, when restricted to each simplex $K \in \mathcal{T}_h$, are linear. We have a similar definition for $S_H$. Then $S_H \subset S_h$, and the containment is proper. Finally, $V_h$ and $V_H$ are, respectively, the subspaces of $S_h$ and $S_H$ that consist of functions that vanish on $\partial\Omega$.

We will consider the minimization of energy $E$ restricted to $V_h$, which is a subspace of $H_0^1(\Omega)$, that is,

$$\min_{v \in V_h} E(v).$$

Thus, for our axiomatic theory, we set $\mathcal{V} = V_h$ with norm

$$\|\cdot\|_{\mathcal{V}} = |\cdot|_{H^1(\Omega)} = \|\cdot\|_{H_0^1(\Omega)}.$$

Notice that (E1) still holds, as $V_h \subset H_0^1(\Omega)$.

Next, we describe a two-level space decomposition of $\mathcal{V}$. Recall that the Lagrange nodal basis $\mathcal{B}_h$ was introduced in Definition 7.4.2. We will use the decomposition

$$\mathcal{V} = \sum_{i=0}^{n_h} \mathcal{V}_i, \tag{13.2.1}$$

where $\mathcal{V}_0 = V_H$, and $\mathcal{V}_i = \operatorname{span}\{\psi_{h,i}\}$, $1 \leq i \leq n_h$.

We now show that this is a stable decomposition. First, we present some properties of $Q_H$, the $L^2$-projection onto $V_H$.

**Definition 13.2.1** ($L^2$-projection onto $V_H$). Denote by $Q_H : L^2(\Omega) \to V_H$ the **$L^2$-projection onto $V_H$**. This operator is defined as follows: for each $v \in L^2(\Omega)$, $Q_H v \in V_H$ is the unique function satisfying

$$\int_\Omega (v - Q_H v) w_H \,\mathrm{d}\mathbf{x} = 0, \quad \forall\, w_H \in V_H.$$ ■

**Lemma 13.2.2** (Properties of $Q_H$). *There is a constant $C > 0$, independent of $H$, such that*

$$\|v - Q_H v\|_{L^2(\Omega)} \le CH|v|_{H^1(\Omega)}, \quad \forall\, v \in H_0^1(\Omega). \tag{13.2.2}$$

*In addition, the $L^2$-projection onto $V_H$ is $H_0^1$-stable, that is, there is another constant $C > 0$, independent of $H$, for which*

$$|Q_H v|_{H^1(\Omega)} \le C|v|_{H^1(\Omega)}, \quad \forall\, v \in H_0^1(\Omega). \tag{13.2.3}$$

*Proof.* To prove this result, we must use an interpolant different from the Lagrange one. Namely, such that it is a projection, $L^2$- and $H^1$-stable, and has suitable approximation properties. In other words, we must be able to produce $\wp_H : L^2(\Omega) \to V_H$ such that $\wp_H^2 = \wp_H$ and, for positive constants $C$, independent of $H$,

$$\|\wp_H v\|_{L^2(\Omega)} \le C\|v\|_{L^2(\Omega)}, \quad \forall\, v \in L^2(\Omega), \tag{13.2.4}$$

$$|\wp_H v|_{H^1(\Omega)} \le C|v|_{H^1(\Omega)}, \quad \forall\, v \in H_0^1(\Omega), \tag{13.2.5}$$

$$\|v - \wp_H v\|_{L^2(\Omega)} \le CH|v|_{H^1(\Omega)}, \quad \forall\, v \in H_0^1(\Omega). \tag{13.2.6}$$

The construction of such an interpolant is beyond the scope of our introductory text. We will only mention that the so-called Scott–Zhang interpolant, developed in [32], satisfies all these requirements. See also [4, 10].

With such an operator at hand, we let $v \in H_0^1(\Omega)$ and estimate

$$\begin{aligned}
\|v - Q_H v\|_{L^2(\Omega)} &\le \|v - \wp_H v\|_{L^2(\Omega)} + \|\wp_H v - Q_H v\|_{L^2(\Omega)} \\
&\le \|v - \wp_H v\|_{L^2(\Omega)} + \|Q_H \wp_H v - Q_H v\|_{L^2(\Omega)} \\
&\le 2\|v - \wp_H v\|_{L^2(\Omega)} \\
&\overset{(13.2.6)}{\le} CH|v|_{H^1(\Omega)},
\end{aligned}$$

which is (13.2.2).

We next establish (13.2.3). Once again, we let $v \in H_0^1(\Omega)$ and estimate

$$\begin{aligned}
|Q_H v|_{H^1(\Omega)} &\le |\wp_H v|_{H^1(\Omega)} + |Q_H v - \wp_H v|_{H^1(\Omega)} \\
&\overset{(13.2.5)}{\le} C|v|_{H^1(\Omega)} + |Q_H v - \wp_H v|_{H^1(\Omega)} \\
&\overset{\text{Lem. 7.3.4}}{\le} C|v|_{H^1(\Omega)} + CH^{-1}\|Q_H v - \wp_H v\|_{L^2(\Omega)}
\end{aligned}$$

$$
\begin{aligned}
&\leq C|v|_{H^1(\Omega)} + CH^{-1}[\|Q_H v - v\|_{L^2(\Omega)} + \|v - \wp_H v\|_{L^2(\Omega)}] \\
&\overset{(13.2.2)}{\leq} C|v|_{H^1(\Omega)} + CH^{-1}\|v - \wp_H v\|_{L^2(\Omega)} \\
&\overset{(13.2.6)}{\leq} C|v|_{H^1(\Omega)}.
\end{aligned}
$$

□

**Theorem 13.2.3** (Stable decomposition). *The decomposition of the finite element space $V_h$ described in* (13.2.1) *satisfies Assumption (*NS1*).*

*Proof.* Let $Q_H : L^2(\Omega) \to V_H$ be the $L^2$-projection onto $V_H$:

$$(Q_H v, w)_{L^2(\Omega)} = (v, w)_{L^2(\Omega)}, \quad \forall w \in V_H.$$

For $v \in V_h$, we denote by

$$\tilde{v} = (I - Q_H)v \in V_h$$

the projection error. Suppose that

$$\tilde{v} = \sum_{i=1}^{n_h} \tilde{v}_i$$

is the nodal decomposition of the error in $V_h$, that is, $\tilde{v}_i \in \operatorname{span}\{\psi_{h,i}\}$. By the inverse inequality and the stability of nodal decompositions in the $L^2$-norm (see Lemma 7.3.3) we have

$$
\begin{aligned}
\sum_{i=1}^{n_h} |\tilde{v}_i|^2_{H^1(\Omega)} &\overset{\text{Lem. 7.3.4}}{\leq} C\sum_{i=1}^{n_h} h^{-2}\|\tilde{v}_i\|^2_{L^2(\Omega)} \\
&= Ch^{-2}\sum_{i=1}^{n_h} |[\tilde{\boldsymbol{v}}]_i|^2\|\psi_{h,i}\|^2_{L^2(\Omega)} \\
&\leq Ch^{d-2}\sum_{i=1}^{n_h} |[\tilde{\boldsymbol{v}}]_i|^2 \\
&\overset{\text{Lem. 7.3.3}}{\leq} Ch^{-2}\|\tilde{v}\|^2_{L^2(\Omega)} \\
&\leq C|v|^2_{H^1(\Omega)}.
\end{aligned}
$$

This, together with (13.2.3), shows that (NS1) holds, that is, the decomposition

$$
\begin{aligned}
v &= Q_H v + (v - Q_H v) \\
&= Q_H v + \sum_{i=1}^{n_h} \tilde{v}_i
\end{aligned}
$$

is stable in $H^1_0$. □

**Theorem 13.2.4** (Strengthened Cauchy–Schwarz). *Let the energy $E$ be defined in* (13.1.2), *and let $V_h$ be decomposed as in* (13.2.1). *Then Assumption (*NS2*) holds.*

*Proof.* Let $u_0 \in V_h$, and let $\mathcal{B} = \mathcal{B}(u_0)$ its energy sublevel set. For $i = 1, \ldots, n_h$, assume that $u_i, v_i, w_i \in \mathcal{V}_i$ are as in the hypotheses of (NS2). By Proposition 11.1.32 and the mean value theorem we find $z_{i,j} \in \mathcal{B}$ for which

$$\begin{aligned}
\sum_{i=0}^{n_h} \sum_{j=i+1}^{n_h} \langle \mathrm{D}E(v_{j-1} + u_j) - \mathrm{D}E(v_{j-1}), w_i \rangle \\
&= \sum_{i=0}^{n_h} \sum_{j=i+1}^{n_h} \langle \mathrm{D}^2 E(z_{i,j}) u_j, w_i \rangle \\
&\le \sum_{i=0}^{n_h} \sum_{j=i+1}^{n_h} \big| (p-1)(|z_{i,j}|^{p-2} u_j, w_i)_{L^2(\Omega)} + D_o(\nabla u_j, \nabla w_i)_{L^2(\Omega)} \big|.
\end{aligned}$$

Define

$$\begin{aligned}
T_1 &\coloneqq \sum_{i=0}^{n_h} \sum_{j=i+1}^{n_h} \big| (|z_{i,j}|^{p-2} u_j, w_i)_{L^2(\Omega)} \big|, \\
T_2 &\coloneqq \sum_{i=0}^{n_h} \sum_{j=i+1}^{n_h} \big| (\nabla u_j, \nabla w_i)_{L^2(\Omega)} \big|,
\end{aligned}$$

so that

$$\sum_{i=0}^{n_h} \sum_{j=i+1}^{n_h} \langle \mathrm{D}E(v_{j-1} + u_j) - \mathrm{D}E(v_{j-1}), w_i \rangle \le (p-1) T_1 + D_o T_2.$$

Recall that $u_i, v_i, w_i \in \mathcal{V}_i$. Thus their support is contained in

$$S_i = \bigcup \{ K \in \mathcal{T}_h \mid \boldsymbol{N}_{h,i} \in K \}.$$

For $i = 1, \ldots, n_h$, we define

$$\mathcal{N}(i) \coloneqq \{ j \in \{1, \ldots, n_h\} \mid j > i,\ S_j \cap S_i \ne \emptyset \}.$$

By quasiuniformity, $\#\mathcal{N}(i) \le C_{\mathcal{T}}$ is much smaller than and independent of $n_h$. Thus we have

$$\begin{aligned}
T_2 &= \sum_{i=0}^{n_h} \sum_{j \in \mathcal{N}(i)} |(\nabla u_j, w_i)_{L^2(S_i \cap S_j)}| \\
&\le \sum_{i=0}^{n_h} \sum_{j \in \mathcal{N}(i)} \|\nabla u_j\|_{L^2(S_i \cap S_j)} \|\nabla w_i\|_{L^2(S_i \cap S_j)}
\end{aligned}$$

$$\leq \left[\sum_{i=0}^{n_h}\sum_{j\in\mathcal{N}(i)} \|\nabla u_j\|^2_{L^2(S_i\cap S_j)}\right]^{1/2} \left[\sum_{i=0}^{n_h}\sum_{j\in\mathcal{N}(i)} \|\nabla w_i\|^2_{L^2(S_i\cap S_j)}\right]^{1/2}$$

$$\leq \left[C_{\mathcal{T}}\sum_{i=0}^{n_h} \|\nabla u_i\|^2_{L^2(\Omega)}\right]^{1/2} \left[C_{\mathcal{T}}\sum_{i=0}^{n_h} \|\nabla w_i\|^2_{L^2(\Omega)}\right]^{1/2}.$$

Now, since $z_{i,j} \in \mathcal{B}$, it satisfies the bound (13.1.5). This, together with a similar reasoning as before, implies that

$$(|z_{i,j}|^{p-2}u_j, w_i)_{L^2(\Omega)} \leq C_1^{p-2}\|u_j\|_{L^p(S_i\cap S_j)}\|w_i\|_{L^p(S_i\cap S_j)}.$$

Consequently,

$$T_1 \leq C_1^{p-2}\left[\sum_{i=0}^{n_h}\sum_{j\in\mathcal{N}(i)} \|u_j\|^2_{L^p(S_i\cap S_j)}\right]^{1/2} \left[\sum_{i=0}^{n_h}\sum_{j\in\mathcal{N}(i)} \|w_i\|^2_{L^p(S_i\cap S_j)}\right]^{1/2}$$

$$\leq C_1^{p-2}\left[C_{\mathcal{T}}\sum_{i=0}^{n_h} \|u_i\|^2_{L^p(\Omega)}\right]^{1/2} \left[C_{\mathcal{T}}\sum_{i=0}^{n_h} \|w_i\|^2_{L^p(\Omega)}\right]^{1/2}$$

$$\leq C_1^{p-2}C_{\mathcal{T}}C_{p,\Omega}^2\left[\sum_{i=0}^{n_h} \|\nabla u_i\|^2_{L^2(\Omega)}\right]^{1/2} \left[\sum_{i=0}^{n_h} \|\nabla w_i\|^2_{L^2(\Omega)}\right]^{1/2}.$$

Upon defining

$$C_S = (p-1)C_1^{p-2}C_{\mathcal{T}}C_{p,\Omega}^2 + D_o C_{\mathcal{T}},$$

we realize that

$$\sum_{i=0}^{n_h}\sum_{j=i+1}^{n_h} \langle DE(v_{j-1}+u_j) - DE(v_{j-1}), w_i\rangle$$
$$\leq C_S\left[\sum_{i=0}^{n_h} \|\nabla u_i\|^2_{L^2(\Omega)}\right]^{1/2} \left[\sum_{i=0}^{n_h} \|\nabla w_i\|^2_{L^2(\Omega)}\right]^{1/2}.$$

In particular, assumption (NS2) holds. □

## 13.3 A one-dimensional implementation

In this section, we provide implementation details for the FASD method applied to a specific one-dimensional nonlinear finite element problem. We consider the energy functional

$$E(v) = \int_0^1 \left( \frac{1}{2}|v'|^2 + \frac{1}{4}|v|^4 - fv \right) dx, \tag{13.3.1}$$

for all $v \in H_0^1(0,1)$. The first Fréchet derivative of this energy is

$$\langle DE(v), w \rangle = \int_0^1 (v'w' + v^3 w - fw)\, dx,$$

for all $w \in H_0^1(\Omega)$. The function $u \in H_0^1(0,1)$ is the unique energy minimizer iff it is the unique solution of the following variational problem: find $u \in H_0^1(0,1)$ such that

$$\mathcal{A}(u,v) + (u^3, v)_{L^2(0,1)} = (f, v)_{L^2(0,1)}, \quad \forall\, v \in H_0^1(0,1), \tag{13.3.2}$$

where, as usual,

$$\mathcal{A}(u,v) = (u', v')_{L^2(0,1)}, \quad \forall\, u, v \in H_0^1(0,1).$$

This variational problem is the weak formulation of the nonlinear boundary value problem

$$\begin{cases} u(0) = 0, \\ -u'' + u^3 = f \quad \text{in } \Omega = (0,1), \\ u(1) = 0. \end{cases} \tag{13.3.3}$$

To demonstrate basic implementation aspects, as in Chapter 4, we consider a grid comprised of $n_1$ equally sized intervals

$$K_{1,i} := (x_{1,i-1}, x_{1,i}), \quad i = 1, \dots, n_1 + 1,$$

where the set of grid points $\mathcal{G}_{h_1}$ is defined in (4.1.1). Specifically, the grid spacing

$$h_1 = x_{1,i} - x_{1,i-1}, \quad i = 1, \dots, n_1 + 1,$$

is assumed invariant for simplicity. We define the fine-grid finite element space as follows:

$$V_1 := \{ v \in C^0([0,1]) \mid v(0) = v(1) = 0,\ v|_{K_{1,i}} \in \mathbb{P}_1,\ 1 \le i \le n_1 \}. \tag{13.3.4}$$

The coarse-grid finite element subspace $V_0$ is defined analogously. See the discussion surrounding equation (4.6.1) in Chapter 4. Importantly, $V_0$ is a nested subspace of $V_1$, and both are finite-dimensional subspaces of $H_0^1(0,1)$.

The respective bases $\mathcal{B}_\ell$, $\ell = 0, 1$, are defined as usual:

$$\mathcal{B}_\ell = \{\psi_{\ell,i}\}_{i=1}^{n_\ell} \subset V_\ell \subset H_0^1(0,1), \quad \ell = 0, 1,$$

is the standard $\mathbb{P}_1$ Lagrange nodal basis for $V_\ell$. We have, precisely, $\dim(V_\ell) = n_\ell$, $\ell = 0, 1$.

The corresponding finite element approximation is defined as follows: find $u_1 \in V_1$ such that

$$\mathcal{A}(u_1, v_1) + (u_1^3, v_1)_{L^2(0,1)} = (f, v_1)_{L^2(0,1)}, \quad \forall\, v_1 \in V_1. \tag{13.3.5}$$

Writing $u_1 = \sum_{j=1}^{n_1} u_{1,j}\psi_{1,j}$ and defining the vectors

$$\boldsymbol{u}_1 = \begin{bmatrix} u_{1,1} \\ u_{1,2} \\ \vdots \\ u_{1,n_1} \end{bmatrix} \in \mathbb{R}^{n_1}, \quad \boldsymbol{f}_1 = \begin{bmatrix} (f, \psi_{1,1})_{L^2(0,1)} \\ (f, \psi_{1,2})_{L^2(0,1)} \\ \vdots \\ (f, \psi_{1,n_1})_{L^2(0,1)} \end{bmatrix} \in \mathbb{R}^{n_1},$$

we have the equivalent nonlinear problem: given $\boldsymbol{f}_1 \in \mathbb{R}^{n_1}$, find $\boldsymbol{u}_1 \in \mathbb{R}^{n_1}$ such that

$$\mathsf{A}_1\boldsymbol{u}_1 + \boldsymbol{g}_1(\boldsymbol{u}_1) = \boldsymbol{f}_1. \tag{13.3.6}$$

Here $\mathsf{A}_1$ is the $n_1 \times n_1$ SPD stiffness matrix defined in (4.1.9), and $\boldsymbol{g}_1(\boldsymbol{u}_1)$ is the nonlinear vector-valued function with components

$$[\boldsymbol{g}_1(\boldsymbol{u}_1)]_i = (u_1^3, \psi_{1,i})_{L^2(0,1)} = I_{1,i}^L + I_{1,i}^R, \quad 1 \le i \le n_1,$$

where

$$I_{1,i}^L := \int_{x_{1,i-1}}^{x_{1,i}} (\psi_{1,i-1}u_{1,i-1} + \psi_{1,i}u_i)^3 \psi_{1,i}\, \mathrm{d}x$$

and

$$I_{1,i}^R := \int_{x_{1,i}}^{x_{1,i+1}} (\psi_{1,i}u_{1,i} + \psi_{1,i+1}u_{1,i+1})^3 \psi_{1,i}\, \mathrm{d}x.$$

Here we use the simplifying convention that

$$u_{1,0} = u_{1,n_1+1} = 0.$$

In other words, we add degrees of freedom at the boundary points and then zero them out. This greatly simplifies the implementation.

By a simple change of variable $z = \frac{x}{h_1}$ we obtain

$$I_{1,i}^L = h_1 \int_0^1 (u_{1,i-1}(1-z) + u_{1,i}z)^3 (1-z)\,dz$$

and

$$I_{1,i}^R = h_1 \int_0^1 (u_{1,i}(1-z) + u_{1,i+1}z)^3 z\,dz.$$

By a direct computation we find

$$I_{1,i}^L = h_1\left(\frac{u_{1,i-1}^3}{5} + \frac{3u_{1,i-1}^2 u_{1,i}}{20} + \frac{u_{1,i-1}u_{1,i}^2}{10} + \frac{u_{1,i}^3}{20}\right)$$

and, symmetrically,

$$I_{1,i}^R = h_1\left(\frac{u_{1,i}^3}{20} + \frac{u_{1,i}^2 u_{1,i+1}}{10} + \frac{3u_{1,i}u_{1,i+1}^2}{20} + \frac{u_{1,i+1}^3}{5}\right).$$

Therefore we can express the nonlinear term as

$$(u_1^3, \psi_{1,i})_{L^2(0,1)} = h_1 g(u_i, u_{i-1}, u_{i+1}), \tag{13.3.7}$$

where

$$\begin{aligned} g(u_i, u_{i-1}, u_{i+1}) :={}& \frac{u_{1,i}^3}{10} + \frac{u_{1,i}^2}{10}(u_{1,i-1} + u_{1,i+1}) \\ &+ \frac{3u_{1,i}}{20}(u_{1,i-1}^2 + u_{1,i+1}^2) + \frac{u_{1,i-1}^3 + u_{1,i+1}^3}{5}. \end{aligned} \tag{13.3.8}$$

The scheme can now be written in component/stencil form as follows: for $1 \le i \le n_1$,

$$\frac{-u_{1,i-1} + 2u_{1,i} - u_{1,i+1}}{h_1} + h_1\, g(u_{1,i}, u_{1,i-1}, u_{1,i+1}) = f_{1,i}, \tag{13.3.9}$$

with the convention $u_{1,0} = 0 = u_{1,n_1+1}$.

For the FASD method, we use the space decomposition

$$V_1 = \operatorname{span}\{\psi_{1,1}\} + \cdots + \operatorname{span}\{\psi_{1,n_1}\} + V_0, \tag{13.3.10}$$

where $V_0$ is the coarse-grid finite element space. We have already seen that this decomposition is stable and the energy satisfies the strengthened Cauchy–Schwarz inequality (NS2) with respect to this decomposition. See Theorems 13.2.3 and 13.2.4. This can be proved by elementary means; see Problem 13.2.

For our implementation, we will use SSO steps for the first $n_1$ subspaces. Recall that SSO steps are generalized by FASD steps, as we have pointed out in Remark 12.4.3. This translates to performing precisely one lexicographic sweep of a nonlinear Gauss–Seidel iteration (see Problem 13.3), which we now describe. We begin with some initial approximation

$$v_1^0 = (v_{1,0}^0, v_{1,1}^0, \dots, v_{1,n_1}^0, v_{1,n_1+1}^0),$$

where we continue the convention

$$v_{1,0}^0 = 0 = v_{1,n_1+1}^0.$$

For nonlinear Gauss–Seidel, we will use the superscripting convention, typical for both the nonlinear and linear versions, even though this will deviate from the convention we established in the last chapter when defining the SSO and FASD methods. The nonlinear Gauss–Seidel update is defined as follows: having obtained $v_{1,1}^1, \dots, v_{1,i-1}^1$ in lexicographic order, find $v_{1,i}^1 \in \mathbb{R}$ such that

$$\frac{-v_{1,i-1}^1 + 2v_{1,i}^1 - v_{1,i+1}^0}{h_1} + h_1\, g(v_{1,i}^1, v_{1,i-1}^1, v_{1,i+1}^0) = f_{1,i}. \tag{13.3.11}$$

We need to solve this nonlinear equation for $v_{1,i}^1$. For each fixed $i$, $1 \le i \le n_1$, we use Newton's method with $u_{1,i}^{1,0} := u_{1,i}^0$:

$$v_{1,i}^{1,m+1} = v_{1,i}^{1,m} - \frac{F_{1,i}(v_{1,i}^{1,m}, v_{1,i-1}^1, v_{1,i+1}^0)}{\frac{\partial F_{1,i}}{\partial x}(v_{1,i}^{1,m}, v_{1,i-1}^1, v_{1,i+1}^0)}, \tag{13.3.12}$$

where

$$F_{1,i}(x,y,z) := \frac{-y + 2x - z}{h_1} + h_1 g(x,y,z) - f_{1,i},$$

and therefore

$$\frac{\partial F_{1,i}}{\partial x} = \frac{2}{h_1} + h_1\left\{\frac{3x^2}{10} + \frac{2x}{10}(y+z) + \frac{3}{20}(y^2+z^2)\right\}.$$

Because of the inherent convexity in the problem, we are guaranteed that Newton's method will converge unconditionally: for each fixed $i$, $1 \le i \le n_1$,

$$\lim_{m\to\infty} v_{1,i}^{1,m} = v_{1,i}^1.$$

Note that no orthogonalization step is needed after each coordinate update for the SSO steps. One full nonlinear Gauss–Seidel sweep is complete when we have stepped lexicographically through the $n_1$ coordinates, and we write

$$v_1^1 = \mathrm{NLGS}(v_1^0).$$

The next step is to compute the update with respect to the subspace $V_0$. For obvious reasons, we will refer to this step as the coarse-grid correction. For the subspace $V_0$ energy, we use

$$E_0(w) = \frac{1}{2}\|w\|^2_{H_0^1(0,1)} = \frac{1}{2}\|w'\|^2_{L^2(0,1)}, \quad \forall\, w \in V_0.$$

The $\tau$-perturbation is, precisely,

$$\tau_0^1 \coloneqq \mathrm{D}E_0(Q_0 v_1^1) - R_0 \mathrm{D}E(v_1^1) \in V_0',$$

where $v_1^1 \in V_1$ is the approximation obtained after a full lexicographic sweep of nonlinear Gauss–Seidel is completed, $Q_0 : L^2 \to V_0$ is the $L^2$-projection, and $R_0 : V_1' \to V_0'$ is the restriction operator.

We must now solve

$$\langle \mathrm{D}E_0(\eta_0^1), w_0\rangle = \langle \tau_0^1, w_0\rangle, \quad \forall\, w_0 \in V_0, \tag{13.3.13}$$

where $\eta_0^1 = s_0^1 + Q_0 v_1^1 \in V_0$. Because $E_0$ is quadratic, the last equation is equivalent to

$$\langle \mathrm{D}E_0(s_0^1), w_0\rangle = \langle -R_0 \mathrm{D}E(v_1^1), w_0\rangle = -\langle \mathrm{D}E(v_1^1), w_0\rangle,$$

for all $w_0 \in V_0$. This problem, in turn, may be expressed equivalently (see Problem 13.4) as follows: for all $w_0 \in V_0$,

$$\mathcal{A}(s_0^1, w_0) = (f, w_0)_{L^2(0,1)} - \mathcal{A}(v_1^1, w_0) - ((v_1^1)^3, w_0)_{L^2(0,1)}. \tag{13.3.14}$$

The object on the right-hand side of (13.3.14) is a residual. If we write

$$\boldsymbol{r}_1^1 = \boldsymbol{f}_1 - \mathsf{A}_1 \boldsymbol{v}_1^1 - \boldsymbol{g}_1(\boldsymbol{v}_1^1),$$

which is the fine-grid residual vector for

$$v_1^1 \in V_1 \overset{\mathcal{B}_1}{\longleftrightarrow} \boldsymbol{v}_1^1 \in \mathbb{R}^{n_1},$$

then the coarse-grid problem is expressed in vector form as

$$\mathsf{A}_0 \boldsymbol{s}_0^1 = \mathsf{R}_0 \boldsymbol{r}_1^1,$$

where

$$s_0^1 \in V_0 \overset{\mathcal{B}_0}{\longleftrightarrow} \boldsymbol{s}_0^1 \in \mathbb{R}^{n_0},$$

and $\mathsf{A}_0$ is the coarse-grid stiffness matrix obtained by the strong Galerkin condition. Of course, the last problem is uniquely solvable, because $\mathsf{A}_0$ is SPD.

Next comes the orthogonalization step. Recall that

$$\alpha_0^1 = \operatorname*{argmin}_{\alpha\in\mathbb{R}} E(v_1^1 + \alpha s_0^1),$$

which is equivalent to

$$\alpha_0^1 = \operatorname*{argzero}_{\alpha\in\mathbb{R}} \langle DE(v_1^1 + \alpha s_0^1), s_0^1\rangle. \tag{13.3.15}$$

Again, because of convexity, the last problem is uniquely solvable. Note that the term $DE(v_1^1 + \alpha s_0^1)$ is just $-1$ times the residual functional. In any case, following the FASD algorithm, once $\alpha_0^1$ is found, we just need to set

$$v_1^2 := v_1^1 + \alpha_0^1 s_0^1,$$

which completes FASD update associated with the coarse space, $V_0$.

To finish, let us examine a practical method for computing $\alpha_0^1$. It is easy to show (see Problem 13.5) that

$$\alpha_0^1 = \operatorname*{argzero}_{\alpha\in\mathbb{R}} G(\alpha), \tag{13.3.16}$$

where $G : \mathbb{R} \to \mathbb{R}$ is

$$G(\alpha) := (\boldsymbol{s}_0^1)^\top \mathsf{R}_0 \boldsymbol{r}_1(\boldsymbol{v}_1^1 + \alpha \mathsf{P}_0 \boldsymbol{s}_0^1),$$

and the residual vector function is

$$\boldsymbol{r}_1(\boldsymbol{w}_1) := \boldsymbol{f}_1 - \mathsf{A}_1\boldsymbol{w}_1 - \boldsymbol{g}_1(\boldsymbol{w}_1) \in \mathbb{R}^{n_1}, \quad \forall\, \boldsymbol{w}_1 \in \mathbb{R}^{n_1}.$$

The derivative of $G$ with respect to $\alpha$ is somewhat complicated. For simplicity and computational efficiency, we seek an alternative to Newton's method for solving $G(\alpha) = 0$. In our implementation, we use the secant method to find $\alpha_0^1$ with the initial data $\alpha_0^{1,-1} = \frac{8}{10}$ and $\alpha_0^{1,0} = \frac{9}{10}$: for $m \geq 0$,

$$\alpha_0^{1,m+1} = \alpha_0^{1,m} - \frac{(\alpha_0^{1,m} - \alpha_0^{1,m-1})G(\alpha_0^{1,m})}{G(\alpha_0^{1,m}) - G(\alpha_0^{1,m-1})}.$$

This method requires only one new function evaluation every iteration and still gives relatively rapid convergence. Once again, unconditional convergence is guaranteed by convexity.

We can now summarize the complete FASD algorithm. Given the approximation $u_1^k \in V_1$, compute

$$u_1^{k+1} = \mathrm{FASD}(u_1^k)$$

according to the following steps:
1. Set $v_1^0 := u_1^k \in V_1$.
2. Use the nonlinear Gauss–Seidel algorithm to update $v_1^0$:

$$v_1^1 := \mathrm{NLGS}(v_1^0) \in V_1.$$

3. Compute the search direction $s_0^1 \in V_0$ by solving

$$\mathcal{A}(s_0^1, w_0) = (f, w_0)_{L^2(0,1)} - \mathcal{A}(v_1^1, w_0) - ((v_1^1)^3, w_0)_{L^2(0,1)},$$

   for all $w_0 \in V_0$.
4. Compute $\alpha_0^1 \in \mathbb{R}$ by solving

$$\alpha_0^1 = \underset{\alpha\in\mathbb{R}}{\mathrm{argzero}} \langle DE(v_1^1 + \alpha s_0^1), s_0^1\rangle.$$

5. Apply the coarse-grid correction $\alpha_0^1 s_0^1 \in V_0$:

$$v_1^2 := v_1^1 + \alpha_0^1 s_0^1 \in V_1.$$

6. Set

$$u_1^{k+1} := v_1^2 \in V_1.$$

## 13.4 Some computational experiments

To conclude this chapter, we perform some computational experiments to confirm the effectiveness of the FASD method for the model nonlinear finite element problem. To keep the presentation simple, we will only consider the one-dimensional setting described in the previous section. The interested reader is referred to [13], where an implementation in two dimensions and interesting computational results are provided using an implementation based on `iFEM`.

We will solve (13.3.5), which is equivalent to (13.3.6). The source term $\boldsymbol{f}_1$ in (13.3.6) is manufactured to correspond to a known exact solution. In our experiments, we specify the exact solution to the discretized system for error measurement:

$$[\boldsymbol{u}_1^{\mathrm{E}}]_i = u_{1,i}^{\mathrm{E}} = \exp(\sin(3\pi x_{1,i})) - 1.$$

The initial approximation $\boldsymbol{u}_1^0$ is chosen to be the zero vector. The FASD algorithm terminates when the norm of the error falls below a prescribed tolerance, for example,

$$\|\boldsymbol{u}_1^{\mathrm{E}} - \boldsymbol{u}_1^k\|_1 \le 10^{-9}.$$

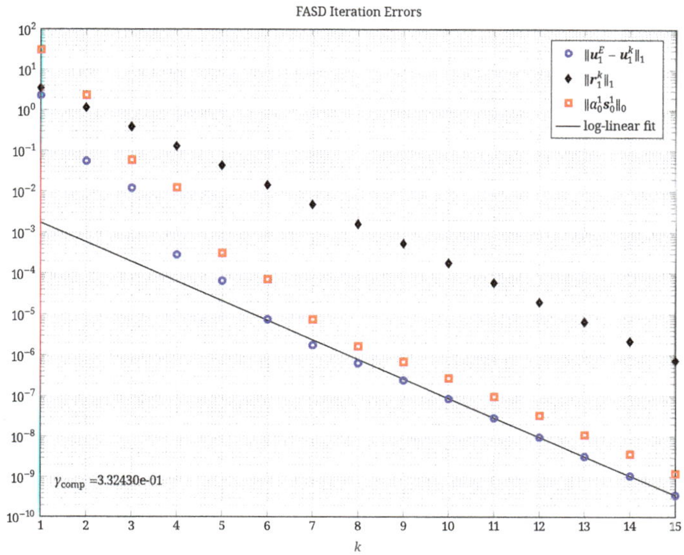


**Figure 13.1:** Error reduction for the FASD solver to the model nonlinear problem. The computed error reduction factor $\gamma_{\text{comp}}$ is roughly 0.332. The factor $\gamma_{\text{comp}}$ is computed using a log-linear fit of the last four exact error values.

In the experiment, we use choose $n_1 = 999$. The results are presented in Figure 13.1. The exact error $\|\boldsymbol{u}_1^{\text{E}} - \boldsymbol{u}_1^k\|_1$, the residual $\|\boldsymbol{r}_1^k\|_1$, where

$$\boldsymbol{r}_1^k := \boldsymbol{r}_1(\boldsymbol{u}_1^k) := \boldsymbol{f}_1 - \mathsf{A}_1\boldsymbol{u}_1^k - \boldsymbol{g}_1(\boldsymbol{u}_1^k) \in \mathbb{R}^{n_1},$$

and the coarse-grid correction $\|\alpha_0^1 \boldsymbol{s}_0^1\|_0$ are plotted against the iteration number $k$. The error contraction factor $\gamma_{\text{comp}}$ is computed using a log-linear fit of the last four error values $\|\boldsymbol{u}_1^{\text{E}} - \boldsymbol{u}_1^k\|_1$. The computed value $\gamma_{\text{comp}} \approx 3.32 \times 10^{-1}$ demonstrates effective convergence of the FASD method. This value is nearly invariant over a wide range of grid sizes; see Table 13.1.

**Table 13.1:** Computed two-grid contraction factors $\gamma_{\text{comp}}$ for various grid sizes $n_1$. The factor $\gamma_{\text{comp}}$ is computed using a log-linear fit of the last four error values $\|\boldsymbol{u}_1^{\text{E}} - \boldsymbol{u}_1^k\|_1$.

| $n_1$ | $\gamma_{\text{comp}}$ |
|---|---|
| 249 | $3.28 \times 10^{-1}$ |
| 499 | $3.31 \times 10^{-1}$ |
| 999 | $3.32 \times 10^{-1}$ |
| 1999 | $3.37 \times 10^{-1}$ |
| 3999 | $3.65 \times 10^{-1}$ |

## 13.5 Problems

13.1 Verify that the first and second Fréchet derivatives of the energy defined in (13.1.2) are indeed given by (13.1.3) and (13.1.4), respectively.

13.2 Prove that the one-dimensional decomposition defined in (13.3.10) is stable using elementary means, that is, by not appealing to the sophisticated abstract techniques in the proof of Theorem 13.2.3. To built this "elementary" argument, establish a one-dimensional version of estimate (13.2.2), in the setting where the appropriate meshes/grids covering $[0, 1]$ are uniform.

13.3 In Section 13.3, we have claimed that $n_1$ SSO steps using the decomposition

$$V_1 = \operatorname{span}\{\psi_{1,1}\} + \cdots + \operatorname{span}\{\psi_{1,n_1}\}$$

are exactly equivalent to the nonlinear Gauss–Seidel algorithm NLGS(·). Prove that this is indeed the case.

13.4 Prove that problems (13.3.13) and (13.3.14) are equivalent.

13.5 Prove that problems (13.3.15) and (13.3.16) are equivalent.

13.6 Consider the boundary value problem

$$\begin{cases} u(0) = 0, \\ -u'' + 2u^5 = f \quad \text{in } \Omega = (0,1), \\ u(1) = 0. \end{cases} \tag{13.5.1}$$

(a) Show that the functional

$$E(v) := \int_0^1 \left( \frac{1}{2}|v'|^2 + \frac{1}{3}|v|^6 - fv \right) \mathrm{d}x, \quad v \in H_0^1(0,1),$$

is twice Fréchet-differentiable, coercive, and strictly convex.

(b) Conclude that (13.5.1) has a unique weak solution, which is also the unique minimizer of $E$.

(c) Modify the one-dimensional MATLAB© implementation from the listing to solve (13.5.1). Verify numerically that the modified FASD solver converges with a comparable error-reduction factor.

# Listings

./FASDFEM/
- FASD1D.m
- FASDSolver.m
- NLOperator.m
- getResidual.m
- prolongation.m (see Listing 4.7)
- restriction.m (see Listing 4.8)
- NLGS.m
- secantSearch.m

**Listing 13.1:** FASD1D.m: Driver code for the FASD nonlinear FEM solver.

```
%
% FASD1D.m
%
% Calling script for solving the 1D nonlinear elliptic
% equation
%
%   -u'' + u^3 = f
%
% on the unit interval [0,1] with homogeneous Dirichlet
% boundary conditions using the FASD
% (Fast Subspace Descent) method.
%
%% Initialization and Parameter

% Clear command line window and workspace.
clear; clc; close all;

% Tolerance for the norm of the error.
tol = 1.0e-09;

% Number of interior grid points on the fine level.
% nf+1 must be even.
nf = 999;

% Fine grid mesh size.
hf = 1.0/(nf+1);

% Define the exact discrete solution on the fine level.
```

```
xE = linspace(0,1,nf+2)';
x = xE(2:nf+1);
uExact = exp(sin(3.0*pi*x))-1.0;

% Maximum number of iterations.
maxIter = 100;

%% Setting up the Solver

% Check nf.
if mod(nf+1,2) ~= 0
  error('nf+1 must be even.');
end
fprintf('Using nf = %d interior points.\n',nf);

% Manufacture the source term vector from uExact.
ff = NLOperator(uExact,hf);

% Initial Guess.
uf = zeros(nf,1);

%% Calling the Nonlinear Solver

tic;
[uf,errVals,kStop] = FASDSolver(uf,ff,hf,uExact,...
                               maxIter,tol);
solveTime = toc;
fprintf('Solver finished in %.4f seconds.\n',solveTime);

% Calculate final error.
finalErrL2 = norm(uf-uExact,2);
fprintf('Final L2 error = %g\n',finalErrL2);

%% Plotting

% Add boundary points (zeros) back for plotting solution.
ufE = [0;uf;0];

% Get exact solution on full fine grid for plotting.
uExactE = [0;uExact;0];

figure;
```

```
plot(xE,uExactE,'r-','LineWidth',2,'DisplayName',...
  'Exact Solution');
hold on;
plot(xE,ufE,'b--','LineWidth',1.5,'DisplayName',...
  'Computed Solution');
hold off;
xlabel('x');
ylabel('u(x)');
title(sprintf('Computed vs Exact (nf = %d)',nf));
legend('show','Location','best');
grid on;

% Plot convergence history.
figure;

semilogy(...
  1:kStop,errVals(1:kStop,1),'bo','LineWidth',1.5,...
  'DisplayName',...
  '$\| \mathbf{u}_1^E - \mathbf{u}_1^k \|_1$');
hold on;
semilogy(...
  1:kStop,errVals(1:kStop,2),'kd','LineWidth',1.5,...
  'DisplayName','$\|  \mathbf{r}_1^k \|_1$');
hold on;
semilogy(...
  1:kStop,errVals(1:kStop,3),'rs','LineWidth',1.5,...
  'DisplayName','$\|  \alpha_0\mathbf{s}_0 \|_0$');

% Estimate the rate of contraction based on exact error.
rateError = 1.0;
if kStop>=4
  kv = kStop-3:kStop;
  le = log(errVals(kStop-3:kStop,1));
  p1 = polyfit(kv,le,1);
  rateError = exp(p1(1));
  p1k = polyval(p1,1:kStop);
  semilogy(1:kStop,exp(p1k),'-k','DisplayName',...
    'log-linear fit');
end

xlabel('$k$','Interpreter','latex');
title('FASD Iteration Errors');
```

```
legend('show','Location','North East',...
  'Interpreter','latex');
grid on;

% Add contraction factor at bottom left.
text(1,errVals(kStop-1,1),...
  strcat('$\gamma_{\rm comp} = $ ',...
  num2str(rateError,'%.5e')),...
  'FontSize',12,'Interpreter','latex');
hold off;

% Print contraction factor.
fprintf('Error contraction factor = %g\n',rateError);
```

**Listing 13.2:** `FASDSolver.m`: FASD solver implementing SSO smoothing and coarse-grid correction. The auxiliary prolongation and restriction functions are provided in Listings 4.7 and 4.8.

```
function [uf,errVals,kStop] = FASDSolver(uf,ff,...
                                          hf,uExactf,...
                                          maxIter,tol)

nf = length(uf);

nc = (nf+1)/2-1;
hc = 1.0/(nc+1);

errVals = zeros(maxIter,2);

for kIter = 1:maxIter

  % SSO sweep on the fine grid.
  uf = NLGS(uf,ff,hf);

  % Fine grid residual after smoothing.
  rf = getResidual(uf,ff,hf);

  % Manufacture the coarse grid -Laplacian matrix.
  onesCoarse = ones(nc,1);
  Ac = spdiags([-onesCoarse,2*onesCoarse,-onesCoarse],...
               -1:1,nc,nc);
  Ac = Ac/hc;
```

```
  % Solve for coarse grid correction direction.
  s0 = Ac\restriction(rf);

  % Secant line search for optimal step size.
  alphaOpt = secantSearch(uf,ff,hf,s0);

  % Compute coarse correction.
  cGc = alphaOpt*prolongation(s0);

  % Update fine grid solution.
  uf = uf+cGc;

  % Convergence bookkeeping.
  errVals(kIter,1) = norm(uf-uExactf,2);
  errVals(kIter,2) = norm(getResidual(uf,ff,hf),2);
  errVals(kIter,3) = norm(cGc,2);

  fprintf(['Iter %3d: residual = %.3e, ',...
    'true error = %.3e, alpha0 = %.4f\n'],...
    kIter, errVals(kIter,2),errVals(kIter,1), alphaOpt);

  if errVals(kIter,1)<tol
    fprintf('Converged after %d iterations.\n',kIter);
    kStop = kIter;
    return
  end
end

fprintf(['Solver did not converge within %d ',...
         'iterations.\n'],maxIter);
kStop = kIter;

end
```

**Listing 13.3:** `NLOperator.m`: Nonlinear discrete operator.

```
function operatorResult = NLOperator(uf,hf)

nf = length(uf);
```

```

% Zero boundary extension.
uE = [0.0;uf(:);0.0];

for i = 1:nf
  um = uE(i);
  ui = uE(i+1);
  up = uE(i+2);
  ui2 = ui*ui;
  ui3 = ui2*ui;
  up2 = up*up;
  up3 = up2*up;
  um2 = um*um;
  um3 = um2*um;

  operatorResult(i) = (-um+2.0*ui-up)/hf...
                      +hf*(0.10*ui3+0.10*ui2*(um+up)...
                      +0.15*ui*(um2+up2)+0.20*(um3+up3));
end

end
```

**Listing 13.4:** `getResidual.m`: Code to compute the residual vector.

```
function residual = getResidual(uf,ff,hf)

% Residual of the nonlinear equation.
residual = ff-NLOperator(uf,hf);

end
```

**Listing 13.5:** `NLGS.m`: Nonlinear Gauss–Seidel smoother with Newton's method.

```
function uf = NLGS(uf,ff,hf)

% Tolerance for Newton's method.
newtonTol = 1e-12;

nf = length(uf);

uE = [0.0;uf(:);0.0];
```

```
for i = 2:nf+1
  um = uE(i-1);
  up = uE(i+1);
  ui = uE(i  );
  ui2 = ui*ui;
  ui3 = ui2*ui;
  up2 = up*up;
  up3 = up2*up;
  um2 = um*um;
  um3 = um2*um;

  % Newton iteration for this node.
  for locIt = 1:7
    uiPrev = ui;

    % g(u_i, u_{i-1}, u_{i+1}) - exact coefficients.
    gVal = (0.10*ui3+0.10*ui2*(um+up)...
           +0.15*ui*(um2+up2)+0.20*(um3+up3));

    % Residual F_i(ui).
    FVal = (2.0*ui-um-up)/hf+hf*gVal-ff(i-1);

    dgVal = (0.30*ui2+0.20*ui*(um+up)+0.15*(um2+up2));

    dFVal = 2.0/hf+hf*dgVal;

    delta = -FVal/dFVal;
    ui = ui+delta;

    if abs(delta)<newtonTol
      break
    end
  end

  uE(i) = ui;
end

uf(1:nf) = uE(2:nf+1);

end
```

**Listing 13.6:** `secantSearch.m`: Secant search algorithm to solve $G(\alpha) = 0$.

```
function alphaOpt = secantSearch(uf,ff,hf,s0)

% Tolerance for Secant method
secantTol = 1.0e-06;
% Maximum number of iterations.
maxSecantIter = 5;

% Initial guesses for alpha
alpham2 = 0.8;
alpham1 = 0.9;

% Pre-calculate the value of G at the two alpha values.
Gm2 = s0'*restriction(getResidual(...
  uf+alpham2*prolongation(s0),ff,hf));
Gm1 = s0'*restriction(getResidual(...
  uf+alpham1*prolongation(s0),ff,hf));

% Secant iteration loop
for j = 1:maxSecantIter

  % Secant update for alpha
  delta = -(alpham1-alpham2)*Gm1/(Gm1-Gm2);
  alphaOpt = alpham1+delta;

  if abs(delta) < secantTol
    break;
  end

  % Update alpha values for next iteration
  alpham2 = alpham1;
  Gm2 = Gm1;
  alpham1 = alphaOpt;
  Gm1 = s0'*restriction(getResidual(...
    uf+alpham1*prolongation(s0),ff,hf));
end



end
```

# Bibliography


[1] iFEM: an Integrated Finite Element Methods Package in MATLAB — lyc102.github.io. https://lyc102.github.io/ifem/. [Accessed 02-05-2025].

[2] R. E. Bank, T. F. Dupont, and H. Yserentant. The hierarchical basis multigrid method. *Numer. Math.*, 52(4):427–458, 1988.

[3] P. Bjorstad and W. Gropp. *Domain Decomposition: Parallel Multilevel Methods for Elliptic Partial Differential Equations*. Cambridge University Press, 2004.

[4] D. Braess. *Finite Elements: Theory, Fast Solvers, and Applications in Solid Mechanics*. Cambridge University Press, third edition, 2007.

[5] D. Braess and W. Hackbusch. A new convergence proof for the multigrid method including the v-cycle. *SIAM J. Numer. Anal.*, 20:967–975, 1983.

[6] J. H. Bramble. *Multigrid Methods*. Longman Scientific and Technical, 1993.

[7] J. H. Bramble, J. E. Pasciak, and J. J. Xu. Parallel multilevel preconditioners. *Math. Comp.*, 55:1–22, 1990.

[8] J. H. Bramble and X. Zhang. The analysis of multigrid methods. In P. G. Ciarlet and J. L. Lions, editors, *Solution of Equations in* $\mathbb{R}^n$ *(Part 3); Techniques of Scientific Computing (Part 3)*, volume 7 of *Handbook of Numerical Analysis*, pages 173–415. Elsevier, 2000.

[9] A. Brandt. Multi-level adaptive solutions to boundary value problems. *Math. Comp.*, 31:333–390, 1977.

[10] S. C. Brenner and L. R. Scott. *The Mathematical Theory of Finite Element Methods*. Springer Verlag, third edition, 2008.

[11] H. Brezis. *Functional Analysis, Sobolev Spaces and Partial Differential Equations*. Universitext. Springer, New York, 2011.

[12] W. L. Briggs, V. E. Henson, and S. F. McCormick. *A Multigrid Tutorial*. SIAM, second edition, 2000.

[13] L. Chen, X. Hu, and S. M. Wise. Convergence analysis of the fast subspace descent methods for convex optimization problems. *Math. Comp.*, 89:2249–2282, 2020.

[14] P. G. Ciarlet. *Introduction to Numerical Linear Algebra and Optimization*. Cambridge University Press, 1989.

[15] P. G. Ciarlet. *Linear and Nonlinear Functional Analysis With Applications*. Society for Industrial and Applied Mathematics, Philadelphia, PA, 2013.

[16] F. Clarke. *Functional Analysis, Calculus of Variations and Optimal Control*, volume 264 of *Graduate Texts in Mathematics*. Springer, London, 2013.

[17] B. Dacorogna. *Direct Methods in the Calculus of Variations*, volume 78 of *Applied Mathematical Sciences*. Springer, New York, second edition, 2008.

[18] I. Ekeland and R. Temam. *Convex Analysis and Variational Problems*. North–Holland, Amsterdam, 1976.

[19] R. E. Ewing, R. D. Lazarov, and P. S. Vassilevski. Local refinement techniques for elliptic problems on cell-centered grids, i. error analysis. *Math. Comp.*, 56:437–461, 1991.

[20] W. Feng, A. J. Salgado, C. Wang, and S. M. Wise. Preconditioned steepest descent methods for some nonlinear elliptic equations involving p-Laplacian terms. *J. Comput. Phys.*, 334:45–67, 2017.

[21] P. L. George. Automatic mesh generation and finite element computation. In *Handbook of Numerical Analysis, Vol. IV*, volume IV of *Handb. Numer. Anal.*, pages 69–190. North-Holland, Amsterdam, 1996.

[22] P. Grisvard. *Elliptic Problems in Nonsmooth Domains*, volume 69 of *Classics in Applied Mathematics*. Society for Industrial and Applied Mathematics (SIAM), Philadelphia, PA, 2011. Reprint of the 1985 original [MR0775683], With a foreword by Susanne C. Brenner.

[23] W. Hackbusch. *Multi-Grid Methods and Applications*. Springer Verlag, 1985.

[24] A. Loseille. Unstructured mesh generation and adaptation. In *Handbook of Numerical Methods for Hyperbolic Problems*, volume 18 of *Handb. Numer. Anal.*, pages 263–302. Elsevier/North-Holland, Amsterdam, 2017.

[25] Y. Nesterov. Efficiency of coordinate descent methods on huge-scale optimization problems. *SIAM J. Optim.*, 22(2):341–362, 2012.

[26] Y. Nesterov. *Introductory Lectures on Convex Optimization: A Basic Course*. Springer Verlag, 2013.


https://doi.org/10.1515/9783111354880-014

[27] R. H. Nochetto, K. G. Siebert, and A. Veeser. Theory of adaptive finite element methods: an introduction. In *Multiscale, Nonlinear and Adaptive Approximation*, pages 409–542. Springer, Berlin, 2009.
[28] M. E. G. Ong. Hierarchical basis preconditioners in three dimensions. *SIAM J. Sci. Comput.*, 18(2):479–498, 1997.
[29] L. F. Pavarino. Additive Schwarz methods for the $p$-version finite element method. *Numer. Math.*, 66(4):493–515, 1994.
[30] Y. Saad. *Iterative Methods for Sparse Linear Systems*. Society for Industrial and Applied Mathematics, Philadelphia, PA, second edition, 2003.
[31] A. J. Salgado and S. M. Wise. *Classical Numerical Analysis: A Comprehensive Course*. Cambridge University Press, 2023.
[32] L. R. Scott and S. Zhang. Finite element interpolation of nonsmooth functions satisfying boundary conditions. *Math. Comp.*, 54(190):483–493, 1990.
[33] J. Stoer and R. Bulirsch. *Introduction to Numerical Analysis*. Springer, third edition, 2002.
[34] X. C. Tai. Rate of convergence for some constraint decomposition methods for nonlinear variational inequalities. *Numer. Math.*, 93:755–786, 2003.
[35] X. C. Tai and J. Xu. Global and uniform convergence of subspace correction methods for some convex optimization problems. *Math. Comp.*, 71:105–125, 2001.
[36] A. Toselli and O. Widlund. *Domain Decomposition Methods—Algorithms and Theory*, volume 34 of Springer Series in Computational Mathematics. Springer-Verlag, Berlin, 2005.
[37] U. Trottenberg, C. Oosterlee, and A. Schüller. *Multigrid*. Academic Press, 2001.
[38] P. Wesseling. *Multigrid Methods*. R.T. Edwards, 2004.
[39] H. Yserentant. On the multilevel splitting of finite element spaces. *Numer. Math.*, 49(4):379–412, 1986.
[40] H. Yserentant. Hierarchical bases. In *ICIAM 91 (Washington, DC, 1991)*, pages 256–276. SIAM, Philadelphia, PA, 1992.

# Index

$L^2$ projection 427
Q-class 330
T-matrix 311, 314
$\alpha$-weak approximation property 157
$\tau$-perturbation 405, 418

Additive GLIS 22
– symmetrized 24
Adjoint matrix
– $A_1$- 76
– level-1 75
Affine Sets 57
Aliasing 109, 112
Approximation property 98, 138, 252
– $\alpha$-weak 157, 252
– strong 157, 201, 206
– weak 157, 207
Arithmetic-Geometric Mean inequality 116
Aubin, J.P. 204

Backward sweep 19
Balanced symmetric pair 70, 146
Banach, S. 366
Banach space 366
Basis
– canonical
  – $\mathbb{R}^n$ 326
– hierarchical 327, 333, 335
– Lagrange nodal 190, 202, 203
Bilinear form 187
Bisection 188
Boundary value problem 1
– elliptic 187
– nonlinear 392
– one-dimensional 31
Braess, D. 174
Braess–Hackbusch Theorem 174
Bramble, J. 349
Bulirsch, R. 81

Calculus of Variations 381
– direct method 381
Canonical basis
– $\mathbb{R}^n$ 326
Canonical splitting 10, 238
Cauchy, A.-L. 393
Cauchy–Schwarz inequality
– strengthened 345, 346, 350, 393
Céa J. 201
Céa's Lemma 201
Cell-centered 230
– discrete Laplacian 265
– Ritz projection 253
Cell-centered finite difference 230, 261
– error estimate 234, 253
– method 234
– stiffness matrix 235
– two dimensions 261
Cell-centered grid 230, 262
Cell-centered points 230, 262
Characteristic function 232
Chebyshev, P.L. 56
Chebyshev polynomial 56
Classical solution 258, 261, 270
Closure of a set 371
– weak 371
Coarse grid correction 73, 147
– exact 148
– two-grid 73
Coarse-grid Ritz projection matrix 149
Coefficient
– diffusion 188
– reaction 188
Coercive functional 375
Compactness
– sequential 372
  – weak 372
Condition number 5, 6, 199, 387
– BPX preconditioner 354
– hierarchical basis preconditioner 348
– spectral 5
– stiffness matrix 88, 199, 237
Conforming triangulation 188
Conjugate gradient method 6
– preconditioned 44
Convergence
– multigrid 258, 270
  – V-Cycle 258
– two-grid method 142
– V-Cycle 208
  – one-sided 322
– W-Cycle 257
  – one-sided 323

https://doi.org/10.1515/9783111354880-015

– symmetric 323
– weak 370
Convergent matrix 7
Convex domain 187
Convex functional 378
– strictly 378
– strongly 379, 392
Convex Minimization 381
Convex set 374
Convolution
– discrete 347
Convolution estimate
– discrete 347
Correction
– coarse grid 73, 147
– exact 148
– two-grid 73
– multilevel 312

Damped Jacobi method 14, 90
Dampening
– Jacobi
– optimal 97
Decomposition
– hierarchical 327
– subspace 325
– (SS1) 325
Descent direction 46
Diagonally dominant matrix 12
Diffusion coefficient 188, 258, 270
– variable 258, 270
Direct sum 334
Discrete convolution 347
Discrete Laplacian operator 265, 266
– cell-centered 265
– level-$\ell$ 249
Domain
– convex 187
– polytopal 187
Dual space 367
Duality argument 204
Duality pairing 369

Elliptic functional 379
Elliptic problem 187
Energy 392
Energy form 249
Energy inner product 187
– level $\ell$ 146
Energy norm 25
Energy section 395
Epigraph 373
Equation
– linear 1
Error contraction factor
– two-grid 117
Error estimate
– cell-centered finite difference 234, 253
– energy norm 204
– full multigrid 209
– interpolation 203, 341
Error sequence 6
Error smoothing 98
Error transfer matrix 7, 91
– multigrid 149
– presmoothing 75
– smoothing 239
– two-grid 77
Estimate
– convolution
– discrete 347
– trace 340, 345
Euclid 2
Euclidean
– norm 2
Euler, L. 384, 392
Euler–Lagrange Equation 384, 392

FAS 402, 406
FASD 405
– approximate line search 417
FASDALS 417
– quadratic step size 418
Fast Subspace Descent (FASD) 405
– approximate line search (FASDALS) 417
Finite difference method 1, 31, 81, 230
– cell-centered 230
Finite difference stencil 265
Finite element method 1, 82, 187
– piecewise linear 82, 189
Finite element space 189
– piecewise linear 189
First smoothing property 159
Forward sweep 16
Fourier, J.-B. J. 81
Fréchet, R.M. 373
Fréchet differentiable 373
– continuously 374

- twice 374
- twice continuously 375, 424
Full Approximation Storage (FAS) method 402, 406
Full multigrid 208
- error estimate 209
- work estimate 211
Function
- characteristic 232
- grid 231, 263
  - cell-centered 231, 263
- hat 82, 190, 202
  - fine-level 82
  - level-0 99
  - level-1 82
- minimization 45
- piecewise constant 231, 263
- quadratic 44
Functional 367
- coercive 375
- convex 378
  - strictly 378
  - strongly 379, 392
- elliptic 379
- Fréchet differentiable
  - twice 374
  - twice continuously 375
- linear 187, 367
- Lipschitz smooth 380
  - locally 380
- lower semicontinuous 372
- minimizer 381
- quasiconvex 391
- section 386
- weakly lower semicontinuous 372
Fundamental orthogonality (FO) 395

Galerkin, B.G. 45, 70, 201
Galerkin condition 70, 104, 152
- (G2) 155
- imbalanced (G3) 176, 248, 260, 268, 272
- strong (G0) 152
- weak (G1) 153
Galerkin orthogonality 45, 201
Gauss, J.C.F 10
Gauss–Seidel method 10, 15, 17, 33, 34
- backward 17
- forward 15
- nonlinear 395
- symmetric 25
General linear iterative scheme (GLIS) 6
Ghost points 230, 262
GLIS 6
- additive 22
  - symmetrized 24
- multigrid as a 152
- multiplicative 22
  - symmetrized 23
Gradient descent method 387
Grid 31
- cell-centered 230, 262
- fine 81
- function 31, 81
- level-1 81
- points 31, 81
- spacing 81, 189
- two-level 96
  - nested 96
  - uniform 96
Grid function 231, 263
- cell-centered 231, 263
  - homogeneous Dirichlet 231, 263

Hackbusch, W. 174
Hat function 82, 190, 202
- fine-level 82
- level-0 99
- level-1 82
Hierarchical basis 327, 333, 335
Hierarchical decomposition 327
Hilbert, D. 366
Hilbert space 366
Horner, W.G. 356
Horner's method 356
Householder, A.S. 28
Householder–John Theorem 28

Imbalanced Galerkin condition 176, 248, 260, 268, 272
Induced norm 3, 8, 367
Inequality
- Arithmetic-Geometric Mean 116
- Cauchy–Schwarz
  - strengthened 345, 346, 350
- inverse 196, 197, 340
- Poincaré 187
  - mean zero 339
- Young 181
  - convolution 347

Inner product 2
- A- 3
- $A_1$- 76
- $A_\ell$- 196
- energy 187
  - level $\ell$ 146
- Euclidean 2
- level-1 75
- level-$\ell$ 196
Interpolant
- Lagrange nodal 203
- Scott–Zhang 428
Interpolation error estimate 203, 341
Inverse inequality 196, 197, 340
Iterator matrix 6

Jacobi, C.G.J 10
Jacobi method 10, 12, 33, 34
- damped 14, 90
- weighted 14
John, F. 28

Krylov, A.N. 44, 51
Krylov subspace 51

Lagrange, J.-L. 384, 392
Lagrange nodal basis 190, 202, 203
Lagrange nodal interpolant 203
Lax, P.D. 187
Lax–Milgram Theorem 187
Lifting operator 255
Linear functional 187
Linear mapping 367
- bounded 367
- continuous 367
Lipschitz, R.O.S. 380, 392
Lower semicontinuity (LSC) 372

Mapping
- Fréchet differentiable 373
  - continuously 374
- linear 367
  - bounded 367
  - continuous 367
Matrix
- adjoint
  - $A_1$- 76
  - level-1 75
- coarse-grid Ritz projection 149
- convergent 7
- diagonally dominant 12
- error transfer 7, 91
  - multigrid 149
  - presmoothing 75
  - smoothing 239
  - two-grid 77
- finite element stiffness
  - fine-level 85
  - level-1 85
- general symmetric 4
- iterator 6
- level-$\ell$ stiffness
  - cell-centered 235
- multilevel-T 314
- preconditioner 6
- projection
  - coarse-grid Ritz 149
- prolongation 70, 99, 193, 246, 268, 325
  - cell-centered 246, 268
  - hierarchical 337
  - multigrid 146
  - multilevel 311
- restriction 70, 99, 194, 246, 268
  - cell-centered 246, 268
  - multigrid 146
  - multilevel 312
- Ritz projection
  - multilevel 313
- smoothing 73
- SPD 1, 3, 85
- stiffness 70, 267
  - cell-centered 267
  - coarse grid 70
  - fine grid 70
  - finite difference 32, 82
  - level-0 70
  - level-1 70
  - level-$\ell$ 193
- symmetric 3
  - $A_1$- 76
  - level-1 75
Mazur, S.M. 378
Mazur's Lemma 378
Mean Value Theorem 426
Milgram, A.N. 187
Minimization 45
Minimizer 381
Minkowski, H. 377

Minkowski sum of sets 377
Modal analysis
- prolongation 111
- restriction 110
- Ritz projection 112
Mode
- frequency 97
  - high 97
  - low 97
- high frequency 97
- low frequency 97
Model problem 187
Monotone Convergence Theorem 396
Multigrid
- multiplicative 311
- one sided 149
- symmetrized 149
- V-Cycle 149
  - convergence 185
  - general symmetric 174, 175
  - simple symmetric 149, 169, 171
- W-Cycle 149
  - convergence 184
  - one-sided 163, 166
  - symmetric 179, 181, 184
Multigrid as a GLIS 152
Multigrid convergence 270
Multigrid method
- generic 149
Multigrid operator 147
Multigrid principle
- first 98
- second 98
Multilevel correction 312
Multilevel T-matrix 314
Multiplicative GLIS 22
- symmetrized 23

Newton, I. 356, 387, 405
Newton's Method 387, 405
Nitsche, J.A. 204
Nitsche's Trick 204
Norm
- $\ell^p$- 3
- A- 4
- energy 25
- Euclidean 2
- induced 8, 367
- operator 367

Objective 392
Operator
- discrete Laplacian 258, 265, 266
  - cell-centered 258, 260, 265
  - level-$\ell$ 249
- interpolation
  - Lagrange nodal 203
  - Scott–Zhang 428
- lifting 255
- multigrid 147
- norm 367
- prolongation 99, 193
- restriction 99, 194
  - full weighting 107
- smoothing 73
- two-grid 72
Orthogonality
- fundamental 395
- Galerkin 45, 201

Parallelization 325
Pasciak, J. 349
PCG method 50
- residual 51, 57
- search direction 51, 57
- standard 56
- zero-start 50
  - convergence rate 54
Piecewise constant function 231, 263
Poincaré, J.H. 187, 339
Poincaré inequality 187, 425, 426
- mean zero 339
Points
- cell-centered 230, 262
- ghost 230, 262
- vertex 230, 262
Polynomial
- Chebyshev 56
Polytopal domain 187
Postsmoothing 148
Preconditioned conjugate gradient (PCG) method 44, 50
- zero start 50
Preconditioned steepest descent (PSD) method 46
Preconditioner 6
- additive 325
  - subspace 329
- BPX 349
  - Numerical Experiment 357

- hierarchical basis 336, 338
Presmoothing 147
Projection
- $L^2$ 250, 427
- Ritz 205, 253, 352
 - cell-centered 253
 - coarse-grid 74
Prolongation matrix 70, 99, 193, 246, 268, 325
- cell-centered 246, 268
- hierarchical 337
- multigrid 146
- multilevel 311
Prolongation operator 99, 193
Property
- approximation 98, 138, 252
 - $\alpha$-weak 157, 252
 - strong 157, 201, 206
 - weak 157, 207
- smoothing 98, 141
 - first 159
 - second 166
PSD method 46
- convergence 48
- correction 46
- error minimization 47
- residual 46
- search direction 46

Quadratic function 44
Quadratic trap 385
- local 408
Quadrisection 188
Quasiconvex functional 391

Rayleigh, J.W.S. 3rd Baron 198, 333
Rayleigh quotient 198, 333
Reaction coefficient 188
Reciprocal wave number 108
Refinement
- bisection 188
- quadrisection 188
Regularity condition 188
Residual sequence 6
Restriction matrix 70, 99, 194, 246, 268
- cell-centered 246, 268
- multigrid 146
- multilevel 312
Restriction operator 99, 194
- full weighting 107
Richardson, L.F. 10, 158
Richardson method 10, 19, 139, 158
Richardson smoother 139, 158
Riesz, F. 187, 370, 417
Riesz Representation Theorem 187, 370
Ritz, W.H.W. 45, 205, 253
Ritz approximation 45
Ritz projection 205, 253, 352
- cell-centered 253
- coarse-grid 74
Ritz projection matrix
- multilevel 313

Schwarz, K.H.A. 393
Scott, L.R. 428
Scott–Zhang interpolant 428
Second smoothing property 166
Section 386, 395
Seidel, P.L. von 10
Sequence
- error 6
- residual 6
- subsequence 372
Sequential compactness 372
- weak 372
Set
- closure 371
- convex 374
- Minkowski sum 377
- sequentially compact 372
- sublevel 373
- weak closure 371
- weakly sequentially compact 372
Shape-regular triangulation 188
Smoother
- Jacobi
 - damped 239
- Richardson 139, 158, 239, 244
 - $\Lambda$ 239
Smoothing 98
- matrix 73
- operator 73
- post- 73, 148
- pre- 72, 147
Smoothing property 98, 141
- first 159
- second 166
Sobolev, S.L. 425
Sobolev embedding 425

Sobolev space 340
Solution
– classical 258, 261, 270
SOR method 32–34
Space
– finite element 189
  – piecewise linear 189
– Sobolev 340
SPD matrix 1
Spectral Decomposition Theorem 4
– general 4
Spectral radius 8, 270
– damped Jacobi 92
– Richardson smoother 242
Splitting
– canonical 10, 238
Splitting method 10
SSO 394
– corrections
  – lower bound 396
  – upper bound 396
Steepest descent method 46
– preconditioned 46
Stencil
– finite difference 265
Stiffness matrix 70, 267
– cell-centered 267
– coarse grid 70
– condition number 88
– fine grid 70
– finite difference 32, 82
– finite element
  – fine-level 85
  – level-1 85
– level-0 70
– level-1 70
– level-$\ell$ 193
Stoer, J. 81
Strengthened Cauchy–Schwarz inequality 345, 346, 350, 393
Strong approximation property 157
Strong Galerkin condition 152
Strongly convex functional 392
Sublevel set 373
Subsequence 372
Subspace
– Krylov 51
Subspace decomposition 325, 392
– (SS1) 325
Successive over-relaxation (SOR) 31, 32
Successive Subspace Optimization (SSO) 394
Sweep
– backward 19
– forward 16
Symmetric Gauss–Seidel method 25
Symmetric matrix 3
– $A_1$- 76
– level-1 75
Symmetrized GLIS 23
– additive 24
– multiplicative 23

Taylor, B. 87, 374
Taylor's Theorem 374, 375
Trace estimate 340, 345
Triangulation
– conforming 188
– grid spacing 189
– nested family 189
– refinement
  – bisection 188
  – quadrisection 188
– shape-regular 188
Two-grid method 73, 177
– convergence 142, 178
– error contraction factor 117
– one-sided 73, 114, 159, 161
– symmetric 73
Two-grid operator 72

V-Cycle multigrid 149, 258
– convergence 185, 208, 258, 322
– general symmetric 174, 175
– simple symmetric 149, 169, 171
Vector space
– Banach 366
– complete 366
– dual 367
– Hilbert 366
Vertex point
– interior 190
Vertex points 230, 262

W-Cycle multigrid 149
– convergence 184, 257, 323
– one-sided 163, 166
– symmetric 179, 181, 184

Wave number
– reciprocal 108
Weak approximation property 157
Weak closure of a set 371
Weak convergence 370
Weak formulation 82
Weak Galerkin condition 153
Weak lower semicontinuity (wLSC) 372
Weighted Jacobi method 14
Work estimate
– full multigrid 211

Xu, J. 349

Young, W.H. 181, 347
Young inequality 181
– convolution 347

Zhang, S. 428

# List of Figures


Figure 1.1 The canonical splitting of A. —— **11**
Figure 1.2 The curve $\rho(T_R)$ (in solid black) as a function of $\omega$. —— **21**
Figure 1.3 Error reduction rates for classical GLIS, $n = 31$. —— **34**
Figure 1.4 Error reduction rates for classical GLIS, $n = 63$. —— **35**
Figure 2.1 Convergence path for PSD. —— **50**
Figure 2.2 Comparison of CG, SD, and Richardson. —— **59**
Figure 3.1 The creation of the coarse grid matrix via the Galerkin condition. —— **71**
Figure 3.2 Restriction from a fine grid ($\ell = 1$) to a coarse grid ($\ell = 0$). —— **71**
Figure 3.3 Prolongation from a coarse grid ($\ell = 0$) to a fine grid ($\ell = 1$). —— **72**
Figure 3.4 Two dimensional fine and coarse grids. —— **72**
Figure 4.1 Piecewise linear basis (hat) functions in 1D. —— **83**
Figure 4.2 Interacting hat functions in 1D. —— **85**
Figure 4.3 Eigenvalues of $A_1$. —— **86**
Figure 4.4 Upper and lower polynomial bounds for $\cos(x)$, $x \in (0, \frac{\pi}{2}]$. —— **87**
Figure 4.5 The function $f(x) = 1 - \cos(\pi x)$. —— **89**
Figure 4.6 Plots of the eigenvalues of $K_1$. —— **93**
Figure 4.7 A uniform and nested two-level grid with $n_1 = 3$. —— **97**
Figure 4.8 A uniform and nested two-level grid with $n_1 = 7$. —— **97**
Figure 4.9 A piecewise linear function on the coarse grid, where $n_0 = 3$. —— **100**
Figure 4.10 A coarse grid function represented on the fine grid. —— **100**
Figure 4.11 Level-0 hat basis functions. —— **101**
Figure 4.12 Level-1 hat basis functions. —— **101**
Figure 4.13 A fine grid function and three possible restrictions. —— **106**
Figure 4.14 Error contraction for the one-sided two-grid algorithm. —— **118**
Figure 4.15 Errors after presmoothing and after coarse grid correction. —— **119**
Figure 4.16 Errors after presmoothing and after coarse grid correction. —— **120**
Figure 4.17 The error contraction for the one-sided two-grid algorithm. —— **121**
Figure 4.18 The error contraction for the one-sided two-grid algorithm. —— **122**
Figure 4.19 The error contraction for the one-sided two-grid algorithm. —— **123**
Figure 4.20 Errors after presmoothing and after coarse grid correction. —— **124**
Figure 4.21 Errors after presmoothing and after coarse grid correction. —— **125**
Figure 5.1 Eigenvalues of the error transfer matrix. —— **140**
Figure 6.1 W- and V- Cycles. —— **149**
Figure 7.1 Bisecting the triangulation $\mathcal{T}_0$ in one dimension. —— **189**
Figure 7.2 Refining the triangulation $\mathcal{T}_0$ in two dimensions. —— **189**
Figure 7.3 Uniform refinement of a two dimensional triangulation $\mathcal{T}_0$. —— **190**
Figure 7.4 Interior vertices and a basis function. —— **191**
Figure 7.5 A uniform mesh and a Lagrange basis function. —— **191**
Figure 7.6 Shape of the full multigrid algorithm. —— **209**
Figure 7.7 The error reduction for the multigrid V-Cycle algorithm. —— **214**
Figure 7.8 Meshes visualized with the command `showmesh`. —— **215**
Figure 7.9 L-shaped mesh obtained by refinement. —— **216**
Figure 7.10 Error reduction for multigrid. —— **216**
Figure 8.1 A uniform and nested two-level cell-centered grid with $q = 2$. —— **231**
Figure 8.2 A uniform and nested two-level cell-centered grid with $q = 4$. —— **231**
Figure 8.3 A piecewise constant function $v_\ell \in V_\ell$. —— **234**
Figure 8.4 Eigenvalues of the level-$\ell$ stiffness matrix $A_\ell$. —— **237**


https://doi.org/10.1515/9783111354880-016

Figure 8.5 Eigenvalues of $K_\ell$ as functions of $k$. — **242**
Figure 8.6 Spectrum of the error transfer for the Λ-Richardson smoother. — **243**
Figure 8.7 A uniform and nested two-level cell-centered grid. — **247**
Figure 8.8 Setup and notation for the proof of Proposition 8.7.2. — **260**
Figure 8.9 Two-dimensional cell-centered grids and grid points. — **263**
Figure 8.10 Lexicographic enumerations. — **264**
Figure 8.11 Graphical illustration of restriction and prolongation. — **269**
Figure 8.12 Grid points involved in the calculation of $-\Delta_\ell^D$. — **271**
Figure 8.13 Computaton of the diffusion coefficient via averaging. — **272**
Figure 8.14 Grid points used in the proof of Proposition 8.8.5. — **273**
Figure 8.15 Forward and backward Gauss–Seidel updates. — **278**
Figure 8.16 Error reduction for the multigrid V-Cycle algorithm. — **280**
Figure 8.17 Effect of damping on multigrid convergence. — **282**
Figure 8.18 Error reduction for the 2D algorithm with variable diffusivity. — **283**
Figure 10.1 A sample function in the hierarchical space $W_\ell$. — **334**
Figure 10.2 Nodal basis functions in one dimension. — **335**
Figure 10.3 Hierarchical versus standard Lagrange nodal bases. — **337**
Figure 10.4 Flow of information for the prolongation between levels. — **357**
Figure 10.5 Error reduction for BPX-PCG. — **358**
Figure 11.1 The graph of a lower semicontinuous function. — **373**
Figure 12.1 The sequence $\{d_k\}_{k=0}^\infty$ used in the proof of Corollary 12.2.5. — **398**
Figure 12.2 The energy defined in (12.2.9). — **400**
Figure 12.3 The function $f_{i,k}$ defined in (12.4.6). — **410**
Figure 12.4 The function $f_{i,k}$ and its quadratic approximation $q_{i,k}$. — **419**
Figure 13.1 Error reduction for the FASD solver. — **439**

# List of Tables

Table 2.1 Convergence history of the PSD method. —— **49**
Table 4.1 The values of $k'$ as a function of $k$. —— **108**
Table 4.2 Numerical values of $\gamma(m_1)$. —— **117**
Table 4.3 Computed two-grid convergence factors. —— **122**
Table 7.1 Computed multigrid convergence factors. —— **213**
Table 8.1 Computed V-cyle multigrid error contraction factors. —— **281**
Table 13.1 Computed FASD contraction factors. —— **439**

https://doi.org/10.1515/9783111354880-017

# List of Listings


Listing 1.1 The main Gauss–Seidel iteration block. —— 33
Listing 1.2 Code for one pass of SOR iteration. —— 33
Listing 1.3 Code for comparing the various GLIS. —— 38
Listing 1.4 The Jacobi iterative solver. —— 41
Listing 1.5 The Gauss–Seidel iterative solver. —— 41
Listing 1.6 The SOR iterative solver. —— 42
Listing 1.7 Code for estimating the contraction rates. —— 43
Listing 2.1 Driver for Example 2.2.6. —— 61
Listing 2.2 PSD solver code for Example 2.2.6. —— 64
Listing 2.3 Identity preconditioner for the PSD solver. —— 64
Listing 2.4 The PCG method. —— 65
Listing 2.5 Driver for Section 2.6. —— 66
Listing 2.6 Richardson's method. —— 68
Listing 4.1 The main damped Jacobi iteration block. —— 91
Listing 4.2 Element-wise prolongation operator. —— 104
Listing 4.3 Element-wise restriction operator. —— 104
Listing 4.4 Calling script for the two-grid solver —— 127
Listing 4.5 Two-grid solver. —— 129
Listing 4.6 Coarse grid correction function. —— 131
Listing 4.7 Prolongation function. —— 131
Listing 4.8 Restriction function. —— 132
Listing 4.9 Damped Jacobi smoothing function. —— 132
Listing 4.10 Gauss–Seidel smoothing function. —— 133
Listing 4.11 Residual calculation. —— 134
Listing 6.1 Multigrid operator implementation. —— 148
Listing 7.1 Manual creation of a finite element mesh. —— 214
Listing 7.2 Refining meshes with the `iFEM` `uniformrefine` command. —— 215
Listing 7.3 Calling script for the 1D finite element multigrid solver. —— 218
Listing 7.4 1D finite element multigrid solver. —— 221
Listing 7.5 1D finite element multigrid operator. —— 222
Listing 7.6 Demo of mesh creation and refinement with iFEM©. —— 224
Listing 7.7 Main calling script for solving Poisson equation. —— 226
Listing 8.1 A single iteration of Richardson's smoother. —— 240
Listing 8.2 Calling script for the 1D cell-centered multigrid solver. —— 285
Listing 8.3 Multigrid solver for a 1D cell-centered elliptic equation. —— 289
Listing 8.4 Multigrid operator for a 1D cell-centered elliptic equation. —— 290
Listing 8.5 Prolongation operator. —— 292
Listing 8.6 Restriction operator. —— 292
Listing 8.7 Restriction of the diffusion coefficient. —— 292
Listing 8.8 Discrete finite difference operator. —— 293
Listing 8.9 Damped quasi-Jacobi smoothing operator. —— 293
Listing 8.10 Richardson's smoothing operator. —— 294
Listing 8.11 Damped quasi-Gauss–Seidel smoothing operator. —— 294
Listing 8.12 Residual calculation for a 1D elliptic equation. —— 295
Listing 8.13 Function to apply various 1D boundary conditions. —— 296
Listing 8.14 Function to compute a scaled $L^2$ norm. —— 296
Listing 8.15 Calling script for the 2D cell-centered multigrid solver. —— 297


https://doi.org/10.1515/9783111354880-018

Listing 8.16 Multigrid solver for a 2D cell-centered elliptic equation. — 302
Listing 8.17 Multigrid operator for a 2D cell-centered elliptic equation. — 303
Listing 8.18 Prolongation operator. — 304
Listing 8.19 Restriction operator. — 305
Listing 8.20 Restriction of the diffusion coefficient. — 305
Listing 8.21 Finite difference operator. — 306
Listing 8.22 Damped quasi-Jacobi smoothing operator. — 306
Listing 8.23 Damped quasi-Gauss–Seidel smoothing operator. — 307
Listing 8.24 Residual calculation for a 2D elliptic equation. — 308
Listing 8.25 Function to apply various 2D boundary conditions. — 309
Listing 8.26 Function to compute a scaled $L^2$ norm. — 310
Listing 10.1 Driver code for the 1D finite element BPX-PCG solver. — 361
Listing 10.2 Implementation of the BPX preconditioning matrix. — 365
Listing 13.1 Driver code for the FASD nonlinear FEM solver. — 441
Listing 13.2 Main FASD solver script. — 444
Listing 13.3 Nonlinear discrete operator. — 445
Listing 13.4 Residual computation. — 446
Listing 13.5 Nonlinear Gauss–Seidel smoother. — 446
Listing 13.6 Secant search algorithm. — 448

# List of Symbols

The next list describes several symbols that will be commonly used within the text.

| | |
|---|---|
| $(\cdot,\cdot)$ | In the context of subspace decompositions, the Euclidean inner product on $\mathbb{R}^n$, page 329 |
| $(\cdot,\cdot)$ | The Euclidean inner product, page 2 |
| $(\cdot,\cdot)_\ell$ | In the context of subspace decompositions, the Euclidean inner product on $\mathbb{R}^{m_\ell}$, page 329 |
| $-\Delta_\ell^\circ$ | In the context of cell-centered finite difference schemes, the level-$\ell$ discrete minus Laplacian, page 249 |
| $-\Delta_{\ell-1}^{\widetilde{D}}$ | In the context of multidimensional cell-centered finite difference schemes, the discretization of the variable-coefficient second-order operator that satisfies Assumption (G3), page 272 |
| $-\Delta_\ell$ | In the context of cell-centered finite difference schemes, the cell-centered finite difference Laplacian, page 265 |
| $-\Delta_\ell^{D,\circ}$ | In the context of cell-centered finite difference schemes, the discretization of a second-order operator with variable coefficient $D$ and homogeneous Dirichlet boundary conditions, page 260 |
| $-\Delta_\ell^D$ | In the context of cell-centered finite difference schemes, the discretization of a second-order operator with variable coefficient $D$, page 258 |
| $[\mathsf{A}]_{i,j}$ | The element in the $i$th row and $j$th column of a matrix A, page 1 |
| $[\boldsymbol{u}]_j$ | The $j$th component of a vector $\boldsymbol{u}$, page 1 |
| $(\cdot,\cdot)_{\mathsf{A}_\ell}$ | In the context of multigrid methods, the energy inner product at level $\ell$, page 146 |
| $\operatorname{argmin}_{x\in X} f(x)$ | The set of minimizers (if it exists) of the function $f$ over the set $X$, page 198 |
| $\mathcal{A}_\ell$ | In the context of cell-centered finite difference schemes, the level-$\ell$ energy form, page 249 |
| $\mathcal{J}_j$ | In the context of a BPX preconditioner, the bilinear form that encodes the scaled Euclidean inner product on $V_j$ when the functions are identified with vectors, page 349 |
| $\mathcal{Y}_j$ | In the context of a hierarchical basis preconditioner, the bilinear form that encodes the Euclidean inner product on $W_j$ when the functions are identified with vectors, page 338 |
| $\mathcal{B}_\ell$ | In the context of cell-centered finite difference schemes, the level-$\ell$ basis of piecewise constant functions, page 232 |
| $\mathcal{B}_\ell$ | In the context of finite element methods, the Lagrange nodal basis of $V_\ell$, page 190 |
| $\mathcal{B}_\ell^W$ | In the context of hierarchical basis preconditioners, the basis of $W_\ell$, the hierarchical finite element space, page 335 |
| $\chi_A$ | The characteristic function of a set $A$, page 232 |
| $:=$ | Equality by definition, page 1 |
| $\mathrm{D}F(v)$ | The Fréchet derivative of a mapping $F$ at the point $v$, page 373 |
| $\mathrm{D}^2E(v)$ | The second Fréchet derivative of a functional $E$ at the point $v$, page 374 |
| $\frac{\mathrm{d}v}{\mathrm{d}x} = v'$ | The derivative of a function $v \in H_0^1(0,1)$, page 82 |
| $\mathsf{E}_\ell$ | In the context of multigrid methods, the error transfer matrix at level $\ell$, page 149 |
| $\mathsf{E}_1$ | The error transfer matrix for a two-grid algorithm, page 77 |
| $(\cdot,\cdot)_\ell$ | In the context of multigrid methods, the canonical inner product on level $\ell$, page 145 |
| $\mathrm{epi}(E)$ | The epigraph of a functional $E$, page 373 |
| FASDALS $(\cdot)$ | The FASD method with approximate line search, page 417 |
| FASD $(\cdot)$ | The fast subspace descent algorithm, page 405 |

https://doi.org/10.1515/9783111354880-019

| | |
|---|---|
| $\mathrm{FMG}(\cdot)$ | The full multigrid operator, page 208 |
| $\mathcal{H}_\ell$ | In the context of additive subspace preconditioners, the hierarchical basis of $V_\ell$, page 335 |
| $H_0^1(0,1)$ | The subspace of $L^2(0,1)$ of functions that have square-integrable weak derivatives and vanish at $x=0$ and $x=1$, page 82 |
| $\hat{\Pi}_\ell$ | $\frac{1}{2}\tilde{\Pi}_\ell$, page 176 |
| $h_\ell$ | In the context of cell-centered finite difference schemes, the level-$\ell$ grid spacing, page 230 |
| $(\cdot,\cdot)_{\mathsf{A}_\ell}$ | The $\mathsf{A}_\ell$ inner product, page 196 |
| $(\cdot,\cdot)_\ell$ | The level-$\ell$ Euclidean inner product, page 196 |
| $(\cdot,\cdot)_{\mathsf{A}}$ | For an SPD matrix A, the A-inner product, page 3 |
| $\kappa(\mathsf{A})$ | The spectral condition number of a matrix A, page 5 |
| $\kappa_\star(\mathsf{A})$ | The $\\|\cdot\\|_\star$-condition number of a matrix A, page 5 |
| $\mathsf{K}_1$ | For a two-level algorithm, the presmoothing error transfer matrix, page 75 |
| $\mathcal{K}_m(\mathsf{M},\boldsymbol{z})$ | The Krylov subspace of degree $m$ of $\mathsf{M}\in\mathbb{R}^{n\times n}$ and $\boldsymbol{z}\in\mathbb{R}^n$, page 51 |
| $\mathsf{K}_\ell$ | In the context of multigrid methods, the error transfer of the smoother at level $\ell$, page 150 |
| $\langle\cdot,\cdot\rangle_{\mathcal{X}',\mathcal{X}}$ | For a Banach space $\mathcal{X}$, the duality pairing between $\mathcal{X}$ and its dual $\mathcal{X}'$, page 369 |
| $\langle v\rangle_D$ | The average of the function $v$ over the domain $D$, page 339 |
| $\{\psi_{h,i}^+\}_{i=1}^{n_h^+}$ | In the context of Galerkin methods, the Lagrange nodal basis for the space $S_h$, page 203 |
| $\{\boldsymbol{N}_{h,j}\}_{j=1}^{n_h}$ | The set of interior vertices of the triangulation $\mathcal{T}_h$, page 202 |
| $\{\boldsymbol{N}_{h,j}^+\}_{j=1}^{n_h^+}$ | The set of all vertices of the triangulation $\mathcal{T}_h$, page 202 |
| $\mathrm{level}_a(E)$ | The sublevel set at height $a$ of $E$, page 373 |
| $\mathcal{B}^+$ | In the context of the fast subspace descent method, an enlarged sublevel set of $E$, page 408 |
| $\mathcal{B}_h$ | In the context of Galerkin methods, the Lagrange nodal basis functions with respect to $\mathcal{T}_h$, page 202 |
| $\mathcal{B}_h^+$ | In the context of Galerkin methods, the collection of interior and boundary Lagrange nodal basis functions, page 203 |
| $\mathcal{B}_i$ | In the context of the fast subspace descent method, $Q_i\mathcal{B}^+$, page 408 |
| $\mathcal{C}_\ell$ | In the context of cell-centered finite difference schemes, the level-$\ell$ space of cell-centered grid functions, page 231 |
| $\mathcal{C}_{\ell,0}$ | In the context of cell-centered finite difference schemes, the level-$\ell$ space of homogeneous Dirichlet cell-centered grid functions, page 231 |
| $\mathcal{G}_\ell$ | In the context of cell-centered finite difference schemes, the level-$\ell$ canonical basis of grid functions, page 232 |
| $\mathcal{I}_h$ | The Lagrange nodal interpolation operator, page 203 |
| $\mathcal{I}_h^+$ | The extended Lagrange nodal interpolation operator, page 203 |
| $\mathcal{L}$ | In the context of cell-centered finite difference schemes, the lifting operator, page 255 |
| $\mathcal{L}(\mathcal{X},\mathcal{Y})$ | For normed spaces $\mathcal{X}$ and $\mathcal{Y}$, the space of continuous linear mappings $\mathcal{X}\to\mathcal{Y}$, page 367 |
| $\mathcal{R}_h$ | In the context of Galerkin methods, the Ritz projection onto $V_h$, page 205 |
| $\mathcal{R}_{\ell-1}$ | In the context of cell-centered finite difference schemes, the discrete cell-centered Ritz projection, page 253 |
| $\mathcal{X}'$ | The dual space of a normed space $\mathcal{X}$, page 367 |
| $\mathfrak{R}_{\mathcal{V}}$ | For a Hilbert space $\mathcal{V}$, the canonical Riesz map, page 370 |

|  |  |
|---|---|
| $\mathfrak{R}_i$ | In the context of fast subspace descent methods, the Riesz map $\mathfrak{R}_i : \mathcal{V}_i' \to \mathcal{V}_i$, page 417 |
| $\mathrm{clo}_s(K)$ | The closure of a subset $K$ of a Hilbert space, page 372 |
| $\mathrm{clo}_w(K)$ | The weak closure of a subset $K$ of a Hilbert space, page 372 |
| $\mathrm{MG}(\cdot)$ | The multigrid operator, page 147 |
| $m_\ell$ | In the context of multidimensional cell-centered finite difference schemes, the number of grid points in each coordinate direction, page 262 |
| $\mathsf{A}^\top$ | The transpose of a matrix $\mathsf{A}$, page 2 |
| $\mathsf{A}_\ell$ | The level-$\ell$ stiffness matrix, page 193 |
| $\mathsf{C}_{\mathrm{BPX}}$ | The BPX preconditioner, page 349 |
| $\mathsf{C}_{\mathrm{H}}$ | The hierarchical basis preconditioner, page 338 |
| $\mathsf{Q}[\boldsymbol{u}]$ | In the context of subspace decompositions, the Q-class of $\boldsymbol{u} \in \mathbb{R}^n$, page 330 |
| $\mathsf{Q}_j^\ell$ | In the context of subspace decompositions, the hierarchical prolongation matrix, page 337 |
| $\mathsf{T}$ | The error transfer matrix of a GLIS, page 7 |
| $\mathsf{T}_{j,\ell}'$ | In the context of multilevel T-matrices, $\Pi_{j,\ell} - (\mathsf{K}_j^*)^m \Pi_{j,\ell}$, page 316 |
| $\mathsf{T}_{j,\ell}(m)$ | A multilevel T-matrix, page 314 |
| $\mathbb{N}_0$ | $\mathbb{N} \cup \{0\}$, page 26 |
| $(\cdot,\cdot)_{\mathsf{A}_1}$ | For a two-grid algorithm, the $\mathsf{A}_1$-inner product, page 76 |
| $(\cdot,\cdot)_{L^2(0,1)}$ | The inner product in $L^2(0,1)$, page 82 |
| $(\cdot,\cdot)_1$ | For a two-grid algorithm, the level-1 inner product, page 75 |
| $\{\psi_{\ell,i}\}_{i=1}^{n_\ell}$ | In the context of finite element methods, the Lagrange nodal basis of $V_\ell$, page 190 |
| $\{\psi_{\ell,j}\}_{j=1}^{n_\ell}$ | In the context of cell-centered finite difference schemes, the level-$\ell$ basis of piecewise constant functions, page 232 |
| $\{\boldsymbol{N}_{\ell,j}\}_{j=1}^{n_\ell}$ | The set of interior vertices of $\mathcal{T}_\ell$, page 190 |
| $\{\boldsymbol{N}_{\ell,j}^W\}_{j=1}^{m_\ell}$ | The collection of nodes in $\mathcal{T}_\ell$ that are not nodes of $\mathcal{T}_{\ell-1}$, page 335 |
| $\{g_{\ell,j}\}_{j=1}^{n_\ell}$ | In the context of cell-centered finite difference schemes, the level-$\ell$ canonical basis of grid functions, page 232 |
| $\lVert\cdot\rVert_{\mathsf{A}_\ell}$ | The $\mathsf{A}_\ell$-induced norm, page 196 |
| $\lVert\cdot\rVert_\ell$ | The level-$\ell$ Euclidean norm, page 196 |
| $\lVert\cdot\rVert$ | In the context of subspace decompositions, the Euclidean norm on $\mathbb{R}^n$, page 329 |
| $\lVert\cdot\rVert_\ell$ | In the context of subspace decompositions, the Euclidean norm on $\mathbb{R}^{m_\ell}$, page 329 |
| $\lVert\cdot\rVert_{\mathsf{A}}$ | The energy norm of an SPD matrix $\mathsf{A}$, page 25 |
| $\lVert\cdot\rVert_{\ell^p}$ | For $p \in [1,\infty]$, the $\ell^p$-norm in $\mathbb{R}^n$, page 3 |
| $\lVert\mathsf{M}\rVert_{\mathsf{A}}$ | For matrices $\mathsf{M}$ and $\mathsf{A}$, the induced matrix norm of $\mathsf{M}$ that comes from the $\mathsf{A}$-energy norm, page 55 |
| $n_\ell$ | In the context of cell-centered finite difference schemes, the number of level-$\ell$ grid points, page 230 |
| $\mathsf{P}_{\ell-1}$ | In the context of multigrid methods, the prolongation matrix on level $\ell$, page 146 |
| $\Pi_{\ell-1}$ | $\mathsf{A}_{\ell-1}^{-1}\mathsf{R}_{\ell-1}\mathsf{A}_\ell$, page 150 |
| $\Pi_{j,\ell}$ | $\mathsf{A}_j^{-1}\mathsf{R}_{j,\ell}\mathsf{A}_\ell \in \mathbb{R}^{n_j \times n_\ell}$, page 312 |
| $\mathsf{P}_0$ | The prolongation operator of the two-grid method, page 99 |
| $\mathbb{P}_1(K)$ | The space of polynomials of degree at most one over the set $K$, page 189 |
| $\mathbb{P}_k^\star$ | The space of polynomials of degree at most $k$ that evaluate to 1 at the origin, page 56 |
| $\prod_{j=\alpha}^{\beta} \mathsf{B}_j$ | For a family of matrices (operators) $\{\mathsf{B}_j\}_{j=\alpha}^{\beta}$, this is inductively defined as $\mathsf{B}_\alpha \prod_{j=\alpha+1}^{\beta} \mathsf{B}_j$, page 311 |
| $\mathsf{P}_{j,\ell}$ | The multilevel prolongation matrix, page 311 |

|  |  |
|---|---|
| $\mathsf{R}_{\ell-1}$ | In the context of multigrid methods, the restriction matrix on level $\ell$, page 146 |
| $\mathbb{R}^{m\times n}$ | The set of all real matrices with $m$ rows and $n$ columns, page 1 |
| $\mathbb{R}^n$ | The set of all real vectors of length $n$, page 1 |
| $\mathbb{R}^n_\star$ | The nonzero elements of $\mathbb{R}^n$, page 1 |
| $\mathsf{R}_0$ | The restriction operator of the two-grid method, page 99 |
| $\mathsf{R}_{j,\ell}$ | The multilevel restriction matrix, page 312 |
| $\mathsf{S}_\ell$ | In the context of multigrid methods, the smoother at level $\ell$, page 150 |
| $\mathrm{SSO}(\cdot)$ | One iteration of the successive subspace optimization method, page 394 |
| $\frac{1}{2}\mathsf{R}_0$ | The full weighting restriction matrix, page 107 |
| $\frac{\partial v}{\partial n}$ | The normal derivative of the function $v$ on the boundary of a domain, page 281 |
| $\mathrm{TG}(\cdot)$ | The two-grid operator, page 72 |
| $\tilde{\mathsf{R}}_0$ | A different type of restriction operator, page 106 |
| $\tilde{\mathsf{T}}'_{j,\ell}$ | In the context of multilevel T-matrices, $\mathsf{P}_{j,\ell}\mathsf{T}'_{j,\ell}$, page 316 |
| $\tilde{\Pi}_1$ | For a two-level algorithm, the coarse-grid Ritz projection, page 74 |
| $\tilde{\Pi}_\ell$ | In the context of multigrid methods, the coarse-grid Ritz projection matrix at level $\ell$, page 149 |
| $\tilde{\Pi}_{j,\ell}$ | The multilevel Ritz projection matrix, page 313 |
| $u_\ell \in V_\ell \overset{\mathcal{B}_\ell}{\longleftrightarrow} \boldsymbol{u}_\ell \in \mathbb{R}^{n_\ell}$ | This indicates that the finite element function $u_\ell$ has the vector $\boldsymbol{u}_\ell$ as a coordinate vector with respect to the Lagrange nodal basis, page 192 |
| $\lvert\cdot\rvert_{H_0^1(\Omega)}$ | The seminorm on $H_0^1(\Omega)$, page 201 |
| $\lvert\cdot\rvert_{H^2(\Omega)}$ | The seminorm of $H^2(\Omega)$, page 188 |
| $V_0$ | In the context of two-grid methods, the space of continuous piecewise linear functions on the level-0 grid that vanish at $x = 0$ and $x = 1$, page 99 |
| $\boldsymbol{v}_\ell \in \mathbb{R}^{n_\ell} \overset{\mathrm{lex}}{\longleftrightarrow} v_\ell \in V_\ell$ | The lexicographic correspondence for a multidimensional finite difference cell-centered discretization, page 264 |
| $V_\ell$ | In the context of cell-centered finite difference schemes, the level-$\ell$ space of piecewise constant functions, page 231 |
| $v_\ell \in \mathcal{C}_{\ell,0} \overset{\mathcal{G}_\ell}{\longleftrightarrow} \boldsymbol{v}_\ell \in \mathbb{R}^{n_\ell}$ | This indicates that the cell-centered grid function $v_\ell$ has the vector $\boldsymbol{v}_\ell$ as a coordinate vector with respect to the canonical basis, page 233 |
| $v_\ell \in V_\ell \overset{\mathcal{B}_\ell}{\longleftrightarrow} \boldsymbol{v}_\ell \in \mathbb{R}^{n_\ell}$ | This indicates that the piecewise constant function $v_\ell$ has the vector $\boldsymbol{v}_\ell$ as a coordinate vector with respect to the canonical basis, page 233 |
| $\wp_H$ | In the context of FASD methods, the Scott–Zhang interpolation operator onto $\mathcal{T}_H$, page 428 |
| $\{\psi_{1,i}\}_{i=1}^{n_1}$ | In the context of two-grid methods, the collection of level-1 hat basis functions, page 82 |
| $\{\mathcal{T}_h\}_{h>0}$ | A (not necessarily nested) shape-regular family of triangulations of $\Omega$, page 201 |
| $\{\boldsymbol{v}_0^{(k)}\}_{k=1}^{n_0}$ | The collection of eigenvectors of the level-0 stiffness matrix, page 105 |
| $\{\lambda_1^{(k)}\}_{k=1}^{n_1}$ | The collection of eigenvalues of the level-1 stiffness matrix, page 85 |
| $\{\lambda_\ell^{(k)}\}_{k=1}^{n_\ell}$ | In the context of one-dimensional cell-centered finite difference schemes, the collection of eigenvalues of the level-$\ell$ stiffness matrix, page 236 |
| $\{\lambda_\ell^{(s,t)}\}_{s,t=1}^{m_\ell}$ | In the context of cell-centered finite difference schemes, the eigenvectors of the two-dimensional discrete Laplacian, page 269 |
| $\{\lambda_0^{(k)}\}_{k=1}^{n_0}$ | The collection of eigenvalues of the level-0 stiffness matrix, page 105 |
| $\{\mu_\ell^{(k)}(\omega)\}_{k=1}^{n_\ell}$ | In the context of one-dimensional cell-centered finite difference schemes, the collection of eigenvalues of the level-$\ell$ Richardson smoother $\mathsf{K}_\ell(\omega)$ with parameter $\omega$, page 241 |

|  |  |
|---|---|
| $\{\psi_{0,i}\}_{i=1}^{n_0}$ | In the context of two-grid methods, the collection of level-0 hat basis functions, page 99 |
| $\{Q_j\}_{j=0}^{L}$ | In the context of subspace decompositions, a collection of ordered prolongation matrices, page 325 |
| $\{\mathcal{T}_\ell\}_{\ell=0}^{L}$ | A nested family of conforming triangulations, page 189 |
| $\{\boldsymbol{e}^k\}_{k=0}^{\infty}$ | The error sequence $\boldsymbol{e}^k = \boldsymbol{u} - \boldsymbol{u}^k$ of a GLIS, page 6 |
| $\{\boldsymbol{r}^k\}_{k=0}^{\infty}$ | The residual sequence $\boldsymbol{r}^k = \boldsymbol{f} - \mathsf{A}\boldsymbol{u}^k$ of a GLIS, page 6 |
| $\{\boldsymbol{v}_1^{(k)}\}_{k=1}^{n_1}$ | The collection of eigenvectors of the level-1 stiffness matrix, page 85 |
| $\{\boldsymbol{v}_\ell^{(k)}\}_{k=1}^{n_\ell}$ | In the context of one-dimensional cell-centered finite difference schemes, the collection of eigenvectors of the level-$\ell$ stiffness matrix, page 236 |
| $\{w_\ell^{(s,t)}\}_{s,t=}^{m_\ell}$ | In the context of cell-centered finite difference schemes, the eigenfuctions of the two-dimensional discrete Laplacian, page 269 |
| $A + B$ | The Minkowski sum of two subsets of a vector space, page 377 |
| $C^0(\overline{\Omega})$ | The space of continuous functions on the closure of the domain $\Omega$, page 189 |
| $C_\ell$ | In the context of one-dimensional cell-centered finite difference schemes, the set of level-$\ell$ cell-centered grid points, page 230 |
| $C_{A0}$ | The approximation property constant, page 137 |
| $C_{S1}$ | The smoothing constant, page 141 |
| $E_i$ | In the context of the fast subspace descent method, a local energy $E_i : \mathcal{V}_i \to \mathbb{R}$, page 405 |
| $f(x) = \Theta(x^r)$ | Big Theta notation, see Definition 4.2.4, page 90 |
| $f_{i,k}$ | In the context of the fast subspace descent method, a one-dimensional energy section, page 410 |
| $H_0^1(\Omega)$ | The space of functions on $\Omega$ that have weak derivatives in $L^2(\Omega)$ and vanish on the boundary $\partial\Omega$, page 187 |
| $H^2(\Omega)$ | The space of functions on $\Omega$ that are weakly twice differentiable, with derivatives in $L^2(\Omega)$, page 188 |
| $H^{-1}(\Omega)$ | The dual of $H_0^1(\Omega)$, page 187 |
| $I_i$ | In the context of the fast subspace descent method, the natural inclusion $I_i : \mathcal{V}_i \hookrightarrow \mathcal{V}$, page 405 |
| $k'$ | The wave number reciprocal to $k$, page 108 |
| $K_{\ell,i}$ | In the context of one-dimensional cell-centered finite difference schemes, a grid cell, page 230 |
| $L$ | In the context of multigrid methods, the maximum number of levels, page 145 |
| $L^2(0,1)$ | The space of square-integrable functions on the interval $(0,1)$, page 82 |
| $L^2(\Omega)$ | The space of square-integrable functions on the domain $\Omega$, page 187 |
| $m_1$ | In the context of multigrid methods, the number of presmoothings at every level, page 147 |
| $m_2$ | In the context of multigrid methods, the number of postsmoothings at every level, page 147 |
| $n_\ell$ | In the context of multigrid methods, the dimension of the space on level $\ell$, page 145 |
| $p$ | In the context of multigrid methods, the number recursive calls to the previous level, page 147 |
| $Q_\ell$ | In the context of cell-centered finite difference schemes, the $L^2$-projection onto $V_\ell$, page 250 |
| $Q_H$ | In the context of finite element methods, the $L^2$-projection onto the space $V_H$, page 427 |
| $Q_i$ | In the context of the fast subspace descent method, a projection $Q_i : \mathcal{V} \to \mathcal{V}_i$, page 405 |

| | |
|---|---|
| $R_B(\cdot)$ | The Rayleigh quotient for an SPD matrix B, page 198 |
| $R_i$ | In the context of the fast subspace descent method, the natural restriction of functionals $R_i = I_i^\top : \mathcal{V}' \to \mathcal{V}_i'$, page 405 |
| $S_\ell$ | In the context of finite element methods, the space of continuous piecewise linear functions on the level-$\ell$ grid, page 189 |
| $S_h$ | In the context of Galerkin methods, the space of continuous piecewise linear functions on the grid $\mathcal{T}_h$, page 201 |
| $V_1$ | In the context of two-grid methods, the space of continuous piecewise linear functions on the level-1 grid that vanish at $x = 0$ and $x = 1$, page 82 |
| $V_\ell$ | In the context of finite element methods, the space of continuous piecewise linear functions on the level-$\ell$ grid that vanish on $\partial\Omega$, page 189 |
| $V_h$ | In the context of Galerkin methods, the space of continuous piecewise linear functions on the grid $\mathcal{T}_h$ that vanish on $\partial\Omega$, page 201 |
| $v_n \rightharpoonup v$ | The sequence $\{v_n\}_{n=1}^\infty$ converges weakly to $v$, page 370 |
| $W^{m,q}(D)$ | For a domain $D \subset \mathbb{R}^n$, $m \in \mathbb{N}$, and $q \in [1, \infty]$, the space of functions that, together with its derivatives up to order $m$, belong to $L^q(D)$, page 340 |
| $W_\ell$ | In the context of subspace decompositions, the level-$\ell$ hierarchical finite element space, page 333 |
| $x_{\ell,i+1/2}$ | In the context of one-dimensional cell-centered finite difference schemes, a vertex point, page 230 |
| $x_{\ell,i}$ | In the context of one-dimensional cell-centered finite difference schemes, a grid point, page 230 |
| (A0) | In the context of multigrid methods, strong approximation property, page 157 |
| (A1) | In the context of multigrid methods, the weak approximation property, page 157 |
| (A2) | In the context of multigrid methods, the $\alpha$-weak approximation property, page 158 |
| (E0) | In the context of the abstract method of subspace corrections, the assumption that the objective is Fréchet differentiable and strongly convex, page 392 |
| (E1) | In the context of the abstract method of subspace corrections, the assumption that the objective satisfies (E0) and is locally Lipschitz smooth, page 392 |
| (E2) | In the context of the FASD method, the set of assumptions on the local energies, page 408 |
| (G0) | In the context of multigrid methods, the strong Galerkin condition, page 153 |
| (G1) | In the context of multigrid methods, the weak Galerkin condition, page 153 |
| (G2) | In the context of multigrid methods, Assumption (G2), page 155 |
| (G3) | In the context of multigrid methods, the imbalanced Galerkin condition, page 176 |
| (NS1) | In the context of the abstract method of subspace corrections, the assumption that the subspace decomposition is stable, page 393 |
| (NS2) | In the context of the abstract method of subspace corrections,the strengthened Cauchy–Schwarz inequality, page 393 |
| (S1) | In the context of multigrid methods, the first smoothing property, page 159 |
| (S2) | In the context of multigrid methods, the second smoothing property, page 166 |
| (SS1) | In the context of space decompositions, this indicates that a set of ordered prolongation matrices supports a subspace decomposition of $\mathbb{R}^n$, page 325 |
| (E2b) | In the context of the fast subspace descent method, the assumption that the local energies are strongly convex, page 408 |
| (E2c) | In the context of the fast subspace descent method, the assumption that the derivatives of the local energies are Lipschitz continuous, page 408 |
| (E2a) | In the context of the fast subspace descent method, the assumption that the local energies are Fréchet differentiable, page 408 |

www.ingramcontent.com/pod-product-compliance
Lightning Source LLC
Chambersburg PA
CBHW061151240926
51846CB00006B/987

* 9 7 8 3 1 1 1 3 5 4 1 9 4 *